COMPANION TO THE COSMOS

'In a masterly Introduction Gribbin gives a clear account of how it all began
... If you want to know, you will find it here'
Lewis Wolpert, *The Times*

'John Gribbin's *Companion to the Cosmos* is going to become a firm friend
... Not only is it packed full of a universe of accessible information; it also
includes many mini-biographies and a timeline'
New Scientist

'There are already many fine books on astronomy and cosmology but
Companion to the Cosmos is a distinctive addition to a crowded field ... He
presents, in accessible prose, an extraordinary range of fascinating material
which will help set in context whatever astronomical advances hit the
headlines in the next few years'   Martin Rees, *Sunday Times*

'Gribbin's powers of explanation ... are massive and have never been used
to greater effect'   Martin Ince, *The Times Higher Education Supplement*

IN THE BEGINNING

'A fascinating book [from] a world-class science writer ... Read this book for
a superb overview of the very latest thinking on the universe'
Heather Couper, *Sunday Times*

'One of the world's leading popularisers of science, John Gribbin has here
used his wide-ranging research to make a major contribution at the cutting
edge of knowledge ... One of the most stimulating books I have read in a
long time ... It is a book to challenge the mind. I would recommend it to
anyone and everyone who is – quite literally – interested in life, the universe and everything'   Nigel Henbest, *The Times Educational Supplement*

'Any shortlist of the world's best popular science writers must include John
Gribbin. His gift lies not just in eloquent and lucid presentation of complicated ideas, but in writing about them as if they were adventures. From the
first page of one of his tours of the universe the reader is gripped'
Anthony Grayling, *Financial Times*

THE OMEGA POINT

'Displays the author's enviable and customary lucidity ... No one writes a better book on fundamental aspects of the universe'
Roy Herbert, *New Scientist*

'The most accurate popular presentation of modern cosmology I have ever seen'   Robert Schaefer, Bartol Research Institute, University of Delaware

IN SEARCH OF THE EDGE OF TIME

'Gribbin writes entertainingly and makes difficult ideas sound simple'
Michael Rowan-Robinson, *The Times Educational Supplement*

'John Gribbin's book is accessible, lucid and fun ... Reading Gribbin on the subject [of astrophysics] makes this reviewer feel like a child allowed to stay up late'   John Wilkes, *Los Angeles Times*

SCHRÖDINGER'S KITTENS

'A superb work of popular science'   Nicholas Lezard, *Guardian*

'Gribbin ... likes to surprise and excite his audience. This gives his writing a delightful vitality, and graphic imagery that makes difficult ideas accessible. The whole parable of "Schrödinger's kittens" is brilliant'
Danah Zohar, *Independent on Sunday*

'The best attempt yet to make the inexplicable explicable ... The philosophical implications of a theory that turns both commonsense and the classical certainties upside down hold a mesmerising sway'
Ian Critchley, *Sunday Times*

'Gribbin proves yet again that he is an outstanding communicator of difficult ideas'   Robin Blake, *Independent*

John Gribbin has a Ph.D. in astrophysics from Cambridge, and is now Visiting Fellow in Astronomy at the University of Sussex and consultant to *New Scientist*. His many bestselling books include *In Search of Schrödinger's Cat*, *The Omega Point*, *In Search of the Big Bang*, *In the Beginning* and *Schrödinger's Kittens*. His books have been translated into many languages, and have won awards in both Britain and the United States. He also writes science fiction.

John Gribbin lives in Sussex with his wife, Mary, also a science writer, and their two sons.

ALSO BY JOHN GRIBBIN

Galaxy Formation
White Holes
Timewarps
Genesis: The Origins of Man and the Universe
The Monkey Puzzle (with Jeremy Cherfas)
Spacewarps: Black Holes, White Holes, Quasars and the Universe
The Redundant Male (with Jeremy Cherfas)
In Search of Schrödinger's Cat: Quantum Physics and Reality
In Search of the Double Helix: Quantum Physics and Life
Weather (with Mary Gribbin)
In Search of the Big Bang: Quantum Physics and Cosmology
The Hole in the Sky
The Stuff of the Universe (with Martin Rees)
Winds of Change (with Mick Kelly)
Hothouse Earth: The Greenhouse Effect and Gaia
The Cartoon History of Time (with Kate Charlesworth)
The Matter Myth (with Paul Davies)
Stephen Hawking: A Life in Science (with Michael White)
Too Hot to Handle?: The Greenhouse Effect (with Mary Gribbin)
In Search of the Edge of Time
In the Beginning: The Birth of the Living Universe
Albert Einstein: A Life in Science (with Michael White)
Being Human (with Mary Gribbin)
Time and Space (with Mary Gribbin)
Schrödinger's Kittens and the Search for Reality
Charles Darwin: A Life in Science (with Michael White)
Fire on Earth (with Mary Gribbin)
Watching the Weather (with Mary Gribbin)
Richard Feynman: A Life in Science (with Mary Gribbin)

FICTION

The Sixth Winter (with Douglas Orgill)
Brother Esau (with Douglas Orgill)
Father to the Man
Ragnarok (with D.G. Compton)
Innervisions

Companion to the
COSMOS

....

JOHN GRIBBIN

edited by Mary Gribbin
illustrations by Jonathan Gribbin
timelines by Benjamin Gribbin

A PHOENIX GIANT PAPERBACK

First published in Great Britain by Weidenfeld & Nicolson in 1996
This paperback edition published in 1997 by Phoenix, a
division of Orion Books Ltd,
Orion House, 5 Upper St Martin's Lane, London WC2H 9EA

A CIP catalogue record for this book is available
from the British Library.

ISBN: 1 85799 891 X

Printed and bound in Great Britain by
Butler & Tanner Ltd, Frome and London

Picture acknowledgements are given on page 519

CONTENTS

....

There is something fascinating about science.
One gets such wholesale returns of conjecture out of
such a trifling investment of fact.

Mark Twain, Life on the Mississippi

INTRODUCTION

....

Where do we come from?

Everybody is intrigued by the story of our origins – how the Universe came
into being, how it got to be the way it is, and why it is a suitable home for
life forms like ourselves. The question 'where do we come from?' is the most
profound question it is possible to ask, and the ability to provide a reasonably
complete answer to that question arguably ranks as the greatest achievement
of human thought.

Virtually all of our information about the Universe at large comes from
studies of electromagnetic radiation – light, radio waves, X-rays and other
variations on the theme – all of which travels at the speed of light, 30,000
million cm per second. Although this is a very large speed, the Universe itself
is very large, so that light and other forms of electromagnetic radiation take
a long time to reach us from other stars and galaxies. Even for a relatively
nearby star, light spends years on its journey to Earth, so that we see the star
as it was years ago, when the light left it. Farther away across the Universe,
we can detect light from galaxies and quasars so remote that the light has
been millions, hundreds of millions or even, in some cases, thousands of
millions of years on its journey across space to us, and we see these objects
as they were that long ago, when the Universe was correspondingly younger.

The disadvantage of this is that light from such distant objects is very faint
by the time it reaches us, and can be analysed only with the aid of powerful
telescopes and sensitive electronic detectors. But to a large extent this disadvantage is offset by the great advantage that we are literally looking farther
back in time as we look farther out into space, seeing the Universe as it used
to be and getting some idea of how it has evolved. The key discovery in all
of cosmology is that the Universe we see around us is indeed evolving – that
it is different now from the way it used to be, and that it had a definite origin
at a certain moment in time. But, in fact, you do not need either large
telescopes or sensitive electronic detectors to work that out – all you need is
the evidence provided by your own two eyes.

The most fundamental astronomical observation is that night follows day.
Although the importance of this observation was not appreciated until the

18th century, and the explanation was not forthcoming at all until the 19th century, and not widely understood until the 1980s, this single observation is enough to tell us that the Universe had an origin at a definite time in the past, and that it has not always been as we see it today.

The fundamental question concerns the darkness of the night sky – how can a dark Universe be full of bright stars? The question is now known as Olbers' Paradox, after the German astronomer Heinrich Wilhelm Olbers, although he was not, in fact, the first person to puzzle over it. Put at its simplest, the puzzle is that if the Universe is infinite in extent, going on forever in all directions, and if every region of the Universe is similar, in an average sense, to the region we live in, then in every direction we look our line of sight should intersect the surface of a star. Every point on the night sky should be bright!

We now know that stars are grouped into galaxies, like our own Milky Way, islands in the Universe which may each contain hundreds of billions of stars; but the 'paradox' can easily be rephrased to take account of that. We also know that, even if the Universe is not infinite, it is certainly big enough for the puzzle to hold with full force – *if* galaxies like those we see today have existed everywhere in the Universe, forever.

The resolution of the puzzle is straightforward, but required a revolution in the way people thought about the Universe. It is simply that the stars and galaxies have not existed forever – that, if you like, there has not been enough time since the birth of the Universe to fill up all the dark spaces between the stars with light. The darkness of the night sky is alone enough to tell us that the Universe had a definite beginning.

The answer seems obvious, to modern generations reared on the idea of a Universe born in a Big Bang. But it is a sign of how revolutionary the idea was that proper discussions of Olbers' Paradox only took place decades after the discovery that the Universe is expanding forced astronomers to reject the idea of an eternal, unchanging cosmos and to begin thinking about the evolution of the Universe itself.

The discovery of the expansion of the Universe came only in the late 1920s, when the American astronomer Edwin Hubble and his colleagues established that galaxies are moving apart from one another. Modern cosmology really began only with that discovery, and the time that has elapsed from the discovery of the expanding Universe to the publication of this book is almost exactly one human lifetime, the standard 'three score years and ten' referred to in the Bible. Just one lifetime ago, the idea that the Universe was eternal and unchanging was an 'obvious' scientific truth, so unquestioned that when Albert Einstein first developed his general theory of relativity, and discovered that the simplest version of the equations required the Universe to be expanding, he added an extra term to the equations to hold them still. He later described this as the 'biggest blunder' of his career.

The lesson to be drawn from all this is not that we are so much more clever and insightful than the astronomers of seventy years ago, but that if even Einstein could make such a cosmic blunder we should, perhaps, be wary of taking too much of what we think we know about the Universe at face value. What seems obvious and commonsense to us just may, in another 70 years or so, seem as laughably out of date as we find the notion of an eternal, unchanging Universe. But that is not to say that we should not take any of the present understanding of the Universe and its origins seriously. Stars and galaxies and the structure of the Universe today are very well understood. The question is how far out into the Universe (and correspondingly back in time) we can push our good understanding of the Universe, and where (and when) speculation begins to play a major part. The boundary lies farther back in time, and under much more extreme conditions, than you might guess.

If galaxies are moving apart today, that must mean that they used to be closer together than they are now. One important point about the expanding Universe is that the galaxies are not moving through space, like fragments of bomb from a great explosion, but space itself is stretching, and carrying the galaxies along for the ride (this was the prediction from the general theory of relativity that Einstein himself initially refused to accept). Long ago, there was no space between what are now the galaxies, which must have overlapped each other; before that, there was no space between what are now the stars, which must have touched; and before that, there must have been a time when there was no space between atoms, which merged into one another.

Astronomers know a lot about stars and galaxies. Physicists know a lot about atoms. Astrophysicists have no trouble describing the behaviour of a soup of matter and radiation so thick that individual atoms merge into one another, with electrons from the outer parts of the atoms being dislodged, leaving the nuclei at the hearts of the atoms exposed. Such a soup of nuclei and electrons, together with radiation, is called a plasma. But even this is not the limit of our sound understanding of matter and radiation under extreme conditions. Indeed, experiments carried out at particle accelerator laboratories like the ones at CERN, in Geneva, or Fermilab, in Chicago, provide insight into the behaviour of atomic nuclei themselves, and the protons and neutrons of which they are made. Physicists tell us – and back their claims up with convincing evidence – that they even understand the behaviour of matter, space, time and energy under conditions so extreme that atomic nuclei themselves are packed together cheek by jowl, and are broken up into their constituent parts.

Physicists also make much more extravagant claims about understanding what goes on under much more extreme conditions than this, but those claims are not always so well backed up by a solid weight of evidence. It is at that point that speculation starts to play a part, initially modest, in their cosmological musings, a part which grows in importance as they consider ever more extreme conditions. We know about atomic nuclei, protons and neutrons because all those things exist in the Universe today and can be studied directly in experiments of different kinds. And it is, therefore, not hopelessly unrealistic to believe that physicists really can tell us what things were like when the entire Universe was as dense as the nucleus of an atom is today, and that they can also tell us how it evolved from this hot, dense state (the Big Bang itself) into the collection of galaxies, stars, planets and people that we see today. Indeed, many physicists would argue that we are being cautiously conservative, in the second half of the 1990s, in setting our sights so low, and in claiming that our really good understanding of the Universe 'only' extends from the time when it had nuclear density up to the present day. But that's OK; in this context, it is better to err on the side of conservatism. So – *when* was the entire Universe in such a dense, hot state? When was the Big Bang?

If we imagine 'winding back' the present expansion of the Universe, it would mean that everything in the Universe as we know it – space, time, matter and energy – emerged from a point of infinite density and zero volume (a singularity) about 15 billion years ago. The exact time is not known, because details of the expansion of the Universe are hard to measure and interpret, but this does not matter. What matters is that the expansion tells us that there was a superdense state, perhaps a bit more than 15 billion years ago, perhaps a bit less, and that taken to extremes the superdense state seems to have started from a singularity. All this is borne out by the equations of the general theory of relativity – but nobody believes that this is really exactly what happened; the effects of quantum physics would dominate the situation close to a singularity, and would ensure that the hypothetical mathematical point was in fact blurred out by a process known as quantum uncertainty.

The question of what exactly went on close to the singularity, and how the quantum processes gave rise to the Big Bang, is one of the most respectable of current cosmological speculations, and attempts to answer that question form the basis of a great deal of research in cosmology today. But there is no need to worry about this just yet. The conditions we are interested in now, as the earliest time and place where the completely solid, securely founded understanding of the physics of objects that exist in our everyday world could be applied, occurred a full tenth of one-thousandth of a second (0.0001 sec) after the time represented by the singularity – sometimes called the 'moment of creation' or the 'birth of the Universe'. In the sense that all of the science involved is well understood, astrophysicists can talk with complete confidence about everything that has happened after the first tenth of one-thousandth of a second; the uncertainties that remain in describing the subsequent evolution of the Universe are simply a result of our imperfect observations of the Universe at large, and our imperfect ability at applying the known laws of physics to describe complicated systems. The interval before that, back to the moment of creation, is still partly a mystery, not only because of our imperfect ability to apply the laws of physics but also because we are not quite sure exactly what the laws of physics are that operate under such extreme conditions.

But by everyday standards, the conditions that existed 0.0001 sec after the moment of creation were extreme enough. At that time, the density of the Universe was 10^{14} grams per cubic centimetre – 100,000 billion times the density of water. The temperature was 1,000 billion degrees above absolute zero (10^{12} K, which for such large numbers is essentially the same as 10^{12} Celsius), and the Universe consisted of a cosmic fireball of hot radiation.

Under such extreme conditions, individual particles (such as protons, neutrons and electrons) do not have much of an independent existence. Individual photons from the fireball radiation ('particles of light') carry so much energy at these temperatures that they are able to convert themselves into pairs of particles, swapping energy for mass in line with Einstein's famous equation $E = mc^2$. The pairs of particles made in this way almost always consist of an everyday particle (such as a proton) and its so-called antimatter counterpart (in this case, an antiproton). When a particle meets up with an equivalent antiparticle, the pair annihilate, giving back in the form of radiation the energy they were made from. In the Big Bang, radiation was constantly being turned into matter, and matter was constantly being turned back into radiation, in a seething maelstrom of activity.

But as the cosmic fireball expanded and cooled, the individual photons in the fireball had less and less energy. Soon, they did not have enough energy to make any more protons and neutrons. If the conversion of radiant energy into matter–antimatter pairs had always been precise, that would have meant that the cooling Universe was left with an exactly equal number of protons and antiprotons, and an exactly equal number of neutrons and antineutrons. Before long, under those extreme conditions, every particle would have met an antiparticle partner and annihilated, leaving nothing but radiation in the cooling Universe. But because of a tiny imbalance in the laws of physics, whose significance was first appreciated in the 1960s by the Soviet physicist Andrei Sakharov, there was a tiny excess of the kind of matter we are made of left over at the end of this process – there was just one everyday particle left over for every billion photons of radiation left in the fireball. Everything that we can see in the Universe today is made out of the one-in-a-billion particles (protons+neutrons) manufactured in this way in the Big Bang fireball.

By one-hundredth of a second after the moment of creation, things were calming down a little. The temperature had dropped to 100 billion K (10^{11} K), and protons and neutrons were no longer being manufactured out of radiation, although they were still being buffeted by the dense sea of photons in which they swam. Initially, there was the same number of neutrons as there was protons. But neutrons, unlike protons, are unstable particles, and left to their own devices they will each spit out an electron (in a process known as radioactive decay) and convert themselves into protons. Today, this process is slow compared with the changes that were going on in the Universe when it was a fraction of a second old. On average, if you have an isolated neutron, it takes more than 10 minutes for it to decay in this way. But the buffeting that neutrons were getting in the cosmic fireball encouraged the change. So by the time the temperature of the Universe had fallen to 30 billion K, just over one-tenth of a second after the moment of creation, the proportion of neutrons to protons had dropped from 50:50 down to 38 per cent neutrons to 62 per cent protons. By the time the Universe had cooled to 10 billion K, 1.1 seconds after the moment of creation, the density was down to 380,000 times that of water and there were only 24 neutrons left for every 76 protons. But, like most of us, the Universe slows down and is less susceptible to change as it gets older; at last, the headlong pace of change in the early Universe had slowed to the point where change could be measured over seconds, rather than in fractions of a second.

After 13.8 seconds, the temperature had dropped to 3 million K and, with correspondingly less energy available to knock them about, the rate at which neutrons were being converted into protons had slowed dramatically. There were still 17 neutrons left for every 83 protons in the Universe, and occasionally nuclei of the isotope deuterium (heavy hydrogen) could form in the fireball as an individual proton and an individual neutron stuck together temporarily before being knocked apart in a collision. Just 3 minutes and 2 seconds after the moment of creation, the temperature of the entire Universe had cooled to the point where it was only 1 billion K, 70 times as great as the temperature at the heart of the Sun today, which is some 15 million K. There were still 14 neutrons around for every 86 protons, and by now the Universe was so old that the natural decay of the neutrons started to be important. Although the average lifetime of a free neutron is more than 10 minutes, as is the way with averages, some live longer and some decay sooner. In every 100 seconds from now on, one in ten of the remaining free neutrons would turn itself into a proton spontaneously. But the neutrons were saved from extinction because just at this time, a little over three minutes into the life of the Universe, conditions had eased to the point where neutrons began to combine with protons to form stable nuclei, first of deuterium and then of helium. The nuclei still collided with each other, and with other particles, but the temperature was now so low that there was not enough energy in these collisions to break the nuclei up. Almost immediately, the remaining neutrons (about 13 for every 87 protons) were locked up in nuclei of helium-4, each of which contains two neutrons and two protons. The proportion of the total mass of neutrons and protons converted into helium was just twice the proportional number of neutrons, which is 26 per cent, and the process was completed by 3 minutes and 46 seconds after the moment of creation.

The exact numbers that emerge from this description of the birth of the Universe are not just pulled out of the hat. They come from a combination of the general theory of relativity, which tells us how fast the Universe was expanding and cooling, and the known facts about the behaviour of particles such as neutrons and protons, and atomic nuclei, deduced from experiments here on Earth. This combination provides the so-called 'standard model' of cosmology, and one of its great triumphs is that it predicts that 26 per cent of the mass of each of the first stars formed in the Universe (which are therefore the oldest stars seen today) should be in the form of helium. This exactly matches the actual amount of helium observed in old stars, using the technique of spectroscopy. The other great prediction of the standard model is that the Universe should be filled with a sea of radiation left over from the fireball. This radiation, at a temperature of about 1 billion K when helium began to form less than 4 minutes after the moment of creation, should have cooled, according to the standard model, all the way down to below 3 K (just under –270°C) today, 15 billion years later. The cosmic microwave background radiation discovered in the 1960s, and predicted by the Big Bang theory, exactly fits the bill. It is hardly any wonder, then, that the standard Big Bang model of the Universe is regarded as one of the jewels in the crown of modern science.

Once helium had formed in the expanding Universe, the processes which led to the formation of galaxies, stars, planets and people seem to have proceeded inexorably on their way. Some of the details of these later processes remain obscure. This is partly because they represent increasingly complex phenomena – one of the strangest aspects to take on board about cosmology is that we know so much about the fireball of the Big Bang because it was a very simple place, running in accordance with detailed laws of physics that are very well understood, and involving a few simple kinds of particle, such as protons, neutrons and electrons. Once you get involved with atomic nuclei, you have more complexity to deal with; atoms themselves interact in accordance with the laws of chemistry, producing another layer of complexity; and life itself involves extremely complex chemistry and the way some complex arrangements of atoms and molecules (such as people) interact with their physical environment. We know less and less about more and more complex systems.

Another source of uncertainty in bringing the story of the Universe up to date concerns the amount of dark matter there is in the Universe, and its nature. Obviously, not everything in the Universe is in the form of a bright star that we can see with our telescopes. Indeed, there is sound evidence that tens of times as much dark matter came out of the cosmic fireball as there was matter in the form of protons and neutrons, the kind of nuclear matter which forms stars, galaxies, planets and people. The influence of this dark matter is seen by the way in which it tugs on the visible matter through gravity. Without the dark matter, nuclear matter would have been spread ever more thinly as the Universe expanded, never clumping together to form stars and galaxies at all; it is only thanks to the gravitational influence of the dark stuff that we exist.

In broad outline, the story of how we got to be here can be picked up about 300,000 years after the Big Bang. At that point, the Universe was still a hot soup, at a temperature of about 5,000 K, a little cooler than the surface of the Sun is today. The nuclear stuff of the Universe was largely in the form of individual protons (nuclei of hydrogen atoms) and helium nuclei, moving in a sea of electrons and dark matter. Until that time, any nucleus that tried latching on to an electron or two and forming an atom would quickly be involved in a collision with an energetic photon, which would rip the electron away from it. Because nuclei each have a positive electric charge and electrons each have a negative charge, and photons like to interact with charged particles, this meant that the Universe was full of charged particles, interacting with photons and making the Universe opaque. No photon could travel far without bouncing off a charged particle and continuing on its journey in a crazy zig-zag path, like the ball in some demented cosmic pinball machine.

Then, quite suddenly by cosmological standards, as the temperature fell the photons no longer had enough energy to disrupt atoms as they tried to form. Each proton captured an electron, each helium nucleus captured two electrons, and all the charged particles were locked up as electrically neutral atoms. There were no longer any charged particles for photons to bounce off, and they streamed through space, around the atoms, essentially unhindered. Overnight, as it were, the Universe became transparent. It is the radiation from that time, a few hundred thousand years after the Big Bang, which has been streaming silently through transparent space ever since, and which we now detect as the background radiation.

When the atoms formed, they were already clumped together in large streamers and sheets of higher than average density, pulled together by the gravitational influence of dark matter in the Universe. Within the great sheets of atomic material formed in this way, even though the Universe as a whole continued to expand, large quantities of gas were pulled together by gravity into sheets surrounding voids of atomic material. As gravity pulled the gas into thinner sheets, clumps formed within the sheets and shrank in their turn, breaking up into smaller fragments, which shrank and fragmented in their turn (actually, not literally 'in their turn' in the sense of one after another; the fragmentation and collapse was going on at all levels simultaneously). The smallest fragments formed in this way became stars, nested inside galaxies, which are themselves nested inside clusters of galaxies within superclusters that form chains, filaments and sheets of bright stuff, forming a frothy distribution of visible material throughout the dark void of the Universe at large.

It was only after the first generations of stars formed in this way and ran through their life cycles that stars like the Sun and planets like the Earth could form. The first stars contained only hydrogen and helium. Heavier elements, including such atoms as those of carbon, oxygen and nitrogen that are essential for life as we know it, were manufactured inside stars by nuclear fusion, and spread through large regions of the young galaxies when the more massive and short-lived of those first-generation stars exploded at the ends of their lives.

The Sun formed much later, only about 5 billion years ago, from the debris of such stellar explosions. The specific collapsing cloud of gas from which the Sun was born probably contained enough matter to make several hundred stars, which formed together as the fragmenting cloud collapsed, but have since gone their separate ways. One glob of matter in that cloud contained a little more mass than the Sun has today, and as the glob collapsed under its own weight, most of the matter formed a hot ball of gas while some formed a ring of material around the embryonic star. The heat of the young star blew away many of the lighter atoms from the ring, leaving a system made up of tiny particles of dust which gradually stuck together and aggregated to form the planets. After that, the story involves geophysics and biology, not cosmology and cosmogony, as life emerged and evolved on the surface of at least one of those planets.

One of the most important features of this story, which still puzzles many astronomers, is that, although it all seems inevitable given what we know about the way the laws of physics work, only very small changes in the way those laws work could have prevented all this happening. *Could* the laws have been different? For example, if the Universe had expanded slightly more slowly, by the time it had cooled to the point where helium nuclei could form there would have been no neutrons left to make helium nuclei with. If the expansion had proceeded only a little more quickly, there would have been so many neutrons left that virtually all of the nuclear matter would have emerged from the Big Bang in the form of helium, with no free protons left to make hydrogen at all. Either way, the Universe would have been a very different place. Stars made entirely of helium, for example, would rapidly run through their life cycles and quickly fade away, perhaps not allowing time enough for life to evolve on any planets that orbited them.

The fact that the Universe contains some helium, but not 100 per cent, depends on a balance between gravity (which determines how fast the Universe expands) and the nuclear forces involved in the formation of helium nuclei, which determine how quickly protons and neutrons combine to make helium. If the balance had been slightly different, we would not exist; so the fact that we exist can be used to work out what some of the properties of the Universe, and the laws of physics, must be. This is an example of what is known as anthropic reasoning, or anthropic cosmology; there is a lively debate about whether this is a mere tautology or whether it can be used to tell us something deeply significant about the way the Universe works.

One exciting possibility is that there may be other universes, in which the laws of physics operate in different ways to the way they operate in our Universe, and life forms like us cannot exist. This brings us right back to the question of what went on before the Big Bang, during the first split-second of the existence of the Universe, and at the moment of creation itself. This is the realm of inflationary cosmology, the most important and dramatic development in cosmology today.

The search for the Big Bang ended in the spring of 1992, when NASA astronomers announced the results of observations of the cosmic background radiation made by the COBE satellite. Those observations left no room to doubt that the Universe as we know it has evolved from a very hot, very dense state – the Big Bang. But they went further. Tiny variations in the strength of the background radiation from place to place on the sky, which soon became known as 'ripples', exactly matched the predictions of the theory of inflation, developed during the 1980s. The combination of inflation and the hot Big Bang was established beyond reasonable doubt as the only good description of the birth of the Universe.

But that does not mean that there is no more work for the cosmologists to do, or that there will be no more exciting stories about the birth of the Universe. With the success of the Big Bang theory, attention has now turned to how and why the Universe got to be in a hot, dense state in the first place – the details of the theory of inflation. There are now many variations on this particular theme, opening up for truly scientific debate one of the last provinces of the philosophers and metaphysicists, and addressing the question 'how did time itself begin?'

At the same time, with the basics of the Big Bang theory firmly in place, cosmologists are increasingly interested in the details of how the Universe got to be the way we see it today. One of the predictions of inflation is that the Universe should contain a great deal of dark matter, much more than we can see in the form of bright stars and galaxies. The presence of at least some of this dark stuff is also revealed, as we have mentioned, by the way that galaxies move. There is a lively debate among the experts about exactly what this dark matter might be, and experimenters on Earth are trying to capture particles of the dark matter in their laboratories.

Another enduring puzzle is the exact value of the Hubble constant, the number which measures how rapidly the Universe is expanding. It is still extremely difficult to measure this parameter, which also indicates the age of the Universe, and current estimates range from about 50 km per second per Megaparsec to about 80 km per second per Megaparsec.

All of these debates about the details of the Big Bang are sometimes presented in news stories as if the uncertainties threaten the Big Bang theory itself. For example, if the Hubble constant is indeed as large as 80, in the usual units, then the simplest version of the cosmological models would tell us that the time that has elapsed since the Big Bang is less than 10 billion years. That would be embarrassing, since many stars are known to be older than that. Obviously, the Universe cannot be younger than the stars it contains! But even if this measurement of the Hubble constant turned out to be correct, that would not spell the death of the Big Bang theory. More subtle versions of the cosmological models can quite easily accommodate such a large value of the constant and keep the age of the Universe greater than that of its oldest stars. It is always nice if the simplest version of a theory turns out to be a good description of reality – but as we all know from everyday experience, life just is not like that, so why should the Universe be so simple? As Richard Feynman has pointed out, the simplest thing would be nothing at all, and nature is far more inventive than that (see Brian Hatfield, ed., *Feynman Lectures on Gravitation*, Addison-Wesley, 1995).

One possibility is that we are being too parochial, and that our corner of the Universe, even though it may be billions of light years across, is not big enough to provide a reliable guide to the Universe at large. Cosmologists baffled by the apparent evidence that the Universe is younger than the stars it contains may simply have been guilty of reading too much into our immediate surroundings in the Universe. According to a group of Chinese researchers, the problem is that we live in a low-density bubble which is not typical of the Universe at large. When the appropriate measurements are made on large enough scales, everything slots into place.

The kinds of scale that cosmologists deal with are much greater than the distances between stars. They are interested in the distances between clusters of galaxies, and regard a whole galaxy of several hundred billion stars, like our Milky Way, merely as a 'test particle' in the Universe at large. Their efforts to measure the scale of the Universe are rather like trying to measure the distribution of island archipelagos across the Pacific Ocean from a base on one of those islands – with the added complication that each archipelago is moving apart from every other archipelago as the Universe expands.

The key question, which has not really been considered much by cosmologists until now, is how typical the region of the Universe over which we can make these measurements is. Just as the hypothetical Pacific islander mapping the known 'universe' may be unaware of the existence of the continents on either side of the ocean, so our local bubble of space may not give us enough information to predict the behaviour of the entire Universe. Xiang-Ping Wu, of the Beijing Astronomical Observatory, and several colleagues, suggested in 1995 that this is indeed the case.

They pointed out that, although this kind of study of the Universe extends out to distances of a few hundred million light years, if the measurements made for clusters at different distances are analysed separately, instead of all being lumped together to give one average figure, they show that the density of matter in the Universe increases the further out we look. On a scale of about 30 million light years, the density is only 10 per cent of the critical value, while on a scale of 300 million light years it may be as much as 90 per cent of the critical value.

The direct implication of this is that, on the scale over which recent measurements of the expansion of the Universe have been made by the Hubble Space Telescope, the expansion rate (given by the Hubble constant) is bigger than the overall average expansion rate by as much as 40 per cent. This means that the age of the Universe has been underestimated by 40 per cent, which is almost exactly the correction needed to boost the age from about 8 billion years to about 12 billion years, matching the ages of the oldest stars. In cosmological terms, it may be that our Pacific islanders have just discovered America.

But whether or not the volume of space we live in is typical of the Universe at large, as we look towards the way cosmology will develop as we enter the 21st century, inflation seems certain to hold centre stage. The marriage between quantum physics and cosmology could be said to have been consummated by the discoveries made by COBE, and since confirmed and refined by several other observations, both from space and from ground-based experiments. The key question that the discovery of the cosmic ripples answered was how irregularities as large as galaxies and clusters of galaxies could ever have grown up in the Universe as it emerged from the Big Bang fireball. The ripples correspond to fluctuations in the temperature of the background radiation from different parts of the sky of only 30 millionths of a degree Kelvin, representing an incredible achievement in measuring them at all. But fluctuations of this size, corresponding to differences in the density of the Universe from place to place at the time when matter and radiation decoupled and went their separate ways, are exactly right to have allowed the growth of the kind of irregularities we see in the Universe today over many billions of years.

The icing on the cake is that inflation can tell us where these irregularities seen at the era of decoupling came from. During the split-second of rapid expansion that we call inflation, quantum processes should, according to the theory, have created tiny distortions in the structure of the Universe. Inflation took these quantum irregularities, and blew them up to the size of clusters of galaxies. The nature of the cosmic ripples measured by COBE exactly matches the kinds of distortion that would have been produced in this way.

This success is all the more impressive because inflation theory was not invented in order to explain where galaxies came from. The motivation for the development of the theory came from two puzzles that astronomers began to focus on in the 1970s. The first is called the horizon problem, and is simply that the Universe looks the same in all directions – in particular, the temperature of the background radiation is the same on opposite sides of the sky. But how do regions on opposite sides of the sky know how to keep in step with each other? After all, there has not been sufficient time since the Big Bang for light (or anything else) to travel across the Universe and back. The second puzzle is related to the presence of dark matter. It is that the Universe is very nearly flat, in the sense that it sits just on the dividing line between expanding forever and one day recollapsing.

The flatness problem can be understood in terms of the density of the Universe. The density parameter is a measure of the amount of gravitating material in the Universe, usually denoted by the Greek letter omega (Ω), and also known as the flatness parameter. It is defined in such a way that if spacetime is exactly flat then $\Omega = 1$. Before the development of the idea of inflation, one of the great puzzles in cosmology was the fact that the actual density of the Universe today is very close to this critical value – certainly within a factor of ten. This is curious because, as the Universe expands away from the Big Bang, the expansion will push the density parameter away from the critical value. If the Universe starts out with the parameter less than 1, Ω gets smaller as the Universe ages, while if it starts out bigger than 1 Ω gets bigger as the Universe ages. The fact that Ω is between 0.1 and 1 today means that in the first second of the Big Bang it was precisely 1 to within one part in 10^{60}. This makes the value of the density parameter in the beginning one of the most precisely determined numbers in all of science, and the natural inference is that the value is, and always has been, exactly 1. One important implication of this is that there must be a large amount of dark matter in the

Universe. Another is that the Universe was made flat by inflation.

It works like this. During the first split-second of the existence of the Universe, everything we can see now was squashed together in a region smaller than the nucleus of an atom today. The hot fireball was so small that there was no problem about photons criss-crossing it at the speed of light and smoothing everything into a homogeneous state. But under these conditions, quantum processes associated with the development of the separate forces of nature that we know about today (gravity, the electromagnetic force and the two forces that operate inside nuclei) acted as a kind of antigravity, forcing the embryonic Universe to expand extremely rapidly. This expansion blew up the tiny fireball to at least the size of a basketball, and perhaps much bigger, in a tiny fraction of a second, doubling the size of the Universe hundreds of times in an interval too short to be measured by any conventional clock. Then, inflation switched off, and the more sedate expansion described by the standard Big Bang model and outlined earlier took over.

The key point is that during inflation opposite regions of the Universe were blasted apart, in a sense, 'faster than light'. Nothing can travel through space faster than light, but it was space itself that expanded in this way, carrying regions that used to be literally in touch with one another out of the range of light communication. That is why opposite sides of the Universe still have the same temperature today – they started out the same, even though they have not 'communicated' with one another since the first split-second after the birth of the Universe. This awesome expansion also ensures that spacetime gets highly flattened, in much the same way that the wrinkly surface of a prune becomes a smooth, flat surface when the prune is placed in water and swells up.

The reason why inflation has become established as the most exciting area of cosmological research today is that the ideas from quantum theory that led to this dramatic resolution of the cosmological puzzles were entirely derived from studies of particle physics, the so-called grand unified theories, or GUTs. The physicists who developed GUTs had no idea that they could be applied to describe the birth of the Universe; the fact that these theories work so well in solving problems they were never designed for is taken as strong evidence that they are telling us something fundamental about the way the Universe works.

Quantum theory can even explain the origins of the tiny primordial seed from which the Universe as we know it inflated – so-called 'quantum fluctuations' allow such minuscule regions of intense energy to appear literally out of nothing at all, making the Universe, in a memorable phrase, 'the ultimate free lunch'. An alternative possibility suggests that new baby universes may be created by inflation where matter collapses into a black hole, forming a singularity like the one in which the Universe was born, from which inflation creates a new universe. On that picture, our own Universe may have formed from a black hole in another universe, and so on, in an interconnected web which had no beginning and will have no end.

The clean simplicity of the original simple picture of inflation has now, inevitably, begun to be obscured by refinements, as inflationary cosmologists add bells and whistles to their models to make them match more closely the Universe we see about us. Some of the bells and whistles, it has to be said, are studied just for fun. Andrei Linde, who now works at Stanford University and has been the chief proponent of inflation in the 1990s, has taken great delight in pushing inflation to extremes, and offering entertaining new insights into how the Universe might be constructed. For example, could our Universe exist on the inside of a single magnetic monopole produced by cosmic inflation? According to Linde, it is at least possible, and may be likely. And in a delicious touch of irony, Linde made this outrageous claim in a lecture at a workshop on the birth of the Universe held in Rome, where the view of creation is usually rather different.

One of the additional reasons why theorists came up with the idea of inflation in the first place (alongside the horizon problem and the flatness problem) was precisely to get rid of magnetic monopoles – strange particles carrying isolated north or south magnetic fields, predicted by many grand unified theories of physics, but never found in nature. Standard models of inflation solve the 'monopole problem' by arguing that the seed from which our entire visible Universe grew was a quantum fluctuation so small that it contained only one monopole. That monopole is still out there, somewhere in the Universe, but it is highly unlikely that it will ever pass our way.

But Linde has discovered that, according to theory, the conditions that create inflation persist *inside* a magnetic monopole even after inflation has halted in the Universe at large. Such a monopole would look like a magnetically charged black hole, connecting our Universe through a wormhole in spacetime to another region of inflating spacetime. Within this region of inflation, quantum processes can produce monopole–antimonopole pairs, which then separate exponentially rapidly as a result of the inflation. Inflation then stops, leaving an expanding Universe rather like our own, which may contain one or two monopoles, within each of which there are more regions of inflating spacetime.

The result is a never-ending fractal structure, with inflating universes embedded inside each other and connected through the magnetic monopole wormholes. Our Universe may be inside a monopole which is inside another universe which is inside another monopole, and so on indefinitely. What Linde calls 'the continuous creation of exponentially expanding space' means that 'monopoles by themselves can solve the monopole problem'. Although it seems bizarre, the idea is, he stresses, 'so simple that it certainly deserves further investigation'.

That variation on the theme really is just for fun, and it is hard to see how it could ever be compared with observations of the real Universe. But most of the modifications to inflation now being made are in response to new observations, and in particular to the suggestion that spacetime may not be quite 'flat' after all. In the mid-1990s, many studies (including observations made by the refurbished Hubble Space Telescope) began to suggest that there might not be quite enough matter in the Universe to make it perfectly flat – most of the observations suggest that there is only 20 per cent or 30 per cent as much matter around as the simplest versions of inflation require. At first sight, the shortfall is embarrassing, because one of the most widely publicized predictions of simple inflation was the firm requirement of exactly 100 per cent of this critical density of matter. But there are ways around the difficulty; and here are two of them to be going on with.

The first suggestion is almost heretical, in the light of the way astronomy has developed since the time of Copernicus. Is it possible that we are living near the centre of the Universe? For centuries, the history of astronomy has seen humankind displaced from any special position. First the Earth was seen to revolve around the Sun, then the Sun was seen to be an insignificant member of the Milky Way Galaxy, then the Galaxy was seen to be an ordinary member of the Universe. But now comes the suggestion that the 'ordinary' place to find observers like us may be in the middle of a bubble in a much greater volume of expanding space.

The conventional version of inflation says that our entire visible Universe is just one of many bubbles of inflation, each doing their own thing somewhere out in an eternal sea of chaotic inflation, but that the process of rapid expansion forces space-time in all the bubbles to be flat. A useful analogy is with the bubbles that form in a bottle of fizzy cola when the top is opened. But that suggestion, along with other cherished cosmological beliefs, has now been challenged by Linde, working with his son Dmitri Linde (of Caltech) and Arthur Mezhlumian (also of Stanford).

Linde and his colleagues point out that the Universe we live in is like a hole in a sea of superdense, exponentially expanding inflationary cosmic material, within which there are other holes. All kinds of bubble universes will exist, and it is possible to work out the statistical nature of their properties. In particular, the two Lindes and Mezhlumian have calculated the probability of finding yourself in a region of this super-Universe with a particular density – for example, the density of 'our' Universe.

Because very dense regions blow up exponentially quickly (doubling in size every fraction of a second), it turns out that the volume of all regions of the super-Universe with twice any chosen density is 10 to the power of 10 million times greater than the volume of the super-Universe with the chosen density. For any chosen density, most of the matter at that density is near the middle of an expanding bubble, with a concentration of more dense material round the edge of the bubble. But even though some of the higher-density material is round the edges of low-density bubbles, there is even more (vastly more!) higher-density material in the middle of higher-density bubbles, and so on forever.

The discovery of this variation on the theme of fractal structure surprised the researchers so much that they confirmed it by four independent methods before venturing to announce it to their colleagues. Because the density distribution is non-uniform on the appropriate distance scales, it means that not only may we be living near the middle of a bubble universe, but the density of the region of space we can see may be less than the critical density, compensated for by extra density beyond our field of view.

This is convenient, since those observations by the Hubble Space Telescope have suggested that cosmological models which require exactly the critical density of matter may be in trouble. But there is more. Those Hubble observations assume that the parameter which measures the rate at which the Universe is expanding, the Hubble constant, really is a constant, the same everywhere in the observable Universe. If Linde's team is right, however, the measured value of the 'constant' may be different for galaxies at different distances from us, truly throwing the cat among the cosmological pigeons. We may seem to live in a low-density universe in which both the measured density and the value of the Hubble constant will depend on which volume of the Universe these properties are measured over. It might sound like the fevered imaginings of an over-enthusiastic theorist – if it were not for those intriguing Chinese observations which suggest that the Universe around us really is like that!

That would mean abandoning many cherished ideas about the Universe, and may, in spite of the observational evidence, still be too much for many cosmologists to swallow. But there is a simpler solution to the density puzzle, one which involves tinkering only with the models of inflation, not with long-held and cherished cosmological beliefs. That may make it more accept-able to most cosmologists – and it is so simple that it falls into the 'why did I not think of that?' category of great ideas.

A double dose of inflation may be something to make the government's hair turn grey – but it could be just what cosmologists need to rescue their favourite theory of the origin of the Universe. By turning inflation on twice, they have found a way to have all the benefits of the inflationary scenario, while still leaving the Universe in an 'open' state, so that it will expand forever.

In the simplest inflation models, the big snag is that after inflation even the observable Universe is left like a mass of bubbles, each expanding in its own way. We see no sign of this structure, which has led to refinements of the basic model to ensure homogeneity. Now, however, Martin Bucher and Neil Turok, of Princeton University, working with Alfred Goldhaber, of the State University of New York, have turned this difficulty to advantage.

They suggest that, after the Universe had been flattened by the original bout of inflation, a second burst of inflation could have occurred within one of the bubbles. As inflation begins (essentially at a point), the density is effectively 'reset' to zero, and rises towards the critical density as inflation proceeds and energy from the inflation process is turned into mass. But because the bubble from which the Universe is expanding has already been homogenized, there is no need to require this bout of inflation to last until the density reaches the critical value. It can stop a little sooner, leaving an open bubble (what we see as our entire visible Universe) to carry on expanding at a more sedate rate.

According to Bucher and his colleagues, an end product looking very much like the Universe we live in can arise naturally in this way, with no 'fine tuning' of the inflationary parameters. All they have done is to use the very simplest possible version of inflation, but to apply it twice. And you do not have to stop there. Once any portion of expanding spacetime has been smoothed out by inflation, new inflationary bubbles arising inside that volume of spacetime will all be pre-smoothed and can end up with any amount of matter from zero to the critical density (but no more). This should be enough to make everybody happy. Indeed, the biggest problem now is that the vocabulary of cosmology does not quite seem adequate to the task of describing all this activity.

The term 'Universe', with the capital 'U', is usually used for everything that we can ever have knowledge of, the entire span of space and time accessible to our instruments, now and in the future. This may seem like a fairly comprehensive definition, and in the past it has traditionally been regarded as synonymous with the entirety of everything that exists. But the development of ideas such as inflation suggests that there may be something else beyond the boundaries of the observable Universe – regions of space and time that are unobservable in principle, not just because light from them has not yet had time to reach us, or because our telescopes are not sensitive enough to detect their light.

This has led to some ambiguity in the use of the term 'Universe'. Some people restrict it to the observable Universe, while others argue that it should be used to refer to all of space and time. If we use 'Universe' as the name for our own expanding bubble of spacetime, everything that is in principle visible to our telescopes, then maybe the term 'Cosmos' can be used to refer to the entirety of space and time, within which (if the inflationary scenario is correct) there may be an indefinitely large number of other expanding bubbles of spacetime, other universes with which we can never communicate. Cosmologists ought to be happy with the suggestion, since it makes their subject infinitely bigger and therefore infinitely more important; and it is in this spirit that we have offered you not merely a guide to the Universe, but a Companion to the Cosmos.

Cross-references to another entry are printed in *italic bold type*. Cross-references are selective and are used only when the other entries are directly relevant to the entry that is being read.

AAT = *Anglo-Australian Telescope.*

aberration An apparent shift in the position of a *star* caused by the finite speed of light and the motion of the Earth in its orbit around the Sun. Over a year, the star seems to move in a small ellipse around its average position. The effect was discovered by James *Bradley* in 1729 and used by him to measure the speed of light.

absolute magnitude The *apparent magnitude* that a star or other bright object would have if it were at a distance of exactly 10 parsecs from the observer (see *magnitude scale*).

absolute zero The lowest temperature that could ever be attained. At absolute zero, atoms and molecules would have the minimum amount of energy allowed by quantum theory. This is defined as 0 on the *Kelvin* (K) scale of temperature; 0 K is −273.15 °C, and each unit on the Kelvin scale is the same size as 1 degree *Celsius*.

absorption line A narrow feature in a *spectrum*, corresponding to absorption of electromagnetic radiation at a well-defined wavelength. The pattern of absorption lines in a spectrum is like a fingerprint, identifying the elements that are absorbing the radiation.

absorption nebula A cold cloud of gas and dust in space that is visible only because it blocks out the light from more distant stars. See *nebulae*.

abundance of the elements See *cosmic abundances*.

accretion Two kinds of accretion are important in the Universe. The first is the process where small particles collide with one another and stick together to make larger objects. Collisions have to be 'just right' for this to occur – if the impacts are too hard, they will break up the objects (*fragmentation*) instead of allowing them to stick together. When the Solar System formed from a cloud of gas and dust in space, collapsing under its own gravitational pull, the young Sun was left surrounded by a disc of material which settled into the equatorial plane. This must have been rather like a grander version of the rings seen around Saturn today. The planets and other objects in the Solar System formed by accretion in this swirling disc of material, starting with tiny grains less than 1 mm across.

The second kind of accretion occurs when a massive object gathers in material from its surroundings by the pull of its *gravitational field*. This happens in a modest way for an ordinary star like our Sun, which is constantly accreting material from interstellar space. Much more dramatic accretion can occur with objects that have stronger gravitational fields, such as neutron stars and black holes. Then, material being sucked down on to the object (perhaps from a nearby companion star in a binary

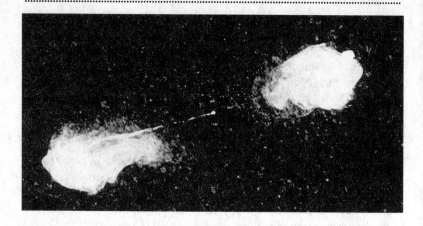

Active galaxy. The active galaxy Cygnus A, 'photographed' at radio wavelengths. The galaxy itself is a small dot at the centre of the image; activity associated with a black hole at the heart of the galaxy emits two jets of material in opposite directions.

system) forms an *accretion disc*. As material gains energy by falling in the gravitational field, and the *atoms* collide with one another in the disc, they can become so hot that they radiate *X-rays*. Processes like this, operating on a large scale at the centres of some *galaxies*, involving black holes many millions of times more massive than our Sun, may provide the power source in *quasars*.

accretion disc   A ring of material surrounding a *star* or other object from which matter spirals inward to fall on the object inside the disc. See *accretion*.

active galaxy   A *galaxy* that is emitting a large amount of energy from its central region, known as the nucleus. This gives such objects an alternative name, active galactic nucleus, usually shortened to AGN. The term covers a variety of energetic galaxy types discovered at different times and given different names, including Seyfert galaxies, N galaxies, BL Lac objects and quasars. It is now thought that these are all powered by essentially the same process, involving the *accretion* of matter on to a supermassive *black hole* at the centre of the active galaxy.

As material from the galaxy falls into the black hole, gravitational energy associated with its mass is released and converted into electromagnetic radiation, including light, X-rays and radio waves. This process is so efficient that 10 per cent or more of the mass of the infalling material can be converted into energy, in line with Einstein's famous equation $E = mc^2$ (see *special theory of relativity*). The central black hole may have a mass of up to 100 million (10^8) solar masses, just 0.1 per cent of the mass of all the bright stars in the surrounding galaxy. It would only need to 'eat' the equivalent of one or two stars like our Sun each year to provide the energy seen in the most powerful active galaxies.

Energy from the central source is often beamed out from the power-house to either side of the galaxy, probably from the 'poles' of the black hole. The energy cannot escape in other directions because it is blocked by the **accretion disc**. Where the beamed radiation interacts with the material in the galaxy and its surroundings, it can produce thin jets or extended regions called lobes which radiate at radio wavelengths (see **radio galaxies, jets**).

Adams, John Crouch (1819–92)   English astronomer who in 1845 pre-dicted the existence of a planet beyond Uranus by studying perturbations in the orbit of Uranus. The French astronomer **Le Verrier** independently made the same prediction, and passed his calculations on to the Berlin Observatory where the planet Neptune was discovered on 23 September 1846.

advance of the perihelion   The orbit of Mercury around the Sun does not trace out the same path every time, but shifts slightly from one orbit to the next. Each orbit is an ellipse, with the Sun at one focus of the ellipse. In each orbit, at the closest approach of Mercury to the Sun (**perihelion**), the ellipse shifts sideways by a tiny amount. This advance of the peri-helion was predicted by Albert Einstein's general theory of relativity, but cannot be explained by Isaac Newton's theory of gravity. See also **binary pulsar**.

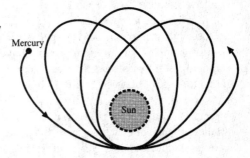

Advance of the perihelion. Greatly exaggerated representation of the way the orbit of Mercury shifts sideways around the Sun.

aerial = *antenna*.

age of the Sun and Solar System   The oldest rocks in the crust of the Earth have been dated by radioactive techniques, and are a little less than 4 billion years old. Samples of Moon rock have been dated in a similar way, and ages range from about 4.1 billion years up to about 4.5 billion

(4.5×10^9) years for the oldest samples from the Moon. Some samples from meteorites are also more than 4 billion years old. Combining all of the evidence suggests that the Solar System is about 4.6 billion years old. Since the Galaxy is about 15 billion years old, this means that the Sun and planets are only one-third as old as the Galaxy.

Although there is no direct way to measure the age of the Sun, its overall appearance as an orange/yellow star on the main sequence of the Hertzprung–Russell diagram is exactly what would be expected for a star with the Sun's mass that is about 4.6 billion years old and halfway through its lifetime on the main sequence (see *stellar evolution*).

age of the Universe   The Universe is expanding, with *galaxies* moving apart from one another as time passes and the space between them stretches. If we imagine winding this process backward in time, it seems that there must have been a beginning, long ago when the galaxies were all piled up in one place. This idea of the origin of the Universe in a *Big Bang* (or *singularity*) is borne out by Einstein's general theory of relativity. The rate at which the Universe is expanding today is given by a number known as *Hubble's constant*, H_0, which can be determined from observations of galaxies. If the Universe had been expanding at the same rate ever since the Big Bang, its age would then simply be given by the inverse of Hubble's constant, $1/H_0$. This is called the Hubble time.

In fact, the gravitational force of attraction between the galaxies tends to slow down the expansion as the Universe ages. Hubble's 'constant' is the same for all galaxies at the same epoch in the life of the Universe, but it decreases as time passes. This means that the age calculated from the present-day value of the constant is always larger than the true age of the Universe, because it was expanding more quickly in the past.

It is difficult to measure the precise value of H_0, but a value widely accepted by astronomers is 55 km per second per Megaparsec. Using this figure, the upper limit on the age of the Universe is 18 billion years. Using standard cosmological *models*, based on Einstein's equations, this translates into a true age of between 12 and 15 billion years.

An indirect way to estimate the age of the Universe is from the ages of stars. Presumably, the Universe must be older than the oldest stars in it. Studies of *globular clusters* suggest that some stars in our Galaxy may be between 14 and 18 billion years old. This is just consistent with the age of the Universe if $H_0 = 55$ km/sec./Mpc. Some cosmologists, however, argue that the constant is much larger, perhaps as great as 100 km/sec/Mpc. That would reduce the 'age of the Universe' below 10 billion years, in which case it would be impossible for it to contain stars 14 billion years old. Some recent studies, though, suggest that the constant may be less than 40 km/sec/Mpc, which would make the Universe comfortably older than the stars it contains.

AGN See *active galaxy*.

Airy, Sir George Biddell (1801–92) English astronomer who became the seventh Astronomer Royal in 1835 and held the post until 1881. On taking up the appointment, he found the *Royal Greenwich Observatory* in a shambolic and inefficient state, but transformed it into the finest observatory of its kind in the world through the ruthless application of his own high standards.

albedo A measure of the reflectivity of an object. A perfectly reflecting surface has an albedo of 1, while a black surface that absorbs all light that falls on it has an albedo of 0. Because it is covered by white clouds, *Venus* has a relatively high albedo, 0.65, whereas *Mercury*, with no atmosphere and a rocky surface, has an albedo of just 0.11. The albedo of the *Earth* is 0.37.

Algol The second brightest *star* in the *constellation* Perseus. The observed brightness of Algol varies in a regular way because light from the star is periodically eclipsed by a fainter companion as it orbits around the star. This was the first *eclipsing binary* to be identified, using spectroscopic techniques, in the 1880s. Algol A, the bright star, is 3.7 times as massive as our Sun, while Algol B, the faint star, has only 0.8 solar masses. They orbit each other with a period of 68.8 hours.

alpha beta gamma theory The explanation of how primordial hydrogen was partly converted into helium in the *Big Bang*, to provide the raw material from which *stars* are made. This theory predicts the existence of the *background radiation* later discovered to fill the Universe.

The calculations on which the theory is based start when the Universe was less than a second old, in an extremely hot, dense state, filled with a mixture of protons, neutrons, electrons and other fundamental particles. In the 1940s, George *Gamow* and his student Ralph *Alpher* showed that, as the Universe expanded from this fireball state and cooled, about 75 per cent of the mass in these particles would remain as protons (hydrogen nuclei), while 25 per cent would be converted into alpha particles (helium nuclei, each containing two protons and two neutrons). This matches the mixture of material observed (by *spectroscopy*) to exist in old stars, formed when the Universe was young, and explains where 99 per cent of the visible material in stars and galaxies comes from (to find out where the other 1 per cent is made, see *nucleosynthesis*).

Although the calculations were published in Alpher's Ph.D. thesis, Gamow decided that they merited wider circulation, and prepared a paper for the journal *Physical Review*. At this point his sense of humour overcame him. As he wrote later, in his book *The Creation of the Universe* (Viking, New York, 1952), 'It seemed unfair to the Greek alphabet to have the article signed by Alpher and Gamow only, and so the name of Dr Hans A. Bethe (*in absentia*) was inserted.' To Gamow's delight, the paper appeared with all three names, coincidentally in the issue of the journal dated 1 April 1948; and to this day it is known as the 'alpha beta gamma' paper, from the names Alpher, Bethe and Gamow. Later in 1948, Alpher and Robert **Herman** extended the theory to predict that the Universe today must be filled with **background radiation** at a temperature of about 5 K.

Alpha Centauri   The second closest star to the Sun, Alpha Centauri is actually a *binary star* with two components, prosaically named A and B, that orbit each other once every 80.1 years. A has a mass slightly greater than that of the Sun, while B has a mass about 90 per cent that of the Sun. The double star is just 1.33 *parsecs* away from the Solar System.

alpha decay   The process in which the nucleus of an atom emits an alpha particle and is transformed into a nucleus with four fewer atomic mass units and two fewer units of charge, corresponding to a different element. See *nucleosynthesis*.

alpha particle   Two protons and two neutrons held together by nuclear interactions to make a very stable *nucleus*. The alpha particle is essentially the nucleus of a helium atom which has lost its two electrons. Because it is so stable, the alpha particle is a fundamental building block in the process of *nucleosynthesis*, by which heavier elements are manufactured inside stars.

Alpher, Ralph Asher (1921–)   American physicist who worked with George Gamow and Robert Herman in the 1940s. Together, they made the first calculations of how elements could have been made in the *Big Bang*, and predicted the existence of the cosmic microwave *background radiation*, although this prediction was forgotten for 25 years. See *alpha beta gamma theory*.

Ambartsumian, Victor Amazaspovich (1908–96)   Georgia-born Armenian astrophysicist and pioneering cosmologist. He worked as an astronomer in the Soviet Union from 1981 onward and was a pioneer, in the mid-1950s, of the idea that the energetic events observed in *radio galaxies* were not a result of collisions between galaxies, but were caused by violent explosions taking place at the centres of individual galaxies.

amino acid   A complex molecule, chiefly (in most cases, entirely) composed of carbon, hydrogen, oxygen and nitrogen (see *CHON*), that is essential for life as we know it, and is a building block of protein. The first detection of an amino acid (glycine) in an *interstellar cloud* was reported in 1994.

Anaxagoras of Clazomenae (about 500–428 BC)   The first astronomer to give the correct explanation for eclipses of the Sun and Moon. Anaxagoras noticed that meteorites fell to Earth as lumps of red-hot iron, and reasoned that the Sun and stars were also balls of red-hot iron. Thinking that the

Earth was flat, he calculated the height of the Sun above the Earth as 4,000 miles, and the Sun's diameter as 35 miles. Using the same calculation but assuming that the Earth is round, Eratosthenes was able to calculate that it is actually the radius of the Earth that is about 4,000 miles. Pericles was one of Anaxagoras' pupils.

Anaximander of Miletus (611–547 BC)   The first philosopher to suggest that the surface of the Earth is curved. Anaximander, a pupil of Thales, thought that the surface of the Earth was like the surface of a cylinder curved in the north–south direction. He was also the first person to suggest that the Earth might float unsupported in space.

Andromeda galaxy   The nearest large galaxy to our own Milky Way Galaxy, Andromeda is a *disc galaxy* at a distance of about 700 kiloparsecs. It is the most distant object that can be seen with the naked eye, as a faint patch of light (a *nebula*) in the constellation Andromeda.

As long ago as the 18th century, Immanuel *Kant* had suggested that such nebulae might be complete star systems beyond the Milky Way, but even at the beginning of the 20th century this had not been proved. An alternative suggestion, that the nebulae were regions where stars were forming from clouds of gas within our Galaxy, also had supporters. The question was resolved in the 1920s, when Edwin *Hubble* used the new 100-inch (254 cm) telescope at Mount Wilson Observatory to identify individual stars in the outer parts of the nebula.

Some of these stars turned out to be Cepheid variables. Because the variations of these stars are related to their absolute magnitude, their apparent brightness enabled Hubble to work out the distance to the Andromeda 'nebula', proving that it was indeed another galaxy in its own right (see *cosmic distance scale*).

The distance estimated by Hubble has since been revised upward, chiefly through the work of Walter *Baade*, but Hubble's work established that our Milky Way is just one galaxy among many and that the Universe extends far beyond the borders of the Milky Way. At a distance of 700 kpc, the Andromeda galaxy (also known by its catalogue numbers as M31 or NGC224) is 60 kpc in diameter, about twice as large as our Milky Way Galaxy, and contains about 400 billion stars.

Anglo-Australian Observatory   Site of the *Anglo-Australian Telescope* and UK Schmidt Telescope, both at Siding Spring Mountain in Australia. See also *Mount Stromlo and Siding Spring Observatory*.

Anglo-Australian Telescope (AAT)   A 3.9 m reflecting telescope sited at Siding Spring in Australia (the *Anglo-Australian Observatory*) at an altitude of 1,150 m. Jointly funded by the UK and Australian governments, the AAT can make observations both in optical light and in the infrared.

ångström   An obsolete unit of length often used in the past to measure

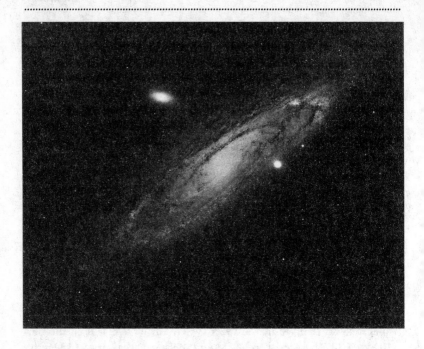

The Andromeda Galaxy (also known as the Andromeda nebula) is the nearest large disc galaxy to our own. Our Milky Way would look something like this if viewed from outside.

wavelengths of *light* and other forms of radiation. 1 Å is one-tenth of a billionth of a metre, or 0.1 nanometres. The unit is named after A. J. Ångström, a Swedish pioneer of *spectroscopy*, who lived from 1814 to 1874.

antenna In astronomy, the component of a *radio telescope* which receives radio waves from space, equivalent to the antennas that receive TV signals for domestic use and pass them on to the TV set. Antennas can also be used, as with TV transmissions, to transmit radio waves or other electromagnetic radiation. Some large radio telescope antennas have been used in this way to bounce radar signals off other planets.

anthropic principle The idea that the existence of life in the Universe (specifically, human life) can set constraints on the way the Universe is now, and how it got to be the way it is now.

The power of anthropic reasoning is best seen by an example. In order for us to exist, there has to be one star (the Sun), orbited at the appropriate distance by one planet (the Earth), made of the right mixture of chemical elements (particularly including carbon, nitrogen, oxygen and the primordial hydrogen left over from the Big Bang). These elements play a key role in life processes (see **CHON**). At first sight, it may seem that the existence of the rest of the Universe, containing millions of galaxies scattered across billions of light years of space, is irrelevant to our existence.

But where did the elements of which we and the Earth are made come from? The **Big Bang** produced only hydrogen, helium and traces of a few light elements. Carbon and other heavy elements were manufactured inside stars (see **nucleosynthesis**), which had to run through their life cycles and explode, scattering the heavy elements across space to form clouds of material from which later generations of stars, including the Sun, and their attendant planets could form. This took billions of years. The evolution of life on a suitable planet to the point where intelligent beings could notice their surroundings and wonder about the size of the Universe took more billions of years. All the while, the Universe was expanding. After billions of years, it is inevitably billions of light years across. So the fact that we are here to ask questions about the size of the Universe means that the Universe must contain many stars, and must be billions of years old and billions of light years across.

This almost (but not quite) tautological argument ('we are here because we are here') seems to have been first expressed in a cosmological context by Robert **Dicke**, who pointed out, in a paper published in 1957, that the size of the Universe is 'not random but conditioned by biological factors' (quoted by Barrow and Tipler, see below). By then, Fred **Hoyle** had already used what can be seen with hindsight to be a genuine prediction based on anthropic reasoning to explain how the elements were made. He argued that the existence of carbon in our bodies required a certain **nuclear fusion** reaction to take place inside stars; experiments were carried out to test his prediction, and the reaction was found to occur exactly as Hoyle had predicted (see **B^{2}FH**, **triple alpha reaction**). This powerful application of anthropic reasoning has not always received the credit it deserves.

Interest in the anthropic principle among cosmologists did not really take off until 1974, when the British researcher Brandon Carter drew a distinction between the 'weak anthropic principle' and the 'strong anthropic principle'. These variations on the theme were later defined by John Barrow and Frank Tipler (see below) as follows:

Weak anthropic principle: *The observed values of all physical and cosmological quantities are not equally probable, but they take on values restricted by the requirement that there exist sites where carbon-based life can evolve and by the requirement that the Universe be old enough for it to have already done so.*

Strong anthropic principle: *The Universe must have those properties which* *allow life to develop within it at some stage in its history.*

The weak version of the principle suggests that in some sense the Universe had a 'choice' about how it emerged from the Big Bang. For example, the strength of the **gravitational interaction** might have been different from the value we know. Suppose gravity were much stronger than it is in our Universe. Then, other things being equal, stars would be smaller than in our Universe and would burn their nuclear fuel more quickly in order to hold themselves up against the pull of gravity. Make gravity strong enough, and stars would burn out before there was enough time for complex life forms like ourselves to evolve.

On this picture, an infinite universe might be separated into 'domains' which have different laws of physics. These domains might be separated from one another in space, beyond the range of our telescopes, or in time – in some sense, perhaps, 'before' the Big Bang. Or they might exist in some extradimensional superspace, connected by **wormholes**. Life forms like us would exist only in domains where stars lived long enough for complex creatures to evolve, and where other conditions were, like baby bear's porridge, 'just right'. This is sometimes referred to as the Goldilocks effect.

The strong version of the anthropic principle suggests that the Universe had no choice about how it emerged from the Big Bang, and is in some sense 'tailor made' for humankind. Some physicists, notably John **Wheeler**, have linked this to ideas from quantum physics, which suggest that nothing is real unless it is observed – that is, that the physical reality of our Universe depends on the presence of intelligent observers who notice its existence, and that this ensures that the fundamental interactions and the constants of nature, such as the strength of gravity, have the values we know.

Others see the 'coincidences' that permit the existence of life in the Universe as evidence of a designer at work. Hoyle has said that 'the laws of nuclear physics have been deliberately designed with regard to the consequences they produce inside stars' (*Galaxies, Nuclei and Quasars*, Heinemann, London, 1965), although few cosmologists agree with him.

At this level, the debate about anthropic cosmology is a variation on the old argument from design used to 'prove' the existence of God by pointing out that living things are far too complicated to have arisen by chance, an argument put forward forcefully by William Paley (1743–1805) in the 18th century. The counter-argument is that evolution by natural selection has produced the complexity of living things on Earth, and has fitted them to their environment, without any need for the hand of God.

Intriguingly, this counter-argument has now been extended into the cosmological realm, largely through the work of the American mathematical physicist Lee Smolin. He has argued that when **baby universes** bud off from their parents through black holes, the laws of physics in the 'new' universe may be slightly different from the laws of physics in the 'old' universe. This variation in the laws of physics may provide the raw material on which natural selection can operate at the level of the universes themselves, with universes that are more efficient at producing black holes, and thereby producing more universes like themselves, winning out in some cosmological struggle for space.

According to this argument, laws of physics that favour the conversion of matter into many black holes will be favoured by the selection process. Smolin argues that our Universe is a very probable end product of such an evolutionary process, and that the laws of physics which seem to us to be so nicely fitted to our existence are in fact fine-tuned for the manufacture of black holes and more baby universes. Our existence might then be seen as a parasitic consequence of the fact that those laws happen to allow the existence of carbon and the other elements on which life as we know it is based.

Further reading: John Barrow and Frank Tipler, *The Anthropic Cosmological Principle*; John Gribbin and Martin Rees, *Cosmic Coincidences* (also known as *The Stuff of the Universe*, Penguin, London, 1995); John Gribbin, *In the Beginning*.

antimatter A form of matter in which key properties of the individual particles have been reversed. For example, the electron has a counterpart in nature, known as an antiparticle, which carries a positive electric charge equal in magnitude to the negative charge on an electron. For this reason, the 'anti-electron' is also known as the positron. Even particles that do not carry charge, however, can have antiparticle counterparts, because other properties associated with the **fundamental interactions** are reversed.

Antiparticles occur naturally in **cosmic rays** and can be manufactured in particle accelerators ('atom smashing' machines) like those at CERN, in Geneva, and Fermilab, in Chicago. In any process where sufficient energy is available, some of the energy (E) can be converted into a particle–antiparticle pair (such as an electron plus a positron) in which the combined mass of the two particles (m) is given by the equation $E = mc^2$ (see **special theory of relativity**). Whenever a particle meets an equivalent antiparticle, both particles are annihilated, releasing the energy equivalent of their combined mass as radiation.

In principle, it is possible for antimatter atoms to exist (such as antihelium, in which a nucleus of two antiprotons and two antineutrons is surrounded by two positrons), and to combine to form antimatter objects including planets and people. But any antimatter that came into contact with everyday matter would be destroyed in a blast of energy. Although there has been speculation that the existence of everyday matter in the stars and galaxies of the visible Universe might be balanced in some sense by the existence of stars and galaxies made of antimatter in far distant regions of space, there is no evidence that any of the visible galaxies contain significant quantities of antimatter.

antiparticle   See *antimatter.*

aperture   The diameter of the objective lens in a *refracting telescope*, the main mirror in a *reflecting telescope*, or the dish of a *radio telescope*.

aperture synthesis   A trick used in *radio astronomy* to make radio telescopes simulate the power of much larger instruments. Several different radio dishes, each antenna perhaps 25 m in diameter, are linked electronically to observe the same part of the sky. As the Earth rotates, the telescopes in the array are carried round with it, each sweeping out an arc that corresponds to part of the surface of a non-existent 'super telescope' with an aperture several kilometres in diameter. The information received by all the telescopes in the array is recorded and then combined using a computer to produce an image of the sky (a synthesis) corresponding to the view that such a super telescope would have. In practice, the simulation of the super telescope is never perfect, but the result can provide far more information than any of the instruments in the array on its own.

Aperture synthesis. The Ryle telescope, just outside Cambridge, uses the principle of aperture synthesis to mimic the power of a single very large dish using several smaller dishes.

The technique was developed in England and in Australia in the 1950s, and the first important instrument of this kind was the Cambridge 'one mile' telescope (later extended to 3 miles, or 5 km, and now known as the *Ryle telescope*), which used individual dishes spaced along a straight railway line. Some of the dishes could be moved along the track from one day to the next, to mimic different parts of the super telescope. The *Very Large Array* and the *Australia Telescope* both use the principle of aperture synthesis.

aphelion   The point in its orbit at which a planet or other object is at its furthest distance from the Sun. The Earth is at aphelion on 3 July each year.

APM   An Automatic Plate-measuring Machine developed at the University of Cambridge, and used to extract astronomical information from photographic plates, especially those obtained by *Schmidt cameras*. One of the tasks of the APM is to identify quasar candidates from analysis of hundreds of thousands of star images.

Apollo asteroids   A group of *asteroids* whose orbits all have *perihelia* inside the orbit of the Earth and *aphelia* outside the Earth's orbit, so that the asteroids cross the Earth's orbit as they move around the Sun. They are named after the asteroid Apollo-1862, which was discovered when it came within 0.07 astronomical units of the Earth in 1932. Apollo-1862 itself is about 1.4 km across. The impact of such an object with the Earth would cause widespread devastation. See *Doomsday asteroid*.

apparent magnitude   The brightness of a star, measured on a standard *magnitude scale*, as it appears from Earth. Because stars are at different distances from us, and objects that are the same brightness will look fainter if they are further away, the apparent magnitude cannot be used on its own to tell us how bright a star really is. See *absolute magnitude*.

arc minute   A measurement of angle. There are 360 degrees in a circle, 60 arc minutes in a degree, and 60 arc seconds in an arc minute. The angle covered by the Moon on the sky (its angular diameter) is just over 30 arc minutes (half a degree).

arc second   A measurement of angle. 60 arc seconds make up one arc minute, 60 arc minutes make up one degree, and 360 degrees make up a circle. So there are 1,296,000 (that is, 1.296×10^6) arc seconds in a circle.

Arecibo radio telescope   The world's largest dish telescope, the Arecibo instrument is situated in a natural hollow in the ground in Puerto Rico. It is operated by Cornell University, in the United States, and has a diameter of 305 m. Because the telescope has been made by covering the natural hollow with reflecting panels, it cannot be steered in the same way as instruments like the famous *Jodrell Bank* telescope, but scans a strip of sky as the Earth rotates.

Aristarchus of Samos (early third century BC)   The first person to attempt to work out the relative distances to the Sun and Moon after it was realized that the Earth is round. An accurate estimate was impossible with the knowledge available at the time (Aristarchus was a contemporary of Archimedes), but this represented the first step towards an appreciation of the size of the Universe, and showed that the Sun is much larger than the Earth. Aristarchus realized that the apparent movement of the stars across the sky is caused by the Earth's rotation, and taught that the Sun was at the centre of the Universe with the Earth and other heavenly bodies moving in circles around it – anticipating the ideas of *Copernicus* by some 1,900 years.

Aristotle (384–322 BC)   Among his many interests, Aristotle wrote about *cosmology*. Building on the ideas of his predecessors, he came up with the model of the Universe as a series of concentric spheres, centred on the Earth and rotating about it. The outermost sphere carried the fixed stars, while inner spheres carried the planets Saturn, Jupiter and Mars, followed by the Sun's sphere and by the spheres for Venus, Mercury and (closest to the Earth) the Moon. In spite of the ideas put forward by *Aristarchus*, this 'geocentric model' held sway until the time of *Copernicus*, in the 16th century. Aristotle also proved that the Earth is spherical. He pointed out that it casts a circular shadow on the Moon, and that individual stars appear to move higher or lower above the horizon as you travel north–south.

Arizona Meteor Crater = *Barringer Crater*.

Arp, Halton Christian ('Chip') (1927–)   American astronomer who has made a systematic study of unusual galaxies and suggests that the usual interpretation of the cosmological *redshift* as a distance indicator is not always applicable.

   Arp was born on 21 March 1927, in New York City. He studied at Harvard University and at the California Institute of Technology, where he obtained his PhD in 1953. He spent most of his career as an observational astronomer in California, at the *Mount Wilson and Palomar Observatories* and at the *Hale Observatory*, but worked at the University of Indiana from 1955 to 1957.

   As an expert observer capable of getting the best out of the telescopes available to him, Arp developed two strands to his career. One followed a conventional path, notable for studies of *novas* and *variable stars* which have helped to establish the *cosmic distance scale*. In particular, his studies of novas in the Andromeda galaxy established a link between the absolute magnitude of a nova and the rate at which its apparent brightness declines. This means that, by studying the *light curve* of a nova in a galaxy, astronomers can infer its absolute magnitude, and by comparing this with its apparent magnitude they can work out the distance to the nova, and therefore to the galaxy.

But in spite of his considerable achievements in these conventional studies, Arp is best known for his unconventional ideas about redshifts and the activity of galaxies. Throughout his career, he has collected photographic examples of unusual galaxies which seem to be ejecting jets of material or to be physically connected to other galaxies or quasars. In 1965 he published an *Atlas of Peculiar Galaxies*. Arp claimed that in many cases faint bridges of stellar material could be seen connecting a galaxy to a neighbouring object, and he found that in such cases the two objects sometimes have very different redshifts.

According to the conventional understanding of cosmological redshifts, this would mean that the two objects are at very different distances from us, and just happen to lie next to each other on the sky. But if they are physically connected by a bridge of stars, they must be at the same distance from us. In that case, at least one of the redshifts could not be a guide to distance.

Arp's peculiar galaxies are very much a minority of those studied by astronomers, and there is no doubt that the redshift–distance relationship works well for galaxies as a whole. Few astronomers accept that any of the objects with 'discrepant' redshifts discovered by Arp are actually physically connected. There is, however, still a possibility that some objects with high redshifts identified as quasars have actually been ejected from relatively nearby galaxies, like a shell being fired from a gun, and that their high redshifts are produced by some other process, not by the expansion of the Universe. This echoes some of the ideas of Victor **Ambartsumian**.

Arrhenius, Svante August (1859–1927)   Swedish physical chemist who received the Nobel Prize for Chemistry in 1903 and was director of the Nobel Institute of Physical Chemistry, in Stockholm, from 1905 to 1927. He suggested that the Earth might have been 'seeded' with life from space, in the form of micro-organisms riding on dust particles (see **panspermia**). Similar ideas have been taken up recently by Sir Fred **Hoyle** and Chandra Wickramasinghe, and by Francis Crick, co-discoverer of the structure of DNA.

arrow of time   One of the greatest mysteries in science is the distinction between the past and the future. At a subatomic level, neither the old ideas of *classical mechanics* nor the modern theory of *quantum mechanics* distinguish between the past and the future. In a typical interaction involving subatomic particles, two particles may come together and interact in some way to produce two different particles, which then separate. The laws of physics say that almost every such interaction can run equally well in reverse, with the 'final' two particles coming together and interacting to make the 'original' two particles. At this level, there is no way to distinguish the past from the future simply by looking at each pair of particles.

But at the macroscopic level of our human senses, the distinction between the past and the future is obvious. Things wear out; people get older. In the equivalent of the particle interaction, we can imagine a wine glass balanced precariously on the edge of a table, then falling to the floor and smashing. We never see smashed glasses reassembling themselves, even though every interaction involving the atoms of the wine glass as it smashes is, according to the known laws of physics, reversible. If we were shown two still photographs, one of the glass on the table and the other of the smashed glass on the floor, we would have no difficulty saying which one was taken first and which one was taken later in time. There is an inbuilt arrow of time, pointing from the past to the future, when we are dealing with complex systems which contain many particles.

It is, however, important to distinguish between an arrow which *points* into the future and one which *moves* into the future. The correct analogy is with the needle of a compass, which points to the north, but does not have to be moving north (or anywhere else) at all. If we had a movie of the glass falling off the table, instead of just two 'before and after' pictures, and if the individual frames of the film were cut up and mixed together, we would still be able to sort them out into the right order. The film does not have to be running through a projector for the distinction between the past and the future to be clear.

Some scientists (and philosophers) argue that our impression of time passing may be no more than an illusion, as our minds scan the events of our own personal histories, like the movie being run through a projector and displayed on a screen. Underlying reality, both in the past and the future, may still be there, like the separate frames of film in the movie, even though our attention is forced to follow the story sequentially, one frame of the movie at a time. Whether or not this is true (and it is a highly contentious issue), it is still true that there is a distinction between the past and the future, which can be represented by an arrow pointing from the past into the future.

This distinction can be expressed mathematically. The science of thermodynamics is based on analysis of the way things change as we 'move' from the past into the future. The key insight is that the amount of disorder in the Universe is always increasing – glasses break, but do not reassemble themselves. Physicists measure disorder in terms of a quantity called entropy; the most fundamental law of physics is that the entropy of a closed system always increases (the **second law of thermodynamics**).

You can get round this law in an open system, which has an external source of *energy*. The second law seems to be violated on Earth, because living things grow and people can, for example, take a pile of bricks and turn it into a much more ordered structure, in the form of a house. But all of this depends on an input of energy, originally from the Sun. The decrease in entropy taking place on Earth is much less than the increase in entropy associated with the *nuclear fusion* reactions going on inside the Sun and the way it radiates heat out into space. The entropy of the whole Universe increases as time passes – that is, states of the Universe with higher entropy correspond to the future direction compared with states of lower entropy.

This same arrow of time is built into the structure of the Universe in another way. The Universe is expanding (see *redshift*), so galaxies are moving further apart. States of the Universe in which galaxies are further apart lie in the future direction compared with states of the Universe with galaxies closer together. The ultimate arrow of time is provided by the *Big Bang* itself – wherever and whenever you are in the Universe, the Big Bang always lies in the past direction of time. Somehow, the Universe emerged from the Big Bang with a low enough entropy to permit stars, planets and people to form; it has been running down ever since. Heat cannot flow from a colder object into a hotter one (another way of expressing the second law), so there is a one-way flow of energy out of the bright stars and into the cold Universe. If and when all the stars and other sources of energy in the Universe have given up their heat, the entire Universe will be in a state of uniform temperature in which nothing ever changes. It will have suffered a 'heat death'.

This highlights another way of looking at the arrow of time and the concept of entropy. The amount of energy in a closed system (or in the whole Universe) cannot change – this is the first law of thermodynamics. Even when mass is converted into energy in line with Einstein's equation $E = mc^2$, mass is regarded as a form of stored energy, so no 'new' energy is created. What the second law then tells us is that, in any interaction in a closed system, the amount of 'useful' energy decreases.

Useful energy is energy that can do work. For example, when the glass falls from the table it could, in principle, be connected to a pulley system that turns a generator and converts the gravitational energy associated with the falling glass into electrical energy. But when the glass is falling freely, this potentially useful gravitational energy is converted into energy of motion (kinetic energy). When the glass hits the floor and shatters, the kinetic energy is turned into heat and dissipated as the atoms and molecules of the glass and the floor are shaken up and vibrate more rapidly. This heat energy is ultimately turned into infrared radiation and dissipated into space, and can never be made to do useful work. We never see radiation coming in from space to make the atoms and molecules of the

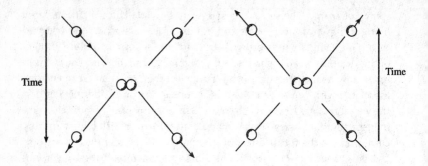

Arrow of time. The laws of physics do not identify a flow of time. An event in which two balls collide and then move apart, for example, can be described just as well whichever way the arrow of time points.

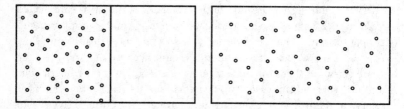

And yet, we all know which way time flows. In this experiment, when the partition is removed, gas from the left-hand side of the box expands to fill the box. We never see gas spontaneously gathering itself in just one half of the box.

floor and the broken bits of glass jiggle about in just the right way to stick the glass together and make it leap up on to the table.

Even if the falling glass had been made to do useful work by turning a generator, some of the energy would have been lost in friction and turned into heat. No energy conversion process is perfect, which is why you can never make a perpetual motion machine (for example, by using the electricity generated by the falling glass to drive a motor to lift the glass back on to the table).

But we are still faced with the puzzle that when the glass falls and shatters every interaction involving a pair of atoms or molecules is, in principle, reversible. Why does it never happen in practice? One possibility is that it is not absolutely *impossible* for the process to 'go backwards', but only extremely unlikely that this should happen.

Forget about the falling glass, for now, and think about a simpler system – a box divided into two halves by a partition, with gas on one side of the partition and a vacuum on the other. If you remove the partition, the gas will spread out to fill the whole box (and, incidentally, cool down a little as it does so – this is the principle on which the refrigerator works). No matter how long you sat and watched the box, you would never expect to see all the atoms and molecules of gas move back into one-half of the container, leaving empty space in the other half. And yet, every collision between two of the particles in the box is, in principle, reversible. If we could wave a magic wand and reverse the motion of every particle, the gas would have to go back where it came from.

In the 19th century, the French physicist Henri Poincaré (1854–1912) showed that such an 'ideal' gas, trapped in a box, must eventually pass through every possible arrangement of particles that is allowed by the laws of thermodynamics. As the atoms and molecules bounce around, sooner or later they will take up any permitted arrangement, including one with all the gas in one-half of the box. If we wait long enough, the system will return to its starting point, and time will seem to run backwards.

But 'long enough' is the key term here. The time it would take for all the particles to pass through all possible arrangements is called the Poincaré cycle time, and it is related to the number of particles in the box. Even a small box of gas might contain 10^{22} atoms, and it would take that many atoms a time much longer than the age of the Universe to pass through every possible arrangement of atoms. Typical Poincaré cycle times have more zeros in the numbers than there are stars in all the known galaxies put together, representing the odds against any particular pattern occurring while you are watching the box of gas, or waiting for the glass to leap on to the table.

So the standard 'answer' to the puzzle of why the world is reversible on the microscopic scale but irreversible on the macroscopic scale (why the arrow of time points only one way) is that the law of increasing entropy is a statistical law; a decrease in entropy is not so much forbidden as extraordinarily unlikely.

This led the Austrian physicist Ludwig Boltzmann (1844–1906), also in the 19th century, to suggest that the Universe might be a gigantic statistical freak. If we imagine a situation in which the heat death has occurred and everything is uniform, then according to Boltzmann's interpretation of Poincaré's work, from time to time it will just happen, by chance, that all the particles in one part of the Universe will be moving in just the right way to create stars, or galaxies, or a Big Bang. In effect, time would temporarily run backwards in such a region of the Universe, creating order out of disorder. Then, the low-entropy bubble 'unwinds' as it returns to a more probable state.

This idea is not taken seriously by cosmologists today, and has been superseded by the Big Bang model, although it does have some similarities to variations on the *Steady State* model. But it still offers intriguing insights into the nature of time, as has been pointed out by Paul Davies, now of the University of Adelaide.

If the arrow of time always points in the direction of increasing entropy, it may not make sense to describe time as 'flowing backwards' as the Boltzmann bubble grows. An intelligent observer in the interesting region of the Universe would still experience an arrow of time pointing towards the high-entropy, heat death state. In other words, even if the Universe were 'really' collapsing instead of expanding, moving towards a *singularity* instead of away from one, intelligent observers like ourselves might still perceive 'the future' as being the time when galaxies are further apart.

This is more than merely philosophical hair-splitting, because some variations on the Big Bang model suggest that the expansion of our Universe will one day halt and then go into reverse. Will time itself run backwards if and when this happens? And if it does, will intelligent observers notice, or will they think they are living in an expanding Universe even though it is contracting? Perhaps we do live in a contracting Universe, and have not noticed!

Further reading: Paul Davies, *The Runaway Universe*; Philip K. Dick, *Counter-Clock World*; Ilya Prigogine and Isabelle Stengers, *Order Out of Chaos*.

asteroid Rocky object, smaller than a planet, in orbit around the Sun. Most asteroids congregate in orbits between those of Mars and Jupiter, where there are estimated to be a million objects bigger than 1 km across. The largest, Ceres, has a diameter of 933 km. Asteroids are probably left-over cosmic rubble from the formation of the Solar System, and may represent the kind of material that planets like the Earth formed out of. See *minor planets*.

Astronomers Royal See *Royal Greenwich Observatory*.

astronomical distance scale See *cosmic distance scale*.

astronomical unit (AU) A measure of distance defined as the average distance between the Sun and the Earth over one orbit (one year). 1 AU is equal to 149,597,870 km (= 499.005 light-seconds).

astrophysics The application of physics to an understanding of the workings of everything in the Universe, including (but not exclusively) stars, and of the Universe itself. Astrophysics began in the 19th century with the application of *spectroscopy* to the stars, which led to measurements of their temperatures and compositions. Astrophysicists are able to study matter in the Universe under extreme conditions (of temperature, pressure and density) that cannot be achieved in laboratories on Earth.

Atkinson, Robert D'escourt (1898–1982) Welsh astrophysicist who graduated from Oxford University in 1922 and obtained a PhD from the University of Göttingen in 1928. He subsequently worked at Rutgers University in New Jersey. Atkinson's chief contribution to science was to show, with Fritz Houtermans, how energy could in principle be produced inside stars by sticking nuclei together (*fusion*). Their calculations, at the end of the 1920s, showed that the *tunnel effect* proposed by George *Gamow* could allow nuclei (such as protons) to come close enough together to combine, in spite of the repulsion caused by their electric charge. See *proton–proton reaction*.

atom The smallest component of an element, the unit that takes part in chemical reactions. An atom consists of a tiny central nucleus, which has positive electric charge, surrounded by a cloud of electrons, which each have negative charge. The overall charge is zero. The proportions of the size of the nucleus to the size of the electron cloud are roughly those of a grain of sand in the middle of the Albert Hall. The largest atom, caesium, is just 0.0000005 mm (that is, 5×10^{-7} mm) across; it would take 10 million atoms side by side to stretch across the gap between two of the points on the serrated edge of a postage stamp.

atomic clock General name for any of a variety of timekeeping devices which are based on regular vibrations associated with atoms. The first atomic clock was developed in 1948 by the US National Bureau of Standards, and was based on measurements of the vibrations of atoms of nitrogen oscillating back and forth in ammonia molecules, at a rate of 23,870 vibrations per second. It is also known as an ammonia clock.

The standard form of atomic clock today is based on caesium atoms. The *spectrum* of caesium includes a feature corresponding to radiation with a very precise frequency, 9,192,631,770 cycles per second. One second is now defined as the time it takes for that many oscillations of the radiation associated with this feature in the spectrum of caesium (see *atomic time*). This kind of atomic clock is also known as a caesium clock; it is accurate to one part in 10^{13} (one in 10,000 billion), or 1 second in 316,000 years.

Even more accurate clocks have been developed using radiation from hydrogen atoms. They are known as hydrogen maser clocks, and one of these instruments, at the US Naval Research Laboratory in Washington, DC, is estimated to be accurate to within 1 second in 1.7 million years. In principle, clocks of this kind could be made accurate to one second in 300 million years.

atomic mass unit (amu) The unit of mass used to measure masses of atoms and molecules. Defined as one-twelfth of the mass of a carbon atom, this is roughly the same as the mass of a proton, 1.66×10^{-27} kg. In round terms, the atomic mass of an atom in amu is equal to the total number of protons and neutrons in its nucleus.

atomic number Number denoting the number of protons in the nucleus of each atom of a particular element (different *isotopes* of the same element still have the same atomic number). Also equal to the number of electrons associated with each atom of that element. The atomic number of hydrogen is 1; that of iron 26.

atomic time Time measured by the oscillations of an *atomic clock*. Since 1967, the second has been defined in terms of atomic time, as the time taken for 9,192,631,770 oscillations of the radiation corresponding to a particular line in the spectrum of atoms of caesium-133. International Atomic Time is measured in seconds from 1 January 1958 (that is, from the astronomical moment of midnight, *Greenwich Mean Time*, on the night of 31 December 1957/1 January 1958).

Australia telescope A *radio telescope* which combines eight antennae at three different sites in New South Wales, using the principles of *aperture synthesis* and long baseline *interferometry*. In some configurations, the instrument simulates the *resolving power* of a telescope with a diameter of 300 km.

axion A hypothetical *elementary particle* required by some *grand unified theories*, but not yet detected in experiments in laboratories on Earth (allegedly named after the washing powder). If axions exist, they each have a very small mass, less than one-hundred-thousandth of an *electron Volt* (10^{-5} eV); but there may be so many of them in the Universe, left over from the *Big Bang*, that they provide the *cold dark matter* which some cosmological models require to make *spacetime* flat. Proof that axions exist would therefore be powerful evidence that our understanding of the laws of physics is correct at both extremes, on the smallest and largest scales. See *inflation*.

B^{2}FH Short-hand reference to the names of Geoffrey and Margaret *Burbidge*, Willy *Fowler* and Fred *Hoyle*. In 1957, this team of astrophysicists published one of the classic scientific papers of all time, describing how all the naturally occurring varieties of *nuclei* except primordial hydrogen and helium (see *alpha beta gamma theory*) are built up inside stars by *nucleosynthesis*. Fowler alone later (in 1983) received a Nobel Prize largely for this work, although Hoyle was the leader of the team. In the words of the Swedish Academy of Science's announcement of Fowler's prize, this paper (also known as 'B^{2}FH') 'is still the basis of our knowledge in this field, and the most recent progress in nuclear physics and space research has further confirmed its correctness'. It literally explains how the elements in your body were put together inside stars. See also *triple alpha process*.

Baade, (Wilhelm Heinrich) Walter (1893–1960) German-born astronomer who spent most of his career in the United States, and was responsible for a major revision in the *cosmic distance scale* in the 1940s and early 1950s.

Baade was born on 24 March 1893, at Shröttinghausen. He studied at Münster and Göttingen universities, obtaining a PhD in 1919. After working at the Bergedorf Observatory of Hamburg University for eleven years, he emigrated to the United States, where he joined the staff of the *Mount Wilson Observatory* in 1931. In 1948 he made the short move to the *Mount Palomar Observatory*, where he stayed until 1958, when he reached the statutory retiring age for his post. He then returned to Göttingen, where he died on 25 June 1960.

At Mount Wilson, Baade worked with Fritz *Zwicky* and Edwin *Hubble* on studies of supernovae and the distances to other *galaxies*. At the time, the 100-inch (2.5 m) telescope at Mount Wilson was the largest in the world, and in great demand for many kinds of astronomical research. But after the United States entered the Second World War, many astronomers were inducted into military research or into the armed forces. Baade, as a German national, was not considered a suitable person to participate in the war effort directly, and by 1943 he was one of the few active astronomers left at Mount Wilson, with almost unlimited access to the telescope. He also benefited from the wartime blackout of nearby Los Angeles, which reduced the light pollution and enabled him to push the telescope to its limits, taking long-exposure photographs that, for the first time, resolved individual stars in the inner part of the Andromeda galaxy.

The observations showed Baade that there are two very different types of star in our neighbouring galaxy. The first type, which he called *Population I*, are mainly young stars. They are hot and blue, and lie in the *spiral arms*. The second type, *Population II*, are older, cooler and redder, and lie in the central part of the galaxy and the *globular clusters* of the *halo*. This distinction is now known to be characteristic of *disc galaxies*.

After the war, even with the lights of Los Angeles switched back on, Baade was able to continue with his work at the new 200-inch (5 m) telescope at Mount Palomar. He found that each of the two populations of stars has its own distinctive population of *Cepheid variables*. Both Population I Cepheids and Population II Cepheids have characteristic *period–luminosity relations*, but the two relations differ from one another. The period–luminosity relation used by Hubble in the first attempt to determine an extragalactic distance scale had been developed from studies of what were now known to be Population II Cepheids in our Galaxy; but it had then been applied to what Baade had found to be the brighter Population I Cepheids of the Andromeda galaxy.

When Baade recalculated the distance to the Andromeda galaxy using the correct period–luminosity relation, in 1952 he came up with a figure of 2 million light years (just over 600,000 parsecs), compared with Hubble's estimate of 800,000 light years (just under 250,000 parsecs). Because the distance estimate for the Andromeda galaxy had been a key step in Hubble's calculation of the distances to galaxies based on *redshifts*, this meant that the estimated distances to all galaxies with measured redshifts were more than doubled at a stroke.

Because the relationship between redshifts and distances (Hubble's Law) is also an indicator of the age of the Universe, the estimated time since the *Big Bang* was also more than doubled by Baade's revision of the Cepheid distance scale, from 2 billion to 5 billion years. This came as a great relief to astronomers, since in 1952 it was already clear from geological evidence that the Earth is more than 4 billion years old, and it is highly unlikely that the Earth could be older than the Universe.

Continuing improvements in observational techniques have since pushed the estimate of the age of the Universe out to as much as 20 billion years, compared with the Earth's age of about 4.5 billion years.

baby universes   Regions of *spacetime* connected to one another by *wormholes*. According to some interpretations of the equations of the general theory of relativity, when an object in our Universe collapses to form a black hole, it can pass through the singularity at the heart of the black hole and expand into a different region of spacetime. This region of spacetime expanding away from a singularity would be exactly equivalent to the expansion of our Universe away from the Big Bang singularity. Because of *inflation*, such a baby universe could become as large as our own Universe, even if the material that went into the original black hole was only a few times more than the mass of our Sun.

It is possible that our Universe was produced in this way from the collapse of a black hole in another region of spacetime, and that the overall structure of spacetime (the 'Metauniverse') is a series of interconnected bubbles, resembling froth on a glass of beer, with no beginning and no end. One way to picture this is to imagine the spacetime of our Universe represented by the skin of an expanding balloon. A baby universe would correspond to a piece pinched off from the balloon and expanding in its own right, with new baby universes budding off from its own skin, and so on indefinitely.

Bizarre though these ideas may seem, they have been taken seriously and investigated mathematically by researchers including Stephen *Hawking* in Britain, Sidney Coleman and Lee Smolin in the United States, and Andrei Linde from Russia. They are discussed in detail in *In the Beginning*, by John Gribbin (Penguin, 1993).

background radiation   The Universe is filled with a sea of radiation at a temperature of just over 2.7 *Kelvin*, detectable at microwave radio fre-

Baby. A new universe can 'pinch off' from the fabric of our spacetime like a bubble pinched off from the skin of an expanding balloon.

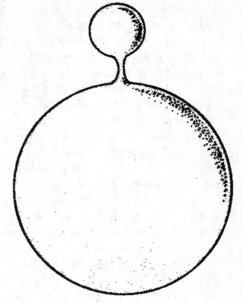

Babies. The process can repeat indefinitely, so that our Universe may be just one bubble in an infinite array of universes.

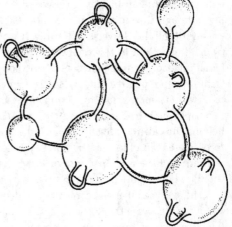

quencies both by Earth-based radio telescopes and by instruments on board artificial satellites. This is interpreted as direct evidence of the **Big Bang** fireball in which the Universe was born. The discovery of the background radiation is therefore the most important observation made in cosmology since the discovery by Edwin *Hubble* that the Universe is expanding; but the discovery was not made easily.

The first person to attempt to describe the conditions in the Big Bang quantitatively was George *Gamow*, in the 1940s. He applied the developing understanding of *quantum physics* to investigate the kind of *nuclear interactions* that would have occurred at the birth of the Universe, and found that primordial hydrogen would have been partly converted into helium (see *alpha beta gamma theory*).

The amount of helium produced in this way depended, according to the calculations, on the temperature of the Big Bang at the time these interactions were taking place. It would have been filled with a fireball of hot, short-wavelength *black body radiation*, in the form of X-rays and gamma rays. Gamow's team realized that the hot radiation corresponding to this fireball would have thinned out and cooled as the Universe expanded, but should still exist in a highly redshifted form, as radio waves.

Because there is nowhere 'outside the Universe' for the radiation to escape into, it always fills the Universe, just as the gas inside a balloon always fills the balloon. If we stretched the balloon to make it bigger, without letting any more gas in, the density of the gas inside the balloon would get less. In the same way, as the Universe expands, the density of the radiation that fills it gets less. This corresponds to a decrease in temperature and an increase in wavelength of the radiation – a redshift. But although the radiation had cooled, like the gas filling the balloon it would still fill the Universe uniformly. It would shine down on the Earth from all directions in space. And the amount by which the waves of the radiation had been stretched by the expansion of the Universe would, Gamow's team pointed out, give their temperature today.

Two of Gamow's students, Ralph *Alpher* and Robert *Herman*, calculated, in a paper published as early as 1948, that in order to make the amount of helium 'cooked' in the Big Bang match the amount revealed by *spectroscopy* to be present in old stars, the relict radiation from the Big Bang fireball would now have a temperature of just 5 K. Gamow himself published a slightly higher figure in 1952, in his book *The Creation of the Universe*.

The exact number you come up with depends on detailed assumptions about the physics of the Big Bang, and estimates of the age of the Universe. But as a rule of thumb, the temperature of the background radiation in Kelvin is given by taking the number 10^{10} (a 1 followed by ten zeros) and dividing it by the square root of the age of the Universe in seconds. So one second after the beginning of time, the temperature was 10 billion degrees, after 100 seconds it was 1 billion degrees, and after an hour it was just 170 million degrees. For comparison, the temperature at the heart of our Sun is about 15 million degrees.

But neither Gamow nor his colleagues realized that the technology to 'take the temperature' of the Universe already existed in the early 1950s. They did not press radio astronomers to make the observations that would have revealed the presence of background radiation, and no radio astronomer seems to have noticed the paper predicting the existence of such radiation. Curiously, though, observations which indicated that the Universe has a temperature close to 3 K had been made in the 1930s, using spectroscopy.

These observations, of the spectrum of a compound known as cyanogen (CN), revealed the temperature of clouds of interstellar material in our Galaxy. In 1940, Andrew McKellar, of the Dominion Astrophysical Observatory in Canada, interpreted these observations as indicating a temperature for the interstellar clouds of about 2.3 K, and by 1950 this measurement was enshrined in standard textbooks. But not even Gamow made the connection with the temperature predicted for the background radiation. One reason was that Gamow's own estimates gave a temperature rather higher than either the one published by McKellar or the estimate made by Alpher and Herman.

In an article published in *New Scientist* in 1981, Fred **Hoyle** recounted how McKellar's estimate came up in a conversation with Gamow in 1956. Hoyle was an ardent supporter of the **Steady State** hypothesis, and did not believe there had been a Big Bang, so (at the time) he argued that there was no background radiation. Gamow believed that there should be background radiation with a temperature rather higher than 5 K. Hoyle recalls that he pointed out to Gamow that McKellar had set an upper limit of 3 K for any such background, and that therefore Gamow was wrong. Neither of them made what seems with hindsight the fairly modest leap of the imagination required to realize that there really was a sea of background radiation, with a temperature lower than Gamow had predicted.

Even more bizarrely, at the same time that Gamow's team were developing their idea in the 1940s, a team of radio astronomers were actually looking for cool radiation from space. Robert **Dicke** and his colleagues used an instrument developed from wartime radar technology to study the sky at microwave frequencies, and found evidence of radiation with a temperature below 20 K, the limit set by their instrument. Their results were published in 1946, in the same volume of the journal *Physical Review* (volume 70, page 340) as the first paper by Gamow's team on **nucleosynthesis** (volume 70, page 572) – but it was almost twenty years before anybody noticed the connection.

By the early 1960s, several groups, including researchers in the United States, Britain and the Soviet Union, had begun thinking about the possibility of detecting left-over radiation from the Big Bang – but the pioneering work of Gamow's team had largely been forgotten, and each team rediscovered the possibility. At Princeton University, a young researcher, P. J. E. **Peebles**, unknowingly repeated the calculations carried out by Alpher and Herman, and realized that the Universe should be filled with a sea of background radiation with a temperature of a few K. His supervisor in this work was Dicke, who, having forgotten his own pioneering efforts in the 1940s, then set two other researchers, P. G. Roll and D. T. Wilkinson, the task of setting up a small radio telescope to search for this radiation.

In 1965, just when they were ready to roll, Dicke received a phone call from Arno **Penzias**, at the Bell Research Laboratories, just 30 miles away, in Holmdel, New Jersey. Penzias and his colleague Robert **Wilson** had been preparing a 20-foot horn **antenna**, initially designed for work with the Echo communications satellites, for use in radio astronomy. They had found a persistent source of interference, microwave radio noise coming uniformly from all over the sky, and they wondered whether Dicke and his colleagues had any idea what it might be.

It was, of course, the background radiation. At last the theory and the observations had come together, and two and two were quickly put together to make four.

The Princeton team soon confirmed the observations, and the two groups published papers on the discovery alongside one another in the *Astrophysical Journal*. Over the next twenty years or so, more and more observations, using different instruments at a variety of wavelengths, confirmed the existence of the background radiation, pinned down its temperature as 2.7 K, and showed that it was perfect black body radiation. Penzias and Wilson received the Nobel Prize for their accidental discovery in 1978. It was the discovery and explanation of the background radiation that convinced most astronomers that there really had been a Big Bang, and which made cosmology a thriving scientific discipline.

But by the 1980s, there was a problem with the background radiation. It was almost *too* smooth and perfect, coming from every direction in space with exactly the same temperature.

The Big Bang theory, by now well established, said that the radiation should have been unchanged (except for being redshifted and cooled) since a time about 300,000 years after the birth of the Universe. This corresponds to the time when the whole Universe had cooled to a temperature of about 6,000 K, roughly the temperature at the surface of the Sun today. At that temperature, individual electrons and nuclei could get together to form stable atoms, and atoms do not have any net electric charge. Because atoms are electrically neutral, they do not interact strongly with electromagnetic radiation, and so the background radiation has been undisturbed since that time.

If the Universe was perfectly smooth 300,000 years after its birth, as the smoothness of the background radiation seemed to imply, then where did things like galaxies and stars and people come from? In order for us to exist, there must have been some irregularities in the Universe by the time it was 300,000 years old – clouds of gas in space that would soon clump together under their own gravity to collapse and form galaxies and stars.

The theory said that, as a result of the presence of these irregularities, there ought to be ripples in the background radiation, slight differences in temperature depending on which part of the sky the instruments were pointed at. The predicted differences were so small that they could be measured only from space, above the disturbances caused by the Earth's atmosphere. In April 1992, NASA announced that the satellite COBE (from COsmic Background Explorer) had found ripples exactly the right size, as predicted by the standard Big Bang model. The discovery was hailed as the ultimate triumph of the Big Bang theory, confirmation that the Universe really did begin at a definite moment in time in a hot fireball of radiation.

So one result of the way that the Universe was born is that it is filled with microwave radiation today, just like the microwaves in a microwave oven, but with the rather low cooking temperature of a bit less than – 270°C. And you do not even need a radio telescope to detect it. Because the background radiation is everywhere in the Universe, some of it gets picked up by any ordinary TV antenna. If you tune your TV set to a frequency between those of the stations broadcasting programmes, you will see the screen covered in dancing white dots, and hear a hiss of noise, sometimes referred to as static. About 1 per cent of the incoming 'signal' causing those dancing dots and that hissing noise is, in fact, cosmic microwave background radiation, broadcasting direct from the Big Bang to your living room.

Further Reading: Jeremy Bernstein, *Three Degrees Above Zero*; Marcus Chown, *The Afterglow of Creation*; George Gamow, *The Creation of the Universe*.

Bahcall, John (1934–) An astrophysicist who has played a leading role in the investigation of the *solar neutrino problem*. Bahcall was born in Shreveport, Louisiana, and obtained his Ph.D. from Harvard University in 1961. He worked at Indiana University and at CalTech in the 1960s, and has been associated with the Institute for Advanced Study in Princeton since 1968, and simultaneously with Princeton University from 1971. In 1962, at the instigation of Willy *Fowler*, he joined forces with Ray *Davis* to investigate the possibility of capturing *neutrinos* from the Sun with detectors on Earth.

Bahcall is a theorist who has been concerned with the calculations of how the Sun works and therefore the number of neutrinos expected to arrive at Earth, while Davis is an experimenter who headed the team trying to capture those neutrinos.

Bahcall's other interests have included the possibility that there may be a non-cosmological influence on the redshifts of distant galaxies and quasars, and the nature of the *dark matter* in the Universe. This is also an interest of his wife, Neta, an observational astronomer whose studies of the *rotation curves* of galaxies pinpointed the presence of dark haloes in galaxies like our own.

Barnard's Star   The star with the largest known *proper motion*. Discovered by the American astronomer E. E. Barnard in 1916, Barnard's Star moves so fast that it would take only 180 years to cover a distance of half a degree on the sky (the angular diameter of the Moon as seen from Earth) relative to the background stars. The star is 1.8 parsecs (about 6 light years) away from us, the fourth nearest known star to the Solar System, and is a *red dwarf*, too faint to be seen with the naked eye. It is one of the faintest stars yet detected, with an *absolute magnitude* just one-hundredth as bright as the Sun. There are slight wobbles in the path of Barnard's Star across the sky, which may be caused by the gravitational influence of planets in orbit around it.

Barred spiral. An example of a barred spiral galaxy. This is NGC 5383, which lies in the direction of (but far beyond) the constellation Canes Venatici.

barred spiral A type of *spiral galaxy* in which the spiral arms are joined to a straight bar of stars across the centre of the galaxy.

Barringer crater A large crater in the Arizona desert, 1.2 km across and 180 m deep, formed by the impact of a *meteorite* with the Earth about 50,000 years ago. The meteorite was largely made of iron, and was probably a few tens of metres across. The crater is named after Daniel Barringer, a mining engineer who was the first person to suggest, early in the 20th century, that it was caused by a meteorite impact. It was more than 50 years before this suggestion became generally accepted.

The meteorite would have had a mass of a few million tonnes, entering the Earth's atmosphere with an impact velocity of between 10 and 20 km per second. It would have released the equivalent of the explosion of about 10 megatonnes of TNT as it struck (kinetic energy converted into heat by the impact), equivalent to a large nuclear bomb. A similar impact occurring today could devastate a small city, but this is by no means the largest meteorite impact that the Earth has experienced (see *Doomsday asteroid*).

The only unusual thing about the Barringer crater is that its desert location means that the traces of the impact have been preserved for so long. Impacts occurring in other parts of the world (for example, in forests or in the seas) would leave little obvious trace after tens of thousands (or even hundreds) of years. On average, the Earth probably experiences one impact this large on land every thousand years, with two impacts in the oceans for every one on land.

baryon Member of the family of *elementary particles* that are affected by the *strong nuclear interaction*. The only stable baryons are the proton and the neutron, and the term 'baryonic material' is often used to refer to everyday atomic matter, which is made of protons, neutrons and the much less massive *electrons*.

baryon asymmetry See *Sakharov, Andrei*.

baryon catastrophe The puzzle that studies of the amount of hot gas in clusters of galaxies suggest that the proportion of *baryons* to *dark matter* in the Universe is too great to allow the possibility of there being exactly the right amount of all kinds of matter put together to match the predictions of the simplest versions of *inflation* and make *spacetime* flat (see *cosmological models*).

It has become firmly established that most of the matter of the Universe is in some invisible form. But while theorists delight in playing with mathematical models that include such exotica as cold dark matter, hot dark matter, WIMPs and mixed dark matter, the observers have slowly been uncovering an unpalatable truth. Although there is definitely some dark matter in the Universe, there may be less to the Universe than some of these favoured models imply.

The standard model of the hot *Big Bang* (incorporating the idea of inflation, which invokes a phase of extremely rapid expansion during the first split-second of the existence of the Universe) says that the Universe should contain close to the 'critical' amount of matter needed to make spacetime flat and to just prevent it expanding forever. But the theory of how light elements formed in the early Universe (see *nucleosynthesis*) limits the density of ordinary baryonic matter (protons, neutrons, and the like) to about one-twentieth of this. The residue, the vast majority of the Universe, consists (on the standard picture) of some kind of exotic particle such as axions. These particles have never been seen directly, although their existence is predicted by the standard theories of particle physics. In the favoured Cold Dark Matter (CDM) model of the Universe, the gravitational influence of the dark particles on the bright stuff gives rise to structures first on small scales, then on successively larger ones as the Universe evolves.

The evidence for dark matter comes from observations on a range of scales. Within our own Galaxy, the Milky Way, there is at least as much unseen matter as that in visible stars. But observations of *gravitational lensing* of stars in the Magellanic Clouds suggest that this particular component of the dark matter may be baryonic, either large planets or faint, low-mass stars known as *brown dwarfs*. There is also evidence from the speed at which stars and gas clouds orbit the outer parts of disc galaxies for more extensive haloes of dark matter, but once again these could be baryonic. As far as individual galaxies are concerned, there is actually no need to invoke CDM at all.

There is no reason to suppose, however, that the contents of galaxies are representative of the Universe as a whole. When a protogalaxy first collapsed, it would have contained the universal mix of baryonic matter (in the form of a hot, ionized gas) plus dark matter. The dark matter is 'cold' in the sense that individual particles move slowly compared with the speed of light, but like the baryonic stuff they have enough energy to produce a pressure which keeps them spread out over a large volume of space. The baryons lose energy by radiating it away electromagnetically, so they cool very quickly; the baryon component of the cloud loses its thermal support and will sink into the centre of the protogalactic halo to form the galaxy that we see today. This leaves the dark matter, which cannot cool (because it does not radiate electromagnetically), spread out over a much larger volume.

To find a more typical mixture of material we must therefore look at larger, more recently formed structures, in which cooling is less efficient. These are clusters of galaxies.

A typical rich cluster may contain a thousand galaxies. These are supported against the attractive force of gravity by their random speeds, which can be more than 1,000 km per second, and are measured from the *Doppler effect* produced by their motion. This shifts features in their spectra either towards the blue or towards the red (this is independent of the *redshift* produced by the expansion of the Universe, which has to be subtracted out from these measurements). By balancing the kinetic energy of the galaxies against their gravitational potential energy, it is possible to estimate the total mass of the cluster. This was first done by Fritz *Zwicky* in the 1930s, and led to the then surprising conclusion that the galaxies comprise only a small fraction of the total mass. This was so surprising that for several decades many astronomers simply ignored Zwicky's findings.

Without the experimental background in particle physics, or the cosmological models which are available today, it would have been natural for those astronomers who did take the observations seriously to identify this missing matter as hot gas. However, this was not done, perhaps because the physical condition of the gas would render it undetectable by any means available at the time. The gas particles are moving at similar speeds to the galaxies, which is equivalent to a gas temperature of about 100 million degrees – this is sufficient to strip all but the most tightly bound electrons from atomic nuclei, leaving behind positively charged *ions*. Such an ionized gas emits mainly at X-ray energies, which are absorbed by the Earth's atmosphere. It was only with the launch of X-ray satellite observatories in the 1970s that clusters were found to be very bright X-ray sources and it was finally realized that the hot gas, or intracluster medium (ICM), cannot be neglected (see *X-ray astronomy*).

The ICM has turned out to be a very important component of clusters of galaxies. Not only does it contain more matter than is present in the galaxies, but its temperature and spatial distribution can be used to trace the gravitational potential and hence the total mass of the cluster in a much more accurate way than from the galaxies alone. To obtain the total mass of gas, one looks at the radiation rate. This radiation is produced in collisions between oppositely charged particles (ions and electrons), and so depends upon the square of the gas density. We observe only the projected emissions, as if the cluster were squashed on the plane of the sky, but assuming spherical symmetry it is relatively easy to invert this to find the variation of density with distance from the centre of the cluster. The gas is found to be much more extended than the galaxies and can in some cases be traced out to several million light years from the cluster centre. Whereas the galaxies dominate in the core of the cluster, there is at least three times as much, and probably a lot more, gas in the cluster as a whole as there is matter in the form of galaxies (it is not the mass of gas which is uncertain, but the mass of the galaxies). But even the combined

Baryon catastrophe. The central core of the Coma cluster of galaxies. The whole cluster is more than 100 Megaparsecs away and half a million parsecs across.

mass of gas and galaxies is less than the total cluster mass, showing that a large amount of dark matter is also present.

The hot gas is supported against gravitational collapse in the cluster by its pressure gradient. To derive this uniquely from the observations, we would have to know the variation of temperature with distance from the cluster centre. Unfortunately, this is not yet possible with present X-ray telescopes (although it is beginning to be so with the Japanese ASCA satellite), and so some simplifying assumptions have to be made. It is usually supposed that the gas is isothermal – the same temperature right across the cluster. This is consistent with both observations and numerical simulations, which show little variation of either random galaxy speeds or gas temperature across the cluster. It is possible that the gas temperature falls in the outer parts of clusters – this would tend to lower the overall mass estimates.

A study by David White and Andy Fabian of the Institute of Astronomy in Cambridge, published in 1995, examined data from the Einstein satellite for nineteen bright clusters of galaxies. They compared the mass of gas with the total cluster mass and concluded that it comprises between 10 and 22 per cent, with an average value of about 15 per cent. These fractions would increase by between 1 and 5 per cent (of the total mass) if the mass of galaxies were included. So the total baryon content of clusters is much greater than the 5 per cent predicted by the standard CDM model for a flat Universe. You still need some dark matter (to the relief of the particle physicists), but only five times as much as there is baryonic matter, not twenty times as much. Since the Big Bang models still say that only 5 per cent of the critical density can be in the form of baryons, this means that, if the distribution of matter in clusters of galaxies is typical of the Universe at large, overall there can only be about 30 per cent of the critical density, even including the dark stuff. If you want to keep the high overall value of the *density parameter*, you have to allow much more than 5 per cent of the total mass of the Universe to be in the form of baryons, but this is forbidden by the rules of primordial nucleosynthesis.

What is the resolution of this problem? There are various uncertainties in the models (for example, the gas may be clumped or may not be isothermal), but these are unlikely to alter the conclusions greatly. One major uncertainty, however, is the distance to the clusters, which is in turn determined by the rate at which the Universe has expanded from the Big Bang to its present size. There is a lively debate among astronomers about the exact value of the parameter which measures the expansion rate, the so-called *Hubble constant*. So far, we have assumed a Hubble constant of 50 km per second per Megaparsec, which is at the lower end of the accepted range and corresponds to a large, old universe. This means that a galaxy 1 Megaparsec away (1 million parsecs, or about 3.26 million light years from us) is receding at a rate of 50 km/sec as a result of the expansion of the Universe, and so on.

In the cosmological models, as the Hubble constant is lowered, the calculated baryon fraction increases. But the predicted baryon fraction from primordial nucleosynthesis increases even faster, so the discrepancy between the two is reduced. By making the Hubble constant low enough, one could reconcile the two, but long before this happens the baryon fraction becomes equal to unity. Since there cannot be more than 100 per cent of the mass of the Universe in the form of baryons, this argument can be reversed to place an absolute lower bound on the Hubble constant of about 14, in the usual units. Very few astronomers would countenance going to such extremes. But it is worth mentioning that a new technique for estimating the Hubble constant (based on the *Sunyaev–Zel'dovich effect*) uses measurements of the influence of the hot cluster gas on the *background radiation* passing through it to determine how fast the Universe is expanding. This technique is in its infancy, but early results from it do suggest a low value of the Hubble constant, perhaps even less than 50.

It would seem, therefore, that one of the cherished foundations of the standard model must be relinquished. Perhaps the least fundamental of these is that the dark matter must be 'cold'. Hot dark matter, made of particles (such as neutrinos) which emerge from the Big Bang with speeds close to that of light, is unable to cluster efficiently due to the large random motions of its particles. At first sight, you might guess that it could fill the space between clusters of galaxies with huge amounts of matter, so that even the clusters are not representative of the stuff of the Universe. However, hot dark matter cannot comprise more than about one-third of the total amount of dark matter because interactions between the hot stuff and ordinary baryonic matter would slow the development of structures such as galaxies and clusters, delaying their formation until later times; this conflicts with the observed number of distant, old radio galaxies and quasars.

There is certainly no way that the baryonic material found so far will go away, and there could be even more of it than we have estimated. If the same analysis is carried out for larger volumes around the clusters, it tends to show an even larger proportion of mass in gas, because the galaxies themselves congregate in the centres of the clusters. In some cases, as much as half the mass of the cluster is in the form of hot gas. In general, heating of the gas will tend to expel it from clusters and exacerbate the baryon discrepancy still further – if there is cold baryonic material outside clusters, there is even more ordinary stuff than the observations suggest. It has been suggested that clusters may contain a surplus of baryons because they have been formed by the aggregation of gas swept up at the edges of large voids produced by huge cosmic explosions. But unfortunately, such models seem to have been ruled out because they would produce excessive distortions in the cosmic microwave background.

People have toyed with the idea of non-standard nucleosynthesis: for example, allowing the baryon abundance to vary from place to place. This allows some relaxation of the upper bound on the baryon fraction, but the models are rather contrived and anyway the models do not work as well as the standard one.

We are left with the simplest explanation, and yet the one which most cosmologists would least like to accept, that the mass density of the Universe is much less than the critical density. If 'what you see is what you get', the Universe could contain as much as 25 per cent baryonic material, with overall about 30 per cent of the critical density, and the baryons themselves about a third in the form of hot cluster gas and about two-thirds in the form of galaxies. The other 75 per cent of the stuff of the Universe would be mainly cold dark matter, perhaps with a smattering of hot dark matter. The Hubble constant could then be rather higher than 50, as some recent observations seem to suggest.

If cosmologists wish to preserve the idea of a spatially flat Universe, as predicated by theories of cosmic inflation, they may have to reintroduce the idea of a cosmological constant.

Whatever the ultimate resolution of this problem, it is sure to lead to a fundamental reshaping of cosmological ideas. Indeed, this is already happening, as some cosmologists have begun to find ways in which their cherished idea of inflation can be tweaked up to produce a Universe in which the density parameter in the entire observable volume of space can have any value between 0 and 1 (see Introduction: Where do we come from?).

Further reading: John Gribbin and Martin Rees, *The Stuff of the Universe*.

baryon fraction   The proportion of the mass of the Universe that is in the form of *baryons*. If the Universe is just closed, so that spacetime is flat (see *cosmological models*), the baryon fraction is about 1 per cent (at most 5 per cent) and by far the bulk of the Universe is in the form of *dark matter* (but see *baryon catastrophe*).

Beijing Observatory   A research institute run by the Chinese Academy of Sciences, which operates five observing stations to carry out astrophysical observations using both optical and radio telescopes.

Bell Burnell, (Susan) Jocelyn (1943–)   Observational astronomer who discovered *pulsars* in 1967, while still a student.

Born on 15 July 1943, in Belfast, Northern Ireland. Jocelyn Bell's father was the architect of the Armagh Observatory, which was near her childhood home. There she was able to nurture her childhood interest in astronomy. After attending school in York, Bell studied at the University of Glasgow, graduating in 1965, then began work for her Ph.D. in Cambridge, under the supervision of Anthony *Hewish*. This involved constructing a special kind of radio telescope which would be sensitive to rapid variations in the brightness of radio sources, called scintillations. These are the radio equivalent of the twinkling of light from a star, and are caused by the passage of radio waves from a distant source, such as a quasar, through clouds of electrically charged material (plasma) in space. This plasma is ejected by the Sun, so the scintillation is more pronounced in daytime.

The telescope was more like an orchard than the conventional image of a radio telescope. A field covering four and a half acres was filled with an array of 2,048 dipole antennae. Each dipole (a long rod aerial) was mounted horizontally on an upright pole, a couple of metres above the field, like the crossed yard-arm of a square-rigged ship. By wiring the antennas up correctly, which was one of Bell's jobs, their signals could be combined to scan a strip of sky as the Earth rotated.

By the summer of 1967, Bell had the system working, and had begun to use it to search for previously undiscovered quasars. On 6 August that year, she found a peculiar signal that could not be explained in terms of scintillation, and repeated observations showed that it was always coming from the same part of the sky, at the same time of night (so the radio waves were not passing through the clouds of plasma associated with the

The radio telescope used by Jocelyn Bell Burnell to discover pulsars. The original covered 4.5 acres; this extended version is twice that area. (The large dish in the background is part of the Ryle telescope.)

Sun). Nor could it be explained as interference from human activities.

By November, it was clear that she had discovered an astronomical source varying with a regular period of just 1.3 seconds, and continued observations refined this to 1.33730113 seconds. This remarkable accuracy led Hewish and his team to consider seriously the possibility that it might be an interstellar beacon, a kind of radio lighthouse, planted by an alien civilization, and among themselves the Cambridge radio astronomers referred to the source as 'LGM 1', the initials standing for 'Little Green Man'.

Just before Christmas 1967, Bell found a second, similar source, with a period of 1.27379 seconds, and soon after two more, with periods of 1.188 seconds and 0.253071 seconds. As the number of similar objects known grew, it became clear that they were a natural phenomenon, and they were given the name pulsating radio sources, soon contracted to pulsars. In February 1968, the discovery of this new kind of astronomical object was announced in a paper in the journal *Nature*.

After being awarded her Ph.D. for this work later in 1968, Bell worked on *gamma ray* astronomy at the University of Southampton, on X-ray astronomy at the Mullard Space Science Laboratory, in infrared, optical and millimetre astronomy while based at the Royal Observatory, Edinburgh, and since 1991 she has been professor of physics at the Open University.

In 1974, Hewish was awarded the Nobel Prize for his part in the discovery of pulsars. The Nobel Committee has made many bizarre decisions in its time, but its failure to include Bell in this award (presumably because she was 'only' a student when she made the discovery) is one of the most striking. Perhaps partly because of the consternation this caused in 1974, when the time came for the award of a Nobel Prize for the discovery of the *binary pulsar*, the student who actually made the discovery did share the award with his supervisor.

Further reading: John Gribbin, *In Search of the Edge of Time*.

bending of light   See *deflection of light*.

Beta Centauri   The second brightest star in the southern *constellation* Centaurus. Beta Centauri is a very bright giant star, 130 parsecs from us, a hundred times more remote than *Alpha Centauri*.

beta decay   The process in which a neutron, either on its own or inside a nucleus, emits an electron (otherwise known as a beta particle) and an anti-neutrino as it transforms into a proton. When this happens inside a nucleus, the charge of the nucleus is increased by one unit, and it becomes a nucleus of a different element.

Beta lyrae variables   A family of *eclipsing binary* stars in which one star has expanded to fill its *Roche lobe* and is spilling matter over on to its companion. The spilled material forms an *accretion disc* around the smaller star. Because one star of the pair fills its Roche lobe, such systems are sometimes said to be 'semi-detached'; some similar semi-detached systems may harbour *black holes*.

Betelgeuse   Bright red star marking the shoulder of the constellation Orion (at the top left of the constellation, as viewed from the Northern Hemisphere). Betelgeuse, also known as Alpha Orionis, is a red supergiant at a distance of 200 parsecs. It has a diameter 800 times that of the Sun, measured directly by *interferometry*.

Bethe, Hans Albrecht (1906–)   German-born American physicist and astronomer, best known for working out the mechanisms by which stars derive their energy from *nuclear fusion* reactions.

Born in Strassburg, Germany (now Strasbourg, in France) on 2 July

1906, his father a professor of physiology, Bethe himself was educated at the universities of Frankfurt and Munich, and was awarded his PhD in 1928. He worked at the universities of Frankfurt, Munich and Tübingen over the next five years, but after the rise of the Nazi Party in Germany he was dismissed from his post (his mother was Jewish) and moved first to Britain and then, in 1935, to the United States, where he was professor of physics at Cornell University from 1935 (assistant professor until 1937) until he retired in 1975.

Bethe has made many important contributions to physics, including nuclear physics and solid state physics. He worked on the wartime development of radar, and on the Manhattan Project (to develop the atomic bomb) at the Los Alamos laboratory in New Mexico, and he later served as a member of the US delegation on negotiations with the Soviet Union on nuclear arms limitation. But he is best known for two contributions to *astrophysics* – one of which he did make, and one of which he did not.

By 1938, astrophysicists knew that the energy of the stars must originate from nuclear processes, but they did not know which nuclear processes. In his book *The Birth and Death of the Sun*, written just after the problem was solved, George *Gamow* describes how Bethe was present at a conference in Washington, DC, in April 1938, where the problem was discussed. According to legend (and Gamow), as Bethe set out on the train journey back to Cornell, he determined to solve the problem before the steward called the passengers to dinner – and did so with seconds to spare.

The process of nuclear energy generation worked out by Bethe on that train journey is now known as the *carbon cycle*, or the carbon–nitrogen–oxygen (CNO) cycle. It was discovered independently by Carl von Weizsäcker (1912–) in Berlin at about the same time, but he lacked an ebullient Gamow to publicize his contribution.

Working with his colleague Charles Critchfield back at Cornell, Bethe later found another way in which stars can obtain their energy, from a process known as the *proton–proton reaction*, or p–p chain. It is now thought that the proton–proton reaction is the mechanism that produces most of the energy inside a star like the Sun, which has a central temperature of about 15 million Kelvin, and cooler stars; the CNO cycle is important inside more massive stars, where the central temperature rises above 20 million K.

A few years later, in the mid-1940s, Bethe was an unwitting participant in Gamow's most famous practical joke, when Gamow added Bethe's name to a paper written by Gamow and Ralph *Alpher*, describing the way primordial hydrogen was partially converted into helium in the *Big Bang*. The name was added simply so that the authors listed on the paper sounded like the first three letters of the Greek alphabet (see *alpha beta*

Hans Bethe (born in 1906).

gamma theory); Bethe played no part in the work at all, but even today some reference books give him credit for it.

Bethe has also worked on theoretical models of supernovae, and was awarded the Nobel Prize in 1967 for his work on the mechanisms of energy production in stars. Although he officially retired in 1975, he continued to be active in research, and hit the headlines in 1986, when he took up and publicized a proposed solution to the *solar neutrino problem* developed by two Soviet researchers, S. P. Mikheyev and A. Yu. Smirnov, on the basis of a suggestion made by a US physicist, Lincoln Wolfenstein. This so-called MSW process is related to the way energy is generated inside the Sun. The human interest of a scientist now in his eighties coming back almost 50 years later to a field of research that he had pioneered caught the imagination of scientists and popularizers alike, and ensured that the MSW idea received a blast of publicity.

Further reading: Jeremy Bernstein, *Hans Bethe;* John Gribbin, *Blinded by the Light,* Bantam, New York, 1991.

Big Bang An overwhelming weight of evidence has convinced most astronomers that the Universe came into being at a definite moment in time, some 15 billion years ago, in the form of a superhot, superdense fireball of energetic radiation. This is known as the Big Bang model of the origin of the Universe, a term actually coined by Fred *Hoyle* at the end of the 1940s, as a mark of derision for a theory he regarded as about as elegant as 'a party girl jumping out of a cake'. Hoyle, one of the originators of the rival *Steady State* model, remains one of the most vociferous opponents of the idea of the Big Bang, but his name has stuck.

Until the 1920s, astronomers thought that the Universe consisted only of what we now know as the Milky Way *Galaxy*, and that it was essentially eternal and unchanging. Individual stars might go through their life cycles and die, but new ones would be born to take their place.

The first clear hint that the Universe might change (evolve) as time passes came when Albert *Einstein* developed his *general theory of relativity*. This is a theory of spacetime, offering a complete mathematical description of the Universe (a model). In 1917, Einstein discovered that when he tried to apply his equations in this way, to describe spacetime as a whole, they did not describe a static, unchanging world. The equations said that the Universe must be either expanding or contracting, but that it could not be standing still. Since there was no astronomical evidence for either expansion or contraction, Einstein introduced an extra term into his equations, a fiddle factor called the *cosmological constant*, to hold the models still. He later described this as 'the biggest blunder' of his career.

Other researchers, notably Willem *de Sitter* in Holland and Alexander *Friedman* in the Soviet Union, also found solutions to Einstein's equations that described different models of the Universe, but all with an inherent tendency to evolve. Some models start out small and expand forever; some expand out to a certain size then collapse. One variation on the theme starts out very big, shrinks down to a certain size and then expands again. And one family of solutions follows a repeating cycle of expansion and collapse, with a 'bounce' at very small size.

The relevance of these mathematical models to the real Universe began to become clear in the 1920s, when the work of Edwin *Hubble* and other observers showed not only that our Milky Way is just one galaxy among many in the cosmos, but that galaxies are moving apart from one another as the Universe expands (see *redshift*). In other words, the simplest cosmological model based on Einstein's equations, without the cosmological constant, is in fact a good description of the behaviour of the Universe at large.

By the beginning of the 1930s, it was clear that the Universe was expanding, with galaxies being carried apart as the space between them stretched. The galaxies do not move through space (at least, not as far as this cosmological expansion is concerned), but are carried along for the ride by the expansion of space. A helpful analogy is with the raisins in a loaf of raisin bread. As the dough rises, the raisins are carried further apart from one another, even though they are not moving through the dough.

But the analogy is not exact, because unlike the Universe a loaf of raisin bread has both a centre and an edge. Einstein's equations tell us that the Universe has neither, either because it is infinite or because spacetime is gently curved round upon itself to make the equivalent in four dimensions of the surface of a sphere.

In that case, just as you could set out from New York and travel in a straight line across the surface of the Earth to end up back in New York, so you could set off in a straight line across the Universe and (eventually!) end up back where you started, after circumnavigating the Universe. There is no centre to the Universe any more than there is a centre to the surface of the Earth or to the skin of a soap bubble.

During the 1930s and 1940s, cosmologists began to try to come to terms with these ideas. The most important implication of the new discoveries was that the Universe must have had a definite beginning in time. If you imagine winding back the expansion we see today, there must have been a time when all the galaxies were on top of one another, as the space between them shrank away. Before that, stars must have been touching each other, merging together to make one great fireball as hot as the inside of a star (15 million K).

Einstein's equations actually say that you can go back further still, to a time when all the matter and energy of the Universe emerged from a point of zero size, a singularity. But the Big Bang idea was not pushed to such extremes at first.

The first person to produce a version of what would now be called a Big Bang model was the Belgian astronomer Georges *Lemaître* (who also happened to be an ordained priest) in 1927. Lemaître did not take the relativistic equations all the way back to the singularity, instead describing the birth of the Universe in terms of expansion from a time when the entire contents of the Universe were packed into a sphere about 30 times bigger than our Sun, which he called the 'primeval atom' (also known, less reverently, as the 'cosmic egg'). This, according to Lemaître, then exploded outwards, for reasons unknown, breaking up into fragments that formed the constituents of the Universe we see around us.

Many people misunderstand this idea as implying that the primeval atom was sitting in 'empty space' and exploded outwards like an exploding bomb. But remember that spacetime, as well as matter and energy, was wrapped up in the cosmic egg. There was no 'outside' for the 'bomb' to explode into, and the expansion is caused by the expansion of space itself, stretching as time passes.

The Big Bang idea was taken a step further by George *Gamow* in the 1940s. He showed how *nuclear interactions* taking place in the fireball of the early Universe could have converted hydrogen into helium, explaining the proportions of these elements in very old stars, and predicting the existence of the *background radiation*. By the 1960s, cosmologists were prepared to 'wind the clock back' to consider a time when the density of matter in the entire Universe was about the same as the density of matter today in the nucleus of an atom. They felt that they understood nuclear interactions well enough to calculate how the Universe would have evolved from that time onward, and those calculations became the standard model of the Big Bang.

If we set the moment when the Universe emerged from a singularity, according to a literal interpretation of Einstein's equations (see *Hawking*), as time zero, the standard model of the Big Bang tells the story of everything that has happened since 0.0001 (10^{-4}) of a second after this moment of creation. At that time, the temperature of the Universe was 10^{12} K (1,000 billion degrees) and the density was the density of nuclear matter, 10^{14} grams per cubic centimetre (1 gram per cubic centimetre is the density of water).

Under these conditions, the photons of the 'background' radiation carry so much energy that they are interchangeable with particles, in line with Einstein's equation $E = mc^2$. Photons create pairs of particles and antiparticles, such as electron–positron pairs, proton–antiproton pairs and neutron–antineutron pairs, and these pairs annihilate one another to make energetic photons in a constant interchange of energy. There were also many neutrinos present in the fireball. Because of a tiny asymmetry in the way the fundamental interactions work, slightly more particles were produced than antiparticles – about a billion and one particles for every billion antiparticles.

When the Universe cooled to the point where photons no longer had the energy to make protons and neutrons, all the paired particles and antiparticles annihilated, and the one in a billion particles left over settled down as stable matter.

One-hundredth of a second after time zero, with the temperature down to 100 billion K (10^{11} K), only the lighter electron–positron pairs still interacted in the dance with radiation. Protons and neutrons had settled out of the maelstrom. At that time, there were as many neutrons as protons, but as time passed interactions with energetic electrons and positrons shifted the balance steadily in favour of protons. One-tenth of a second after time zero, the temperature was down to 30 billion K, and there were only 38 neutrons for every 62 protons. About a third of a second after time zero, neutrinos ceased to interact with ordinary matter, except

Cosmic expansion

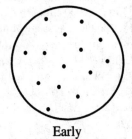

Early

Big Bang. As the Universe expands, all clusters of galaxies move further apart from all other clusters of galaxies, like spots moving further apart on the surface of an expanding balloon. The clusters are not moving through space to cause this increased separation; rather, the space between the clusters is expanding.

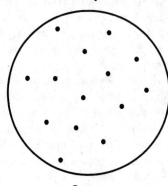

Later

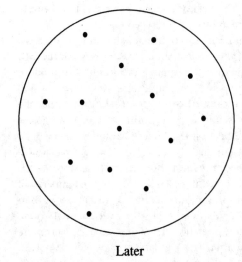

Later

....

(possibly) through gravity (see *dark matter*), and 'decoupled'.

By the time the Universe had cooled to 10^{10} K (10 billion K), 1.1 seconds after time zero, its density was down to just 380,000 times the density of water, neutrinos had decoupled, and the balance between protons and neutrons had shifted further, with 24 neutrons for every 76 protons. By the time the Universe had cooled to 3 billion K, 13.8 seconds after time zero, nuclei of *deuterium*, each containing one proton and one neutron, began to form, but they were soon knocked apart by collisions with other particles. Only 17 per cent of the *nucleons* were now left in the form of neutrons.

Three minutes and two seconds after time zero, the Universe had cooled to 1 billion K, only 70 times hotter than the centre of the Sun is today. The proportion of neutrons was down to 14 per cent, but they were saved from disappearing entirely from the scene because the temperature had at last fallen to the point where nuclei of deuterium and helium could be formed and stick together in spite of collisions with other particles.

It was at this epoch, during the fourth minute after time zero, that the reactions outlined by Gamow and his colleagues in the 1940s, and refined by Fred Hoyle and others in the 1960s, took place, locking up the remaining neutrons in helium nuclei. The proportion of the total mass of nucleons converted into helium is twice the abundance of neutrons at the time, because each nucleus of helium (helium-4) contains two protons as well as two neutrons. Four minutes after time zero, the process was complete, with just under 25 per cent of the nuclear material converted into helium, and the rest left behind as lone protons – hydrogen nuclei.

A little more than half an hour after time zero, all of the positrons in the Universe had annihilated with almost all of the electrons – again with one in a billion left over, matching the number of protons – to produce the background radiation proper. The temperature was down to 300 million K, and the density was only 10 per cent of that of water. But the Universe was still too hot for stable atoms to form; as soon as a nucleus latched on to an electron, the electron was knocked away by an energetic photon of the background radiation.

This interaction between electrons and photons continued for 300,000 years, until the Universe had cooled to 6,000 K (roughly the temperature at the surface of the Sun) and the photons were becoming too weak to knock electrons off atoms. At this point (actually, over about the next 500,000 years), the background radiation decoupled, and had no more significant interaction with matter. The Big Bang was over, and the Universe was left to expand relatively quietly, cooling as it does so, and expanding ever more slowly as gravity tries to pull it back together.

All of this is well understood in terms of the general theory of relativity, a tried and tested theory of gravity and spacetime, and our tried and tested understanding of *nuclear interactions*. The standard model of the Big Bang is solid, respectable science, but it does leave some questions unanswered.

It was about 1 million years after time zero that stars and galaxies could begin to form, processing hydrogen and helium inside stars to make heavy elements (see *nucleosynthesis*) and eventually giving rise to the Sun, the Earth and ourselves. But astrophysicists still do not have a completely satisfactory theory of how galaxies form.

Apart from the question of the origin of the Universe, the big question left unanswered by the standard model of the Big Bang in the 1970s concerned the ultimate fate of the Universe. Would it expand forever (an 'open' model), or would it one day slow to a halt and then collapse back into a *Big Crunch* (a 'closed' model)? Either possibility is allowed by Einstein's equations. The fate is determined by the amount of matter the Universe contains, and therefore how strong the gravitational force trying to halt the expansion is.

There is certainly not enough visible matter, in the form of stars and galaxies, to make the Universe closed, but it is known that there is also a great deal of dark matter in the Universe. By the middle of the 1980s, cosmologists also had a developing understanding of what happened in the interval between time zero and 0.0001 seconds (the very early Universe), in the form of a theory known as *inflation*; this suggests that the Universe sits almost precisely on the dividing line between being open and closed (that it is nearly 'flat'), although just on the closed side.

Further reading: John Gribbin, *In Search of the Big Bang*; Steven Weinberg, *The First Three Minutes*.

Big Bang nucleosynthesis See *nucleosynthesis*.

Big Crunch Counterpart to the *Big Bang*. If the Universe contains sufficient mass to make *spacetime* closed, gravity will one day halt its present expansion and then cause a collapse into a *singularity*, the Big Crunch, at the end of time. See *Omega Point*.

billion 1,000 million, 10^9.

binary pulsar A binary pulsar exists when two *neutron stars*, one of which is a *pulsar*, are in orbit around one another, forming a binary star system. The term is also used to refer to a pulsar in orbit about any other star: for example, a white dwarf. More than twenty binary pulsars are now known, but astronomers reserve the term '*the* binary pulsar' for the first one to be discovered, which is also known by its catalogue number, as PSR 1913 + 16. This pulsar has provided the most accurate test yet of Albert Einstein's *general theory of relativity*, and is the most accurate clock yet discovered.

The binary pulsar was discovered in 1974 by Russell *Hulse* and Joseph *Taylor*, of the University of Massachusetts, working with the *Arecibo radio telescope* in Puerto Rico. Hulse, a research student at the time, was responsible for the day-to-day running of a search for pulsars using the telescope, while Taylor, his research supervisor, was in overall charge of the project and made regular visits from Amherst, Massachusetts, to Arecibo during the summer of 1974. What they found that summer was so important that in 1993 the pair received the Nobel Prize for their work on the binary pulsar.

The first hint of the existence of the binary pulsar came on 2 July, when the instruments recorded a very weak signal. Had it been just 4 per cent weaker still, it would have been below the automatic cut-off level built into the computer programme running the search, and would not have been recorded. The source was especially interesting because it had a very short period, only 0.059 seconds, making it the second fastest pulsar known at the time. But it was not until 25 August that Hulse was able to use the Arecibo telescope to take a more detailed look at the object.

Over several days following 25 August, Hulse made a series of observations of the pulsar and found that it varied in a peculiar way. Most pulsars are superbly accurate clocks, beating time with a precise period measured to six or seven decimal places; but this one seemed to have an erratic period which changed by as much as 30 microseconds (a huge 'error' for a pulsar) from one day to the next. Early in September 1974, Hulse realized that these variations themselves follow a periodic pattern, and could be explained by the *Doppler effect* caused by the motion of the pulsar in a tight orbit around a companion star.

Taylor flew down to Arecibo to join the investigation, and together he and Hulse found that the orbital period of the pulsar around its companion (its 'year') is 7 hours and 45 minutes, with the pulsar moving at a maximum speed (revealed by the Doppler effect) of 300 km per second, one-tenth of the speed of light, and an average speed of about 200 km/sec, as it zipped around its companion. The size of the orbit traced out at this astonishing speed in just under 8 hours is about 6 million km, roughly the circumference of the Sun. In other words, the average separation between the pulsar and its companion is about the radius of the Sun, and the entire binary pulsar system would just fit inside the Sun.

All pulsars are neutron stars; the orbital parameters showed that in this case the companion star must also be a neutron star. One of the key tests of the general theory of relativity is the *advance of the perihelion* of Mercury, an orbital shift predicted by Einstein's theory but not by Isaac *Newton's* theory of gravity. The researchers calculated that the equivalent effect in the binary pulsar (the shift in the 'periastron') would be about 100 times stronger than for Mercury, and whereas Mercury orbits the Sun only four times a year, the binary pulsar orbits its companion 1,000 times a year, giving that much more opportunity to study the effect. It was duly measured, and found to conform exactly with the predictions of Einstein's theory – the first direct test of the general theory of relativity made using an object outside the Solar System. By feeding back the measurements of the periastron shift into the orbital data for the system, the total mass of the two stars in the system was eventually determined to unprecedented accuracy, as 2.8275 times the mass of our Sun.

But this was only the beginning of the use of the binary pulsar as a gravitational laboratory in which to test and use Einstein's theory. Extended observations over many months showed that, once allowances were made for the regular changes caused by its orbital motion, the pulsar actually kept time very precisely. Its period of 0.05903 seconds increased by only a quarter of a nanosecond (a quarter of a billionth of a second) in a year – equivalent to a clock that lost time at a rate of only 4 per cent in a million years.

The numbers became more precise as the observations mounted up: period, 0.059029995271 seconds; rate of increase, 0.253 nanoseconds per year; orbital period, 27906.98163 seconds; rate of change of periastron, 4.2263 degrees of arc per year. Because the pulsar period is actually changing, numbers given to this accuracy refer to a specific date, or 'epoch'; these are for 1 September 1974.

The accuracy of the observations soon made it possible to carry out more tests and applications of the theory of relativity. One involves the **time dilation** predicted by the **special theory of relativity**. Because the speed of the pulsar around its companion is a sizeable fraction of the speed of light, the pulsar 'clock' is slowed down, according to our observations, by an amount which depends on its speed. Since the speed varies over the course of one orbit (from a maximum of 300 km/sec down to 'only' 75 km/sec), this will show up as a regular variation of the pulsar's period over each orbit. And because the pulsar is moving in an elliptical orbit around its companion, its distance from the second neutron star varies. This means that it moves from regions of relatively high gravitational field to regions of relatively low gravitational field, and that its timekeeping mechanism should be subject to a regularly varying gravitational redshift.

The combination of these two effects produces a maximum measured variation in the pulsar period of 58 nanoseconds over one orbit, and this information can be fed back into the orbital calculations to determine the ratio of the masses of the two stars. Since the periastron shift tells us that the combined mass is 2.8275 solar masses, the addition of these data reveals that the pulsar itself has 1.42 times the mass of our Sun, while its companion has 1.40 solar masses. These were the first precise measurements of the masses of neutron stars.

But the greatest triumph of the investigation of the binary pulsar was still to come. Almost as soon as the discovery of the system had been announced, several relativists pointed out that in theory the binary pulsar should be losing energy as a result of *gravitational radiation*, generating ripples in the fabric of spacetime that would carry energy away and make the orbital period speed up as the binary pulsar and its companion spiralled closer together as a result.

Even in a system as extreme as the binary pulsar, the effect is very small. It would cause the orbital period (about 27,000 seconds) to increase by only a few tenths of a millionth of a second (about 0.0000003 per cent) per year. The theory was straightforward, but the observations would require unprecedented accuracy. In December 1978, after four years of work, Taylor announced that the effect had been measured, and that it exactly matched the predictions of Einstein's theory. The exact prediction of that theory was that the orbital period should decrease by 75 millionths of a second per year; by 1983, 9 years after the discovery of the binary pulsar, Taylor and his colleagues had measured the change to a precision of 2 millionths of a second per year, quoting the observed value as 76±2 millionths of a second per year. Since then, the observations have been improved further, and show an agreement with Einstein's theory that has an error of less than 1 per cent. This is the most spectacular and comprehensive test of the general theory ever carried out, and has effectively ruled out any other theory as a good description of the way the Universe works. The accuracy of the test is so precise, and the agreement with the predictions of Einstein's theory are so good, that the general theory of relativity can now be regarded as one of the two most securely founded theories in the whole of science, alongside *quantum electro-dynamics*.

In principle, the binary pulsar, and some of the other systems like it, provide a way to measure the passage of time more accurate than any clock on Earth, including the most accurate *atomic clock*. If we only had atomic clocks with which to measure the variations of a single binary pulsar, we could never prove this. But by comparing the signals from at least three binary pulsars with those of atomic clocks and each other, it ought to be possible to establish a way of keeping time using the pulsars (calibrated against each other) to improve on the timekeeping of the atomic clocks. It is not inconceivable that, just as the second is now defined in terms of the behaviour of caesium atoms, not by the rotation of the Earth, one day the second may be defined in terms of the period of the binary pulsar.

Further reading: Clifford Will, *Was Einstein Right?*

binary stars A pair of *stars* that are physically associated with one another and orbit around one another as a result of their mutual gravitational attraction to form a binary system. The stars in a binary system usually move in elliptical orbits around the *centre of mass* of the system, obeying *Kepler's laws*. Most stars are in binary or more complicated multiple star systems.

Binary stars are of key importance in astrophysics because analysis of their orbits provides the only way to determine the masses of stars directly. The first binary was identified in 1650, when the Jesuit astronomer Joannes Riccioli (1598–1671) discovered, with the aid of a telescope, that the star Zeta Ursae Majoris is actually a double star. But it was not until 1767 that John *Michell* suggested that stars which appear close together on the sky are really physically associated in space, and not the result of a chance juxtaposition of two stars at quite different distances along the line of sight.

Sometimes, only one star in a binary is visible, but the presence of the companion can be inferred from the way the visible star moves. In close binary systems, the components cannot be *resolved*, but *spectroscopic* studies show changes in the light from the system as the stars orbit one another. Stars in close binary systems are often distorted by tidal forces. If gas is being drawn off one star and on to the other by these forces, the system is said to be semi-detached; if the two stars are touching one another they form a contact binary. If the companion in a semi-detached system is a compact neutron star or black hole, the energy released by matter falling on to it from the larger star can produce large amounts of X-rays and other high-energy radiation. See *symbiotic stars.*

binary system See *binary star.*

Birr Castle Site, in central Ireland, of the 72-inch (183 cm) reflecting telescope built by the Earl of Rosse in 1845. The telescope was the most powerful in the world for many years, and was used by Rosse to reveal the spiral structure of many *nebulae.*

black body radiation An object that absorbed all the electromagnetic radiation that fell on it would be a perfectly black body. Black body radiation is the radiation that would be produced by such a hypothetical object when it was warmed.

The closest equivalent to a genuine black body on Earth is provided by making a small hole in a large container, and shining radiation into the container through the hole. When the walls of the container are heated, the radiation that comes out of the hole is black body radiation.

The nature of this radiation depends on the temperature of the black body, with more energy in higher-frequency radiation at higher temperatures. This is why a warm lump of iron radiates invisible infrared radiation, a slightly hotter lump of iron glows red, a hotter lump still glows white hot, and so on.

Although a black body is an idealized concept, the radiation from many astronomical objects can be approximately described in terms of the equivalent black body radiation. The Sun, for example, radiates rather like a black body with a temperature of about 6,000 K. The ultimate black body is the Universe itself, which is filled with perfect black body radiation left over from the *Big Bang*, now cooled to a temperature of 2.7 K and detectable at microwave radio frequencies. See *background radiation*.

black dwarf   A cold, dead *star* which no longer radiates any light. See *white dwarf*.

black hole   A concentration of matter which has a gravitational field strong enough to curve spacetime completely round upon itself so that nothing can escape, not even light, is said to be a black hole. This can happen either if a relatively modest amount of matter is squeezed to very high densities (for example, if the Earth were to be squeezed down to about the size of a pea), or if there is a very large concentration of relatively low mass material (for example, a few million times the mass of our Sun in a sphere as big across as our Solar System, equivalent to about the same density as water).

The first person to suggest that there might exist 'dark stars', whose gravitation was so strong that light could not escape from them, was John *Michell*, a Fellow of the Royal Society whose ideas were presented to the Society in 1783. Michell based his calculations on Isaac *Newton*'s theory of gravity, the best available at the time, and on the corpuscular theory of light, which envisaged light as a stream of tiny particles, like miniature cannon balls (now called *photons*). Michell assumed that these particles of light would be affected by gravity in the same way as any other objects. Ole *Rømer* had accurately measured the *speed of light* a hundred years earlier, and Michell was able to calculate how large an object with the density of the Sun would have to be in order to have an *escape velocity* greater than the speed of light.

If such objects existed, light could not escape from them, and they would be dark. The escape velocity from the surface of the Sun is only 0.2 per cent of the speed of light, but if you imagine successive larger objects with the same density as the Sun, the escape velocity increases rapidly. Michell pointed out that such an object with a diameter 500 times the diameter of the Sun (roughly as big across as the Solar System) would have an escape velocity greater than the speed of light.

The same conclusion was reached independently by Pierre *Laplace*, and published by him in 1796. In a particularly prescient remark, Michell pointed out that, although such objects would be invisible, 'if any other luminiferous bodies should happen to revolve about them we might still

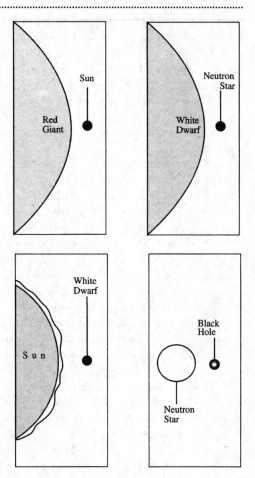

Black holes. Relative sizes of stars and black holes. *Top left*, the Sun compared with a red giant. *Bottom left*, a white dwarf compared with the Sun. *Top right*, a neutron star compared with a white dwarf. *Bottom right*, a black hole compared with a neutron star.

perhaps from the motions of these revolving bodies infer the existence of the central ones'. In other words, he suggested that black holes would most easily be found if they occurred in *binary systems*. But the notion of dark stars was forgotten in the 19th century and only revived in the context of Albert Einstein's *general theory of relativity*, when astronomers realized that there was another way to make black holes.

One of the first people to analyse the implications of Einstein's theory was Karl *Schwarzschild*, an astronomer serving on the eastern front in the First World War. The general theory of relativity explains the force of gravity as a result of the way spacetime is curved in the vicinity of matter. Schwarzschild calculated the exact mathematical description of the geometry of spacetime around a spherical mass, and sent his calculations to Einstein, who presented them to the Prussian Academy of Sciences early in 1916. The calculations showed that for *any* mass there is a critical radius, now called the Schwarzschild radius, which corresponds to such an extreme distortion of spacetime that, if the mass were to be squeezed inside the critical radius, space would close around the object and pinch it off from the rest of the Universe. It would, in effect, become a self-contained universe in its own right, from which nothing (not even light) could escape.

For the Sun, the Schwarzschild radius is 2.9 km; for the Earth, it is 0.88 cm. This does not mean that there is what we now call a black hole (the term was first used in this sense only in 1967, by John **Wheeler**) of the appropriate size at the centre of the Sun or of the Earth. There is nothing unusual about spacetime at this distance from the centre of the object. What Schwarzschild's calculations showed was that *if* the Sun could be squeezed into a ball less than 2.9 km across, or *if* the Earth could be squeezed into a ball only 0.88 cm across, they would be permanently cut off from the outside Universe in a black hole. Matter can still fall into such a black hole, but nothing can escape.

For several decades this was seen simply as a mathematical curiosity, because nobody thought that it would be possible for real, physical objects to collapse to the states of extreme density that would be required to make black holes. Even *white dwarf* stars, which began to be understood in the 1920s, contain about the same mass as our Sun in a sphere about as big as the Earth, much more than 3 km across. And for a time nobody realized that you can also make a black hole, essentially the same as the kind of dark star envisaged by Michell and Laplace, if you have a very large amount of matter at quite ordinary densities. The Schwarzschild radius corresponding to any mass M is given by the formula $2GM/c^2$, where G is the constant of gravity and c is the speed of light.

In the 1930s, Subrahmanyan **Chandrasekhar** showed that even a white dwarf could be stable only if it had a mass less than 1.4 times the mass of the Sun, and that any heavier dead star would collapse further. A few researchers considered the possibility that this could lead to the formation of *neutron stars*, typically with a radius only one-seven-hundredth of that of a white dwarf, just a few kilometres across. But the idea was not widely accepted until the discovery of *pulsars* in the mid-1960s showed that neutron stars really did exist.

This led to a revival of interest in the theory of black holes, because neutron stars sit on the edge of becoming black holes. Although it is hard to imagine squeezing the Sun down to a radius of 2.9 km, neutron stars with about the same mass as the Sun and radii less than about 10 km were now known to exist, and it would be a relatively small step from there to a black hole.

Theoretical studies show that a black hole has just three properties that define it – its mass, its electric charge and its rotation (angular momentum). An uncharged, non-rotating black hole is described by the *Schwarzschild solution* to Einstein's equations; a charged, non-rotating black hole is described by the *Reissner–Nordstrøm* solution; an uncharged but rotating black hole is described by the *Kerr solution*; and a rotating, charged black hole is described by the *Kerr–Newman solution*. A black hole has no other properties, summed up by the phrase 'a black hole has no hair'. Real black holes are likely to be rotating and uncharged, so that the Kerr solution is the one of most interest.

Both black holes and neutron stars are now thought to be produced in the death throes of massive stars that explode as supernovae. The calculations showed that any compact supernova remnant with a mass less than about three times the mass of the Sun (the *Oppenheimer–Volkoff limit*) could form a stable neutron star, but any compact remnant with more than this mass would collapse into a black hole, crushing its contents into a singularity at the centre of the hole, a mirror-image of the *Big Bang* singularity in which the Universe was born. If such an object happened to be in orbit around an ordinary star, it would strip matter from its companion to form an *accretion disc* of hot material funnelling into the black hole. The temperature in the accretion disc might rise so high that it would radiate *X-rays*, making the black hole detectable.

In the early 1970s, echoing Michell's prediction, just such an object was found in a binary system. An X-ray source known as Cygnus X-1 was identified with a star known as HDE 226868. The orbital dynamics of the system showed that the source of the X-rays, coming from an object smaller than the Earth in orbit around the visible star, had a mass greater than the Oppenheimer–Volkoff limit. It could only be a black hole. Since then, a handful of other black holes have been identified in the same way, and in 1994 a system known as V404 Cygni became the best black hole 'candidate' to date when it was shown to be made up of a star with about 70 per cent as much mass as our Sun in orbit around an X-ray source with about twelve times the Sun's mass. But such confirmed identifications may be much less than the tip of the proverbial iceberg.

Such 'stellar mass' black holes can be detected only if they are in binary systems, as Michell realized. An isolated black hole lives up to its name – it is black, and undetectable (but see *gravitational lens*). But very many stars should, according to astrophysical theory, end their lives as neutron stars or black holes. Observers actually detect about the same number of good black hole candidates in binary systems as they do *binary*

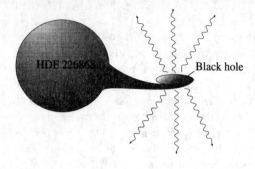

HDE 226868

Black hole

The archetypal X-ray binary system, HDE 226868, is made up of a large star orbited by a black hole. The black hole is stripping material from the star by gravity and swallowing it up. On its way into the black hole the material forms a hot, swirling disk from which X-rays are emitted.

pulsars, and this suggests that the number of isolated stellar-mass black holes must be the same as the number of isolated pulsars. This supposition is backed up by theoretical calculations.

There are about 500 active pulsars known in our Galaxy today. But theory tells us that a pulsar is only active as a radio source for a short time, before it fades into undetectable silence. So there should be correspondingly more 'dead' pulsars (quiet neutron stars) around. Our Galaxy contains 100 billion bright stars, and has been around for thousands of million of years. The best estimate is that there are around 400 million dead pulsars in our Galaxy today, and even a conservative estimate would place the number of stellar mass black holes at a quarter of that figure – 100 million. If so, and the black holes are scattered at random across the Galaxy, the nearest one is probably just 15 light years away. And since there is nothing unusual about our Galaxy, every other galaxy in the Universe must contain a similar profusion of black holes.

They may also contain something much more like the kind of 'dark star' originally envisaged by Michell and Laplace. These are now known as 'supermassive black holes', and are thought to lie at the hearts of active galaxies and quasars, providing the gravitational powerhouses which explain the source of energy in these objects. A black hole as big across as our Solar System, containing a few million solar masses of material, could swallow matter from its surroundings at a rate of one or two stars a year. In the process, a large fraction of the star's mass would be converted into energy, in line with Einstein's equation $E = mc^2$. Quiescent supermassive black holes may lie at the centres of all galaxies, including our own.

In 1994 observers using the Hubble Space Telescope discovered a disc of hot material, about 150,000 parsecs across, orbiting at speeds of about 2 million kilometres per hour (about 3×10^7 cm/sec, almost 1 per cent of the speed of light) around the central region of the galaxy M87, at a distance of about 15 million parsecs from our Galaxy. A jet of hot gas, more than 1 kiloparsec long, is being shot out from the central 'engine' in M87. The orbital speeds in the accretion disc at the heart of M87 is conclusive proof that it is held in the gravitational grip of a supermassive black hole, with a mass that may be as great as 3 billion times the mass of our Sun, and the jet is explained as an outpouring of energy from one of the polar regions of the accretion system.

It was also in 1994 that astronomers from the University of Oxford and from Keele University identified a stellar-mass black hole in the binary system known as V404 Cygni. As we have mentioned, the orbital parameters of the system enabled them to 'weigh' the black hole accurately, showing that it has about twelve times as much mass as our Sun and is orbited by an ordinary star with about 70 per cent of the Sun's mass. This is the most precise measurement so far of the mass of a 'dark star', and is therefore the best individual proof that black holes exist.

A more speculative suggestion is that tiny black holes, known as *mini black holes* or primordial black holes, may have been produced in profusion in the Big Bang and could provide a significant fraction of the mass of the Universe. Such mini black holes would typically be about the size of an atom and each have a mass of perhaps 100 million tonnes (10^{11} kg). There is no evidence that such objects really exist, but it would be very hard to prove that they do not exist.

See also *time travel, wormholes, Hawking radiation*.

Further reading: John Gribbin, *In Search of the Edge of Time*; Kip Thorne, *Black Holes and Time Warps*.

black-widow pulsar A *binary pulsar* in which the intense beam of *radiation* from the pulsar (like a fire hose turned on a pile of sand) is eating away at its companion, which is losing mass as a result.

blazar A term coined from the names *BL Lac object* and *quasar* to refer to a class of highly energetic variable *galaxies*. Many active galaxies are expelling jets of material from their centres; a likely explanation of blazars is that in these cases we are viewing such a jet end on, pointing straight towards us.

Bliss, Nathanial (1700–64) The fourth Astronomer Royal, from 1762 until his death. He had previously worked with James *Bradley* at the *Royal Greenwich Observatory*.

BL Lac objects Unusual, bright, compact sources of intense energy lying at the hearts of some galaxies. Probably closely related to *quasars*.

The first of these objects to be identified, BL Lacertae, had been known for decades as a variable 'star', but in 1968 it was discovered to be a strong source of radio noise as well. This attracted attention to BL Lac itself, which turned out not to be a star after all, and led to the discovery of a handful of similar objects (about 100 are now known). Like quasars, BL Lac objects lie at the hearts of galaxies, and are probably associated with supermassive *black holes*. But they are extremely variable, and can brighten up by a factor of 100 in a few weeks before fading back to their normal brightness.

Unlike quasars, BL Lac objects have featureless spectra, devoid of emission or absorption lines (see *spectroscopy*), and so it is very difficult to measure a redshift and determine their distances. But during their quieter phases, some BL Lac objects become faint enough for light from the surrounding galaxy to be analysed, instead of being lost in the glare of the central source. For the few BL Lac objects where this is possible, the inferred distances are comparable to those of relatively low-redshift quasars.

Most BL Lac objects seem to lie in the centres of elliptical galaxies rather than disc galaxies. One plausible explanation for their variability and their featureless spectra is that their central black holes are so big (containing more than 100 million times the mass of our Sun) that they can swallow whole stars intact.

This would produce a bright flare of energy – a large fraction of the mc^2 corresponding to the mass of the star being swallowed (see *special theory of relativity*), compared with perhaps one-tenth of 1 per cent of this energy normally released in the entire lifetime of the star. And because the star is swallowed whole, instead of being torn apart by tidal forces and swallowed piecemeal, there is no debris left to help form a cloud of material around the black hole, where spectral lines could be produced as light from the central object makes its way outward. See *active galaxies*.

blue sheet *Radiation* falling on to a very dense object will experience an extreme *blueshift*, gaining energy from the object's *gravitational field*. It has been suggested that *white holes*, a hypothetical counterpart to *black holes*, would be smothered in a resulting blue sheet of radiation and could not expand outwards into the Universe. But more sophisticated calculations suggest that there are ways around this difficulty. See *wormholes*.

blueshift The opposite of *redshift*, a blueshift can be produced by the reverse of any of the processes that cause redshift – as a *Doppler effect* for an approaching object, by *radiation* gaining energy as it falls in a *gravitational field*, or (in principle) by the contraction of space in a collapsing universe.

blue stragglers Stars, often found in globular *clusters*, that lie on the blue side of the appropriate *main sequence* 'turn-off point'. See *Hertzprung–Russell diagram*.

Bode, Johannes Elert (1749–1826) German mathematician and astronomer who was director of the Berlin Observatory from 1786 to 1825. He is best known for the law named after him, although he did not discover

the law but popularized the earlier work of Johann Titius (1729–96). See
Bode's Law.

Bode's Law   A mathematical relationship which relates the distances of the
planets from the Sun to a simple numerical sequence. As is often the case
in science, the law does not carry the right name – it was first noticed by
Johann *Titius*, but was popularized by Johann *Bode* in the 1770s.

In the form presented by Bode, the key sequence of numbers is 0, 3,
6, 12, 24 and so on, with each number after the '3' twice the preceding
number. Then, you add 4 to each number. If the distance of Earth (the
third planet) from the Sun is set at 10 units, then the distances of Mercury,
Venus and Mars are all at the equivalent distances given by the series 4, 7,
10, 16. There is a gap at 28, with Jupiter at 52 and Saturn at 100 units from
the Sun, as 'predicted' by the law.

The discovery of Uranus in 1781, close to 196 units from the Sun, was
regarded as a confirmation of the accuracy of Bode's Law, and in the early
19th century the *asteroids* were discovered in a belt between Mars and
Jupiter, corresponding to the 'missing' planet.

Although Neptune and Pluto, discovered later, do not have orbits
which match the predictions of Bode's Law, there has recently been a
revival of interest in the possibility that some simple rule of thumb of this
kind must apply to the spacing of the orbits of planets in any planetary
system, and to the spacing of moons in systems like those around Jupiter
and Saturn.

The discovery of three planets orbiting a pulsar known as PSR
B1257 + 12 has provided a boost for these ideas by revealing a system with
properties that almost exactly match those of the inner Solar System, made
up of Mercury, Venus and the Earth. The similarities are so striking that it
seems there may be a law of nature which ensures that planets always
form in certain orbits and have certain sizes.

PSR B1257 + 12 is a rapidly spinning neutron star, containing slightly
more matter than our Sun packed into a sphere only about 10 km across.
As the star spins, it flicks a beam of radio noise around, like the beam of a
lighthouse, producing regularly spaced pulses of radio noise detectable on
Earth. It can only have been produced in a supernova explosion, long ago,
which would have disrupted any planetary system the star possessed at
the time. So the present planets associated with the pulsar are thought to
have formed from the debris of a companion star disrupted by the pulsar.

The three planets cannot be seen directly, but are revealed by the way
in which they change the period of the pulsar's pulses as they orbit around
it. There is enough information revealed in the changing pulses to show
that the three planets have masses roughly equal to 2.8 times the mass of
the Earth, 3.4 times the mass of the Earth, and 1.5 per cent of the mass of

the Earth. And they are spaced, respectively, at distances from the pulsar equivalent to 47 per cent of the distance from the Earth to the Sun, 36 per cent of the Sun–Earth distance, and 19 per cent of the Sun–Earth distance.

Tsevi Mazeh and Itzhak Goldman, of Tel Aviv University, have pointed out that the ratio of these distances, 1: 0.77: 0.4, is extremely close to the ratio of the distances of the Earth, Venus and Mercury from the Sun, which is 1: 0.72: 0.39. And the masses of the three inner planets in the Solar System are 1 Earth mass, 82 per cent of the mass of the Earth, and 5.5 per cent of the mass of the Earth. In each case, two outer planets with roughly the same mass have an inner companion with a much smaller mass.

So Bode's Law also works for the planets of pulsar PSR B1257 + 12, and the indications are that there is a universal mechanism for the formation of planets around stars. If it works for systems as diverse as a pulsar and our Sun, the chances are that it works for all stars, and that 'Solar' Systems very like our own may be the rule, rather than the exception, among the stars of the Milky Way.

Bok, Bart (1906–1983) Dutch astrophysicist who studied at the universities of Leiden and Groningen before moving to the United States in 1929. He has also worked in Australia. In 1947 Bok discovered small, dark, circular clouds of material in space, which show up against the background of stars or luminous clouds. These *Bok globules* are thought to be stars like our Sun in the process of forming.

Bok globules Clouds of dark material in space, first discovered by Bart Bok in 1947, which are thought to be on the brink of collapsing to form stars. Such globules may be only 0.04 parsecs (8,000 astronomical units) across or ten times bigger, and may contain a tenth as much mass as our Sun or a few tens of solar masses of material in the form of gas and dust.

bolometric luminosity See *luminosity*.

Bondi, Sir Hermann (1919–) Austrian-born mathematician and astronomer who was one of the three original proponents of the *Steady State* model of the Universe, along with Tommy *Gold* and Fred *Hoyle*. Bondi was born in Vienna, but studied mathematics at Cambridge University. He was briefly interned as an 'enemy alien' in 1940, and met Gold during his internment. He worked on radar for the British Admiralty in 1942, and met Hoyle as a result of this work. Bondi became a British citizen in 1947, and has held several posts in the administration of British and European science, as well as being professor of applied mathematics at King's College in London. He was elected a Fellow of the Royal Society in 1959 and knighted in 1973; in 1983 he became Master of Churchill College, in Cambridge.

BOREXINO A detector being built to study *neutrinos* from the Sun, in the Gran Sasso Laboratory under the Apennine mountains in Italy. The detector, which operates on the same basic principle as the *Kamiokande* detector, uses 5 tonnes of pure organic liquid contained in a nylon ball 2 m across, in which neutrinos produce flashes of light as they collide with electrons associated with the atoms of the fluid. The light flashes are recorded by an array of detectors around the sphere. The assembly is immersed in a tank 11 m on each side, containing 1 million litres of pure water; the whole thing weighs 100 tonnes. BOREXINO should be able to detect neutrinos produced by *nuclear fusion* reactions involving boron-8 and beryllium-7 that take place in the heart of the Sun. See *solar neutrino problem*.

bosons   *Elementary particles* that are not conserved during particle interactions, and which behave in accordance with statistical rules developed in the 1920s by Satyendra Bose (1894–1974) and Albert *Einstein*, known as 'Bose–Einstein statistics'. The archetypal boson is the *photon*, the particle of light; photons are created in copious quantities every time a light is turned on.

Using Bose–Einstein statistics to describe electromagnetic radiation as a 'gas' of photons predicts all of the properties of *black body radiation*, without using any description of the radiation in terms of waves. This is an example of the wave–particle duality of the quantum world, which says that entities such as photons or electrons can be described either in terms of waves or as particles. Bosons are the particles which are responsible for the transmission of forces in *quantum theory*. The electromagnetic force, for example, can be described in terms of the exchange of photons between two charged particles, such as an electron and a proton, like two footballers exchanging passes.

In the language of quantum physics, the key property of a boson is its spin. All bosons have a spin which is either zero or a whole number – 1, 2, 3 and so on. This is analogous to the spinning of a child's top – but not exactly like the spinning of a child's top, because a particle such as an electron with half-integer spin has to 'rotate' *twice* in order to get back to where it started.

See *fermions*.

bottom-up model   See *galaxy formation*.

bounce   See *oscillating universe models*.

Bradley, James (1693–1762)   English astronomer who discovered the *aberration* of starlight and used this to determine the speed of light, arriving at a figure equivalent to 308,300 km/sec., close to the modern value of 299,792 km/sec. This figure was published in 1729 and confirmed the accuracy of Ole *Rømer*'s measurement, based on a different technique, published in 1679. Bradley was the third Astronomer Royal, succeeding *Halley* to the title in 1742.

Brahe, Tycho (1546–1601) Danish astronomer (sometimes referred to only by his first name), who made accurate measurements of the positions of stars and the movements of *planets*, paving the way for *Kepler* to discover the laws of planetary motion. He was the greatest observational astronomer of the pre-telescopic era.

Tycho came from an aristocratic family, and was born in Knudstrup, now part of southern Sweden but then ruled by Denmark, on 14 December 1546. His father's brother, who was childless, literally stole him away when he was a year old, and was allowed to keep the boy, whom he sent to study philosophy and law at Copenhagen (in 1559) and Leipzig (1562–5), intending him for a diplomatic career. But Tycho had observed a *solar eclipse* in 1560, and had become fascinated by astronomy as a result. So while officially reading law in Leipzig, he was also studying astronomy on his own initiative.

In 1563 he observed a conjunction of Jupiter and Saturn (when the two planets are seen close together on the sky), and noticed that it had occurred a month earlier than predicted. He realized that the astronomical tables used at the time were inaccurate, and began a long series of observations to provide his own, more accurate, tables.

Tycho's uncle died in 1565, leaving him free to follow his own career inclinations. He studied at the University of Rostok, graduating in 1566; while he was there, he fought a duel over a point of mathematics, and had a large part of his nose sliced off. Afterwards, he wore a false nose made of silver.

After further studies in Basle and Augsburg, Tycho returned home because his father was ill. On 11 November 1572, he noticed a bright new star in the constellation Cassiopeia, and made a series of observations of it, using instruments he had made himself, until it faded from view in March 1574. From his description of the star, published as *De Stella Nova* (not to be confused with the book of the same title published by *Kepler*), we know that this was a *supernova*.

The book made Tycho famous, and as at about the same time he had upset his aristocratic family by marrying a peasant girl, he was happy to take up offers he now received to lecture in Copenhagen and Germany. He considered settling in Switzerland, but in 1576 the King of Denmark, Frederick II, gave Tycho the island of Hven as the site for a new observatory, and promised him an income for life (a pension).

Tycho's observatory became a centre of excellence for astronomy, visited by scholars from across Europe as his reputation grew. He realized that the elongated orbits of *comets*, crossing those of the planets, posed difficulties for the notion of 'heavenly spheres' dating from the time of *Aristotle*, but he did not see all of the implications, and never came to terms with the ideas of *Copernicus*. Instead of accepting that the Sun was at the centre of the Solar System, Tycho suggested a compromise in which the Sun orbits around the Earth while all the other planets orbit around the Sun.

In spite of his successes over a period of 21 years, the observatory was never completed to Tycho's satisfaction, and in 1597 Frederick's successor withdrew both Tycho's funding and his 'ownership' of the island. He left Denmark and worked for a time in Germany before settling in Prague in 1599, at the invitation of the Emperor, Rudolph II, who gave him a castle to observe from and a pension to live on.

Many of Tycho's instruments were moved from Hven to Prague, where he was joined in 1600 by the young Johannes *Kepler*. Their collaboration was short-lived, because Tycho died on 24 October 1601; but Kepler picked up the baton, continuing Tycho's work, completing the astronomical tables and going on to become a great astronomer in his own right.

brown dwarf A star which forms from the gravitational collapse of a cloud of gas in space, but has insufficient mass to trigger *nuclear fusion* reactions in its core. This is possible only if the mass of the star is less than about 8 per cent of the mass of our Sun. Such a star would shine faintly for about 100 million years as gravitational energy is converted into heat (see *Kelvin–Helmholtz timescale*). No brown dwarf had been observed directly by 1994, but such objects may explain *gravitational lensing* events observed in our Galaxy (see *MACHOs*).

Bruno, Giordano (1548–1600) Italian monk who was an early supporter of the proposal made by *Copernicus* that the Earth moves around the Sun. Burned at the stake for heresy in Rome.

Burbidge, Geoffrey (1925–) and Burbidge, (Eleanor) Margaret (1922–) British husband-and-wife team of astrophysicists who have spent most of their working lives in the United States. They are best known for their work with Fred *Hoyle* and Willy *Fowler* on *nucleosynthesis*, which explains how the elements were manufactured inside stars (see *B^2FH*).

Margaret Burbidge was born at Davenport and studied at University College, London, obtaining a PhD in 1947. She worked as assistant director of the University of London Observatory until 1951, when she moved to the United States and had a succession of appointments at Yerkes Observatory, the California Institute of Technology and the University of San Diego. In 1972 she was appointed director of the Royal Greenwich Observatory in England, the first woman to hold the post; but unlike her predecessors there she did not become Astronomer Royal because the two posts were separated at that time. Unhappy with the bureaucratic way in which astronomy in Britain was being run, she resigned from the post in 1973 and returned to her post as professor of astronomy in San Diego.

Since 1979 she has been director of the Center for Astrophysics and Space Sciences there.

Geoffrey Burbidge was born at Chipping Norton, on 24 September 1925, and graduated from Bristol University in 1946 before moving to University College, London, where he completed a PhD in 1950 before going to the United States to work first at Harvard University and then at the University of Chicago. In 1953 he returned to England to work at the Cavendish Laboratories in Cambridge, but in 1955 he moved once again, to the *Mount Wilson and Palomar Observatories*, then on to posts at the California Institute of Technology, University of Chicago, and from 1962 the University of California, San Diego (first as associate professor, then as professor of physics from 1963 to 1978). From 1978 to 1984 Geoffrey Burbidge was director of the *Kitt Peak National Observatory* in Arizona.

Both the Burbidges were regular visitors to Cambridge, England, in the 1970s, when they often spent the summer months at the Institute of Astronomy run by their friend and colleague Fred Hoyle.

As well as their work with Hoyle and Fowler on nucleosynthesis, the Burbidges jointly wrote an important book about *quasars*, published in 1967. They have been involved in some unconventional ideas, including the suggestion (promoted vigorously by Halton *Arp*) that some of the objects identified as quasars are physically associated with galaxies, and cannot, therefore, be at the distances indicated by the cosmological interpretation of their *redshifts*. In 1970 Geoffrey Burbidge was involved in one of the first studies which showed that the bright stars in elliptical galaxies contribute less than a quarter of the mass (see *dark matter*); Margaret Burbidge was involved in several of the spectroscopic studies that established the high redshifts of quasars in the early 1960s, and has continued to make spectroscopic studies of galaxies and quasars.

Burnell, Jocelyn   See *Bell Burnell, Jocelyn*.

bursters   Sources that emit bursts of *X-rays*. They were first discovered in 1973, by astronomers working with the Astronomical Netherlands Satellite. All bursters are located in our *Galaxy*, relatively close to the galactic centre. The bursts are thought to be produced in *binary stars* when *accretion* on to the surface of a *neutron star* triggers a flash of *nuclear fusion*.

butterfly diagram   Pattern made by plotting the *sunspots* observed over an 11-year *solar cycle* of activity. At the start of each cycle, the spots occur well away from the solar equator, but later in the cycle they occur closer to the equator. The pattern looks like the wings of a butterfly.

3C 273   The first *quasar* to be identified, in 1963. 3C 273 has a *redshift* of 0.158, corresponding to a distance of about 500 million parsecs, but is the nearest known quasar. It is also (hardly surprisingly, in view of its relative proximity) the brightest quasar on the sky. The central starlike object is a strong *radio source*, and it has a very long, faint *jet* ending in a bright secondary radio source.

8C 1435+63   The most distant *galaxy* known, with a *redshift* of 4.25.

caesium clock   See *atomic clock*.

Callisto   The faintest and outermost of the four large moons of Jupiter observed by *Galileo*. It has a radius of 2,400 km and a density less than twice that of water, showing that it is largely made of ice.

Cameron, Alastair Graham Walter (1925–)   Canadian-born astrophysicist who has mainly worked in the United States, and became a US citizen in 1963. He independently came up with ideas about stellar *nucleosynthesis* similar to those of the B^2FH team, in the mid-1950s.

Canada–France–Hawaii Telescope   One of several instruments at the *Mauna Kea Observatory* in Hawaii. The reflecting telescope has a mirror with an aperture of 3.6 m and is used for both optical and infrared observations.

cannibalism   Process whereby one *galaxy* swallows up another. Cannibalism is important in the formation and evolution of galaxies like our own, which have been formed by mergers involving smaller components. The largest galaxies, known as cD galaxies, lie at the centres of *clusters of galaxies* and seem to have swallowed several smaller galaxies.

The term is also used to describe the process whereby a star swallows up a companion in a *binary star* system.

Cannon, Annie Jump (1863–1941)   American astronomer who carried out pioneering work in the classification of *stars* using *spectroscopy*.

Born in Dover, Delaware, on 11 December 1863, Cannon, the daughter of a wealthy shipbuilder, was educated at Wellesley College and graduated in 1884. After ten years back in the family home, during which she went deaf as a result of contracting scarlet fever, she returned to Wellesley as an assistant in the physics department and as a postgraduate student. She went on to Radcliffe College in 1895, where she studied astronomy, and became a protégé of Edward *Pickering*, the director of the Harvard College Observatory, where she became an assistant in 1896.

Cannon stayed at Harvard for the rest of her career, working first as a calculator (before the days of electronic computers, Pickering used well-educated women to do the laborious arithmetical calculations involved in astronomical studies), then as the curator of the astronomical photo library, and being appointed William Cranch Bond Astronomer in 1938, at the age of 75. She was the first woman to be awarded an honorary DSc degree by the University of Oxford.

Although Cannon made detailed observations of variable stars and carried out other valuable work in astronomy, her lasting legacy is the classification system for stars, which is still used today. In 1890, Pickering and his colleague Williamina Fleming established a classification system

Annie Jump Cannon (1863–1941).

which assigned stars to one of seventeen spectral types, labelled A to Q, according to the intensity of the absorption lines in their spectra caused by hydrogen. In the early years of the 20th century, Cannon revised this system into a more natural classification based on the surface temperature (and therefore the colour) of the star. In order of decreasing surface temperature, this gave her a ten-step sequence with the letters O, B, A, F, G, K, M, R, N, S, from Pickering's earlier classification scheme.

Most stars – by far the majority – fall into one of the categories between O and M. O, B and A stars are white or blue, F and G stars are yellow, K are orange, and the rest are red. The Sun is an ordinary yellow G star. Cannon's classification system has been only slightly modified since, mainly by subdividing the categories. The sequence has been remembered by generations of astronomy students using the decidedly not 'politically correct' mnemonic 'Oh, Be A Fine Girl, Kiss Me'.

Cannon became so expert at classifying stars according to this system that she could classify three stars a minute by looking at photographs of their spectra. In the *Henry Draper Catalog* (published in nine parts between 1918 and 1924), she classified more than 225,000 stars brighter than tenth *magnitude*, including almost all stars brighter than ninth magnitude; she added a further 130,000 classifications of fainter stars in extensions to the catalog between 1925 (when she was already 62) and 1936, when she was 73. She then took on the classification of another 10,000 faint stars at the request of the Cape of Good Hope Observatory.

This enormous body of painstakingly acquired, meticulous information established that almost all stars can be categorized easily as members of a continuous sequence in terms of brightness and colour; this was an essential prerequisite to the development of a good theory of *stellar evolution*. She died on 13 April 1941.

See also *Hertzsprung–Russell diagram*.

carbohydrates Compounds containing *carbon*, *hydrogen* and *oxygen*, based on the chemical structure $C_x(H_2O)_y$. They are important chemicals for life, and include sugars, starch and cellulose. Fred **Hoyle** and Chandra Wickramasinghe have made the controversial claim that some features of the *spectra* of clouds of interstellar material can be explained by the presence of carbohydrates in those clouds; if they are correct, this has important implications for the origin of life. Not to be confused with *hydrocarbons*.

carbon One of the most common types of atomic material found in the Universe – the fourth most abundant, after hydrogen, helium, and oxygen. It has an *atomic number* of 6, and the most common *isotope* has six protons and six neutrons in each nucleus. Carbon plays an important part in the *carbon cycle*, which provides the energy source in massive stars on the *main sequence*, and is one of the key elements for life (see *CHON*).

carbon cycle (CNO cycle) The process of *nuclear fusion* reactions that provides the energy source inside hot, massive stars, especially those with a *spectral classification* of O, B or A.

At the heart of such stars, the temperature is above 20 million Kelvin. Most of the material there is in the form of nuclei of hydrogen (protons), but there are traces of other nuclei, including those of carbon. The CNO cycle works like this:

First, a proton penetrates a nucleus containing six protons and six neutrons (a nucleus of carbon-12), through the *tunnel effect*. This creates an unstable nucleus of nitrogen-13, which emits a positron and a neutrino, converting itself into a nucleus of carbon-13. If a second proton now tunnels into this nucleus, it becomes nitrogen-14, and the addition of a third proton converts it into oxygen-15, which is unstable and spits out a positron and a neutrino as it transmutes into nitrogen-15. But now, if yet another proton tunnels into the nucleus, it ejects an alpha particle (two protons and two neutrons bound together to form a nucleus of helium-4). This leaves behind a nucleus of carbon-12, identical to the one that started the cycle.

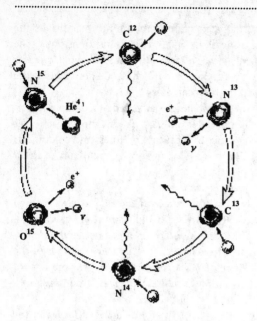

Carbon cycle. The cycle of nuclear interactions that generates heat inside massive stars.

The net effect is that four protons have been converted into one helium nucleus, a couple of positrons and two neutrinos. But the mass of a helium nucleus plus these other particles is less than the mass of four protons put together. The difference in mass has been converted into energy, in line with Einstein's equation $E = mc^2$, keeping the heart of the star hot. Just 0.7 per cent of the mass of each set of four protons is turned into energy every time a helium nucleus is made.

The CNO cycle was worked out in 1938 by Hans **Bethe**, and independently by Carl von Weizsäcker (1912–).

See also *proton–proton chain*.

carbon stars   Cool *red giant* stars which are shown by *spectroscopy* to have more carbon than oxygen in their outer layers. Oxygen in the atmosphere of such a star is locked up in the form of carbon monoxide, with excess carbon forming other compounds with *nitrogen* and hydrogen. Carbon stars are losing mass into space, lacing the *interstellar matter* with carbon, oxygen, nitrogen and traces of other elements produced by the *s-process*.

Cassegrain (about 1650–1700)   The inventor of the system of mirrors used in a *Cassegrain telescope*. Very little is known about Cassegrain – not even his first name – but he worked as an astronomer, physician and sculptor at the court of Louis XIV.

Cassegrain telescope   A *reflecting telescope* in which light from the concave main mirror is reflected up to a small convex secondary mirror and back down through a small hole in the main mirror to a focus. This gives convenient access to the focus of the telescope, particularly useful for large telescopes. It was the second successful design for a reflecting telescope, following the *Newtonian telescope*; *Cassegrain* published his design in 1672, although Isaac *Newton* claimed that Cassegrain was influenced by the earlier work of James Gregory. The design was not put into practice until the 18th century.

Cassini, Giovanni Domenico (1625–1712)   Italian–French astronomer, born in Nice (then part of Italy) and educated in Genoa, who became professor of astronomy at the University of Bologna in 1650. He determined the rotation periods (the 'length of day') for the planets, and calculated tables of the motions of the satellites of Jupiter. In 1675 he discovered a gap in the rings of Saturn, still known as the Cassini division.

Cassiopeia A   A strong *radio source* in our *Galaxy*, in the direction of the constellation Cassiopeia, but far beyond it at a distance of 2,900 parsecs. Thought to be the remains of a *supernova* that must have been visible in the 17th century, but was not recorded by observers on Earth.

cataclysmic variable   A *binary star* system in which a *white dwarf* is in a close orbit around another star, so that matter from the companion is being torn away by the gravity of the white dwarf and falls on to its surface, via an *accretion disc*, creating outbursts of energy. This produces sudden, unpredictable changes in the brightness of the system.

The classic type of cataclysmic variable is a *dwarf nova*, and more than 200 of these stars are known. But the name is also used for *novae* and *recurrent novae*, and for so-called 'nova-like objects', which share all the other properties of novae, but have never been observed to brighten enough to be classified as novae.

About 600 cataclysmic variables are known, and the orbital periods have been determined for about a quarter of them. All except one have

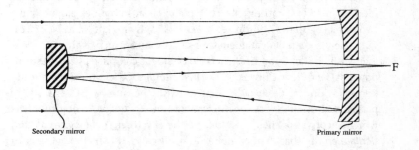

Secondary mirror   Primary mirror

Schematic representation of a Cassegrain telescope.

periods less than 15 hours, but they are all greater than 80 minutes, and there are very few with periods between 2 and 3 hours. The masses of the white dwarfs in these systems are typically about 90 per cent of the mass of our Sun, rather higher than the average for other white dwarfs, about 60 per cent of the mass of the Sun; their companions typically have masses in the range from 10 per cent to 100 per cent of the mass of the Sun, but always less than the mass of the white dwarf.

In most cases, the more frequent the outbursts from a cataclysmic variable are, the less spectacular each outburst is. A nova may brighten by more than 10 *magnitudes* and fade away over several years, but it does not repeat the outburst for thousands of years. A dwarf nova brightens by only a few magnitudes, fading away in a week or so, but it may repeat this pattern of behaviour every few weeks or months.

catalogues Astronomers need to have catalogues which list known astronomical objects and their positions on the sky, both so that they can find them to make observations and so that they know which objects have already been investigated. The first major catalogue of non-stellar objects was compiled by Charles Messier (1730–1817), who was interested in finding new *comets* and wanted to make sure that he did not misidentify known objects, such as the *Crab nebula*, as comets. His first list of 45 objects appeared in 1771, a second list adding 23 more objects in 1780, and his final list of 103 objects in 1781 (published in 1784).

Some of the objects in Messier's catalogue had been discovered by other comet-hunters, and a few objects were added to his list by others. Astronomers still refer to astronomical objects by their numbers in this list, so that the Crab nebula is known as M1 and the *Andromeda galaxy* as M31. Many of the nebulae in the Messier catalogue are now known to be *galaxies*.

The standard list of celestial objects, many of them galaxies, is the New General Catalogue of Nebulae and Clusters of Stars, usually abbreviated to NGC, which was first published in 1888 and later extended to include more than 13,000 entries. Objects which appear in both catalogues have two numbers, so the Andromeda galaxy (M31) is also known as NGC 224.

There are similar catalogues for radio sources, such as the '3C' catalogue, which stands for the third radio catalogue compiled by radio astronomers at Cambridge University. The first *quasar* to be identified was already known as one of these radio sources, with the number 3C 273. Many *pulsars* and other objects of interest are now catalogued using numbers which incorporate their position on the sky (their *right ascension* and *declination*), so that the first pulsar to be discovered (which, having been discovered in Cambridge, might have been called CP 1, following the tradition of Messier) is now known as PSR 1919 + 21.

CCD = *charge coupled device.*

celestial equator   The imaginary circle on the sky that lies directly overhead to a person being carried round on the equator by the Earth's rotation. Tilted at an angle of about 23°26′ to the plane of the *ecliptic.*

Celsius, Anders (1701–44)   Swedish astronomer who had an undistinguished career, but who suggested a temperature scale based on two fixed points: 0 degrees for the boiling point of water, and 100 degrees for the melting point of ice. Shortly after his death, his colleagues at Uppsala University began to use the scale in reverse (so that ice melts at 0° and water boils at 100°). This became known as the Centigrade scale, and is now known as the Celsius scale to avoid confusion with a unit of angle known as the grade. The size of each degree on the Celsius scale is the same as on the *Kelvin* scale.

Centaurus A   A strong source of radio and X-ray emission which lies in the direction of the constellation Centaurus, but far beyond the borders of our *Galaxy*, at a distance of 5 Megaparsecs. This is the closest *active galaxy*, and covers 9 degrees of arc on the sky, eighteen times the apparent diameter of the Moon and equivalent to a physical extent of about 400 kiloparsecs.

Centaurus X-3   A *binary system* that is a source of X-ray emission and lies in our *Galaxy* in the direction of the constellation Centaurus. The X-ray source orbits around its companion star once every 2.09 days, and itself varies with a period of 4.8 seconds. The orbital dynamics suggest that the X-ray source itself is a rotating *neutron star*, making it an X-ray *pulsar.*

centre of mass   The 'balance point' in a system of gravitating objects. Two *stars* in a *binary system*, for example, each *orbit* around their mutual centre of mass, while any object far outside the binary feels the combined gravitational influence of the two stars as if all their mass were concentrated at the centre of mass.

centrifugal force   Outward force felt by an object which is being forced to move in a curved trajectory. It is often mistakenly referred to as a 'fictitious' force, because it is present only for an accelerated object, and does not exist in an *inertial frame*: that is, for an object moving in a straight line at constant speed. But Albert Einstein's *general theory of relativity* allows observers even in a non-inertial frame to regard themselves as at rest, and the forces they feel to be real. Anyone who tells you that centrifugal force is fictitious does not understand the general theory of relativity.

The classic example of centrifugal force is the force you feel when in a car rounding a bend at high speed. The force pushes you to the outside of the car, and a tennis ball left lying on the back shelf of the car will roll in the appropriate direction as it feels the force. An observer standing by the roadside will see that the ball is 'really' trying to continue in a straight line, and that the car is turning; to them, the force is fictitious, but it is not to anyone inside the car.

For an object in *free fall* in an *orbit* (for example, a *satellite* orbiting the Earth), there is no sensation of centrifugal force, because this is exactly balanced by the *centripetal force* of (in this case) the Earth's gravity. This is an example of the *equivalence principle*, which says that the effects of gravity and of acceleration are indistinguishable from one another.

Centrifugal force is the equivalent, for curved motion, of the force felt by observers in an accelerating frame of reference such as a lift, a fast car or a rocket. In both cases the existence of the force is intimately connected with motion that involves acceleration relative to the average distribution of matter across the Universe (see *Mach's Principle*), but there is no consensus on how this relationship works.

centripetal force   The inward-directed force (usually *gravity*) needed to hold an object in a closed *orbit*. When you whirl a stone tied to a piece of string around in a circle, the centripetal force acts along the string and stops the stone flying off at a tangent. See also *centrifugal force*.

Cepheid variables   Stars which vary in a regular way, with periodic variations whose periods are related to the average brightness (*absolute magnitude*) of the star. Measuring the period of a Cepheid reveals its absolute magnitude, so measuring its *apparent magnitude* then indicates how far away the star is. This is one of the key stepping stones in the determination of the *cosmic distance scale*.

Cepheids have periods in the range 1–50 days. The *period–luminosity relation* was discovered in the early part of the 20th century by Henrietta *Leavitt*, working under the supervision of Edward *Pickering* at the Harvard College Observatory. The Cepheid family gets its name from Delta Cephei, which was identified as a variable star by the English astronomer John Goodricke in 1784. Leavitt found that the brighter a Cepheid is, the more slowly it goes through its cycle of variations, and she published this result in 1908. She was then able to work out the exact relationship between brightness and period by studying Cepheids in the Small Magellanic Cloud, which is so far away from us that all the stars in it can be regarded as at roughly the same distance from us, just as everybody in New York City is at roughly the same distance from Trafalgar Square in London.

This meant that she could tell the relative distances to Cepheids in our own Galaxy – that one particular star is twice as far away as another, or one-third as distant, or whatever – but she did not know their actual distances because none of the distances to individual Cepheids had been measured. Finally, in 1913 the distance scale was calibrated when Ejnar *Hertzsprung* measured the distances to several of the nearer Cepheids in our own Galaxy, using a version of the *parallax* technique. It is now known that there are two distinct families of Cepheids, with different period–luminosity relations; this led to a revision of the cosmic distance scale in the early 1950s (see Walter **Baade**).

Cerenkov radiation   *Electromagnetic radiation* produced when electrically charged particles move through a medium at velocities greater than the *speed of light* in the medium. The optical equivalent of a sonic boom.

Cerro Tololo Inter-American Observatory   An observatory near La Serena, in Chile, run by the US Association of Universities for Research in Astronomy. At an altitude of 2,200 m, the observatory operates three main *reflecting telescopes*, with apertures of 4 m, 1.5 m and 1 m.

C-field   The 'creation' field proposed by Fred **Hoyle** as part of his version of the *Steady State* description of the Universe. The idea was first proposed in the 1940s, and was subsequently developed considerably by Hoyle and his colleagues, notably Jayant Narlikar. Although the Steady State model is widely regarded as a failed description of the Universe, and to have lost out to the Big Bang model, the C-field has remarkable similarities to the processes that are thought to power *inflation* in the currently favoured versions of Big Bang theory. It is largely an accident of history and semantics that these inflationary models involving *baby universes* are regarded as a development of the Big Bang idea rather than the Steady State idea; in fact, they are a hybrid of both descriptions of the Universe.

According to Hoyle, the C-field fills the Universe, but is strongest in regions where new matter is being created. The effect of the C-field in the Universe at large is to produce an expansion, balancing the tendency of the newly created particles to make the Universe contract – in effect, the C-field carries negative energy. The C-field is carried by its own *bosons*, analogous to *photons*, which fall into any regions of strong gravity and gain energy as they do so. This, says Hoyle, will 'open the creation tap' to produce a flood of new particles. But as the process continues, there comes a point where the outward pressure associated with the strengthening C-field becomes so strong that it causes an outward explosion.

On this picture, the creation activity can occur on many scales, including at the centres of galaxies and quasars, and in what we usually think of as the Big Bang. A crucial feature of Hoyle's argument is that it does not allow the formation of *black holes* or collapse to a *singularity*. In any situation where gravity tries to compress matter to such extremes, the C-field will reverse the process in an outward blast of new particles. But in the most extreme cases, these are not particles in the everyday sense of the word; instead they are *Planck particles* with enormous mass-energy contained in a region no bigger than the *Planck radius*. The modern version of the Big Bang theory sees the early Universe as bursting out from just such a sea of Planck particles, so that each particle creation in Hoyle's theory is like a Big Bang in itself.

The main difference between the C-field version of the expanding Universe and the standard Big Bang version is that on Hoyle's picture everything happens more slowly. The Universe as we know it expands relatively slowly away from the 'Big Bang', then collapses in a Big Crunch which leads to another Big Bang, and so on. But the whole process proceeds so slowly that the Galaxy is more like 300 billion years old than the 15 billion years allowed by the standard version of the Big Bang.

There are considerable difficulties in reconciling the complete picture painted by Hoyle with the accepted interpretation of observations of the Universe. Nevertheless, the idea of the Universe as we see it – as one bubble among many in a much larger region of spacetime, with its early expansion powered by a field associated with the conditions of extreme *pressure*, *density* and gravity that existed at the *Planck time* – has much in common with inflation.

Further reading: Fred Hoyle, *Home is Where the Wind Blows*.

chain reaction   See *nuclear fission*.

Chandrasekhar, Subrahmanyan (1910–95)   Indian-born American astrophysicist best known for his theoretical investigations of *stars* at the end point of their lives, especially *white dwarf* stars, and the mathematical theory of *black holes*.

Chandrasekhar was born on 19 October 1910, in Lahore, then in India, now part of Pakistan. He studied at the University of Madras, graduating in 1930. While officially studying an old-fashioned physics syllabus, Chandrasekhar taught himself the new physics of *quantum theory* by reading the scientific papers of Niels Bohr, Werner Heisenberg and Erwin Schrödinger in research journals in the college library, and from Arnold Somerfeld's book *Atomic Structure and Spectral Lines*. He actually published two research papers while still an undergraduate, and on the strength of this work won a scholarship to study in England.

In July 1930, on the boat trip to England, Chandrasekhar used his knowledge of quantum physics to carry out some calculations which showed that a white dwarf star could be stable only if it had a mass less than 1.4 times the mass of our Sun. Any star at the end point of its evolution with more mass than this (now known as the *Chandrasekhar limit*) would collapse indefinitely.

Chandrasekhar's result was not regarded as particularly important by his research supervisor in Cambridge, Ralph Fowler, but the young Indian research student was allowed to publish it, in the *Astrophysical Journal*, where it appeared in 1931. He completed his PhD studies in 1933, just before he turned 23, and became a Fellow of Trinity College, in Cambridge.

The claim that old stars with masses greater than the Chandrasekhar

Subrahmanyan Chandrasekhar
(1910–95).

limit must collapse indefinitely (into what we now call black holes) brought Chandrasekhar into conflict with Sir Arthur *Eddington*, the grand old man of British astronomy in the 1930s, who ridiculed the idea at a meeting of the Royal Astronomical Society in 1935. Partly as a result of that conflict, Chandrasekhar left Cambridge in 1936 to work at the University of Chicago, deciding 'that there was no good in my fighting all the time, claiming that I was right and that the others were all wrong. I would write a book. I would state my views. And then I would leave the subject.'

True to his word, Chandrasekhar wrote the book (*An Introduction to the Study of Stellar Structure*) and moved on to other things, setting the pattern for his working life. He would spend several years working in a particular area of research, write a comprehensive book on the subject, then move on to pastures new. This led him through the study of stellar dynamics, stellar atmospheres and other topics to major research on the application of the *general theory of relativity* to astrophysics in the 1960s, and studies of the mathematical theory of black holes in the 1970s and 1980s. The wheel had turned full circle; when Chandrasekhar received the Nobel Prize for his work in 1983, it was for both his latest black hole studies and his first major investigation, into the stability of white dwarfs –

for, of course, his calculations on that boat trip had long since been borne out by further studies.

Although there is an intermediate stage between a white dwarf and a black hole (a *neutron star*), Chandrasekhar had been right in the 1930s when he said that a dead star with more than a certain mass must collapse indefinitely.

Chandrasekhar worked at the University of Chicago and its associated *Yerkes Observatory* throughout his career after 1936, and became an American citizen in 1953. He became editor of the *Astrophysical Journal* in the same year, and ran it until 1971. And in spite of his confrontation with Eddington in the 1930s, he wrote an affectionate memoir of the great astronomer, whom he described as 'the most distinguished astrophysicist of his time'.

He died of a heart attack on 21 August 1995.

Further reading: Subrahmanyan Chandrasekhar, *Eddington*; John Gribbin, *In Search of the Edge of Time*; Kip Thorne, *Black Holes and Time Warps*.

Chandrasekhar limit   The maximum possible mass that a *white dwarf star* can have without collapsing under its own weight to become either a *neutron star* or a *black hole*. Almost exactly 1.4 times the mass of our Sun. See also *Chandrasekhar, Oppenheimer–Volkoff limit*.

chaos   Unpredictable and irregular patterns of behaviour that occur in systems that are actually governed by simple and exact laws. Chaos is caused by the extreme sensitivity of such a system to its starting conditions, so that a small variation in those conditions leads to very different outcomes. It is an intermediate stage between completely predictable behaviour and completely random behaviour.

A simple example of this sensitivity to initial conditions occurs if you try to balance a pencil on its point. It will always fall down, but you can never predict which way it will fall (as long as you are honest about trying to balance it perfectly), even though it is obeying the precise law of gravity.

Weather systems are often very sensitive to initial conditions, and the classic example of chaos at work is the so-called butterfly effect, where it is suggested that the flap of a butterfly's wings in Brazil might make a difference to the development of a storm system in Africa.

Because of chaos, it is impossible to predict how some systems will develop without knowing the *exact* values of all the properties of the system at the start – for example, the precise position and velocity of every object in the system. In practice, this is impossible – in the case of weather forecasting, it would mean knowing the position and velocity of every *molecule* in the air, among other things (this is actually impossible in principle, not just because of the deficiencies of our measuring devices; see **quantum theory**).

If the errors do not make much difference to the behaviour of the system, it becomes predictable. Weather is sometimes predictable and sometimes chaotic, which is why weather forecasting is a mixture of art and science.

In astronomy, chaos is especially important in calculating the orbits of objects in the Solar System. It makes it impossible to predict the exact orbit of an object: for example, a **comet** which is perturbed by the gravitational influence of other objects, such as the **giant planets**. Orbital calculations can be carried out for as far into the future as you like, but they will deviate more and more from the actual orbit as the effects of errors in the initial settings of the calculation build up.

charge   Originally, the name given to a property of some **elementary particles** that causes an electric force between them. Charge comes in two varieties, arbitrarily labelled 'positive' and 'negative'. Two particles with the same kind of charge (such as two **electrons**) repel one another, while two with opposite kinds of charge (such as an electron and a **proton**) attract. The name has since been extended to refer to other properties of elementary particles, such as colour charge. (See **quark**.)

charge coupled device (CCD)   The electronic equivalent of a photographic plate, a CCD is an electronic detector which is sensitive to electromagnetic radiation, including visible light but extending into the infrared and the ultraviolet and low-energy X-ray parts of the **spectrum**.

CCDs are smaller than photographic plates, and typically cover a few square centimetres, but they are much more efficient recorders of electromagnetic energy (photons) than photographs, so, when attached to telescopes, they obtain more information about astronomical objects in less time than photographic techniques.

CCDs consist of a flat array of very small **pixels** (picture elements), on a thin wafer of a semiconductor such as silicon. An individual pixel may be 15 micrometres (15 millionths of a metre) across, or even smaller. The array contains several thousand rows and an equivalent number of columns of pixels, and each pixel responds to photons falling on it, producing a build-up of electrons proportional to the amount of radiation the pixel has received. An individual pixel can have the capacity to hold more than half a million electrons in this way.

The charge on each pixel is 'read' electronically and converted into a digital form for analysis in a computer. After processing, the digital data can be used to produce images in the form of photographs, or displays on TV-type monitor screens, or analysed directly within the computer.

This requires considerable computer power. Standard CCD formats today include arrays of 1,024 by 1,024 pixels and 2,048 by 2,048 pixels, while 4,096 by 4,096 arrays are available. Even a 2,048 by 2,048 array contains more than 4 million pixels, and each pixel carries 2 bytes of information. So a computer memory of 8 Megabytes is needed to store each 'photograph' taken with a CCD, even before any processing is carried out.

CCDs were invented at the Bell Laboratories in the United States in 1970; they are now, thanks to the development of high-speed electronic computers, the most commonly used detectors in astronomy.

Charon Satellite of the planet Pluto, discovered by James Christy of the US Naval Observatory in 1978. Charon orbits Pluto once every 6.4 days at a distance of 19,400 km, and like Pluto is made of water ice and frozen methane. About 1,300 km in diameter, Charon is more than half the size of Pluto. Not to be confused with *Chiron*.

Chiron A super-*comet*, or *ice dwarf*, discovered by Charles Kowal in 1977, that orbits between the orbits of Saturn and Uranus. Chiron takes 50 years to orbit around the Sun once, and has a diameter of at least 200 km. It may be one of the biggest and brightest members of a belt of icy objects in similar orbits. Not to be confused with *Charon*.

CHON Acronym for Carbon–Hydrogen–Oxygen–Nitrogen, the four most important *elements* in living organisms.

Experiments in which ultraviolet radiation or electric sparks are passed through sealed flasks containing a mixture of water (a compound of hydrogen and oxygen), carbon dioxide and ammonia (a compound of hydrogen and nitrogen) have produced more complex compounds called *amino acids*, which are themselves the building blocks of proteins and therefore only one step removed from molecules of life. This is a simple demonstration of the fact that the four elements present in those flasks dominate the structure of living things.

Every living person is made up of about 65 per cent water, but even the other third is made up mainly of the same hydrogen and oxygen atoms that are present in water, plus carbon and nitrogen. A full 96 per cent of your body is made up of CHON. Hydrogen was left over from the *Big Bang*, and is everywhere in the Universe; it can hardly be a coincidence that carbon, nitrogen and oxygen are among the more abundant products of nucleosynthesis inside stars. Life has evolved by making use of the available materials, and some steps towards life may have taken place in *interstellar clouds*, where the same ingredients that are in those flasks are available, together with ultraviolet light.

See *cosmic abundances, panspermia hypothesis*.

Christie, Sir William Henry Mahoney (1845–1922) British astronomer who became the eighth Astronomer Royal in 1881 and retired from the post in 1910. He extended the work of the *Royal Greenwich Observatory* beyond its traditional role of positional measurements, upgraded the equipment used by the observatory by adding a 28-inch (71.1 cm) aperture *reflecting telescope*, and other instruments, and initiated daily observations of *sunspots* and *spectroscopic* studies of stars. Christie transformed the observatory into a centre for *astrophysics*, as well as timekeeping and navigation.

chromosphere See *Sun*.

Circinus X-1 A source of *X-rays*, lying within our *Galaxy* in the direction of the constellation Circinus, but at an unknown distance. The X-ray source is orbiting a companion star once every 16.6 days and is probably a *neutron star*.

Clarke, Arthur C. (1917–) See *geosynchronous orbit*.

classical mechanics Essentially the laws of motion that we learned in school, based on Isaac *Newton*'s three laws of motion – that an object stays still or moves in a straight line at constant speed unless it is acted on by a force, that the rate at which its speed changes (acceleration) is proportional to the force applied to it and inversely proportional to its own *mass*, and that there is an equal and opposite reaction to the force exerted on one object by another.

All of Newton's laws of motion apply extremely accurately to objects that are moving at speeds much less than the *speed of light*. For speeds that are a significant fraction of the speed of light, the laws have to be modified to take account of the way the mass of an object increases as it moves faster (see *special theory of relativity*). This modification to Newtonian mechanics is called *relativistic mechanics*.

Even relativistic mechanics, however, is often regarded as a classical theory, because it deals with continuous space and time. The real distinction is with quantum mechanics (see *quantum theory*), which says that the properties of a system cannot change continuously, but only in distinct steps, called *quanta*. These steps are very small, so the quantum theory does not have to be used when we are dealing with objects that are much larger than atoms. So the strict domain of classical mechanics is the world of objects much larger than atoms, moving much slower than light, and it works perfectly for tasks such as designing bridges or calculating the flight of a thrown baseball, or calculating the orbit of a planet around a star.

classical physics See *classical mechanics*.

Clifford, William Kingdon (1845–79) British mathematician who was one of the first people to suggest that non-*Euclidean* geometry discussed by *Riemann* might be the correct description of the Universe in terms of curved space. He died the year Albert *Einstein* was born. See *general theory of relativity*.

closed universe   See *cosmological models*.

clusters of galaxies   *Galaxies* tend to occur in groups, called clusters, that
may contain several thousand individual galaxies. Our own Galaxy is a
member of a small cluster called the Local Group, which has only about
40 members, most of them *dwarf galaxies*, but including the *Andromeda
galaxy*. Clusters are held together by gravity.

The nearest large cluster of galaxies to us is the Virgo cluster, a huge
congregation of at least 2,500 galaxies, three-quarters of which are *disc
galaxies*. The exact distance to the Virgo cluster is a matter of debate (see
cosmic distance scale), but a widely accepted figure is 15 million parsecs; if
the exact distance can be determined, perhaps from studies of Cepheid
variables using the Hubble Space Telescope, it will provide an important
calibration of the cosmic distance scale. First results of these measurements
by the HST, announced in 1994, suggest a relatively high value of about
80 kilometres per second per Megaparsec for the *Hubble constant*, but these
are not yet definitive.

Clusters of galaxies. Part of the Virgo cluster of galaxies, which altogether contains at least two
thousand galaxies spread across a volume 3 Megaparsecs across, about 15 Megaparsecs away
from us.

One of the most important members of the Virgo cluster, the second brightest galaxy in the cluster (M87), is a giant elliptical galaxy which has been detected both as a radio source (Virgo A) and as an X-ray object (Virgo X-1).

Clusters of galaxies are themselves grouped together to make *super-clusters*, and the Local Group is part of the same supercluster as Virgo, sometimes called the Local Supercluster. Although the Local Group is being carried further away from the Virgo cluster as space stretches in the expanding Universe (see *Big Bang*), the gravitational attraction of the Virgo cluster is strong enough to cancel out part of this cosmological expansion, reducing the measured recession velocity of Virgo by about 250 km/sec.

This still means that we are moving away from Virgo at about 1,000 km/sec, but, confusingly, astronomers often refer to the amount by which this recession velocity has been decreased as our 'infall' *towards* Virgo.

The distribution of clusters and superclusters across the Universe is important evidence for the existence of *dark matter.*

clusters of stars   Groups of stars that are physically associated with one another, not just chance aligments on the sky of objects at different distances. See *globular cluster, open cluster.*

CNO cycle   See *carbon cycle.*

COBE   NASA *satellite*, launched in 1989, which discovered ripples in the cosmic microwave background and confirmed the accuracy of the *Big Bang* model of the Universe. See *background radiation.*

cold dark matter (CDM)   Hypothetical form of non-baryonic *dark matter* made up of particles that move slowly compared with the speed of light, and may comprise up to 99 per cent of the gravitational mass of the Universe. Also known as weakly interacting massive particles (WIMPs). See also *mixed dark matter.*

collapsar   Generic name for any of the three kinds of 'collapsed star' that may form when a star reaches the end of its life – *white dwarf, neutron star* or *black hole.* Sometimes used specifically to refer only to black holes.

colour   The colour of a star depends on the temperature of its surface (see *black body radiation*), and is an important indicator of its underlying nature. Shorter wavelengths of light correspond to higher temperatures. The coolest stars radiate mainly red light, and for increasing temperatures the colours are orange, yellow, white and blue.

But the colour of a star is not measured by astronomers at a single wavelength. Instead, they measure the brightness of the star at two wavelengths at least, and make a comparison of these measurements to determine its temperature (see *colour index*).

The first standard system of measuring colour in this way was established by Edward *Pickering* in the 1890s, and was based on measurements in blue light and in yellow light. In the 1950s, this was largely superseded by an improved system which takes measurements at three wavelengths, in the ultraviolet (at 360 nanometres), blue (420 nm), and visual (540 nm) parts of the spectrum. This UBV system provides more information by offering a comparison between the UB index and the BV index, which shows how much of the light from the star has been absorbed by dust in space on its way to us (interstellar reddening), and can distinguish between *dwarf* and *giant* stars, as well as providing information about the chemical composition of the star.

colour index   The brightness of a *star* or *galaxy* is measured at different wavelengths, corresponding to different colours of *light*: for example, blue and red. The difference between the two measured brightnesses is a colour index, and is related to the temperature of the star.

colour–magnitude diagram   See *Hertzsprung–Russell diagram*.

Coma cluster   A very large group of *galaxies* at a distance of about 100 Megaparsecs from our own *Galaxy*. The *cluster* contains at least 1,000 bright galaxies, held together by their mutual gravitational attraction and moving through space as a unit, like a swarm of bees. It is receding from us at a rate of about 6,700 km/sec because of the expansion of the Universe (see *redshift*).

comet   One of the minor constituents of the Solar System, a comet is a lump of icy material and dust (perhaps several lumps moving together), which becomes visible if it approaches the Sun. The heat of the Sun makes material evaporate from the comet, forming a cloudy coma around the icy nucleus, and a streaming tail of tenuous material which always points away from the Sun, because of the pressure of the *solar wind*. This gives comets their name, from the Greek *kometes*, meaning a long-haired star. The 'dirty snowball' model was proposed by Fred *Whipple* in 1949, and has been confirmed by visits of unmanned spaceprobes to comets.

Comets are thought to originate in a spherical shell or halo, beyond the *orbits* of the *planets* and about halfway to the nearest star (tens of thousands of *astronomical units* from the Sun). Comets may have been stored in this *Oort cloud* since the formation of the Solar System; a rival theory suggests that the Oort cloud is renewed by 'new' comets picked up by the Solar System when it passes through giant *molecular clouds*. The Oort cloud may contain 100 billion comets. From time to time, the gravitational influence of a passing star will disturb the Oort cloud and send comets in towards the Sun, where the gravitational influence of Jupiter and the other *giant planets* may capture them into relatively short-period orbits.

An intermediate ring of comets and other cosmic debris, called the

Comet. Daniel's comet, photographed on 6 August 1907, showing the characteristic extended tail that develops when a comet approaches the Sun.

Kuiper Belt, lies beyond the orbits of Pluto and Neptune, between about 35 and 1,000 astronomical units from the Sun. It contains perhaps 100 million comets, some of which may have fed into the belt from the Oort cloud. Whatever their origin, objects in the belt can eventually feed into the planetary part of the Solar System. *Chiron* may have recently been captured into its present orbit from the Kuiper Belt, but long-period comets are thought to fall in directly from the Oort cloud.

The solid nucleus of a typical comet is quite small – *Halley's Comet*, for example, has a nucleus about 15 km by 10 km by 10 km – but the surrounding coma may be hundreds of thousands of kilometres across and the tail may stretch for 100 million km. The material to make the coma and tail all comes from evaporation of the nucleus, so a comet nucleus gets smaller each time it passes near the Sun, and eventually fades to leave a trail of orbiting dust particles, which cause *meteor* showers when the Earth runs through the stream.

Comets are arbitrarily divided into two classes, long period and short period. Short-period comets have orbits that take less than 200 years to complete, and lie more or less within the orbit of Neptune; long-period comets have orbits that take longer than 200 years (some may take millions of years) to complete, and which carry them far out beyond the orbits of the planets. About 150 short-period comets are known, and more are discovered every year. Halley's comet is the brightest of these regular visitors to the inner Solar System, and has a period of 76 years; Encke's comet has the shortest period, 3.3 years.

compact galaxy   Obsolete term coined by Fritz *Zwicky* to refer to a class of galaxies with very bright *nuclei*, so that they look like stars. Now known to be *starburst galaxies* or *active galaxies* of other kinds. Not to be confused with *dwarf galaxies*.

compactification   See *Kaluza–Klein models*.

Compton Gamma Ray Observatory   *Satellite* launched in 1991, by NASA, to study the Universe in the *gamma ray* part of the *spectrum*. Named after the physicist Arthur Compton (1892–1962), who carried out pioneering experiments with *X-rays*. The satellite has a mass of 16 tonnes and observes in the energy range from 15 keV to 30 GeV.

conjunction   An alignment such that two objects in the Solar System are lined up on the sky so that they have the same *right ascension* (longitude) as seen from Earth. Inferior conjunction is when a *planet* or other body lies between the Sun and the Earth, superior conjunction is when the planet or other body is directly behind the Sun, as seen from Earth. A planetary conjunction occurs when there is a close alignment on the sky between a planet and another planet or some other object which is not the Sun. See *opposition*.

constellations   Originally, constellations were rather arbitrary groupings of stars, with vaguely defined boundaries, whose patterns on the sky might be imagined to mark the outlines of mythological heroes or strange creatures. The system of dividing up the sky in this way that we use today is based on Greek astronomy, but had earlier, now lost, roots. Other ancient civilizations (for example, the Chinese) had their own systems. These patterns have no astrophysical significance, and the stars in a constellation may be at very greatly different distances from us along the line of sight – the two 'pointers' to the Southern Cross, in the constellation of Centaurus, for example, look side by side on the sky, but Alpha Centauri is just over one *parsec* away from us, while Beta Centauri is more than 100 parsecs away.

Ptolemy listed 48 constellations (all of them, of course, visible from the Northern Hemisphere), and many more were added in the 16th, 17th and 18th centuries, especially when explorers began to visit the Southern Hemisphere. In 1933 the International Astronomical Union rationalized the system by dividing the whole sky up into 88 areas, given the historical Latin names of constellations. Any object in the sky must lie in one of the 88 constellations defined by the IAU (some extended objects stretch across constellation boundaries), and the approximate location of any star, *galaxy* or other celestial object can be indicated by saying that it 'lies in' a certain constellation.

Even after this rationalization, the constellations vary enormously in size, from 72 square degrees for Equuleus (The Foal – one of Ptolemy's original 48 constellations) to 1,303 square degrees for Hydra (The Water-snake, also one of Ptolemy's 48). The system is far from perfect, but astronomers are used to it and it works.

contact binary   See *equipotential surfaces*.

continuous creation   See *Steady State hypothesis*.

convection   When heat is transported upwards in a fluid medium (liquid or gas) by bulk movements of the fluid – summed up in the phrase 'hot air rises'. A bubble of gas that is hotter than its surroundings rises and gives up its heat to cooler layers above before sinking down again under the pull of *gravity*. Convection is important in some layers inside *stars*.

Copernican system   The Sun-centred *model* of the Solar System proposed by Nicolaus *Copernicus* in the 16th century and published in his book *De Revolutionibus*. The mode still used the idea of circular *orbits* and epicycles carried over from the *Ptolemaic system*, but placed the Earth as one planet among others orbiting the Sun.

Copernicus, Nicolaus (Mikolaj Kopernigk) (1473–1543)   Polish astron-omer (and doctor) who set out the idea that the Sun, and not the Earth, is at the centre of the Solar System.

Copernicus was born at Torun in Prussia, then ruled by Poland, on 19 February 1473. His father, a wealthy merchant, died in 1483, and Mikolaj (he only Latinized his name later) was looked after by his uncle, who later became Bishop of Ermeland. He studied mathematics and classics at the University of Krakow between 1491 and 1494, then travelled to Italy where he studied astronomy at the University of Bologna. Although he was not in Poland, his uncle's influence was sufficient for him to be made Canon at the cathedral of Fromberk in 1497, even though he never took holy orders. The income from this position, which was almost a sinecure, allowed Copernicus to follow his interest in astronomy as much as he wished.

But he was no wastrel. He studied medicine in Padua between 1501 and 1505, and was made a doctor of canon law by the University of Ferrara in 1503. He returned to Poland in 1506, where he worked for his uncle as both his physician and his private secretary, carried out his fairly nominal duties at the cathedral, and served on several diplomatic missions. His interest in astronomy was far from being an overriding passion, and he made only a few observations, preferring to make his calculations of the motions of the *planets* using data long since collected by others.

Nicolaus Copernicus (1473–1543).

Nevertheless, in the early 1510s he realized that the Earth-centred cosmology of **Ptolemy** was unsatisfactory, and became convinced that the Sun was at the centre of the Universe, with the Earth and other planets in **orbit** around it.

Copernicus was well aware of the revolutionary and possibly heretical nature of such a claim. Although he wrote down a brief outline of his ideas in 1512 or 1513, and a more complete version in 1530, he did not risk formal publication, and only circulated these works privately, to trusted friends. One of these friends, Georg Joachim von Lauchen (also known as Rheticus) eventually persuaded Copernicus to set out a full presentation of his ideas in book form, and this became *De Revolutionibus Orbium Coelestrum* (*On the Revolution of the Celestial Spheres*).

The book had probably been essentially complete in 1530 and, in spite of the author's reluctance to publish, news of its contents spread widely (it was even referred to, not unkindly, by the Popes Leo X and Clement VII), but Copernicus still delayed publication until pressed by Rheticus. By now, in the early 1540s, Copernicus was old and ill, and Rheticus took on the task of copying the manuscript and taking it to Nuremburg to oversee the printing. But before the job was completed he had to leave to take up a post in Leipzig.

In a not too subtle public relations exercise, Copernicus dedicated the book to the Pope, Paul III; without the authorization of Copernicus the book also carried a preface offering the idea of a Sun-centred Universe as a mere hypothesis, rather than claiming that the Universe really was like that. This was added anonymously by the Lutheran minister Andreas Osiander, who supervised the printing of *De Revolutionibus* after Rheticus left for Leipzig. The book was finished in March 1543, and Copernicus saw a copy on the day he died, 24 May.

At first, the idea of a heliocentric Universe was neither seen as obviously being an improvement on Ptolemaic ideas, nor (probably because it was not widely accepted) strongly opposed by the Church. A key problem with the Copernican *model* was that it still dealt with circular orbits, and therefore still required the complicated use of *epicycles*, so that it did not at first sight look much simpler than the older model. Many people also simply could not accept that the Earth was flying through space.

But as the ideas in *De Revolutionibus* gained support, the book was seen as more of a threat, and it was placed on the Catholic index of forbidden books in 1616, staying there until 1835, when the Catholic Church grudgingly accepted the possibility that the Earth might be moving around the Sun. The work of Copernicus had by then long since revolutionized astronomy through its influence on Tycho *Brahe*, Johannes *Kepler*, *Galileo* Galilei and Isaac *Newton*.

Coriolis force   Because the Earth rotates from west to east, an object at the equator is moving eastward at high speed. At the poles, there is no such motion, with intermediate speeds in between. If an object at the equator is pushed north or south (a shell fired from a gun, or a wind blowing towards the poles), its excess eastward motion will carry it sideways, as if it has been pushed by a force. That force is known as the Coriolis force, after Gustave Coriolis (1792–1843).

corona   The tenuous outer layer of the atmosphere of a *star*. The corona of the Sun is best observed when the bright disc of the Sun is covered by the Moon during a *solar eclipse*. It extends out to many times the radius of the Sun, and gradually fades away into interplanetary space. See *Sun*.

Cos B   A satellite which studied *gamma rays* from space, launched by the European Space Agency in 1975. It made observations in the energy range from 70 MeV to 5000 MeV, and operated until 1982.

cosmic abundances   Relative amounts of each *element* in the Universe.

Although *hydrogen* and *helium* were produced in the *Big Bang*, almost all of the heavier elements were manufactured inside stars later in the evolution of the Universe (see *nucleosynthesis*) and are present in much smaller quantities. The standard measure of cosmic abundances is based on studies of the Sun, the Earth and the other objects in the Solar System. In terms of the number of atoms of each element, the solar abundance is 90.8 per cent hydrogen, 9.1 per cent helium and 0.1 per cent for everything else put together. This is similar to the proportions for other stars, determined by *spectroscopy*, although old stars, formed when the Universe was younger, have even less of the heavy elements.

Including everything in the Solar System, the abundances of the most common elements can be expressed in terms either of mass or number of atoms. Because hydrogen is the lightest element, it contributes only 70.13 per cent of the mass of the Solar System, while helium contributes 27.87 per cent and oxygen, the third most common element by mass, contributes just 0.91 per cent. Most astronomers, however, prefer to measure cosmic abundances in terms of numbers of atoms. On this scale, sulphur is the tenth most common element. For every atom (strictly speaking, every nucleus) of sulphur in the Universe (at least, in our part of the Universe), there are roughly: 1 atom of iron; 2 atoms each of neon and magnesium; 3 atoms of silicon; 4 atoms of nitrogen; 20 atoms of *carbon*; 30 atoms of oxygen; 3,000 atoms of helium; and 50,000 atoms of hydrogen.

Apart from this top ten, five other elements (aluminium, argon, calcium, nickel and sodium) have abundances in the range from 10 per cent to 50 per cent of the abundance of sulphur. Everything else is much rarer; for every 10 million sulphur atoms, for example, there are just 3 atoms of gold. Elements heavier than iron are rare because they can only be made in *supernovae*.

See also *CHON*.

cosmic censorship   The hypothesis that there must be a law of physics, as yet undiscovered, which ensures that every *singularity* is hidden behind an *event horizon*, and that therefore (among other things) *time travel* is impossible.

In the mid-1960s, Roger *Penrose* showed that according to the *general theory of relativity* any object which contracts within its event horizon must collapse all the way to a singularity, a point of infinite density and zero volume, where the laws of physics break down and literally anything can happen. This did not concern physicists too much, since nobody outside the event horizon of a *black hole* can ever see what is going on inside, so a singularity hidden in this way can have no influence on the outside Universe.

If a naked singularity did exist, it would behave more like a *white hole* than a black hole, spewing matter and energy out into the Universe. Literally anything could emerge from a naked singularity – hydrogen gas, a stream of frozen TV dinners, or a million replicas of Stephen *Hawking*. It is much more likely that what would emerge would be simple entities, the basic constituents of matter, *protons* and *neutrons*. Indeed, Hawking and Penrose proved that the expansion of the Universe shows that it was born out of a singularity at the beginning of time in just such a process. But what comes out of a naked singularity is produced entirely at random, so it really could be anything at all.

Penrose has speculated that we could be saved from this worrying situation if there is no such thing as a naked singularity, but as relativist Clifford Will has summed up the situation: 'There is no convincing proof of the Cosmic Censorship Hypothesis. There is not even general agreement on how to formulate the vague notion of censorship in terms that can be translated into mathematics.' Indeed, since we know that the Universe itself emerged out of a singularity, such evidence as there is suggests that the cosmic censorship hypothesis is wrong.

In the 1990s, computer simulations of the way non-spherical objects (such as spindles) collapse suggested that they may indeed form singularities that are not concealed behind event horizons. And even singularities that are hidden in this way may lose their cloak of respectability eventually if the black hole in which they reside 'evaporates' through the emission of *Hawking radiation*.

See also *baby universes, wormholes*.

cosmic distance scale   The first step in measuring the scale of the Universe uses triangulation, the same technique used by surveyors on Earth, although astronomers usually call the technique *parallax*. You can see how it works by holding up a pencil at arm's length, and closing each of your eyes in turn. Viewed first by one eye then the other, the pencil seems to move against the background. This is because your eyes are seeing the pencil from slightly different angles along a short 'baseline'; it is exactly because of this that the two eyes give you stereoscopic (three-dimensional) vision and you are able to judge distances.

If astronomers at two widely separated observatories observe the Moon at the same time, it will seem to be in a different place against the background of distant stars from each observatory, because of parallax. By knowing the distance between the observatories (the baseline) and measuring the parallax effect, astronomers can calculate the distance to the Moon (about 400,000 km). (See diagram, page 368.)

The same technique works for the nearer *planets*. The distance to Mars was first measured reasonably accurately in 1671, when a team of French astronomers observed the position of the planet on the sky from Cayenne, in French Guiana, while a team back in Paris noted its position at the same time. When the Guiana team returned home, they could compare notes with the Paris team and work out the distance to Mars.

By combining this kind of parallax measurement with *Kepler's laws* of planetary motion, astronomers were able to work out the distances of the Earth and other planets from the Sun. This gave them a new baseline. The average distance from the Earth to the Sun is 149.6 million km so the diameter of the Earth's orbit is about 300 million km. It takes 1 year for the Earth to move around the Sun once in its orbit, so observations made 6 months apart from the same observatory on Earth will be made from opposite ends of a baseline 300 million km long, across a diameter of the Earth's orbit.

A few stars are so close to our Solar System that they show a measurable parallax from observations made in this way, but the parallaxes are very small – a fraction of an *arc second*. The technique led to the definition of a new unit of distances, the *parsec*; a star that was exactly 1 parsec (pc) away would show a displacement of 2 arc seconds over the 300 million km baseline of the Earth's orbit (that is, it would show a parallax of 1 arc second if we could make measurements from the Sun and Earth simultaneously, over a baseline 150 million km or 1 *astronomical unit* long). One parsec is about 3.26 *light years*, or just under 206,265 times the distance from the Earth to the Sun – and no star is close enough to us to show a parallax as large as even 1 arc second, which is why it took until the 1830s for the first stellar parallaxes to be measured.

The first clutch of stellar parallax measurements gave astronomers their first true guide to the scale of the Universe. They found that the star 61 Cygni has a parallax of 0.3136 arc seconds, corresponding to a distance of 3.4 pc; Alpha Lyrae has a parallax of 0.2613 arc seconds, corresponding to 8.3 pc; and Alpha Centauri, now known to be the nearest star system to our Sun, has a parallax of 0.76 arc seconds corresponding to a distance of 1.3 pc, just 4.3 light years. The nearest star system is 7,000 times further from the Sun than the most distant planet in our Solar System, Pluto.

At the beginning of the 20th century, parallaxes had been measured for just 60 stars. But photographic techniques extended the use of parallax measurements, the use of *charge coupled devices* improved the technique further, and the *Hipparcos* satellite, launched in 1989, measured the positions of more than 100,000 stars to an accuracy of 0.002 arc seconds. Even this accuracy, however, only takes the range of parallax measurements out to a range of a few hundred parsecs. And that is the limit of direct measurements of astronomical distances; all other measurements, beyond the range of parallax, rely on indirect measurements and a chain of reasoning, which has led to considerable debate about the accuracy of the cosmic distance scale.

Three techniques have been particularly valuable in extending the range of distance measurements within our Galaxy. The first involves measurements of the colours of stars, and *spectroscopic* studies of their light. Family resemblances between stars make it possible to say that individual stars that have similar colours and spectroscopic properties to one another have roughly the same *absolute magnitudes*. So if one star of a particular type is close enough for its distance to be measured by parallax, we can estimate the distances to similar stars from their **apparent magnitudes**, compared with the apparent magnitude of the star whose distance is known.

The other two techniques depend on the way stars move across the sky. The speed with which a star is moving towards or away from us can be measured using the **Doppler effect**, which produces a **redshift** or a **blueshift** in its light. The velocity of a star across the line of sight can be measured directly (if it is close enough and moving fast enough), and the two velocities can be added together to give its true velocity through space.

One trick using these observations works for **clusters of stars**, moving together through space and not too far from the Sun. A group of stars moving in the same direction are, in effect, running along parallel lines, like a railway track. And just as the lines of a railway track seem to converge on a point in the distance, so the motions of the stars in such a cluster, measured over many years, will seem to converge on a point on the sky. This tells astronomers which direction the stars are moving in. Knowing how fast the stars are moving, and in which direction, makes it possible to work out how far away they must be to produce the observed motion across the line of sight.

The technique is called the moving cluster method, and it works out to a distance of a few tens of parsecs. One important application of the technique, in the first decade of the 20th century, was the determination of the distance to the **Hyades cluster**, a group of more than 200 stars about 46 parsecs away. Because all the stars in the cluster are at roughly the same distance from us, this enabled astronomers to calibrate the brightnesses of several family types.

The other important technique for measuring distances to stars sounds too bizarre to work, but it does. If we take any large number of stars, chosen at random, all at roughly the same distance (as far as we can tell) and just close enough to measure their **proper motions** across the line of sight, we can guess that on average there is just as much chance of a star in the group moving one way as another. Since the Galaxy is neither collapsing nor flying apart, all the random motions must more or less cancel out. So if we add up the velocities of all the stars in our random sample along the line of sight, measured by the Doppler effect, and take the average, we would expect the average velocity of this group of stars *across* the line of sight (or in any other direction) to be much the same. Assuming this is so, it is possible to assign an 'average distance' to all the stars by comparing this guesstimate with their measured proper motions.

This technique is called statistical parallax, and it turns out to be a reasonable guide to distances if you have a large enough group of stars to play with (we know this because, of course, distances to some stars can also be measured by other techniques, such as parallax). And, crucially, it proved possible, using this technique, to measure the distances to a group of stars that includes a couple of *Cepheid variables*. Because the variations of these stars are related to their absolute magnitudes, knowing the distances to a few Cepheids made it possible to estimate distances to all other Cepheids by measuring the periods of their variations (see *Leavitt, Henrietta*).

The scale of our whole Galaxy has been determined from observations of Cepheids. It is a flattened disc about 4 kiloparsecs thick in the middle (much thinner at the rim) and 30 kpc across, with the Sun about 9 kpc out from the centre, very much in the galactic backwoods. The disc is embedded in a sparse spherical halo of *globular clusters*, about 150 kpc in diameter.

Distances to other *galaxies* have been estimated by a variety of techniques, but the most important is the use of Cepheid variables. This made it possible to calibrate the relationship between redshift and distance in the expanding Universe (see *Hubble's Law*), so that redshift can now be used in its own right as a measure of distance for the most remote objects known, such as the *quasars*. But the whole edifice of distance measurements outside our immediate stellar neighbourhood rests upon the knowledge of directly measured distances to a handful of Cepheids using statistical parallax.

To put all of this in perspective, think in terms of something more homely, like an aspirin. If our Sun were the size of an aspirin, the nearest star would be represented by another aspirin 140 km away. This is fairly typical of the distances between stars – the distance from one star to its neighbour is tens of millions of times the diameter of the star itself (except, of course, in binary and multiple star systems). Galaxies like our own Milky Way contain hundreds of billions of stars, spread over correspondingly huge volumes, all held together by *gravity*, orbiting around the centre of the galaxy.

In order to get a feel for the spacing between galaxies, change the scale so that now one aspirin represents the entire Milky Way. Now, the nearest large galaxy to us, the *Andromeda galaxy* (M31) is represented by another aspirin just 13 cm away. This is slightly misleading, because both our Galaxy and the Andromeda galaxy are members of the *Local Group*, a system of galaxies held together by gravity. But the distance to the nearest similar small group of galaxies, the Sculptor Group, is still only 60 cm on the aspirin scale. Just 3 m away, on this scale, we find the *Virgo Cluster*, a huge congregation of about 2000 aspirin-sized galaxies spread over the volume of a basketball.

We can extend the analogy further. Some 20 m away there is another large cluster of galaxies, the *Coma cluster*, and further out still there are even larger clusters, some themselves 20 m or so across. The powerful radio galaxy Cygnus A is 45 m distant, the brightest quasar in the night sky, *3C 273*, is 130 m away. And the entire visible Universe can be contained in a sphere just 1 km across, on the scale where an aspirin represents our Galaxy.

Clearly, extragalactic space is far richer in galaxies, far more crowded, than galactic space is in terms of stars. If Alpha Centauri were as close to the Sun in relative terms as the Sculptor Group is to the Milky Way, it would actually be slightly closer to the Sun than we are! While if galaxies were as far apart, relatively speaking, as the spacing between stars, then the nearest galaxy would lie about 100 times further away from us than the most distant object seen in the real Universe. We would not even know that there was anything beyond our own Local Group. The only reason cosmologists are able to study the Universe at large, the way matter is spread through the Universe, and how that distribution has changed as the Universe has evolved, is because it is a crowded place, jammed full of galaxies.

Further reading: John Gribbin, *In Search of the Big Bang*; Michael Rowan-Robinson, *The Cosmological Distance Ladder*.

cosmic dust   Material between the stars in the form of small particles of matter. Interstellar dust grains may be as large as 10 micrometres (10 millionths of a metre) across, or as small as 0.01 micrometres in diameter. Their presence is revealed by the way in which they absorb and scatter the blue part of visible light and ultraviolet radiation, which makes stars seen through the dust look redder – this is exactly equivalent to the way in which dust in the atmosphere of the Earth scatters blue light, leaving red to produce spectacular sunsets and sunrises.

The interstellar *reddening* alters the colour of stars, and has to be taken account of in many astronomical observations. Within our Galaxy, this interstellar absorption of light reduces the brightnesses of stars by about 1 *magnitude* for every 1,000 *parsecs* of space the light travels through.

Spectroscopic studies show that most of the grains of interstellar dust are made of graphite (a form of carbon) and silicates. The grains may be coated in layers of frozen water or ammonia ice (ammonia is a compound of hydrogen and nitrogen), or solid carbon dioxide. They probably form from material escaping from the atmospheres of cool stars, and about 2 per cent of all the mass of interstellar clouds is dust. This adds up to 200 million times the mass of our Sun in the form of dust spread through the disc of our Galaxy.

The icy grains of soot and ammonia in interstellar clouds provide surfaces on which chemical reactions can take place, building up complex molecules. Sir Fred *Hoyle* and Chandra Wickramasinghe have proposed that this could include molecules of life (see *panspermia*), a speculation given credence by the discovery of an *amino acid* (*glycine*) in an interstellar cloud in 1994.

There is no dust between the *galaxies*, although there is definitely *intergalactic matter* in the form of gas, and possibly also *dark matter*.

See also *CHON, interstellar chemistry*.

cosmic microwave background   See *background radiation*.

cosmic rays   Actually particles from space, almost all of them carrying electric *charge*, which strike the atmosphere with high energies – up to 10^{20} *electron volts* (eV) per particle. This is much more than the highest energies achieved in particle accelerators on Earth. Low-energy cosmic rays originate from the Sun, those with energies from 10^9 to 10^{19} eV from sources in our *Galaxy*, such as *supernovae*, and the highest-energy particles probably from beyond our Milky Way Galaxy.

cosmic string   Hypothetical material left over from the *Big Bang*, in the form of tubes of energy much narrower than an atom, but possibly stretching across the entire Universe.

Cosmic string is a by-product of the era of the Big Bang itself, and the best way to think of it is as a piece of the Big Bang 'frozen' and trapped inside a tube with a diameter of just 10^{-14} that of an atomic *nucleus*. Because the string contains the energy density of the Universe about 10^{-35} seconds after the moment of creation, even though it is so narrow each centimetre of cosmic string would contain the equivalent of 10 trillion tonnes of mass. A loop of cosmic string 1 metre long would weigh as much as the Earth.

Cosmic string cannot have ends (if it did, the energy inside would leak out), so it can only exist in the form of closed loops, or pieces that stretch right across the Universe. A closed loop of cosmic string is like a stretched elastic band, under tension, and will 'twang' like a plucked guitar string. The twanging occurs as fast as possible, at nearly the speed of light, so a loop of cosmic string 1 *light year* across vibrates roughly once a year. This will produce *gravitational radiation*, draining energy out of the string until the loop shrinks away to nothing.

There is no proof that cosmic strings exist, or ever have existed, but such objects could have provided the 'seeds' on which galaxies grew when the Universe was young. The gravitational influence of the loops of string would make clouds of gas clump together, and by the time the strings had twanged themselves out of existence, the clouds of gas would be big enough to carry on the job of galaxy formation unaided.

If a strand of cosmic string passed through your room, you would not be aware of its mass in the usual gravitational sense (this is only 'visible' when loops of string are viewed from a distance). But if the string were moving at close to the speed of light, it would distort space as it passed. If the string passed right through the room (and you) horizontally, you would feel nothing as it passed, but behind the string the floor and ceiling of the room (not to mention your head and your feet) would rush together at a speed of several kilometres per second.

Few people take the idea of cosmic string seriously, but many enjoy playing with the equations that describe such bizarre possibilities. They are discussed in *The Stuff of the Universe* (John Gribbin and Martin Rees).

Cosmic string should not be confused with the kind of 'strings' invoked by some theories of particle physics – see *string theory*.

cosmic year The time taken for the Sun and Solar System to *orbit* once around the centre of our *Galaxy*; equal to about 225 million years.

cosmogony The study of the origin and evolution of the contents of the Universe, rather than of the Universe itself (see *cosmology*). It was originally restricted to the origin of the Solar System, but now extends to *stars* and *galaxies*, and even to studies of how matter emerged from the *Big Bang*.

cosmological constant A parameter introduced into the description of the Universe in terms of the *general theory of relativity* by Albert *Einstein*, in order to make those *models* static. At that time, in 1917, the expansion of the Universe had not been discovered, and Einstein had been disconcerted to find that the solutions to his equations always suggested that space was either expanding or contracting, but could not be static.

When Edwin *Hubble* and others found that the Universe is expanding, in the late 1920s, there was no need for the cosmological constant in its original form, and Einstein later described his invention of the parameter as the 'biggest blunder' of his career. Nevertheless, some cosmologists still liked to play with the equations including a cosmological constant, since by choosing the value of the constant they could 'create' a wide range of 'toy' universes.

Recently, the idea of a cosmological constant has been revived in the context of theories of *inflation* and in connection with the problem of galaxy formation. Because the constant affects the rate at which the Universe expands, it can be chosen to *increase* the expansion rate, acting as a kind of antigravity, or an energy of empty space (energy of the vacuum). This is exactly what happened during the era of inflation, which can be seen as an era in which there was a suitably large cosmological constant, which later faded away to zero.

It certainly must have faded to very close to zero, because observations of distant *galaxies* show no influence of any cosmological constant bigger than 10^{-66} (in the units used by Einstein) on the expansion of the Universe today. Even a very small cosmological constant could have an influence on how the Universe got to be the way it is, however, and some theorists have toyed with such models. In particular, adding in some cosmic repulsion makes the Universe older and allows more time for features like galaxies and *clusters of galaxies* to evolve. But these models appear ugly and unnatural, requiring very careful 'fine tuning' of the models to make them match reality.

cosmological models Cosmologists cannot make physical *models* of the Universe, but they can set up systems of mathematical equations that describe the behaviour of different kinds of possible universe. These universes (lower-case 'u') are cosmological models, which may or may not tell us something about the real Universe (upper-case 'U').

Some of the equations are relatively easy to solve, and the behaviour of these models can be investigated using nothing more than a pencil, some paper, and a little brain power. This was how Albert *Einstein* discovered that the equations of the *general theory of relativity* predict that the Universe is expanding, and how researchers such as Alexander *Friedmann* developed the first understanding of the different kinds of universe allowed by the general theory.

Other cosmological models incorporate more details and more complicated interactions, and the relevant equations can be solved only with the aid of high-speed electronic computers. But it is a curious and presumably significant feature of our Universe that it seems to be very well described by the simplest solutions of Einstein's equations – or, as Einstein himself put it, 'the most incomprehensible thing about the Universe is that it is comprehensible'.

The three simplest families of cosmological model, based on Einstein's equations with no *cosmological constant*, are distinguished by the ultimate fate of the universes they describe. An open universe is one which starts in a *Big Bang* and expands forever; a closed universe is one which starts in a Big Bang, expands to a certain size, and then collapses into a *Big Crunch*; and a flat universe sits exactly on the dividing line between the two, so that it expands forever but at a gradually decreasing speed, 'hovering' for all time at a final state, without ever collapsing. A variation on the closed model is a universe that expands and collapses repeatedly, with a 'bounce' instead of a Big Crunch. Our Universe is indistinguishable from a flat universe, although it could be either just open or just closed.

By adding in a cosmological constant, more complicated models can be created. One starts at infinite size, contracts to a finite size and then expands again; another starts expanding away from a Big Bang, then the expansion slows to a halt, so that the model stays the same size for an arbitrarily long time before expanding again. And there are other possibilities. But these exotic models are not thought to have much relevance to the real Universe.

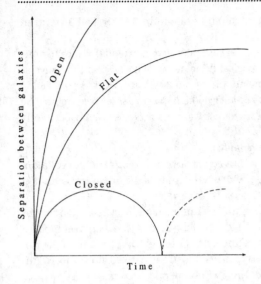

Cosmological models. The three main possibilities. Either the Universe is open, and will expand forever, or it is flat and will gradually come to a halt, or it is closed and will one day recollapse (perhaps 'bouncing' into another cycle of expansion).

See also *deceleration parameter, Steady State hypothesis.*

cosmological principle   The statement that there is no preferred place in the Universe – that the overall features of the Universe will look the same wherever you are located in the Universe.

The most obvious example of this is the nature of the expansion of the Universe, with the rate of recession of galaxies proportional to their distances from us (see *redshift*). This recession law (redshift proportional to distance) will apply from whichever galaxy you happen to be sitting on. If you imagine a piece of rubber with ink dots on it in a line, each dot 1 cm from its neighbours, then if you stretch the rubber so that it is twice as long, every dot will seem to recede from every other dot in accordance with this law.

Two dots that were 2 cm apart become 4 cm apart, while two dots that were 4 cm apart become 8 cm apart, and so on. A dot that starts out twice as far from any chosen dot as another dot will have moved twice as far as the other dot in the same time. In other words, the Earth is not at the centre of the Universe, even though we see galaxies receding from us uniformly in all directions.

Spectroscopic studies of the light of distant objects such as *quasars* show that the laws of physics operate in the same way everywhere in the observable Universe, another example of the cosmological principle. This can also be expressed as the principle of terrestrial mediocrity – that the

Earth is an ordinary planet orbiting an ordinary star and lies in an extremely ordinary part of the Universe.

In the 1940s, some cosmologists attempted to extend this idea to time, proposing a 'perfect cosmological principle' which said that the Universe should look much the same not only from any place within it, but at any time. This led to the *Steady State hypothesis*, but has since been refuted by evidence that the Universe is changing as time passes and almost certainly originated in a *Big Bang* a finite time ago.

cosmological redshift   See *redshift*.

cosmology   The study of the Universe at large, its origins and its evolution. It is distinct from *cosmogony*, which is the study of the origin and evolution of objects within the Universe, such as galaxies.

Although the roots of cosmology can be traced back to ancient myths and legends, and the Greek study of the way the planets move, modern cosmology is essentially a mathematical description of the behaviour of *spacetime* on the largest scale, using equations derived from Albert Einstein's *general theory of relativity*. So the date on which modern cosmology was born can be set very precisely, as 1917, the year Einstein first applied those equations to describe the Universe at large.

Although some theorists have developed alternative theories of gravity and spacetime, which have led to *cosmological models* different from those derived from Einstein's theory, these alternative cosmologies have now been ruled out by observations (see *binary pulsar*). Within the framework of Einstein's theory, there have been two great cosmological hypotheses, the Big Bang and the Steady State.

Both these models conform to the observational evidence that the Universe is expanding as spacetime stretches, but in the Big Bang model this is seen as evidence that the Universe was born out of a *singularity* a finite time ago, while in the Steady State model it was required that new matter should be created continuously to fill the gaps between the galaxies as they moved apart, so that the overall appearance of the Universe stayed the same.

The simple Steady State hypothesis has been found to be incorrect, because there is now clear observational evidence that the Universe changes as time passes. This leaves a variety of possible Big Bang models which can be considered as possible descriptions of the real Universe. To within the limits of observational accuracy, our Universe is indistinguishable from a flat model which obeys the laws of *Euclidean geometry*. This is the simplest possible universe allowed by Einstein's equations.

COSMOS   An automatic system used by the Royal Observatory in Edinburgh to scan photographic plates obtained by telescopes (especially the *UK Schmidt Telescope*) and extract astronomical information from them.

Cosmos Term used historically as synonymous with the term *Universe*, but which we suggest might now be used to refer to the 'super-Universe' required by the theory of *inflation*, within which our Universe is just one bubble among many.

Coudé telescope A *reflecting telescope*, incorporating a system of mirrors, which focuses light from a *star* or other astronomical object at a fixed point in the observatory, even though the telescope moves to track the apparent passage of the object across the sky caused by the rotation of the Earth. This makes it possible to carry out long observations of the object using bulky equipment which cannot move with the telescope.

Crab nebula Astrophysicists have been known to quip that their trade could be divided into two roughly equal categories – the study of the Crab nebula and its contents, and the study of everything else in the Universe. The quip has lost some of its force in recent years with the opening up of the Universe to observations across the electromagnetic *spectrum* and the discovery of new kinds of astrophysically interesting object; but it is still true that the Crab contains something of interest to almost any astrophysicist.

The Crab nebula itself is a glowing cloud of gas and dust in the constellation Taurus, about 2 *kiloparsecs* away from us. It is also known as Taurus A, M1 and NGC1952. It has so many names because it appears in almost every observation of the sky at different wavelengths – the Crab was one of the first three radio sources to be identified with known objects, it was the second X-ray source to be discovered, and it is the second brightest source of gamma rays visible from Earth.

The Crab is the remnant of a *supernova* explosion that was observed by Chinese astronomers in AD 1054, and was temporarily brighter than Venus, being visible in daylight for 23 days. The cloud of debris produced in that explosion has been expanding ever since, and the material in the nebula is still moving outwards at a speed of about 1,500 km per second, so that its appearance has changed significantly since it was first studied telescopically by the English amateur astronomer John Bevis (1693–1771).

The cloud of material contains long, thin filaments that were first observed by Lord *Rosse* in 1844. His drawings of the filaments in the nebula vaguely resembled the pincers of a crab, which is how the Crab nebula got its name.

The rate at which the filaments are moving outwards (revealed in the 20th century both by the *Doppler effect* and by direct measurements on photographs taken many years apart) ties in nicely with the age of the Crab based on the Chinese supernova observations – steady expansion at this rate would make a cloud just the right size in about 900 years. But this posed a puzzle, since if the cloud were merely debris moving outward from

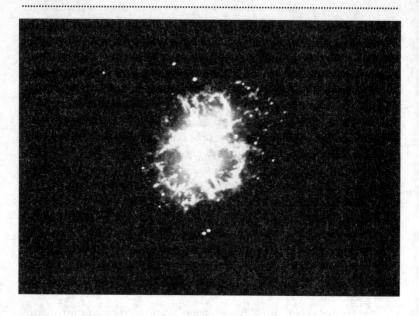

The Crab nebula. One of the most spectacular sights in the heavens through a telescope, this nebula is also known as Taurus A, M1 and NGC 1952.

the site of a long-ago explosion, it ought to be slowing down as it ploughed into tenuous clouds of interstellar matter.

This puzzle was linked to other features of the Crab. The filaments in the nebula are chiefly composed of hydrogen, which is hot enough to glow with a strong red colour; the overall colour of the nebula is a distinctive and unusual pale yellow, and its overall appearance in astronomical photographs has been described as like a ball of yellow cotton wool wrapped in red cotton strands.

The yellow glow is caused by a process known as **synchrotron emission**, which occurs when electrons, freed from their atoms, move in spirals in a strong magnetic field. This often produces radio emission, but there is so much energy available in the Crab that the electrons also radiate visible light. Investigations of the Crab in the 1950s provided the first proof that synchrotron emission occurs naturally in the Universe. And unlike most **nebulae**, the Crab is brightest at its centre, which tells astronomers that there must be a source of energy still at work in the heart of the Crab, keeping its nebula hot and driving the gases outward.

The total energy output of the Crab nebula in the form of synchrotron radiation is 3×10^{38} ergs per second, which is 75,000 times more than the total energy output of the Sun, 4×10^{33} ergs per second. And this from a star which 'died' 900 years ago! A few astronomers had an idea where the energy might be coming from. Back in the 1930s, Fritz **Zwicky** had suggested that supernova explosions might leave behind stellar remnants in the form of **neutron stars**, and although few people took the idea seriously, Walter **Baade** had pointed out that the best place to find a neutron star would be in the Crab nebula. He had even suggested a particular star that might be a candidate, sometimes known as Baade's Star. And in 1966, a year before the discovery of **pulsars**, John **Wheeler** and the Italian theorist Franco Pacini had speculated that the source of the energy emitted by the Crab nebula might be a spinning neutron star.

But their proposal for the source of this energy was not taken seriously until 1968 (to be fair, there was scarcely time for anyone to take much notice of the idea). Then, following the discovery of the first pulsars a year earlier, radio astronomers found that there is a pulsar almost at the centre of the Crab nebula. At the time, this was the most rapid pulsar known, spinning 30 times a second; it is electrons emitted by the pulsar that are responsible for the synchrotron emission and which transfer energy outward from the pulsar to the nebula.

As a result of this energy loss, the **Crab pulsar** (also known as NP 0532) is slowing down so rapidly that its rotation rate will be halved in just 1,200 years. It may sound like a modest slowdown, but the spinning neutron star at the heart of the Crab stores so much energy that this loss, increasing the period of its pulses by just 3×10^{-8} seconds each day, is enough to account for the entire energy output of the nebula.

After the Crab pulsar had been identified at radio frequencies, it was also found to flash at the same rate in visible light and in X-rays. Baade's Star is indeed a neutron star, and if anyone in the early 1960s had considered the ludicrous possibility of a star that flashes once every 33 milliseconds and designed an experiment to look for such flashes from Baade's Star, they could have discovered pulsars using optical astronomy years before the radio astronomers made the discovery. More seriously, though, it is because Baade's Star was already recognized (if only by Baade) as a neutron star candidate, in exactly the right place at the heart of a supernova remnant, that its identification as a pulsar convinced most astronomers in 1968 that all pulsars are indeed neutron stars produced in supernovae.

All of the activity associated with the star is a sign that the pulsar is very energetic, and this linked to its youth – it is the youngest pulsar known.

The filaments in the Crab are interesting in their own right, and **spectroscopy** shows that they contain about seven times as much helium, in proportion to hydrogen, as is usually seen in the surface layers of stars.

This is because helium was produced by *nuclear fusion* processes inside the parent star before it exploded (see *nucleosynthesis*), and the helium-rich interior became mixed with the surface layer during the explosion.

The mass of luminous gas in the filaments alone is about equal to that of the Sun, and there must be much more material that was ejected in the explosion, but which we cannot see because it is not luminous. The original star must have been many times more massive than the Sun, although the pulsar remnant (the neutron star) is probably only about half as massive as the Sun. The ejection of enriched stellar material that we can see going on in the Crab nebula is the process which spreads *heavy elements* across space to form the raw material for later generations of stars and planets.

Crab pulsar A *pulsar*, also known as NP 0532, in the *Crab nebula*. This is the youngest pulsar known, formed in the *supernova* explosion which created the Crab nebula, seen on Earth in 1054. It has a rotation period of 0.0331 seconds, and has been detected using *optical*, *radio* and *X-ray* astronomy. NP 0532 was discovered in 1967.

craters Bowl-shaped depressions seen on the surface of the Moon, on the *terrestrial planets*, on Earth itself and on many of the *moons* of the *giant planets*. A few craters may have been produced by volcanic activity, but almost all are the result of the impacts of solid objects (*meteoroids*) striking the planet or moon from space.

The size of a crater depends on the size of the impacting object. Some are only 1 m or so across, the largest are more than 1,000 km in diameter. The meteoroids that formed the craters were essentially debris left over from the formation of the Solar System, which has been steadily swept up by encounters with the planets and moons. So cratering was more frequent when the Solar System was young and there was more debris around, but it continues today.

On Earth and Venus, the presence of an active atmosphere (and geological activity) has caused severe erosion, so that only recently formed craters are clearly visible (see *Barringer Crater*); even on Mars, which has a thin atmosphere, the oldest craters have been obliterated by erosion. But on airless bodies such as the Moon and Mercury, the full effect of cratering can be seen, and their faces are covered in overlapping craters caused by repeated impacts over the past 4 billion years, with most of them occurring between 3 billion and 4 billion years ago.

On Earth, the atmosphere acts as a shield, slowing down objects with less than about 100 tonnes mass, with the heat caused by their passage through the air also burning away much of their material. A meteoroid containing 1,000 tonnes at the top of the atmosphere would be reduced to 300 tonnes by the time it hit the ground, typically at a speed of 5 km/sec. Such an impact would create a crater 150 m across.

Much larger impacts have occurred, and could occur again (see **Doomsday asteroid**). The impact on Earth of an object like the **asteroid** Icarus, nearly 1 km across, would produce a crater more than 20 km wide and 2 km deep. The eroded remains of an ancient crater roughly this size have been found in Germany, while even larger 'fossil' craters have been found in Canada, South Africa and Siberia.

critical density The density that the Universe would need today in order to be exactly flat, on the borderline between expanding forever and eventual recollapse. It is between 10^{-29} and 2×10^{-29} grams per cubic centimetre, roughly a hundred times greater than the density of bright stars and galaxies in the Universe. See **cosmological models, dark matter**.

curvature of spacetime Distortion of **spacetime** caused by the presence of **mass**, which gives rise to the force of **gravity**.

The idea of curved space emerged in the 19th century through the mathematical investigation of **non-Euclidean geometry**. The first person to suggest that space might be curved in the real Universe, closed around upon itself to make the equivalent in three dimensions of the curved surface of a sphere, was Bernhard **Riemann**, in 1854 – 25 years before Albert **Einstein** was born. The idea was taken up and promoted by the English mathematician William **Clifford**, who realized that, as well as this overall curvature of the entire Universe, there could be local distortions, which he likened to 'little hills'.

Einstein's **special theory of relativity**, published in 1905, united space and time in one mathematical description, and in 1908 Hermann **Minkowski** showed that this was mathematically equivalent to a description in terms of four-dimensional **Euclidean geometry**, so that the combination of space and time in spacetime could be regarded as the four-dimensional equivalent of a flat sheet of paper.

The **general theory of relativity**, published in 1916, describes what happens when this 'flat sheet' is distorted by the presence of matter. Think of it now as like the stretched surface of a trampoline; a heavy weight put on the surface will make a dent, curving spacetime (this is exactly the opposite of Clifford's analogy of little hills). If you roll a marble across the curved surface, it will follow a curved trajectory, as if its orbit has been bent by the action of a force. This, according to Einstein's theory, is how gravity works – the presence of matter tells spacetime how to curve, and the curvature of spacetime tells matter how to move.

The possibility that the overall spacetime of the Universe may be curved is still an open question (see **cosmological models**), but the idea of curved spacetime has been applied with spectacular success to the study of **black holes**.

Further reading: John Gribbin, *In Search of the Edge of Time*.

61 Cygni   A faint star in the constellation Cygnus, of interest only because it is one of the closest stars to the Sun and was therefore the first to have its *parallax* measured (in 1838). The parallax is 0.29 *arc seconds*, placing 61 Cygni (which is a *multiple star* system, with one star in orbit around a *binary*) at a distance of 3.4 *parsecs*. All three stars put together contain only about as much mass as our Sun.

Cygnus A   The brightest *radio source* outside our Galaxy, as 'seen' by radio telescopes on Earth. Some other radio sources are intrinsically brighter, but look fainter because they are further away. Cygnus A is a double source, with two lobes of radio noise, one on either side of a central *galaxy*, which has a *redshift* of 0.057.

Cygnus X-1   An intense source of *X-rays* that lies in the direction of the constellation Cygnus, and almost certainly contains a *black hole*. The X-ray source is in orbit around the star HDE 226868 in a *binary system*, and has a mass of six to fifteen times the mass of our Sun, well above the *Oppenheimer–Volkoff limit*.

Cygnus X-3   An *X-ray* source that lies in the direction of the constellation Cygnus. The source is a *binary system* with an unusually short period of 4.8 hours. It is probably a *neutron star* in orbit around a low-mass *star*.

dalton = *atomic mass unit.*

dark matter   Astronomers know that there is more to the Universe than meets the eye. The bright *stars* and *galaxies* are the obvious components of the Universe to creatures such as ourselves, who have eyes sensitive to visible light, and until the 1980s it was widely accepted that most of the matter in the Universe could be studied by its emission of light or other forms of electromagnetic radiation. But it is now clear that much less than half of the *mass* of the Universe is in the form of bright stuff. And it is possible that most of the dark matter that makes up the bulk of the mass of the Universe may not even be the kind of matter that the Sun and stars, the Earth and ourselves are made of.

Astronomers have known since the 1930s that there is at least some dark matter in our own Milky Way Galaxy. Although most of the stars in the Galaxy orbit in a relatively thin disc about 100,000 light years across but only some 2,000 light years thick (thicker near the centre; thinner at the edge), they bob up and down within the confines of the disc as they orbit the centre of the Galaxy. This motion, rather like the way the needle of a sewing machine bobs up and down through the cloth, is constrained by the amount of matter there is in the disc. The more matter there is, the smaller the amplitude of the bobbing, because gravity holds the stars more tightly in its grip. Statistical studies show that there is at least twice as much matter in the disc as we can see in the form of bright stars.

More recently, studies of the way galaxies like our own rotate (using *spectroscopy* and measurements of the *Doppler effect*) have shown the presence of even more dark matter. Across the entire disc of a galaxy like the Milky Way, the rotation speed is constant. This can only mean that the entire disc of bright stars is embedded in a much bigger halo of dark material, which carries the bright Galaxy around in its gravitational grip. The picture is rather like the way a thin layer of cream swirls around when it is stirred into a cup of dark coffee.

By the middle of the 1980s, it was clear that overall our Galaxy contains up to ten times as much dark matter as the matter we can see in the form of stars. At about the same time, studies of the Universe at large had shown that there is much more dark stuff in the depths of intergalactic space, holding galaxies together in *clusters*.

The speeds with which individual galaxies are moving within a cluster can be found using the Doppler effect. The overall *redshift* for the cluster is caused by the expansion of the Universe, but individual galaxies within the cluster show slightly different redshifts, because of their random motions adding to (or subtracting from) this cosmological redshift. It turns out that the galaxies are moving too fast within the clusters to be held in place by the gravity of the material we can see in the form of galaxies. Since the clusters are held together, or they would not be there, there must be additional dark matter present. There is about ten times more of this cosmological dark stuff than the amount of matter there is in galaxies themselves, *including* the dark component of galaxies.

On the largest scale of all, the Universe as a whole, there may be additional dark matter. If the *inflationary model* of the *Big Bang* is correct, then the spacetime of the Universe must be very nearly flat (see *cosmological models*), and the average density of matter throughout the Universe must be about 5×10^{-27} kg per cubic metre. The amount of bright stuff in the Universe corresponds to only about one-hundredth of this critical density, and even adding in the dark matter required to explain the movement of galaxies within clusters only brings the figure up to 30 per cent of the critical density. In round terms, there is at least 30 times more dark matter than bright stuff in the Universe, possibly 100 times as much. What is it? And where is it?

All the matter that we know from direct observations – stars, planets and people – is made of the same kind of material that we are, so-called baryonic material (see *baryon*). The amount of baryonic material in the Universe was determined by the conditions in the cosmic fireball of the Big Bang, in which the Universe was born. Studies of the *background radiation*, regarded as the afterglow of the Big Bang, and spectroscopic measurements of the amount of helium in very old stars, formed when the Universe was young, provide tight constraints on the amount of

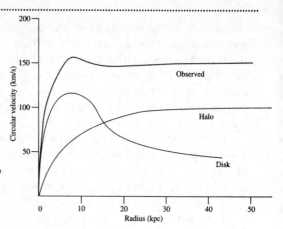

Dark matter. Schematic representation of the typical 'rotation curve' seen in a disk galaxy. The observed rotation curve (top line) cannot be explained solely by the gravity of the distribution of bright stars seen in the disk (middle line). There must be a dark halo as well, adding its influence to the disk's gravity to produce the overall effect.

baryonic material that could have been produced. Overall, there could be up to ten times as much baryonic material as we see in the form of bright stars, but no more.

This means that all the dark stuff in our Galaxy could be made of the same kind of matter as the atoms that make up your body, or planet Earth, or our Sun. But the cosmological dark matter *cannot* be baryonic.

There are two ways in which large amounts of baryonic matter in our Galaxy could be dark. Stars shine because *nuclear fusion* reactions are going on in their interiors, converting light elements (typically hydrogen) into heavier elements (typically helium) and releasing energy in the process. But these fusion reactions can be triggered only if a star has sufficient mass to squeeze the hydrogen in its heart enough to overcome the tendency of the positively charged atomic nuclei to repel one another.

The critical mass for making a star lies somewhere between the size of our Sun and the size of Jupiter, the largest planet in the Solar System, but with only 0.4 per cent of the Sun's mass. An object with mass less than about 8 per cent that of the Sun (twenty times more massive than Jupiter) may still get hot, as it shrinks under the inward tug of gravity, but never hot enough to trigger nuclear burning. It would be a ***brown dwarf***, with a surface temperature never rising above 2,000 K and a brightness never exceeding 1 millionth of the brightness of the Sun. Then, it would fade into complete oblivion as a ***black dwarf***.

The second possibility is more spectacular, but less plausible. Some astronomers have argued that, when our Galaxy was very young, large clouds of material may have collapsed to form superstars, very heavy objects which ran through their life cycles very quickly and then exploded, leaving massive ***black holes*** behind. Such black holes would count as baryonic material for the purposes of this discussion, because they had been made from stars. They could each have a mass millions of times that of the Sun – but if they did exist, they ought to exert a detectable influence on the structure of galaxies (over and above the smoothness of their rotation), which is not seen.

Since these dark objects, of either category, dominate the Galaxy and hold it in a firm gravitational grip, astronomers coined a suitably tough acronym for them – *MACHOs*, from 'massive astronomical compact halo objects'. There is some evidence, from *gravitational lensing*, that MACHOs do exist in the halo of our Galaxy.

But what of the non-baryonic material required to hold clusters together and to make spacetime flat? This is known by more sets of initials. Overall, it is called weakly interacting massive particles, or *WIMPs*. This means that it is a kind of matter that has mass, and therefore interacts through gravity, but otherwise interacts only weakly with ordinary baryonic material. WIMPs come in two (hypothetical) varieties. *Cold dark matter*, or CDM, is the name given to particles that emerge from the Big Bang travelling much more slowly than the speed of light. *Hot dark matter*, or HDM, is the name given to hypothetical WIMPs that emerge from the Big Bang travelling at speeds close to the speed of light.

Intriguingly, the *grand unified theories* of particle physics do require the existence of as yet undetected particles, and these theories allow for the possibility of both CDM and HDM WIMPs. So both cosmology, the study of the Universe on the largest scale, and particle physics, the study of the world of the very small, suggest to physicists that dark matter, in the form of WIMPs, should exist.

If the bulk of the matter in the Universe is in the form of WIMPs, they will have had a profound effect on the way galaxies formed when the Universe was young. CDM particles tend to clump together and produce gravitational 'potholes', which attract baryonic material (hydrogen and helium gas) by their gravity. If the Universe was dominated by CDM, galaxies grew from the 'bottom up', with small concentrations of matter getting bigger as time passed.

Fast-moving HDM particles, on the other hand, tend to blast apart any clouds of gas that start to form in the early Universe, like a cannon ball demolishing a brick wall. As the HDM particles cooled and slowed down, baryonic matter would be left spread in great pancakes across the Universe, and these would break up as a result of gravitational instability, forming galaxies from the fragments in a 'top-down' process.

Neither of these simple models exactly fits the actual pattern of galaxies on the sky. But a combination of studies of the way galaxies cluster today and the ripples in the background radiation observed by the COBE satellite, which indicate the kind of irregularities that existed when the Universe was young, suggests that a mixture of about two-thirds CDM, one-third HDM and just 1 per cent baryons could have produced the distribution of matter that we see in the Universe today out of the Big Bang. This is the 'mixed dark matter', or MDM, model.

That would mean that two-thirds of the mass of the Universe is in the form of kinds of particle that have never been detected, even though they are required by particle theory. They go by names such as axions and gravitinos, and efforts are now being made to find these CDM WIMPs in laboratory experiments.

Up to a third of the mass of the Universe could still be in the form of HDM particles, and there is one known candidate for the role, the *neutrino*. Although it was originally assumed that neutrinos have zero mass, until early 1995 laboratory experiments only set an upper limit on its mass, 20 *electron Volts*. This is ludicrously tiny compared even with the mass of a proton, about 1 billion eV, but there are enormous numbers of neutrinos in the Universe, roughly a billion for every baryon. So the mass for each neutrino required by cosmologists to make neutrinos fit the HDM role in the MDM model is just 5–7 eV, tantalizingly close to the limit set by experiments. Early in 1995, researchers at the Los Alamos National Laboratory claimed that they had experimental evidence for a neutrino mass of a few electron Volts, exactly what the cosmologists would like; but these claims have not yet (May 1995) been confirmed by other experiments.

Most discussions of neutrinos focus on one variety, called the electron neutrino, which is produced in nuclear reactions involving electrons. These are the neutrinos which, theory says, should be produced in huge quantities inside the Sun and stream across space and past the Earth. Experiments have so far failed, however, to detect as many neutrinos from the Sun as theory predicts (see *solar neutrino problem*). There are also two other varieties of neutrino, associated with, respectively, the muon and the tau particle, which are heavier counterparts to the electron. Physicists have long suspected that the deficit in solar electron neutrinos occurs because some of these neutrinos have been converted into the other varieties, which cannot be detected with present-day solar neutrino experiments.

This switching of neutrinos from one variety to another can happen only if they have a very small mass. This was good news for cosmologists, and the recent claims that neutrinos do have mass originally focused on the possibility that one kind of neutrino might have a mass of around 5 eV, one-hundred-thousandth of the mass of an electron. This is exactly right to provide the 20 per cent hot dark matter the cosmologists require. But Joel Primack of the University of California, Santa Cruz, Jon Holtzman of the Lowell Observatory, Anatoly Klypin of New Mexico State University, and David Caldwell of the University of California, Santa Barbara, found an even better way to do the job.

They drew attention to suggestions that the best way to explain the actual observations of neutrinos arriving at the Earth from space is to allow at least two of the varieties of neutrino to have almost the same mass. This encourages the 'oscillations' that change neutrinos from one variety to another. There are two possibilities consistent with the observations of solar neutrinos. Either all three of the neutrino varieties have the same mass, or the muon and tau neutrinos have the same mass, but the electron neutrino is very much lighter and associated with a so-called 'sterile' neutrino which takes no part in nuclear reactions. According to Primack and his colleagues, the experimental data which directly detected neutrino mass early in 1995 can be explained by the second possibility, but not by the first.

This means that instead of the hot dark matter in the Universe being provided by one kind of neutrino, each with a mass of 5 eV, it is in the form of two kinds of neutrino, each with a mass of about 2.4 eV (leaving a little bit for the electron neutrino). And the icing on the cake is that, when the computer simulations of how galaxies grow are run with the 20 per cent hot dark matter in the form of twice as many particles each with half as much mass, they give an even better match with the real Universe than if all the neutrino mass is carried by one variety of the particle.

It should be possible to improve the experiments sufficiently to measure the masses of neutrinos directly, within a few years. If and when experimenters do measure these masses, and if it turns out that they add up to about 5 eV, it will provide the greatest triumph yet of the complete cosmological package based on inflation, the Big Bang and dark matter. Cosmologists will have been able to predict the mass of the lightest particle known (apart from massless particles such as the *photon*), even before particle physicists were able to measure that mass.

Whatever the outcome of those experiments, though, and whatever the fate of the mixed dark matter model, there is no doubt that most of the mass of our own Galaxy is in the form of dark matter, possibly baryonic, and that tens of times more mass than we see in bright galaxies, definitely not baryonic, is holding clusters of galaxies together.

Further reading: John Gribbin, *In the Beginning*; Lawrence Krauss, *The Fifth Essence*, Basic Books, New York, 1989.

dark nebula = *absorption nebula*.

dark night sky   See *Olbers' Paradox*.

Davis, Ray Jr (1914–)   American astrophysicist and chemist who devised and built the first experiment to measure the flux of *neutrinos* coming from the Sun.

Born in Washington, DC, Davis was educated at the University of Maryland and at Yale University, obtaining his PhD in physical chemistry in 1942. After army service during the Second World War, he worked for two years with the Monsanto Chemical Company, but he spent most of his career, from 1948 until he retired in 1984, working as a chemist at the Brookhaven National Laboratory; he then moved to the University of Pennsylvania, where he became professor of astronomy. This rather strange-looking late career move resulted from Davis' long involvement in an experiment which has chemical connections, but which is designed to detect particles arriving at the Earth from the heart of the Sun.

The particles are neutrinos produced in *nuclear fusion* reactions that produce energy in the centre of the Sun. Because neutrinos are extremely reluctant to interact with everyday matter, they would pass out from the centre of the Sun and right through its outer layers essentially unscathed, streaming across space to the Earth. In principle, they might be detected, and provide a direct 'window' on conditions at the heart of the Sun. In 1948, long before anyone dreamed of detecting solar neutrinos, Davis was allowed to choose his own research programme at Brookhaven, and decided that neutrino physics, a new field in which he could use his background in physical chemistry, would be appropriate.

In the 1950s, Davis' main interest was in the design and construction of detectors to capture neutrinos produced in nuclear reactors here on Earth, although he also investigated radioactive *isotopes* found in geological samples and in *meteorites* (this later led to an involvement in the study of Moon rocks brought back by the Apollo missions).

In the late 1950s and early 1960s, astrophysicists such as Willy *Fowler* and John *Bahcall* calculated the strength of the solar neutrino flux. Although it was clear that the reluctance of neutrinos to interact with everyday matter would make building a solar neutrino detector very difficult, the challenge caught the imagination of Davis, who set out to build just such a detector, tailored to the detection of solar neutrinos. It had to be shielded from anything else that might cause interference (*cosmic rays* and the like), so it was built (beginning in 1964) deep underground in a gold mine at Lead, South Dakota. Some 7,000 tonnes of rock had to be removed to make room for the detector, which is a tank the size of an Olympic swimming pool. It contains 40,000 litres of perchlorethylene (a fluid commonly used in so-called dry-cleaning processes), containing a total of about 2×10^{30} chlorine *atoms*. Extremely rare interactions with solar neutrinos would, according to theory, occasionally convert a single chlorine atom into an active form of argon, which could be detected by a combination of chemical techniques and its characteristic *radioactive decay*.

The Davis detector. The tank of dry-cleaning fluid, as large as an Olympic swimming pool, used by Ray Davis and his colleagues to capture neutrinos from the Sun.

First results emerged from the experiment in 1968. Ever since, it has consistently shown that only one-third to one-half of the number of neutrinos predicted by standard *models* of the Sun are arriving at Earth – enough to produce just one active argon atom in the Davis detector every two or three days. This is the *solar neutrino problem*.

At the beginning of the 1990s, new detectors such as *ICARUS* and *SAGE* confirmed that the solar neutrino flux does not match the predictions of the standard model. It is not yet clear why this should be so, but a favoured possibility is that the neutrinos themselves are changing into an undetectable form en route to Earth.

Davis experiment See *Davis, Ray.*

decay See *radioactive decay.*

deceleration parameter The rate at which the expansion of the Universe is slowing down, usually denoted by the symbol q_0. If this is less than 0.5, the Universe will expand forever; if it is greater than 0.5, it will one day collapse in a *Big Crunch* (see *fate of the Universe*). The actual value of the parameter is so close to 0.5, and the measurements of it are so difficult, that observers cannot tell which fate awaits us.

de Chésaux, Jean-Phillippe Loÿs (1718–51)   Swiss astronomer who was the first person to spell out in its modern form the puzzle known as *Olbers' Paradox*.

De Chésaux was a child prodigy who was taught by his grandfather, himself a distinguished scholar, and was responsible, at the age of eighteen, for the construction of the first astronomical observatory in Switzerland. He made his name in the early 1840s with the discovery of two comets, and wrote a book about these observations. It was in an appendix to this book, published in 1744, that he put forward his version of the dark night sky puzzle. De Chésaux suffered from poor health throughout his life, and died at the age of 33, while on a visit to Paris.

declination   One of two coordinates used in astronomy to define the position of an object on the sky (see *right ascension*). Declination (dec) is the angular distance of the object north or south of the equator – equivalent to celestial latitude.

decoupling era   The time, about 300,000 years after the *Big Bang*, when electrons and nuclei of hydrogen and helium combined to form neutral atoms. Because *electromagnetic radiation* interacts only with particles that carry electric charge, matter and radiation 'decoupled' at this point. The *background radiation* is a fossilized imprint of what the Universe looked like at that time, when everything in the Universe was as hot as the surface of the Sun is today.

Deep Space Network   A system of radio *antennas* used by NASA to track and communicate with spacecraft beyond the *orbit* of the Moon. Three standard dish antennas each 70 m across (essentially *radio telescopes*) are spaced around the world so that, as the Earth rotates, at least one of them can receive signals from any spacecraft at any time.

deflection of light   One of the key tests of the *general theory of relativity*, which predicted that light from a distant star passing close by the Sun would be 'bent' by a certain amount. The only way to observe this is during an eclipse, when the bright light of the Sun itself is blocked by the Moon, and stars can be seen on the sky around the eclipsed Sun.

Albert *Einstein* published his paper predicting this effect in 1916, and by a happy coincidence a suitable solar eclipse was due in 1919. The British astronomer Arthur *Eddington* organized an expedition to observe the eclipse from Principe, off the west coast of Africa, while a second team observed it from Brazil. Photographs of the positions of the stars near the Sun on the sky were then compared with photographs of the same part of the sky taken 6 months earlier, when the Earth was on the other side of the Sun in its *orbit* and those stars were visible at night. The comparison showed that the stars photographed during the eclipse seemed to have been shifted sideways slightly by the deflection of light – by exactly the amount that Einstein had predicted.

In principle, this deflection occurs whenever light passes near a massive object, although usually the effect is too small to be measurable. It is caused by the *curvature of spacetime* associated with the *mass*. A more extreme version of the same effect causes the *gravitational lens* phenomenon, and in the ultimate extreme light is trapped completely within a *black hole*.

degenerate matter   Matter at such high density that quantum effects dominate its behaviour, and in particular provide an outward pressure much greater than the pressure appropriate to that density of material according to *classical mechanics*.

Degenerate matter exists in old stars which have undergone *gravitational collapse* after they have exhausted all their nuclear fuel, and can no longer keep themselves hot inside by *nuclear fusion* processes. Under the extreme conditions of temperature and pressure inside a star, electrons are not held tightly to atomic nuclei to form atoms, but move freely among the nuclei in a form of matter known as a *plasma*. As the dying star shrinks under its own weight, the electrons and nuclei are packed more and more tightly together, until quantum effects prevent the electrons from being squeezed any more. At that point, the star becomes a stable *white dwarf*, about the size of the Earth, supported by electron degeneracy pressure – provided it is light enough.

If the star still has more mass than the *Chandrasekhar limit* (a little less than one and a half times the mass of our Sun) at this stage of its evolution, even the pressure of the degenerate electrons cannot prevent further gravitational collapse. The electrons are forced to combine with protons to make neutrons. This allows further collapse, to the point where the same quantum processes that provide an electron degeneracy pressure now make the neutrons degenerate and prevent them coming any closer to one another; the entire star becomes a ball of neutrons a few kilometres across – a *neutron star*. But if the star has more than about three times as much mass as our Sun (the *Oppenheimer–Volkoff limit*) at this stage of its life, even neutron degeneracy cannot hold it up, and it collapses further to become a *black hole*, crushing the matter from which it was made out of existence.

degenerate star   A star made of *degenerate matter*; see also *white dwarf, neutron star*.

Deimos   One of the natural *satellites* (*moons*) of Mars. A small, irregular object (only 15 × 12 × 11 km), orbiting Mars at a distance of 23,460 km once every 1.26 days. Probably a captured *asteroid*.

Democritus of Abdera (about 470–400 BC)   Greek philosopher who proposed that the Universe consists only of empty space and atoms, and that .

133

these atoms are small, hard, eternal and in ceaseless motion. This was a development of even earlier ideas from Leucippus; although the work of Democritus was not forgotten (it was, for example, known to Isaac **Newton**), it had no direct influence on the development of the modern atomic theory at the end of the 18th century.

density The mass of an object divided by its volume. Ordinary water has a density of 1 g/cubic cm (10^6 g/cubic metre). The gas between the stars has a density of about 10^{-20} kg/cubic metre, while the densest form of stable matter, a *neutron star*, has a density of 10^{17} kg/cubic metre.

density parameter A measure of the amount of gravitating material in the Universe, usually denoted by the Greek letter omega (Ω), and also known as the flatness parameter. It is defined in such a way that if *spacetime* is exactly flat (see *cosmological models*) then $\Omega = 1$. Before the development of the idea of *inflation*, one of the great puzzles in *cosmology* was the fact that the actual density of the Universe today is very close to this critical value – certainly within a factor of 10. This is curious because as the Universe expands the expansion will push the density parameter away from the critical value. If the Universe starts out with the parameter less than 1, Ω gets smaller as the Universe ages, while if it starts out bigger than 1, Ω gets bigger as the Universe ages. The fact that Ω is between 0.1 and 10 today means that in the first second of the **Big Bang** it was precisely 1 to within one part in 10^{60}. This makes the value of the density parameter in the beginning one of the most precisely determined numbers in all of science, and the natural inference is that the value is, and always has been, exactly 1 (see *flatness problem*). One important implication of this is that there must be a large amount of *dark matter* in the Universe.

density wave model An explanation of the spiral structure of *disc galaxies* in terms of a wave moving around the galaxy, similar to the ripples spreading outward from a stone dropped in a pond.

Because of the *differential rotation* of a galaxy, with each star moving in its own *orbit*, the overall pattern made by the distribution of stars should be constantly changing as faster-moving stars (those nearer to the centre of the galaxy) overtake slower-moving stars. This could make a spiral pattern, in exactly the same way that cream stirred into black coffee makes a spiral pattern. But differential rotation alone is not enough to explain the spiral structure of disc galaxies, because the pattern persists for too long.

The spiral patterns produced by differential rotation 'wind up' rapidly, and would contain as many turns as the number of times the galaxy has rotated. But although galaxies like our own Milky Way are so old that they must have rotated many times since they were born, the typical spiral pattern shows just one or two arms, traced by lines of bright young stars.

The explanation is that the spiral arms are the visible parts of a density

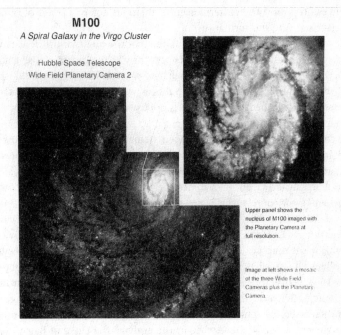

M100

A Spiral Galaxy in the Virgo Cluster

Hubble Space Telescope
Wide Field Planetary Camera 2

Upper panel shows the nucleus of M100 imaged with the Planetary Camera at full resolution.

Image at left shows a mosaic of the three Wide Field Cameras plus the Planetary Camera.

Density wave model. This spectacular Hubble Space Telescope image shows M100, a disc galaxy with a well-developed spiral pattern.

wave, which is moving around the galaxy in the same direction as the stars themselves, but more slowly. The density wave moves at a speed of about 30 km/sec, about three times faster than the speed of sound in the tenuous gas of interstellar space. So it produces a supersonic shock wave, like the shock wave associated with a supersonic aircraft such as Concorde, at the leading edge (the outer, convex part) of the spiral pattern. But stars and clouds of gas and dust are moving through the spiral pattern, overtaking it and ploughing into the shock wave, at speeds of 200 to 300 km/sec. As clouds of gas are squeezed by the shock wave, *gravitational collapse* is triggered, leading to bursts of *star formation*.

Once something has happened to set this process off, it is very largely self-sustaining. A density wave initially created by, perhaps, a gravitational disturbance caused by a close encounter with another galaxy will trigger star formation, and the self-sustaining star formation will help to maintain the density wave and the spiral pattern for much longer than if it were not reinforced in this way. Some disc galaxies have only a very patchy spiral structure, and it may be that these have not been perturbed by close encounters, and the patterns seen in them are solely a result of bursts of self-sustaining star formation.

Some astronomers have speculated that the origin of the density wave may be linked to explosive events occurring at the hearts of disc galaxies, perhaps associated with supermassive *black holes*. This would spread ripples outward, and the ripples could get twisted into spiral patterns by the processes described here. But this is not a widely accepted interpretation of the spiral structure.

Descartes, René (1596–1650) French mathematician and philosopher who invented the techniques of coordinate geometry, also known as Cartesian geometry in his honour. Cartesian geometry is an invaluable tool in the description of *spacetime* in the context of Albert *Einstein*'s *general theory of relativity*.

Descartes, who was the son of a Counsellor in the Rennes Parliament, was born at La Haye, in Touraine, on 31 March 1596. He was a sickly child, who spent much of his time lying in bed, thinking. Educated at a Jesuit college, he studied law at the University of Poitiers and graduated in 1616. Even before he finished his law degree, Descartes began studying mathematics, and this continued after he graduated. He then entered military service, putting his mathematical talents to use as a military engineer. It was while serving in the army of the Duke of Bavaria, lying snug in bed in quarters on the bank of the river Danube on the night of 10 November 1619, that Descartes came up with his revolutionary insight into geometry.

The details were recounted in Descartes' epic book *A Discourse on the Method of Rightly Constructing Reason and Seeking Truth in Science*, published in 1637 and usually referred to simply as *The Method*. He had realized, while lying in bed and watching a fly buzzing around the corner of his room, that the exact position of the fly at any moment in time could be defined simply by three numbers, specifying the distance of the fly from each of the three surfaces (two walls and a ceiling) that met in the corner. He immediately saw this in three-dimensional terms, but it is exactly the same in two dimensions, where each point on a graph can be specified in terms of two numbers (its coordinates) that give its distance from the x and y axes. It is also the same in a modern city – if you give someone instructions to go 'three blocks south and two east', you are giving directions in Cartesian coordinates.

Descartes opened up the possibility of using relationships between sets of numbers – algebraic equations – to study geometry, and this eventually found its application in Einstein's general theory of relativity, including the mathematical description of the *Big Bang* and *black holes* in terms of curved four-dimensional spacetime.

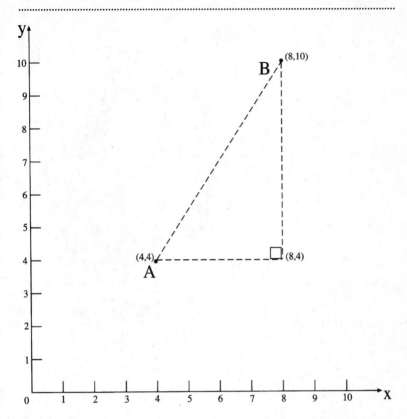

Cartesian geometry. Cartesian geometry makes it possible to work out relationships between points algebraically, using equations. Using Pythagoras' theorem for right-angled triangles, the distance between A and B can be calculated in terms of their Cartesian coordinates. It doesn't matter where the zero point of the coordinates is put, because the calculation only involves the *differences* between the coordinates of the points – that is, (10–4) and (8–4) in this example.

Descartes made many other contributions to astronomy, mathematics and philosophy. He settled in the Netherlands after leaving military service in 1629, but in 1649 could not resist an invitation from Queen Christina of Sweden to become a member of her court in Stockholm and found an academy of sciences there. Unfortunately, it turned out that his duties included visiting the Queen at 5 a.m. each day to give her personal tuition in philosophy.

Now in his fifties, roused at an unearthly hour each day from the familiar comfort of his warm bed, Descartes soon caught a chill in the cold

René Descartes (1596–1650).

RENE DES CARTES

of the Swedish winter. This developed into pneumonia which, with the aid of the enthusiastic bleeding that doctors used to treat the illness, soon finished him off. He died on 11 February 1650, a few weeks short of his 54th birthday.

Further reading: John Gribbin, *In Search of the Edge of Time*; J. F. Scott, *The Scientific Work of René Descartes*.

de Sitter, Willem (1872–1934) Dutch astronomer and cosmologist who was one of the first people to apply Albert *Einstein's* equations of the *general theory of relativity* to provide a mathematical *model* of the Universe.

Born on 6 May 1872 at Sneek in the Netherlands, de Sitter was the son of a judge. He studied mathematics and physics at the University of Groningen, where he became interested in astronomy. After graduating, in 1897 he went to work at the Cape Town Observatory, returning in 1899 to take up a post at Groningen; his PhD dissertation was based on observations he had made at the Cape, and he was awarded the degree in 1901. In 1908 de Sitter was appointed professor of astronomy at the

University of Leiden, and from 1919 onwards (until his death) he was also director of the Leiden Observatory.

De Sitter's observational work was solid but unspectacular. He was, however, one of the few astronomers to appreciate the significance of Einstein's *special theory of relativity* when it was published in 1905, and in 1911 he wrote a paper on the possible implications of the special theory for the orbital motions of the planets. When the general theory of relativity was published in 1916, de Sitter reviewed the theory and developed his own ideas in a series of three papers which he sent to the Royal Astronomical Society in London. The third of these papers included discussion of possible cosmological models – both what turned out to be an expanding universe (the first model of this kind to be developed, although the implications were not fully appreciated in 1917) and an oscillating universe model.

De Sitter's solution to Einstein's equations seemed to describe an empty, static Universe (empty *spacetime*). But in the early 1920s it was realized that, if a tiny amount of matter were added to the model (in the form of particles scattered throughout the spacetime), the particles would recede from each other exponentially fast as the spacetime expanded. This means that the distance between two particles would double repeatedly on the same timescale, so they would be twice as far apart after one tick of some cosmic clock, four times as far apart after two ticks, eight times as far apart after three ticks, sixteen times as far apart after four ticks, and so on. It would be as if each step you took down the road took you twice as far as the previous step.

This seemed to be completely unrealistic; even when the expansion of the Universe was discovered, later in the 1920s, it turned out to be much more sedate. In the expanding Universe as we see it now, the distances between 'particles' (clusters of galaxies) increase steadily – they take one step for each tick of the cosmic clock, so the distance is increased by a total of two steps after two ticks, three steps after three ticks, and so on. In the 1980s, however, the theory of *inflation* suggested that the Universe really did undergo a stage of exponential expansion during the first split-second after its birth. This inflationary exponential expansion is exactly described by the de Sitter model, the first successful cosmological solution to Einstein's equations of the general theory of relativity, dating from 1917.

As well as adding his own contributions, de Sitter provided an important link between Einstein in Germany and astronomers such as Arthur *Eddington* in Britain. Because Holland was neutral during the First World War, Einstein could send copies of his papers to de Sitter, and de Sitter could send them on to Eddington, who was Secretary of the Royal Astronomical

Society at the time and ensured that they received proper publicity.

After observations by Edwin *Hubble* and others in the 1920s had shown that the Universe is indeed expanding, Einstein and de Sitter worked together to develop another model of the Universe, based on Einstein's equations, which was published in 1932. This Einstein–de Sitter model is the simplest one that can be constructed using the basic equations of the general theory. It expands, as those equations require, but at the sedate rate required to match the observations; the space that is expanding is flat (that is, it obeys *Euclidean geometry*), and is essentially the space described by the special theory. The model requires that the Universe was born at a definite moment in time, out of a *singularity*, and (combined with the original de Sitter model) it closely matches both the appearance of the real Universe and the standard model of the Universe developed in the 1980s and 1990s based on the theory of inflation. The 1932 paper even acknowledges the possibility that there may be *dark matter* in the Universe.

This was de Sitter's last significant contribution to cosmology; he died of pneumonia in Leiden on 19 November 1934.

Further reading: John Gribbin, *In Search of the Big Bang*; Edward Harrison, *Cosmology*.

de Sitter expansion   See *de Sitter, Willem*.

de Sitter universe   See *de Sitter, Willem*.

detached system (detached binary)   See *equipotential surfaces*.

deuterium   A form of hydrogen in which each atomic *nucleus* contains both a *proton* and a *neutron*. It is also known as heavy hydrogen. The deuterium nucleus is also known as a deuteron.

deuteron   See *deuterium*.

De Vaucouleurs, Gerard Henri (1918–95)   French-born American astronomer who specialized in investigations of the distribution of *galaxies* and attempts to determine the value of the *Hubble constant*, and therefore the *age of the Universe*.

De Vaucouleurs was born on 25 April 1918 and studied at the University of Paris, graduating in 1936. His career was interrupted by the Second World War, but after 1945 he worked in France and Australia as an observer before moving to the *Lowell Observatory* in Arizona in 1957, the same year that he was awarded a DSc degree by the Australian National University. In 1958 he took up a post at the Harvard College Observatory, and from 1965 onwards he was professor of astronomy at the University of Texas, Austin.

At the time De Vaucouleurs began his research, it was widely accepted that, although galaxies occur in groups known as clusters, the clusters of galaxies themselves are scattered at random across the Universe. During the course of his career, he had shown that this is not the case, and that clusters of galaxies themselves group together to make 'superclusters'. De Vaucouleurs argued that this process continues indefinitely, in a hierarchical structure; the latest observational data do not support this, but show that superclusters of galaxies of the kind first described by De Vaucouleurs are distributed in a filamentary fashion across the Universe, forming chains and sheets which seem to be wrapped around empty regions, known as voids.

De Vaucouleurs' other main interest was attempting to determine the value of the Hubble constant, which is a measure of how fast the Universe is expanding, and therefore of how much time has elapsed since the *Big Bang*. Estimates of the Hubble constant depend on determinations of the distances to remote galaxies, and there is considerable uncertainty in these estimates (see *cosmic distance scale*).

Two schools of thought have emerged from these studies, both using essentially the same observations but coming to quite different estimates for the value of the constant. De Vaucouleurs was the leading proponent (along with Sidney van den Berg) of estimates which lead to a relatively high value of the constant, close to 100 km per second per Megaparsec. This corresponds to an embarrassingly low age for the Universe, less than 10 billion years, which is less than the estimated ages of some stars.

The rival camp, led by Allan *Sandage* and Gustav *Tamman*, favours a lower value for the constant, about 55 km/sec/Mpc, which corresponds to an age of the Universe of about 18 billion years.

Both estimates cannot be correct, and although a compromise figure of 75 km/sec/Mpc is often used for convenience (even though neither camp would accept this as plausible!), it is certain that one of the figures is completely wrong. It is even possible that they both are; in the 1990s some new techniques have suggested a value for the Hubble constant as low as 40 km/sec/Mpc.

De Vaucouleurs died on 7 October 1995.

Further reading: Michael Rowan-Robinson, *The Cosmological Distance Ladder*.

Dicke, Robert Henry (1916–) American physicist who is best known for his investigations of the cosmic *background radiation*.

Dicke was born in St Louis, Missouri, on 6 May 1916. He received his first degree, in physics, from the University of Princeton in 1939, and his PhD from the University of Rochester in 1941. During the Second World War he worked on radar at the Massachusetts Institute of Technology, but in 1946 he returned to Princeton, where he spent the rest of his career, eventually becoming chairman of the department of physics and Albert Einstein Professor of Science.

During his time at MIT, Dicke developed an instrument for measuring radiation in the microwave part of the electromagnetic *spectrum*. The design is known as a Dicke radiometer, and the same principles are incorporated in modern instruments designed to do the same job. With three of his colleagues, Dicke pointed one of these instruments at the sky to see if there was any microwave radiation coming from galaxies; they found nothing, and concluded that if there was any microwave radiation coming from space, it had a temperature of less than 20 K. A short paper announcing this negative result was published in the journal *Physical Review* in 1946.

Dicke and his colleagues were unaware at the time of the possible significance of a microwave *background radiation* filling the Universe, which was predicted at about this time by George *Gamow* and his colleagues. Equally, Gamow's team were not aware that the technology to measure such radiation already existed.

Turning his attention to other areas of astrophysics, Dicke forgot all about these pioneering measurements. Among those other interests, he developed a theory of gravity (in collaboration with Carl Brans) which differs from Albert Einstein's theory, the *general theory of relativity*. The Brans–Dicke theory has now been ruled out by observations, including those of the *binary pulsar*, but Dicke's interest in gravity led him to show that *gravitational mass* and *inertial mass* are equivalent to one another to better than one part in 10^{11}.

In the early 1960s, Dicke became interested in the Big Bang model of the Universe. Still unaware of Gamow's pioneering work, he and Jim *Peebles* independently made essentially the same prediction as Gamow, that the Universe should be filled with a sea of microwave radiation at a temperature of a few K, left over from the Big Bang. In 1964 his team at Princeton were in the late stages of constructing a radio telescope (essentially an improved Dicke radiometer) to search for this radiation when they received a telephone call from Arno *Penzias* asking if Dicke could explain the peculiar microwave radiation from space that he and Robert *Wilson* had detected using an instrument at Holmdel, New Jersey – just 50 km away from Princeton.

Dicke and Peebles continued to be interested in both the microwave background and the nature of the Big Bang. In the 1970s, they were instrumental in drawing the attention of astronomers to the *flatness problem*, the remarkable way in which the Universe is balanced between being open (so that it will expand forever) and closed (so that it will one day recollapse). It was a talk on this topic given by Dicke that made Alan *Guth* aware of the problem, and inspired him to develop the theory of *inflation*.

Further reading: Marcus Chown, *The Afterglow of Creation*.

differential rotation   When different parts of a system rotate at different angular speeds. A solid body like the Earth has to rotate uniformly, but gaseous bodies can rotate differentially. Both Jupiter and the Sun, for example, rotate faster at their equators than at higher latitudes – so, in both cases, their 'day' is longer at higher latitudes.

Digges, Leonard (?1520–59) and Digges, Thomas (?1546–95)   Father and son who in turn invented the telescope (Leonard) and promoted its use (Thomas). Thomas Digges also proposed that the Universe is infinite in extent.

Little is known about the early life of Leonard Digges, but he was educated at Oxford University and became well known as a mathematician, surveyor and author of several books which were, unusually for the time, all written in English. The first of these, *A General Prognostication*, appeared in 1553 and was a best-seller. It included a perpetual calendar and a wealth of information about astronomy, weather lore and other information, including a description of *Ptolemy*'s Earth-centred model of the Universe.

Historical research carried out in the 1990s has shown that Leonard Digges undoubtedly invented a *reflecting telescope* (more than 100 years before Isaac *Newton*), and may also have made a *refracting telescope*. (See Colin Ronan, *Endeavour*, volume 16, page 91; and volume 17, page 177.) But he did not have the opportunity to put them to significant use. In 1554 he was condemned to death for his part in a rebellion led by Sir Thomas Wyatt, and although his sentence was reduced to the confiscation of his estates, he spent the rest of his life struggling to regain his property.

Thomas Digges was about thirteen when his father died in 1559, but had as his guardian John Dee, a mathematician. Dee's library contained more than 1,000 manuscripts, which the younger Digges studied. He published his own first mathematical work in 1571, as well as a posthumous book by his father, *Pantometria*, in which Leonard's invention of the telescope is discussed. Thomas observed the *supernova* of 1572, and some of his observations were used by Tycho *Brahe* in his analysis of that event.

In 1576, Thomas Digges published a revised edition of his father's *A General Prognostication*, by now retitled *Prognostication Everlasting* and including a description of the heliocentric model of the Universe developed by Nicolaus *Copernicus*. In the same book, Thomas Digges stated that the Universe is infinite; this conclusion may have been based partly on his telescopic observations, which predated *Galileo*'s astronomical use of the telescope by at least 35 years.

Thomas Digges also found time to serve as a Member of Parliament on two occasions, as a government adviser, and as Muster-Master General to the English forces in the Netherlands between 1586 and 1593.

dirty snowball   Descriptive term for the core (nucleus) of a *comet*, which is largely made of ice and dust. The close encounter of the spaceprobe *Giotto* with *Halley's comet* confirmed this structure, and showed that a great deal of the dust is in the form of carbon in the comet's nucleus.

disc galaxy   A type of *galaxy* in which a central bulge of cool *stars* (the nucleus) is surrounded by a flattened disc of material which includes stars, gas and dust. Many (but by no means all) disc galaxies show a spiral pattern among the stars of the disc, and the term 'spiral galaxy' is often used loosely to refer to all disc galaxies, whether or not they have spiral arms. The proportions of the nuclear bulge and the flattened disc are roughly those of the yolk and the white of a fried egg, easy over; the thickness of the disc is about one-fifteenth of its diameter.

There is a wide range of disc galaxies. At one extreme, the central bulge is very prominent, and is surrounded by tightly wound spiral arms; at the other, the bulge is inconspicuous, and the visible galaxy consists almost entirely of the disc itself, with loosely wound spiral arms. In some cases, the spiral arms twist outwards from opposite ends of a bar of stars across the centre; in others, they spiral outwards directly from the centre of the galaxy with no sign of a bar.

All disc galaxies are in *differential rotation* (revealed by the *Doppler effect*), and are rich in gas and dust clouds which are the birthplaces of new stars. Star formation is triggered by the passage of density waves through the disc, which initiates *self-sustaining star formation*.

Our *Galaxy* is a typical disc galaxy, and we can see that it is embedded in a spherical *halo* of faint stars and *globular clusters*; other disc galaxies are thought to have similar halos, but these are difficult to see because of their faintness and distance from us. The stars of the disc and spiral arms are typically members of *Population I*, while those of the bulge and halo are older *Population II* stars. In addition, the way in which disc galaxies rotate shows that they are all embedded in much more substantial halos of *dark matter*.

Disc galaxies vary enormously in size. Our near neighbour the *Andromeda galaxy* is a large one, with a diameter of about 40 *kiloparsecs*, containing possibly as many as 1,000 billion (10^{12}) stars (perhaps ten times more than our Galaxy), but most are much less than half this size. About 30 per cent of all the galaxies in the Universe are disc galaxies, with 60 per cent *elliptical galaxies* and 10 per cent *irregular galaxies*.

distances   See *cosmic distance scale*.

Dominion Astrophysical Observatory   Mainly used for *optical astronomy*, located at an altitude of 220 m near Victoria, in British Columbia (Canada). Established in 1917, it had the largest *telescope* in the world for a time (a 1.8 m reflector, still in use), and has a more modern 1.2 m reflector which has been operating since 1962.

Dominion Radio Astrophysical Observatory   Radio counterpart to the *Dominion Astrophysical Observatory*. Situated near Penticton, in British Columbia (Canada), where there is very little interference from human radio sources. Its *radio telescopes* include a 26 m dish *antenna* and an *aperture synthesis* system which uses seven 9 m dishes on a track 600 m long.

Doomsday asteroid   Term used to refer to the possibility that a large *asteroid* might strike the Earth and destroy civilization. This possibility was highlighted by the discovery of geological evidence for a major impact from space at the time of the death of the dinosaurs, some 65 million years ago; such impacts are, however, rare.

Doppler, Christian Johann (1803–53)   Austrian physicist who predicted what is now known as the *Doppler effect* in 1842, when he was a professor at the State Technical Academy in Prague. The prediction was tested and confirmed in 1845 in Holland, using a steam locomotive to haul an open carriage carrying several trumpeters.

Doppler effect   A change in the *frequency* of *light*, or in the pitch of a sound, caused by the motion of the object emitting the light or making the sound.

The Doppler effect, predicted by Christian *Doppler* in 1842, is familiar in everyday life through its effect on sound waves. When a vehicle such as an ambulance is moving rapidly towards you with its siren wailing, the note you hear is higher than when the same vehicle has passed you and is moving away at high speed. This is because when the vehicle is moving towards you, sound waves are squashed together by its motion (making a higher frequency), while when it is moving away the sound waves are stretched out (making a lower frequency). The abrupt change in pitch of the siren as the ambulance passes you is called a 'down Doppler'.

In astronomy, the Doppler effect is important because it operates in exactly the same way for light, and for other *electromagnetic radiation*. The light waves from a star (or other object) that is moving towards you are squeezed together, producing a *blueshift* in its *spectrum*, and the light waves from a star that is moving away from you are stretched to longer wavelengths, producing a *redshift*.

Although the Doppler effect gives no indication of how rapidly objects are moving across the line of sight, it does give a precise measure of their velocity along the line of sight, towards or away from us. Where the transverse velocity of a star can be measured by watching it move across the sky over a long period of time, this can be combined with the Doppler velocity to reveal its actual motion through space; this technique is important in determining the *cosmic distance scale*.

Doppler measurements also show how fast stars in *binary systems* are moving, which indicates what *mass* they must have in order to be in the observed *orbits*; without those measurements, we would have no idea of the mass of any star except our Sun. And they are used to measure how quickly a *galaxy* is rotating, and how rapidly galaxies in *clusters* move relative to one another, which reveals the presence of *dark matter* in the Universe.

The *cosmological redshift* of galaxies associated with the expansion of the Universe is not, however, due to the Doppler effect, because it is caused by space itself stretching, not by galaxies moving through space.

double star   Two stars which lie close together on the sky. Some double stars are *binary systems* physically associated with one another, but others are chance alignments of stars which are actually at different distances from us.

Drake, Frank Donald (1930–)   American astronomer best known for his contributions to the search for extraterrestrial intelligence (*SETI*). The first phase involved listening for signals using a 26 m radio dish in 1960, and was called Project Ozma. In 1975, time on the *radio telescope* at *Arecibo* was allotted to a revived Project Ozma, but no signal suggestive of a communication from another civilization has ever been received. Drake is the originator of the *Drake equation*, which quantifies the chance of other civilizations existing in our Galaxy.

Drake equation   An equation proposed by Frank *Drake* which purports to provide a measure of the number of advanced technical civilizations (defined as any civilization capable of radio astronomy) in our Galaxy. Any such civilization capable of radio astronomy is one that we could make contact with using existing technology on Earth.

Drake's equation links all of the factors that are involved in providing the right conditions for a civilization like our own to exist. It takes the number of stars in the Galaxy (N_*), the fraction of stars that have planetary systems (f_p), the number of planets in each system that are suitable for life (n_e, for Earthlike), the fraction of suitable planets on which life arises (f_l), the fraction of inhabited planets on which intelligence evolves (f_i), the fraction of those planets on which communicative technical civilization emerges (f_c) and the fraction of a planetary lifetime during which the civilization exists (f_L), multiplying them all together to give the number of technological civilizations in the Galaxy at any one time (N). That is,

$$N = N_* f_p n_e f_l f_i f_c f_L$$

which is Drake's equation. Each of the *f*s is a number between 0 and 1, and when all the fractions are multiplied together, they drastically reduce the number of possible civilizations around today, from the total number of stars in the Galaxy, which is about 100 billion (10^{11}).

Just how much this number is reduced depends on how optimistic you are about the various factors, but if a third of all stars have planets, two planets in each planetary system are habitable, a third of these are actually occupied by life, and technological civilization has arisen on just 1 per cent of those planets (that is, $f_i \times f_c = 0.01$), then there are a billion planets in our Galaxy alone on which technological civilization has appeared. How many are around today depends on how long they survive, but it still seems possible that there could be millions of technological civilizations around in the Galaxy today.

Such calculations echo the *Fermi Paradox* – the puzzle of why none of these civilizations has actually made contact. But it may be that this particular choice of numbers to put into the Drake equation is simply wrong, and life (or, at least, technological civilization) is much rarer than those numbers would suggest.

See also *life in the Universe, SETI.*

Draper, Henry (1837–82)   American astronomer who was one of the pioneers of photographic *spectroscopy* (his father, John, had taken the earliest known Daguerreotype of the Moon). After his death, Draper's widow established a fund to continue his work; this eventually led to the publication of the Henry Draper Catalog, which lists stars according to their spectroscopic properties.

The Dumbbell nebula (also known as M27 and as NGC 6853). This cloud of gas blown away by a dying star is expanding at about 27 km per second. It is just under 1 parsec across.

Dumbbell nebula   A *planetary nebula*, also known as M27 and NGC 6853, in the constellation Vulpecula. It was named by Lord *Rosse*, who observed its hour-glass shape using his 1.8 m *reflecting telescope* in the 1840s. (See picture, page 147.)

dust in space   See *cosmic dust*.

dwarf galaxy   A small *galaxy* like the *Magellanic Clouds* that are *satellites* of our own Galaxy. The dividing line between what constitutes a dwarf galaxy and what counts as a normal galaxy has never been defined, but a dwarf may contain only a few million stars, compared with the hundreds of billions of stars in our Galaxy. Not to be confused with *compact galaxies*.

dwarf novae   Faint stars that sometimes increase dramatically in brightness, at intervals of a few weeks or a few months. The increase in brightness is usually between 2 and 5 *magnitudes*, and lasts for a few days. They are *cataclysmic variables*, *binary systems* in which a *white dwarf* is accreting matter from a close companion, a star similar to our Sun, probably through an *accretion disc*.

dwarf planets   See *terrestrial planets*.

dwarf star   Obsolete term for any star on the *main sequence* of the *Hertzsprung–Russell diagram*. The term originated from an early classification of stars into dwarfs and *giant stars*. In this terminology, the Sun is a typical dwarf star. But the term 'main sequence star' is now preferred, to avoid confusion with *white dwarf* stars and other stars much smaller than the Sun.

Dwingeloo 1   A nearby *disc galaxy*, about 10 million *light years* away, that was only discovered in 1994. It had not been identified previously because it is largely obscured by dust in the *Milky Way*. It is probably a member of the same group as the *Maffei galaxies*. Dwingeloo 1 is named after the Dutch *radio telescope* used to discover it.

Dyson, Sir Frank Watson (1868–1939)   British astronomer, born in Ashby-de-la-Zouch, who was the ninth Astronomer Royal (from 1910 to 1933). His main work involved studies of the *proper motions* of stars, and observations of the outer parts of the Sun during *eclipses*. He was the organizer of the two eclipse expeditions in 1919 which confirmed the accuracy of Albert Einstein's *general theory of relativity* by measuring the *deflection of light* passing near the Sun.

Dyson, Freeman John (1923–)   English-born American physicist best known in science for his contributions to the *quantum theory* of *radiation* (quantum electrodynamics). He is an enthusiast for *space travel*, and originator of the idea of the *Dyson sphere*.

Dyson sphere   Hypothetical shell of artificial material around a *star*, created by an advanced civilization.

   The English-born physicist Freeman *Dyson* suggested in 1959 that an intelligent species with an expanding population, confined to a single planetary system like our own Solar System, will eventually rearrange the raw materials provided by those planets to build a hollow sphere around the parent star. This would not be a rigid structure, but would consist of huge numbers of individual 'cities in space', each following its own *orbit* around the star.

The point of building such an artificial habitat is that every city would be at the optimum distance from the parent star for the life forms that lived there, and as well as maximizing the area available for the civilization to spread over, it would enable the civilization to capture almost all of the energy released by the star in the form of electromagnetic radiation. This would make the star invisible to ordinary optical telescopes; but the energy used by the civilization would be re-radiated as infrared radiation (see *second law of thermodynamics*).

Dyson has proposed a search for 'infrared stars' as possible sites of advanced alien civilizations. There is no astronomical evidence (yet) for the existence of Dyson spheres, but the idea has been eagerly taken up by some science fiction writers.

$E = mc^2$ See *special theory of relativity*.

Earth   Our home in space, the third planet out from the Sun in the Solar System.

Earth is the largest of the four *terrestrial planets*, with a diameter of 12,756 km. It *orbits* the Sun at an average distance of 149,597,870 km (1 *astronomical unit*), and it rotates once on its axis every 23 hours 56 minutes and 4 seconds. This is the sidereal day, measured relative to the distant stars; because of the movement of the Earth in its orbit as it rotates, the time from noon one day to noon the next day (the solar day) is a little longer – 24 hours, or 86,400 seconds. It takes 365.24 solar days (1 year) for the Earth to travel once around its orbit, from *equinox* to equinox.

The atmosphere of the Earth is intermediate in density between those of Venus and Mars, its nearest planetary neighbours, and the planet is unique among those of the Solar System in having large oceans of liquid water. The atmosphere today consists of 77 per cent nitrogen, 21 per cent oxygen, 1 per cent water vapour, 0.9 per cent argon and traces of other gases, the most important of which is carbon dioxide (see *greenhouse effect*).

Earth is the only planet known to be geologically active, and has a surface which is very young in geological terms, having been produced by the processes of plate tectonics. These destroy old crust by pushing it below the Earth's surface in deep ocean trenches, while making new crust from molten material spewed out by volcanic activity at the sites of spreading ridges in the oceans.

The Earth. A view of the Earth from space, taken by the astronauts on the Apollo 17 Moon mission.

As well as being geologically young, the surface of the Earth is constantly being changed by erosion caused by wind, water and waves, and by biological activity. This makes it difficult to pick out surface features corresponding to the craters of the Moon, Mars and Mercury (and revealed by radar mapping on Venus), but geological evidence shows that, like the other terrestrial planets, Earth was subjected to intense meteoritic bombardment early in its life.

The mass of the Earth is 5.976×10^{27} grams, and its volume is 1.083×10^{27} cubic cm, so its overall density is just over 5.5 g/cubic cm, five and a half times that of water. The solid crust of the Earth is only 5 km thick under the oceans, and averages about 30 km thickness on the continents. Beneath this skin, a region known as the mantle, subdivided into different layers, extends down to within 5,000 km of the centre of the planet, which is an iron-rich core.

The mantle makes up 82 per cent of the Earth's volume, and is thought to be made of silicate rocks. The solid inner core, with a radius of 1,700 km, is divided from the mantle by a semiliquid outer core; the temperature in

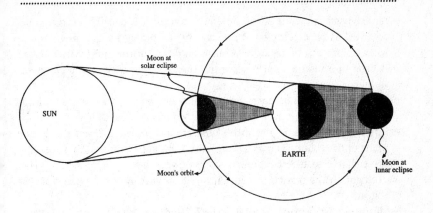

Eclipse. Schematic representation of lunar and solar eclipses (very much not to scale, and the Moon cannot, of course, be in two places at the same time!).

the core is about 3,000 *Kelvin.* The density rises from about 13.5 times that of water in the core to between 5.5 and 3.5 times the density of water in the mantle, and averages about 3 times the density of water in the rocks of the crust.

eclipse   An eclipse occurs when the light or other radiation by which we see an astronomical object is cut off temporarily because of the presence of another object. Usually, this is because the second object has passed in front of the bright object; in the case of lunar eclipses, though, it is because the Moon has moved into the Earth's shadow.

A solar eclipse can occur only close to the time of new Moon, when the Moon is directly in line between the Sun and the Earth, so that the Moon's shadow falls on part of the Earth. It does not happen at every new Moon, because the Moon's orbit is tilted at about 5° to the plane of the *ecliptic*; an eclipse will happen only when the Moon is crossing the ecliptic at the time of new Moon. A solar eclipse exhibits a strange coincidence – although the Sun is 400 times bigger than the Moon, it is also 400 times further away, so the two objects look the same size (about half a degree of arc) on the sky. So during an eclipse, the Moon almost exactly 'fits over' the image of the Sun on the sky.

A lunar eclipse occurs when the Moon passes through the shadow of the Earth, on the other side of the Earth from the Sun. This can only happen at times close to full Moon and, again, only when the full Moon is at the right part of the Moon's tilted orbit.

Eclipses also occur in *binary systems*, when one of the stars in the binary passes in front of the other, as seen from Earth. This can also be regarded as an *occultation* of one star in the binary by the other, and a solar eclipse can be regarded as an occultation of the Sun by the Moon. Eclipses (occultations) also occur in other systems, such as the *moons* of Jupiter.

eclipsing binary   A *binary system* in which one component passes in front of the other, as seen from Earth, as they *orbit* each other.

ecliptic   The plane of the Earth's *orbit* around the Sun. From Earth, the plane of the ecliptic is traced out by the Sun's apparent motion across the sky (relative to the background stars) during the course of a year. The orbits of all the planets, except Pluto, lie very nearly in the plane of the ecliptic.

Eddington, Sir Arthur Stanley (1882–1944)   The first astrophysicist, Eddington was an English theoretical astronomer who carried out the crucial test of Albert Einstein's *general theory of relativity*, developed the application of physics to an understanding of the structure of stars, and was a great popularizer of science in the 1920s and 1930s.

Eddington was born in Kendal, in Cumbria, on 28 December 1882, but moved to Weston-super-Mare, in Somerset, when his father died in 1884. He was brought up as a Quaker, and remained one throughout his life, which was to play a part in his participation in the expedition to test the *deflection of light* predicted by Einstein. He studied at Owens College in Manchester (which later became the University of Manchester), and then at Cambridge University, graduating in 1905. After a short spell of teaching, he became a Fellow of Trinity College in Cambridge and took up a post at the *Royal Greenwich Observatory* (then still at Greenwich). In 1912, aged 30, he became Plumian Professor of Astronomy and Experimental Philosophy at the University of Cambridge, and in 1914 he was appointed Director of the Cambridge Observatories. Along with these duties, he was sent to Malta in 1909 to determine the exact longitude of an observing station there, and in 1912 he was the leader of an expedition sent to Brazil to observe an eclipse of the Sun.

When Einstein announced his general theory of relativity to the Berlin Academy of Sciences in 1915, Britain and Germany were at war. But copies of Einstein's papers went to Willem *de Sitter* in neutral Holland, and de Sitter passed them on to Eddington, who was Secretary of the Royal Astronomical Society at the time. He was ideally placed to present Einstein's work to the scientific community in Britain, and quickly became the leading proponent of the new theory outside Germany.

The Astronomer Royal, Sir Frank *Dyson*, was in favour of organizing two expeditions in 1919 to test Einstein's predictions during an eclipse of the Sun, and Eddington was the obvious person to lead such an expedition.

Sir Arthur Eddington (1882–1944).

The situation was complicated when the draft was introduced in Britain, and Eddington let it be known that as a Quaker he was a conscientious objector and would refuse to serve; after considerable wrangling between the scientific establishment and the Home Office, Dyson came up with a solution to the problem. Eddington's draft was deferred, but only on condition that, if the war ended by May 1919, he would lead the expedition to test the light-bending prediction! Eddington duly confirmed the accuracy of Einstein's theory, and went on to be one of its chief popularizers.

Earlier in his career, Eddington had studied *proper motion* of stars. After that, he went on to apply the laws of physics to the conditions that operate inside stars, explaining their overall appearance in terms of the known laws relating temperature, pressure, density and so on inside them. This was a crucial step towards an understanding of how stars obtain energy from *nuclear fusion*. His book, *The Internal Constitution of the Stars*, was published in 1926 and became a classic.

In the 1930s, Eddington tried unsuccessfully to unite relativity and *quantum theory*, going up a blind alley and making no more significant contributions to astrophysics. He was also strongly opposed to the idea of what we now call *black holes*, publicly ridiculing the suggestion, made by Subrahmanyan *Chandrasekhar*, that a star with more than a certain amount of *mass* must collapse indefinitely at the end of its life. In spite of these minor blemishes on an outstanding career, the best description of Eddington is one made by Chandrasekhar himself: he was 'the most distinguished astrophysicist of his time'. He died in Cambridge on 22 November 1944.

Eddington limit   A limit on the maximum brightness of a star with a particular mass. Stars are held together by gravity, which depends on mass; energy released inside the star by *nuclear fusion* prevents the star from collapsing by exerting a radiation pressure balancing the inward pull of gravity. If the brightness of the star exceeded the Eddington limit, the radiation pressure would be great enough to blow the star apart.

effective temperature   For an object such as a *star*, the temperature of a *black body* which would have the same total *luminosity* as that object.

Effelsberg telescope   A *radio telescope* with a 100 m diameter steerable dish, located 40 km south-west of Bonn, in Germany. It is run by the Max-Planck-Institute for Radio Astronomy, and is one of the largest fully steerable radio telescopes in the world. It is also used in *very long baseline interferometry*.

Einstein, Albert (1879–1955)   German-born physicist who developed both the *special theory of relativity* and the *general theory of relativity*, as well as making major contributions to the development of *quantum theory*.

Einstein was born at Ulm, in Germany, on 14 March 1879, but soon moved with his family to Munich, where he went to school. He was not an outstanding student, and after his family moved to Italy in 1894 (his father, an unsuccessful businessman, repeatedly moved in the hope of finding success), Einstein dropped out of school without completing his courses and spent some months visiting the art centres of Italy. After failing the entrance examination for the Swiss Federal Institute of Technology (ETH) in Zurich in 1895, he had to spend a year in a Swiss secondary school before retaking the examination successfully in 1896.

At the ETH, Einstein did not bother with lectures he considered boring, and enjoyed a typical student life. When the time came for his final examinations, in 1900, he was able to make up for the deficiencies in his studies by swotting up from the carefully kept notes of his friend Marcel Grossman, who, unlike Einstein, had attended all the lectures.

Although Einstein did well in the examinations, he was unable to obtain a post at the ETH or any other university, partly because of his reputation for laziness, and after temporary work as a teacher he settled into a junior post at the Patent Office in Bern in 1902. The work gave him enough financial security to marry his first wife, Mileva (the couple had already had an illegitimate daughter, who had been adopted) and enough free time to continue with his own work in theoretical physics; it was

Albert Einstein (1879–1955).

while working as a patent officer that Einstein completed his PhD in 1905. The same year saw the publication of three great scientific papers, one introducing the special theory of relativity, another explaining the *photo-electric effect* in terms of light *quanta* (later to be called *photons*), and a third explaining the way tiny particles of dust in the air or a liquid are buffeted by atoms (brownian motion).

The importance of the special theory was not fully appreciated until it was explained in terms of the geometry of four-dimensional *spacetime*, by Hermann *Minkowski* in 1908. It is no coincidence that Einstein left the Patent Office to become a professor at the University of Zurich in the following year, 1909.

Over the next few years, Einstein also worked in Prague and Zurich before settling at the Kaiser Wilhelm Institute in Berlin in 1914. It was there that he completed his general theory of relativity, and confirmation of the predictions of the theory by the *eclipse* expedition headed by Arthur *Eddington* in 1919 made Einstein world famous. But his Nobel Prize (awarded in 1922, although it was actually the prize for 1921 held over for a year) was not for his work on relativity, but harked back to his interpretation of the photoelectric effect.

That investigation of the behaviour of photons was just one of Einstein's many contributions to **quantum theory**, and although in later life he disagreed with the philosophical interpretation of quantum theory used by most physicists (the Copenhagen Interpretation), many substantial aspects of quantum theory (for example, the description of lasers) still depend upon Einstein's work.

After the middle of the 1920s, Einstein's work became increasingly isolated from the mainstream of physics, as he tried unsuccessfully to develop a unified theory combining relativity and quantum theory. He moved to the United States after Hitler came to power in Germany in 1933, and worked at the Institute for Advanced Study in Princeton, becoming a US citizen in 1940. Although a lifelong pacifist, he allowed his reputation to be used to draw the attention of the US President to the possibility of the atom bomb, but was later involved in the movement to abolish nuclear weapons. In 1952 he was offered, but turned down, the opportunity to be President of Israel. He died in Princeton on 18 April 1955.

The modern description of the Universe is entirely based on Einstein's theory of **gravity**, the general theory of relativity, which provides the equations that describe the evolution of the expanding Universe and indicate that it was born at a definite moment in time in the **Big Bang**.

Further reading: Ronald Clark, *Einstein: The life and times*; Michael White and John Gribbin, *Einstein: A life in science*.

Einstein cross   An image of a distant **quasar** which has been split into four points of light by **gravitational lensing**, marking the four corners of an imaginary square or the ends of a cross formed by the diagonals of the square. (See picture, p. 215.)

Einstein–de Sitter universe   See *de Sitter*.

Einstein ring   A **gravitational lens** effect, predicted by Albert **Einstein** in the 1930s but not observed until the 1980s, in which light or other electromagnetic radiation from a distant point source (such as a **quasar**) is spread into a ring on the sky by the gravity of an intervening object (such as a **galaxy**) along the line of sight.

Einstein–Rosen bridge   See *wormholes*.

Einstein satellite (Einstein Observatory)   A satellite launched by NASA in 1978 to study the Universe at **X-ray** frequencies. It carried four X-ray **telescopes** into orbit, and operated for more than two years, identifying thousands of previously unknown X-ray sources.

electric force   See *fundamental forces*.

electromagnetic force (electromagnetic interaction)   See *fundamental forces*.

electromagnetic radiation   A form of *radiation* in which energy is carried by oscillating electric and magnetic fields, which move together and travel through space at the speed of light. Electromagnetic radiation is generated by the acceleration of an electric charge – for example, by the movement of *electrons* in an *antenna*.

The electric and magnetic fields can be thought of as oscillating at right angles to one another as they move together through space. In the 19th century, James Clerk *Maxwell* developed a set of equations that describe this behaviour, with the changing electric field producing the changing magnetic field, and the changing magnetic field producing the changing electric field. The equations included a constant, c, which is the speed at which the waves move through empty space. It turned out that this constant is precisely the *speed of light*, proving that light is a form of electromagnetic radiation. Electromagnetic radiation can also travel through some media, at a slightly slower speed than c.

The entire spectrum, from gamma rays to long-wave radio waves, is explained by Maxwell's equations, with the appropriate wavelength for the radiation. *Quantum theory* also describes the radiation in terms of a stream of particles, called photons; each quantum of radiation (photon) with a particular *frequency* f carries hf units of *energy*, where h is *Planck's constant*. Gamma rays are the most energetic, radio waves the least energetic.

In terms of the quantum description, photons behave like particles obeying the Bose–Einstein statistics (*bosons*). Although our everyday experience makes it hard to understand how something can behave as both waves and particles, many experiments have shown that this is just the way the quantum world works, and physicists use whichever description is convenient. *Black body radiation*, for example, can be described perfectly in terms of photons as a 'gas' of bosons.

electromagnetic spectrum   See *spectrum*.

electron   One of the fundamental *elementary particles*, a member of the *lepton* family, with a *rest mass* of 9.1094×10^{-31} kg and a negative electric *charge* of 1.6022×10^{-19} coulombs. The electron has no size and is a 'point-like' particle. Electrons are one of the three basic constituents of *atoms*.

electron Volt (eV)   A unit of *energy*, equal to the amount of energy gained by an *electron* when it accelerates across a potential difference of 1 volt. 1 eV is 1.602×10^{-19} joules. Because *mass* and energy are related by *Einstein's* equation $E = mc^2$, energies can be converted into masses by dividing by c^2. Particle physicists often quote masses of *elementary particles* in terms of electron Volts, by which they mean the Einstein mass equivalent. 1 MeV is a million eV; 1 GeV is 10^9 eV. In these units, the mass of the *proton* is close to 1 GeV.

electroweak interaction   See *fundamental forces, grand unified theories.*

element   A substance composed entirely of *atoms* that all have the same number of *protons* in their *nuclei*, and therefore the same number of *electrons*. Some of the atoms may have different numbers of *neutrons* in their nuclei, making them different *isotopes* of the same element. An element cannot be decomposed into a simpler substance without breaking the atoms apart. There are 92 naturally occurring elements, including *hydrogen, helium, oxygen, carbon* and *nitrogen*. See *cosmic abundances, nucleosynthesis.*

elementary particles   The basic constituents from which all material things are built up; also the particles that carry the *fundamental forces* in *quantum theory*.

Strictly speaking, an elementary particle is one which cannot be broken down into any constituent parts. On this definition, there are only two families of elementary (or fundamental) particles, the *quarks* and the *leptons*. But although both protons and neutrons are composed of quarks, it is impossible to break either of these *baryons* down into its quark constituents, because it is impossible for an isolated quark to exist. So protons and neutrons, and other baryons, are often regarded as elementary particles, even though they are made of quarks.

Until the end of the 19th century, *atoms* were thought to be the fundamental building blocks of matter. Then J. J. Thomson (1856–1940), a pioneering English particle physicist who worked at the Cavendish Laboratory in Cambridge, found that a form of radiation produced by atoms could be explained as a stream of tiny charged particles, now called electrons, broken off from the atoms themselves.

Since electrons carry negative electric charge, while atoms are electrically neutral, it was clear that there must be other particles inside atoms that carry positive charge, to cancel out the negative charge of the electrons. Early in the 20th century, the New Zealand-born physicist Ernest Rutherford (1871–1937), working in Manchester (he later succeeded Thomson as director of the Cavendish Laboratory) showed that this positive charge, together with most of the *mass* of an atom, is concentrated in a tiny central *nucleus*.

At first, it was thought that the nucleus consisted of a mixture of positively charged protons and electrons. It was only in 1932 that James Chadwick (1891–1974), also working at the Cavendish, discovered the neutron, a particle with no electric charge and almost the same mass as the proton. The nucleus was then explained as a collection of protons and neutrons held together by the *strong nuclear interaction*, or strong force.

At the time, it seemed that these three particles – electrons, protons and neutrons – were the only elementary particles, out of which all matter could be constructed. But studies of **cosmic rays** and experiments which involve smashing beams of particles together at high energies in particle accelerators showed that many other kinds of 'subatomic' particle could exist, although these 'new' particles were not stable, and would quickly 'decay' in a shower of other particles, ending up with the familiar electrons, protons and neutrons.

It is important to appreciate that these new particles do not in any sense exist 'inside' the particles (such as protons) that are being smashed together in the particle accelerators. The additional particles are created out of the energy that goes into the accelerators, in line with Einstein's equation $E = mc^2$ (or rather, in this case, $m = E/c^2$). But during their brief lifetimes, they are real particles with properties such as mass and charge. Such particles would have been present in profusion in the high-energy conditions of the **Big Bang**.

Physicists puzzling over how these particles could be fitted into a satisfactory theory of physics were also trying to explain the workings of the fundamental forces which operate between these particles. They did so in terms of another kind of particle, the **mesons**, which carry the forces, using the analogy of the way photons carry the **electromagnetic force** between charged particles. But what were mesons made of?

For a time, the situation was extremely confused. But during the 1960s and 1970s order was restored by the development of the quark theory. On this picture, all of the known particles can be divided into two groups. One group is made of quarks, and 'feels' the strong force, which operates only between quarks. They are called **hadrons**. The other group does not feel the strong force, but does take part in interactions mediated by the so-called **weak force** (or weak interaction), which is responsible, for example, for the process of **radioactive decay** (including **beta decay**). They are called **leptons**. Hadrons can feel the weak force too, as well as getting involved in strong interactions.

The leptons are truly fundamental particles, and are not made of anything else. The archetypal lepton is the electron, which is associated with another lepton called the neutrino (strictly speaking, the electron neutrino). When electrons take part in processes such as radioactive decay, neutrinos are always involved as well.

For reasons which nobody understands, this basic pattern has been duplicated twice in nature, making three 'generations' of leptons. As well as the electron itself, there is also a heavier particle called the muon, which is exactly like an electron except that it is 207 times heavier, and an even heavier counterpart, the tau particle, which has nearly twice as much mass as a proton. Each of these heavy electrons has its own kind of neutrino, so that there are six particles (three pairs) in the lepton family. Although both muons and tau particles can be made out of energy in particle accelerators or produced in cosmic rays, they quickly decay, turning into electrons and neutrinos.

The hadron family itself subdivides into two varieties. Particles that are each made up of three quarks are called baryons, and these are what we would think of in everyday terms as 'material' particles, including protons and neutrons (baryons and leptons are members of the family of *fermions*, effectively another term for everyday material particles). Particles that are made up of pairs of quarks are called mesons, and these are particles that carry the fundamental forces, although there are other mesons as well (these force carriers and other mesons are also known as *bosons*).

Only two kinds of quark (whimsically dubbed 'up' and 'down') are needed to explain the structure of protons and neutrons. A proton is made of two up quarks and one down quark held together by the strong force, and a neutron is composed of two down quarks and one up quark held together by the strong force. The strong force itself is regarded, in particle terms, as involving the exchange of *gluons*, which are themselves made of pairs of quarks and are therefore mesons.

Just as the lepton family is triplicated in nature, so is the quark family. Although only two quarks are needed to explain the nature of protons and neutrons, the pattern is repeated in two more successively heavier generations, the first involving the so-called 'strange' and 'charmed' quarks, and the heaviest the 'bottom' and 'top' quarks. Like the heavy leptons, these particles can be produced in high-energy experiments (and must have existed in profusion in the *Big Bang*), but quickly decay into their lighter counterparts. Although it is impossible to isolate a single quark, particle accelerator experiments have shown direct evidence for all six members of the quark family; the last (top) was identified in 1994 by researchers at Fermilab, in Chicago.

Studies of the masses and other properties of quarks show that there cannot be any more generations, and that there are only three families of quarks and three families of leptons. Happily, the standard model of the Big Bang also says that there can be no more than three generations of particles; if there were, the pressure caused by the extra neutrinos in the very early Universe would have made it expand too quickly, and this would leave it with the wrong amount of helium to match observations of very old stars (see *alpha beta gamma theory*, *nucleosynthesis*). This is one of the neatest pieces of evidence that the standard models of both particle physics and cosmology are telling us something close to a fundamental truth about the way the Universe works.

Apart from the early moments of the Big Bang, however, the second and third generations of particles have played essentially no role in the evolution of the Universe or the behaviour of its contents. Everything we can see in the Universe can be described in terms of two quarks (up and down) and two leptons (the electron and its neutrino); indeed, since individual quarks cannot be isolated, it is still a very good approximation to describe the behaviour of everything we can see in terms of the electron, neutron and proton, with the addition only of the electron neutrino to the list known in 1932, and the four fundamental forces.

But there may be more to the Universe than we can see; there are both observational and theoretical reasons to think that there is much more *dark matter* in the Universe than there is bright stuff. A great deal of this matter may be in the form of fundamental particles which are neither hadrons nor leptons. But that is another story.

Further reading: Frank Close, *The Cosmic Onion*; John Gribbin, *In Search of the Big Bang*.

elliptical galaxy A *galaxy* which looks like an elliptical or circular patch of light on the sky, with no evidence of a surrounding disc of stars. It used to be thought that such galaxies have the shape of an American football, but studies of the orbital speeds of stars in these galaxies (using *spectroscopy* and the *Doppler effect*) have shown that in general all three axes of the ellipsoid (three-dimensional ellipse) have different lengths.

Elliptical galaxies are mainly composed of old, red stars; although they do contain considerable amounts of dust and gas between those stars, there is little sign of active star formation going on in these galaxies today. Because of this, it used to be thought that ellipticals were the oldest galaxies, and have remained largely unchanged since early in the life of the Universe; but more recent studies suggest that many ellipticals (perhaps all the large ones) have formed from collisions and mergers between *disc galaxies* (see *galaxy formation and evolution*). *Starburst galaxies* may show this process at work.

The masses of elliptical galaxies range from about a million times the mass of the Sun in dwarf ellipticals (which resemble the *globular clusters* in our own Galaxy) to as much as 10^{12} times the mass of the Sun in giant ellipticals, the largest galaxies known. Allowing for the many dwarf ellipticals that cannot be seen at great distances, about 60 per cent of all galaxies are ellipticals. They are the most common members of *clusters of galaxies*, and in very rich clusters there is usually a massive, gravitationally dominant giant elliptical at the centre. The most powerful *radio sources* are associated with large elliptical galaxies, which suggests that they may harbour *black holes* in their hearts. (See picture p. 162.)

emission line A narrow feature in a *spectrum*, corresponding to emission of *electromagnetic radiation* at a well-defined wavelength. The pattern of

Elliptical galaxy. A typical elliptical (lens-shaped) galaxy, photographed using the 60-inch reflector on Mount Wilson on Christmas Day, 1911.

emission lines in a spectrum is like a fingerprint identifying the *elements* that are producing the radiation.

emission nebula   A cloud of gas and dust in space which emits light. In most cases, such as the *Orion nebula*, this is because the cloud is being heated by *radiation* from a nearby hot, young star (or stars). See also *planetary nebula*.

Empedocles (about 490–430 BC)   Greek philosopher who was one of the earliest proponents of the idea that everything is made up of four 'elements': fire, air, water and earth.

energy   Everybody knows what energy is, but most definitions seem to go round in circles. To a scientist, the energy of a system measures its capacity to 'do work', which in effect means making a change in some other system.

The most important thing about energy is that it can be neither created nor destroyed (provided we include mass as a form of energy, in line with Einstein's equation $E = mc^2$). It can only be turned into another form of energy.

In terms of an everyday journey in a car, we start with energy stored in a chemical form in the fuel tank. This energy is released to 'do work' when the fuel burns explosively in the cylinders of the engine, driving pistons which in turn drive the wheels and make the car move. In the moving car, the energy is now in the form of kinetic energy. When the car stops, the kinetic energy is turned back into heat energy in the brakes (at a lower temperature than the heat of the burning fuel in the engine), and by the friction in all of the moving parts of the vehicle. This low-grade heat energy is no longer useful for doing work (see *arrow of time*).

In the Universe today, the most important trade-off is between gravitational energy and kinetic energy. A stone dropped from your hand loses gravitational energy and gains kinetic energy as it falls; a rocket moving upward trades chemical energy for kinetic energy for gravitational energy as it moves higher, and it cannot reach *escape velocity* if it does not have enough energy to trade. The trade-off between kinetic energy and gravitational energy determines the way planets and stars move in their orbits, and how gravity holds galaxies, clusters of galaxies and the Universe itself together (see *fate of the Universe*).

Electromagnetic radiation, such as sunlight, is also a form of energy. Indeed, the energy stored in the fuel in your car came originally from sunlight, trapped by plants during photosynthesis. In the very early Universe, the trade-off was between radiant energy and mass energy, and this determined the amount and kind of material particles (chiefly nuclei of hydrogen and helium) that went into the first generation of stars.

epicycle   The *orbits* of the *planets* can be described solely in terms of circles, with a planet moving around a small circle called an epicycle, while the centre of the epicycle moves around the Earth or the Sun in a circle (if necessary, the trick can be extended so that the epicycles themselves follow epicycles). The device was rendered obsolete when *Kepler* showed that it is simpler to describe the orbits of the planets as ellipses, with the Sun at one focus of the ellipse.

Epsilon Aurigae   An *eclipsing binary* which has been studied since 1821 and has been a testbed for astrophysics. The system is about 600 *parsecs* away, in the direction of the constellation Auriga. The binary period is 9,892 days (more than 27 years), and the system consists of a bright F *supergiant* about fifteen times the mass of our Sun, with a large, dark companion of about the same mass, probably a disc of material around a *main sequence* B star. Each *eclipse* lasts for about 610 days.

equinox The moment in the Earth's *orbit* when the Sun seems to cross the *celestial equator*, and the day and night are the same length, everywhere on Earth. The spring (or vernal) equinox occurs on 21 March; the autumn equinox occurs on 23 September (the names were given by chauvinistic astronomers in the Northern Hemisphere; the seasons are reversed in the Southern Hemisphere).

equipotential surfaces The gravitational equivalent of contour lines. Contours are lines of equal height on the surface of the Earth; equipotential surfaces are (imaginary) surfaces where the strength of *gravity* is constant (in fact, this means that contours could also be described as 'equipotential lines', since the strength of the Earth's gravity is indeed the same every-where along a contour line).

For a single *star*, the equipotential surfaces are simply a series of con-centric spherical shells around the star. But in a close *binary system*, the gravitational fields of the two stars interact and make equipotential surfaces with an hour-glass shape. One critical equipotential surface forms two lobes, one around each star, which meet at a point somewhere between the two stars. These are known as the Roche lobes; the point of contact between the two Roche lobes is known as the inner Lagrangian point, and its exact position depends on the masses of the two stars and their distance apart.

If both stars in the binary sit within their Roche lobes, they form a detached system. If one star expands to fill its Roche lobe (see *stellar evolution*), matter will escape through the inner Lagrangian point and fall on to the other star. It is then a semi-detached system. The stream of material falling on to the other star will probably form an *accretion disc*, and will give rise to various kinds of activity, including *dwarf novas* and *X-ray* emission. If each star in the binary fills its own Roche lobe, the system is a contact binary, and matter can escape across the critical equipotential surface into space.

equivalence principle That the effects of acceleration are indistinguishable from the effects of a uniform gravitational field. This equivalence results from the equivalence between *gravitational mass* and *inertial mass*. It led Albert Einstein to the development of his *general theory of relativity*, when he realized that a person falling from a roof would not feel the effects of gravity – the acceleration of their fall would exactly cancel out the feeling of weight.

In modern language, the equivalence is best described in terms of a spaceship being accelerated through space by constant firing of its rocket motors. When the motors are not firing, everything inside the spaceship floats about in *free fall*, just as weightless as the person falling from a roof. In principle, the acceleration of the rocket could be adjusted so that everything inside felt a force exactly as strong as the force of gravity on

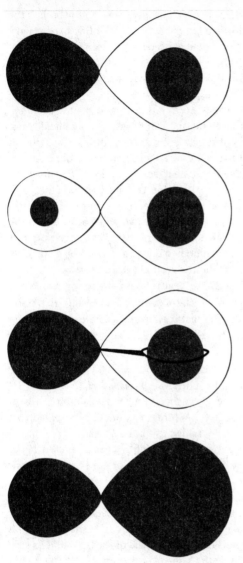

Equipotential surfaces. Four kinds of binary star and their Roche lobes. *Top:* A semi-detached binary, with one star filling its Roche lobe. *Second:* A detached binary. *Third:* Material streaming from a star filling its Roche lobe onto its companion in a semi-detached system. *Bottom:* A contact binary.

Earth (or any other strength you chose), pushing things to the back of the vehicle as it moved forward through space. Any scientific experiments carried out in this accelerating *frame of reference* (for example, studies of the swing of a pendulum, or the way balls roll across the floor) would give exactly the same results as if the spaceship were standing on its launch pad on Earth, and not accelerating at all.

There is one important caveat. Acceleration is equivalent to a *uniform* gravitational field. Strictly speaking, the Earth's gravitational field is not uniform, because it spreads out from a point at the centre of the Earth. If you were sealed in a lift in free fall down a shaft drilled to the centre of the Earth, you would be able to tell that you were not falling freely in space by careful observations of any two objects sharing the lift with you – perhaps a pair of oranges. As you moved closer to the centre of the Earth, the oranges would move closer together, on converging paths. But at the surface of the Earth, the equivalence between gravity and acceleration is very nearly perfect. This is why nobody has ever been hurt by a fall, only by the sudden end to the fall when the ground gets in the way.

Eratosthenes of Cyrene (about 273–192 BC)   A polymath who worked as the chief librarian of the great library in Alexandria and as tutor to the son of Ptolemy III. He carried out the first reasonably accurate calculation of the size of the spherical Earth. He used measurements of the angular height of the Sun above the horizon at different places at the same time to work out that the circumference of the Earth is about 29,000 miles. The modern figure is just under 25,000 miles. See *Anaxagoras*.

EROS   Acronym for Expérience de Recherche d'Objets Sombres, a two-pronged French project to search for *MACHOs* by studying photographs of a large part of the southern sky obtained using the *Schmidt camera* at the *European Southern Observatory*, and also using a *charge coupled device* system concentrating on the dense central region of the *Large Magellanic Cloud*.

escape velocity   The minimum speed needed for an object to keep moving away from a planet or star, instead of either falling back to its surface or entering a closed *orbit* around it. The escape velocity at the surface of a body such as the Earth depends only on the mass of the body and its size. So the escape velocity would be the same (in this case, 11.2 km/sec) for a stone thrown vertically into the air by Superman or for a rocket being launched from Cape Canaveral. An object in orbit around the Earth, or the Sun, or any other massive object could escape if it were given a sufficient boost in speed. The escape velocity at any distance r from the centre of a mass m is given by the square root of $2Gm/r$, where G is the *gravitational constant*.

The same formula, with r set as its radius, gives the escape velocity from the surface of a massive body. Although very large objects such as Jupiter or the Sun contain much more mass than the Earth, and have a stronger gravity overall, this mass is spread across a larger volume, so that the surface of Jupiter or the Sun is much further from its centre than the surface of the Earth is from the centre of the Earth. The force of gravity falls off with increasing distance from the centre of a massive body, so this dilutes the strength of gravity at the surface of a massive object, and tends to reduce the escape velocity (because of the 1/r in the formula). At the same time, the extra mass tends to increase the escape velocity. The mass (*m* in the formula) will be roughly proportional to the volume of the object, assuming it has a roughly uniform density, which is proportional to the radius cubed. The overall effect is that the surface escape velocity is proportional to the radius for different-sized objects with the same density.

The balance of these two effects means that (after allowing for their different densities as well) the escape velocity from the surface of the Moon is 2.4 km/sec, while from the surface of Jupiter it is 61.1 km/sec. The escape velocity from the surface of the Sun is 624 km/sec, just 0.2 per cent of the speed of light. Because the surface escape velocity only increases in proportion to the radius, you would need an object 500 times bigger than our Sun, as big as our Solar System, with the same density as the Sun, before the escape velocity became greater than the speed of light, and you had a *black hole*.

But if you could squeeze an existing massive object to make it smaller, its surface would be closer to its centre, and the escape velocity would be increased even though the mass stayed the same. In this way, the escape velocity would exceed the speed of light and the Sun would become a black hole if it could be squeezed into a ball with a radius of just 2.9 km, its *Schwarzschild radius*. This is not one-five-hundredth of its present radius, but the *square* of one-five-hundredth, or 0.000004, of its present radius.

ether The hypothetical medium through which it used to be thought that light and other electromagnetic radiation must travel, in the same way that sound waves travel through the air. The equations developed by James Clerk *Maxwell* showed that there is no need for an ether, because electromagnetic waves are self-propagating; but it was only after the failure of the *Michelson–Morley experiment* to detect any motion of the Earth through the ether, and the success of the *special theory of relativity* in explaining the behaviour of light, that the notion of the ether was finally discarded.

Euclid Greek mathematician who lived around 300 BC. Very little is known about his life, and he does not seem to have made original contributions to mathematics, but his name is remembered for an epic book in thirteen volumes, called *Elements of Geometry*, in which he gathered together and set down everything known in mathematics at the time in a logical and systematic way.

Euclidean geometry   The kind of geometry you learn in school, where the angles of a triangle always add up to 180° and parallel lines stay the same distance apart forever.   This system of geometry was not invented (or discovered) by *Euclid*, but he gathered the knowledge of his time (about 300 BC) together and wrote it all down.   His writings were translated into Arabic and from Arabic into Latin, surviving centuries of turmoil in Europe to become the archetype for the development of mathematics and then other sciences.   In many places, Euclidean geometry formed the basis of mathematical teaching in schools until well into the 20th century.

Euclidean geometry is not wrong, but it has limited applications. Strictly speaking, it applies only to a flat surface, like a flat piece of paper. On the surface of a sphere, for example (or the surface of the Earth), the angles of a triangle add up to more than 180°, and the bigger the triangle is, the more the angles add up to. But Euclidean geometry can be extended into three or more dimensions, to describe relationships in the three dimensions of ordinary space (called 'flat' space by analogy with a flat surface), or in the four-dimensional flat *spacetime* of the *special theory of relativity*.

In order to describe curved spacetime within the framework of the *general theory of relativity*, however, you need to use *non-Euclidean geometry*. Even here, though, Euclidean geometry is useful when describing small pieces of spacetime in which the curvature is negligible (see *geodesic*).

Europa   One of the four large *moons* of Jupiter.   Smaller than our own Moon, it has a diameter of 3,138 km and a *density* of 2.97 g/cubic cm, nearly three times that of water.   It is thought to be made of a rocky core covered with a thick layer of ice.

European Southern Observatory (ESO)   An organization involving eight European states which runs an observatory at La Silla, near La Serena, in Chile.   The two main *telescopes* have *apertures* of 3.5 m and 3.6 m.   The observatory is at an altitude of 2,400 m in a region of dry climate with clear skies.

European Space Agency (ESA)   An organisation of twelve European states which co-operate in all aspects of the development of *satellites* for scientific and commercial use.   It has centres at Noordwijk in Holland, Darmstadt and Cologne in Germany, and Frascati in Italy, and a launch site at Kouru in French Guiana.   ESA satellites are launched using the Ariane rocket.

evening star   See *Venus*.

event horizon   The imaginary surface surrounding a *black hole* on which

The European Southern Observatory (ESO), at La Silla, in Chile. The tallest building houses a 3.6-metre reflecting telescope.

the *escape velocity* is equal to the *speed of light.* Within the event horizon, the escape velocity would be greater than the speed of light. Because nothing can travel faster than light, nothing (not even light) from within the event horizon can cross this surface and escape into the outside Universe, which is what gives black holes their name. But there is nothing to prevent matter or radiation from the outside Universe crossing the horizon on its way into the black hole; it is not a physical barrier like a wall.

If the black hole is not rotating, the event horizon is a spherical surface with radius equal to the *Schwarzschild radius* of the hole, centred on the *singularity* at the heart of the black hole. If the black hole is rotating, however, the event horizon is distorted, in effect being pushed outward at the 'equator' by the rotation.

exclusion principle See *Pauli exclusion principle.*

exobiology Entirely theoretical (as yet) studies of the kind of living things that might exist elsewhere in the Universe – not on planet Earth. Nobody is a full-time exobiologist, but scientists from many disciplines have dabbled in this area.

One school of exobiological thought concentrates on the possibility of life forms based on the same chemical constituents as ourselves (see

CHON) living on planets rather like the Earth (see *Drake equation*). The fact that we are made of the most common *elements* in the Universe, and the strong likelihood that planets like the Earth exist in profusion in our *Galaxy*, makes this a respectable, if speculative, branch of science.

Some exobiologists, though, lean more towards science fiction (indeed, some of them advise science fiction writers on how to 'create' aliens, and some of them write their own science fiction novels). They point out that the only essential requirement for life to exist is a source of energy which can be used (temporarily) to get round the *second law of thermodynamics* (see *arrow of time*). If this is correct, then in principle it will be possible for life to exist almost anywhere in the Universe, even (at one extreme) in the form of a tenuous cloud of material in space, feeding off the feeble photons of starlight, or (at the other extreme) as tiny, dense creatures on the surface of a neutron star, feeding off the intense magnetic fields.

See also *life in the Universe*; for respectable novelizations of some of these ideas by scientists, see *The Black Cloud*, by Fred Hoyle, and *Dragon's Egg*, by Robert Forward.

EXOSAT   An *X-ray astronomy satellite* launched by the *European Space Agency* in 1983. The satellite weighed 400 kg and made many observations over a three-year lifetime, including studies of *binary stars*, *supernova remnants*, *bursters* and *active galaxies*.

expanding Universe   The Universe in which we live is expanding as the space between clusters of galaxies stretches. This was predicted by the equations of the *general theory of relativity* in 1916, but even Albert Einstein did not accept what the equations were trying to tell him at first. It was the discovery of the *redshift* in the light of galaxies beyond the *Local Group*, by Edwin *Hubble* and his colleagues at the end of the 1920s, which made astronomers realize that the Universe is expanding, and led to the acceptance of the general theory of relativity as a good description of the Universe at large.

Because clusters of galaxies are moving apart now, they must have been closer together in the past. Look back far enough into the past, and they must have been on top of one another. This is the basis of the *Big Bang model* of the Universe.

It is a key feature of the expanding Universe, however, that the expansion is not caused by the galaxies moving through space, like the fragments of a bomb moving through space away from the point at which the bomb has exploded. Space itself is expanding, and carrying galaxies along for the ride.

Although there is no doubt that our Universe is expanding, there are several different models, even within the framework of the general theory of relativity, which describe different kinds of expanding universe. It is not yet clear exactly which of these models best describes our Universe, so although its origin in a Big Bang is well established, its ultimate fate is not known with such certainty (see *fate of the Universe, cosmological models*).

extraterrestrial life   See *life in the Universe.*

Extreme Ultraviolet Explorer (EUVE)   A *satellite* launched by NASA in 1992 to study the Universe at wavelengths between 7 and 76 nanometres, in the extreme ultraviolet part of the *electromagnetic spectrum*. These wavelengths, close to the *X-ray band*, are especially useful in the study of *white dwarfs, cataclysmic variables* and *interstellar matter.*

Faber–Jackson relation   A way of estimating the *luminosity* of an *elliptical galaxy*, which gives an indication of its distance by comparing this with its apparent brightness.

Astronomers are always looking for new ways to estimate distances across the Universe, and thereby to improve their understanding of the *cosmic distance scale*. In 1976 the American researchers S. M. Faber and R. E. Jackson discovered that there seems to be a correlation between the brightness of an elliptical galaxy and the spread in velocities of its stars, indicated by the *Doppler effect*.

The relationship they discovered is a logical example of the trade-off between different kinds of energy in such a galaxy. Brighter galaxies than average contain more stars (which is why they are brighter), and therefore have more mass, so they have more gravitational energy. As a result, they are able to hold on to stars which have more kinetic energy than average, which means higher velocities. So the bigger the measured spread of velocities, the brighter the galaxy must be.

The relation is calibrated for galaxies whose distances have been estimated in other ways, and then used to estimate the distances to other galaxies. The technique is not a precise measure of distances, but it is a useful indicator, especially for clusters of galaxies where some of the distances can be estimated in other ways (see *Tully–Fisher relation*). It gives a distance for the important *Virgo cluster* of about 15 million *parsecs* (15 Mpc).

falling star   See *meteor.*

fate of the Universe   See *cosmological models.*

Fermi, Enrico (1901–54)   Italian-born physicist who made pioneering studies of *radioactivity* and is best remembered as the leader of the group that built the first nuclear reactor (then, in 1942, called an 'atomic pile'). He is known in astronomy for his claim that the fact that the Earth has not been colonized by aliens 'proves' that we are the only civilization in the Galaxy. See *Fermi Paradox.*

fermions   *Elementary particles* that are conserved during particle interactions, and which behave in accordance with statistical rules developed in the 1920s by Enrico *Fermi* (1901–54) and Paul Dirac (1902–84), known as 'Fermi–Dirac statistics'. The archetypal fermion is the **electron**. An electron can never be created or destroyed in isolation in a particle interaction; if an electron is produced (for example, in **beta decay**), it must always be accompanied by an appropriate *antiparticle*. So the total number of fermions in the Universe is always the same, and was determined by conditions that operated in the **Big Bang**.

Fermions are what we are used to thinking of as particles – the electrons, protons and neutrons that make up 'solid matter', together with more exotic unstable particles that can be produced in energetic interactions at particle accelerators or in energetic astronomical sources.

Even these particles can, however, also be described in terms of waves. This is an example of the wave–particle duality of the quantum world, which says that entities such as photons or electrons can be described either in terms of waves or as particles.

In the language of quantum physics, the key property of a fermion is its spin. All fermions have a spin which is 'half-integer' – 1/2, 3/2, 5/2 and so on. This is analogous to the spinning of a child's top – but not exactly like the spinning of a child's top, because a particle such as an electron, for example, has to 'rotate' *twice* in order to get back to where it started.

See **bosons**.

Fermi Paradox   'If they existed, they would be here.' The words used by Enrico *Fermi* to 'prove' that there are no intelligent life forms in our *Galaxy* except ourselves. Although Fermi popularized the 'space travel argument' for the non-existence of other intelligences, it has been used by many other people, and dates back at least to the 17th century, when it was used to prove that there is no intelligent life on the Moon.

The essence of the argument is that any intelligent species which develops to our present level of civilization will, within a few centuries, develop rocket technology sufficient for them to begin to spread across the Galaxy, either as passengers in those rockets or by sending computerized robot systems to do the exploring for them.

Robot spaceprobes would be a cheap and effective means of exploration, because a robot that arrived in a 'new' planetary system could use raw materials from **comet** and **asteroid** belts to build new probes (replicas of itself) to send on to explore other planetary systems. A combination of computers and rockets slightly more sophisticated than we have today would enable any intelligent species to explore the entire Galaxy, even restricted to speeds that are only a fraction of the **speed of light**, in less than about 300 million years. Our Solar System is already nearly 5 billion years old, and the whole Galaxy is at least twice as old. So – 'if they existed, they would be here'.

This is not really a paradox, but an opinion; it can be refuted by, for example, asking whether any intelligent species would bother to look at every planet, or whether they would let us know they were watching us. See *Drake equation*.

field equations/field theory   The idea of a field as the means by which a force is transmitted across space goes back to Michael Faraday's investigations of electricity in the 19th century. Faraday's 'lines of force', familiar from schoolday physics lessons, could be thought of as mathematical lines stretching out from every charged particle in the Universe, every one starting on a charge with one flavour (positive or negative) and ending on an opposite charge (negative or positive). Together, the lines of force make up a field of force around a charged particle, which is described mathematically by field equations.

Like stretched rubber bands, the lines of force tend to pull opposite charges together; but like squeezed elastic blocks, the bunched up lines of force between similarly charged particles tend to keep them apart.

James Clerk *Maxwell* elaborated this idea into the first field theory, describing electricity, magnetism and *electromagnetic radiation*. The fields

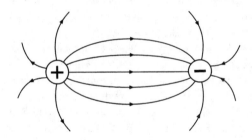

Lines of force. The concept of lines of force (in this case, always pointing outwards from positive charge and in towards negative charge) helps to explain why opposite charges attract and like charges repel.

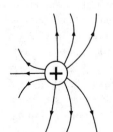
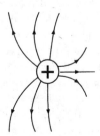

described by Maxwell's equations are 'classical' fields, regarded as smooth and continuous, with no gaps between lines of force and no breaks in the lines of force themselves. The *general theory of relativity* is also a classical field theory, describing gravity in equivalent terms.

The modern version of field theory is quantum field theory, in which *energy* comes in discrete lumps, called quanta. *Photons* are the quanta of the electromagnetic field. Particles such as *electrons* are described by field equations corresponding to waves which fill the entire Universe (a matter field), and the particles (in this case, the electrons) are energetic bits of the quantum field.

By applying the quantum rules once to electrons (or other particles), we arrive at a description in terms of a matter field; by applying the quantum rules again ('second quantization'), we recover the idea of particles as little lumps of mass-energy. In quantum field theory, there is nothing else in the Universe except quantized quantum fields, each described by the appropriate field equations.

fifth force   See *fundamental forces*.

first three minutes   See *Big Bang*.

Fisher–Tully relation   See *Tully–Fisher relation*.

fission and fusion   The processes by which a heavy *nucleus* splits into two or more fragments (fission) and by which light nuclei join together (fusion), each releasing *energy* as they take place.

Both fission and fusion release energy because it happens that the amount of energy per *nucleon* in a nucleus is least for *elements* with the number of *protons* per nucleus near the middle of the range of stability. This corresponds to nuclei of iron-56, which contain 56 nucleons (26 protons and 30 *neutrons*). All natural systems 'like' to be in a state of minimum energy (like water flowing downhill), so it is easier to encourage nuclear reactions which move nuclei towards iron-56 than those which move them away from iron-56.

You can think of this as a valley of stability, with light elements up one side and heavy elements up the other, and iron at the bottom. Each element sits on a little ledge up the sides of the valley. On one side, hydrogen (at the top of the 'valley', with just a single proton in its nucleus) can be encouraged to form helium (two protons and two neutrons) by fusion, jumping down one step towards the valley floor, and other elements are built up successively inside stars by further fusion. On the other side of the valley, some heavy nuclei (originally created in *supernovae*) split spontaneously, jumping down the steps in the valley side to make lighter elements. In other cases (or for the same radioactive elements), the heavy nuclei will be induced to split if they are struck by a suitably energetic stray neutron.

Since fission usually releases two or three additional neutrons, a substance such as uranium-235 (in which each nucleus contains 92 protons and 143 neutrons) can be used to create a chain reaction, in which the fission of one nucleus triggers the fission of one or more additional nuclei, and so on.

See also *nucleosynthesis*.

Fitzgerald, George Francis (1851–1901)   Irish physicist best known for his suggestion that moving objects are shortened in the direction of their motion (the *Fitzgerald contraction*).   See *special theory of relativity* and *Lorentz*.

Fitzgerald contraction   The shrinking of a moving object in the direction of its motion. It was first proposed by the Irish physicist George *Fitzgerald*, to explain the failure of the *Michelson–Morley experiment* to measure any change in the *speed of light*, regardless of which way the light is moving relative to the Earth.   Fitzgerald argued that this could be because the entire experimental apparatus, and the Earth itself, shrinks in the direction of motion.   The shrinkage would give us the illusion that the speed of light is a constant.

A couple of years later, the same idea was put forward independently by Hendrik *Lorentz*; rather unfairly, it is sometimes referred to as the Lorentz–Fitzgerald contraction, in spite of Fitzgerald's priority both alphabetically and historically.   Both Fitzgerald and Lorentz, however, imagined that there was some absolute 'frame of reference', defined by a hypothetical substance known as 'the ether', through which the Earth was thought to move.   Light was imagined as a wave in the ether analogous to waves on the sea, and it was thought to be motion relative to the ether which caused the Fitzgerald contraction.

At the beginning of the 20th century, Albert Einstein showed, with his *special theory of relativity*, that there is no special frame of reference and no need for an ether.   The Fitzgerald contraction occurs for any relative motion, so that any observer (who is entitled to say that they are at rest) will measure this shrinkage for any object moving relative to the observer. The effect only becomes detectable for speeds that are a sizeable fraction of the speed of light, but it has been confirmed by many experiments involving high-speed particles.

See also *Lorentz transformations*.

fixed stars   Old term for stars in general, which were thought to have no motion across the sky, by contrast with the *planets*, which were known as wandering stars.   The term has been known to be incorrect since the early 18th century, when the first *proper motions* were measured, but it is still sometimes used.

Flamsteed, John (1646–1719)   English astronomer who became the first Astronomer Royal in 1675. The *Royal Greenwich Observatory* was set up for

Flamsteed to prepare accurate astronomical tables (important in navigation), which led to a catalogue of nearly 3,000 star positions, measured to an accuracy of 10 *arc seconds*, published six years after Flamsteed died.

flares   Sudden brightenings of small regions of the visible surface of the Sun, which flare up in the space of a few minutes then fade away over about an hour.   As well as releasing energy in the form of light and other electromagnetic radiation, flares produce streams of particles such as protons and electrons, some of which travel across space to the Earth, where they interact with charged layers high in the Earth's atmosphere and can disrupt radio communications.   These particles (part of the *solar wind*) can also affect the orbits of artificial satellites, and could pose a health hazard to any astronauts in space at the time.   The arrival of the charged particles from the flare at the Earth also causes bright auroral displays (the northern and southern lights).

Flare activity is associated with the Sun's magnetic field, and varies during the course of the *solar cycle*; flares are much more common when there are more *sunspots* visible on the face of the Sun.   There is some evidence that flares affect the weather on Earth; unfortunately the data are not conclusive, and most meteorologists are cautious about taking this evidence at face value.

flatness parameter   See *density parameter.*

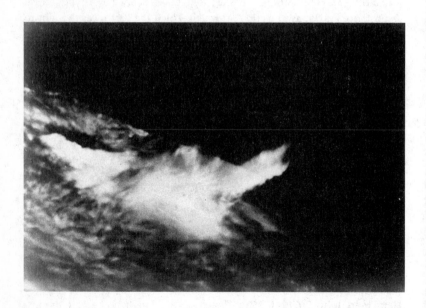

Flares. A flare on the surface of the Sun, ejecting material into space from a region of activity.

flatness problem   The puzzle that the *spacetime* of the Universe is very nearly *Euclidean*, so that the Universe sits just on the dividing line between eternal expansion and eventual recollapse. The problem is resolved by the *inflationary model*. See also *cosmological models*.

forces of nature   See *fundamental forces*.

Foucault's pendulum   Any long pendulum which swings freely enough to demonstrate the rotation of the Earth. It was first demonstrated in this way in 1851, by Jean Foucault, a French physicist who carried out many experiments with light and made the first accurate measurements of the speed of light in the laboratory. It was while he was developing an accurate timekeeping device for use in his light experiments that Foucault noticed that a swinging pendulum stayed swinging in the same plane even when he rotated the apparatus it was attached to. He realized that this persistence of the swing would keep the pendulum swinging in the same plane while the Earth itself rotated underneath it, so that it would seem as if the plane of the pendulum was slowly drifting round relative to the ground.

At the poles, a Foucault pendulum will seem to rotate completely once in 24 hours, or by 15 degrees of arc per hour. At the equator, there is no rotation effect, and there are intermediate rates of change at intermediate latitudes, in between the equator and the poles. So as well as demonstrating that the Earth rotates, a Foucault pendulum could be used to find out what latitude you are at. The pendulum actually keeps swinging in the same plane relative to the *fixed stars* – an example of *inertia*, and a demonstration of the kinds of puzzle addressed by *Mach's Principle*.

Fowler, William Alfred ('Willy') (1911–95)   American physicist who contributed to the development of an understanding of the way *elements* are manufactured inside *stars* (*nucleosynthesis*) and in the *Big Bang*.

Fowler was born in Pittsburgh, Pennsylvania, on 9 August 1911, and graduated from Ohio State University in 1933. He obtained his PhD from the California Institute of Technology in 1936, and stayed there for the rest of his career (apart from wartime work in Washington and elsewhere), studying in the laboratory (using particle accelerators) the kind of *nuclear fusion* reactions that are important inside stars. He measured the actual probabilities (known as 'cross-sections') of the reactions proposed by Hans *Bethe* for the *carbon cycle* and the *proton–proton reaction*, and in the 1950s joined forces with Fred *Hoyle* and Geoffrey and Margaret *Burbidge* in a more comprehensive investigation of processes that go on in inside stars (see *B²FH*). This explained how all the elements more massive than helium can be manufactured inside stars. Fowler later refined the original idea of the proton–proton reaction into its modern form, sometimes called the proton–proton tri-chain.

In the 1960s, Fowler worked with Hoyle and Robert Wagoner on a

Foucault's pendulum. A contemporary engraving of the pendulum used by Foucault in 1851 to demonstrate the rotation of the Earth.

similar investigation of the kind of nuclear reactions that must have occurred in the Big Bang itself, refining the pioneering work of George **Gamow** to show that exactly the amount of helium and other very light elements (such as **deuterium**) seen in old stars should indeed have been produced in the Big Bang.

Although Hoyle was undoubtedly the leader of both the B^{2}FH team and the Wagoner, Fowler and Hoyle team, contributing the key idea of the resonance process by which carbon is made out of helium inside stars, to the astonishment of the astronomical community Fowler alone received a Nobel Prize for this work, in 1983. At the time, Fowler had published more than 200 scientific papers, of which 25 (about 10 per cent) carried Hoyle's name as well; it is unfortunate that the Nobel Committee did not see fit to keep the two names together in 1983, although Fowler certainly deserved this recognition.

Asked about the importance of his prizewinning work, Fowler said, 'If there turn out to be practical applications, that's fine and dandy. But we think it's important that the human race understands where sunlight comes from' (Robert Weber, *Pioneers of Science*, Adam Hilger, 1988). He died on 14 March 1995, at the age of 83.

frame of reference   See *inertial frame.*

Fraunhofer, Josef von (1787–1826)   See *spectroscopy.*

Fraunhofer lines   See *spectroscopy.*

Fred L. Whipple Observatory   Observatory run by the Smithsonian Institution, and formerly known as the Mount Hopkins Observatory. It is located on Mount Hopkins, in Arizona, at an altitude of 2,600 m. The main *telescopes* are the *Multiple Mirror Telescope* (now being converted into a 6.5 m *reflecting telescope*), a 1.5 m reflector and a *gamma ray* detector.

free fall   A state of weightlessness experienced by anything that is moving only under the influence of gravity. A person falling off a tall building on Earth is not really in free fall, because the resistance of the air also affects their motion. But any object in *orbit* around another object (such as the planets moving around the Sun), or moving on an unpowered trajectory between the stars, is in a state of free fall. Astronauts in a spaceship orbiting around the Earth are also weightless and in free fall. Each astronaut is, strictly speaking, in his or her own orbit, and the spaceship itself is in its own orbit, but the astronauts' orbits keep them inside the spaceship.

Some books describe this as not being true weightlessness, because the astronauts (and their spacecraft) are still being tugged on by the Earth's gravity. But those books are wrong. The effect of gravity is precisely cancelled out by the acceleration of a freely falling object (see *equivalence principle*), and orbiting astronauts are both freely falling and literally weightless. The fall does not carry them down to Earth, however, because the acceleration produced by the fall combines with their forward motion to keep them in a closed orbit.

Weightlessness can also be induced briefly inside an aircraft which flies on a carefully planned parabolic trajectory; this technique is used in the training of astronauts to give them a taste of free fall.

free lunch universe   The idea that the Universe may have appeared out of nothing at all, and contains zero *energy* overall, making it, in the words of Alan *Guth* 'the ultimate free lunch'. This is closely linked with the idea of *inflation.*

The idea developed from an unsigned commentary in *Nature* (written by John Gribbin) in 1971 (volume 232, page 440), which suggested that the Universe could be described as the inside of a *black hole*. The speculation was developed further by R. K. Pathria, of the University of Waterloo, Ontario (*Nature*, volume 240, page 298), and then by Edward Tryon, of the City University in New York, who suggested that this black hole Universe might have appeared out of nothing as a so-called vacuum fluctuation, allowed by *quantum theory* (*Nature*, volume 246, page 396).

Quantum uncertainty allows the temporary creation of bubbles of *energy*, or pairs of particles (such as *electron–positron* pairs) out of nothing, provided that they disappear in a short time. The less energy is involved, the longer the bubble can exist. Curiously, the energy in a gravitational field is negative, while the energy locked up in matter is positive. If the Universe is exactly flat (see *cosmological models*) then, as Tryon pointed out, the two numbers cancel out, and the overall energy of the Universe is precisely zero. In that case, the quantum rules allow it to last forever.

If you find this mind-blowing, you are in good company. George *Gamow* told in his book *My World Line* (Viking, New York, reprinted 1970) how he was having a conversation with Albert *Einstein* while walking through Princeton in the 1940s. Gamow casually mentioned that one of his colleagues had pointed out to him that according to Einstein's equations a star could be created out of nothing at all, because its negative gravitational energy precisely cancels out its positive mass energy. 'Einstein stopped in his tracks,' says Gamow, 'and, since we were crossing a street, several cars had to stop to avoid running us down.'

Unfortunately, if a quantum bubble (about as big as the *Planck length*) containing all the mass-energy of the Universe (or even a star) did appear out of nothing at all, its intense gravitational field would (unless something else intervened) snuff it out of existence immediately, crushing it into a *singularity*. So the free lunch Universe seemed like an irrelevant speculation – until the development of the inflationary scenario showed how such a quantum fluctuation could expand exponentially up to macroscopic size before gravity could crush it out of existence. See also *monopole universe*.

frequency   The number of oscillations of a vibrating system that occur in a certain time, usually in 1 second. For a wave, such as *light* or other *electromagnetic radiation*, the frequency is the number of waves that pass a point in each second, and is measured in *Hertz* (= cycles per second). For electromagnetic radiation, the *wavelength* is equal to the frequency multiplied by the *speed of light*.

Friedmann, Aleksandr Aleksandrovich (1888–1925)   Russian mathematician who worked in practical applications such as hydromechanics and meteorology, but who is best remembered for his solutions to Albert Einstein's equations of the *general theory of relativity*, which showed that the Universe need not be static and which provided the basis for the development of *cosmological models*.

Friedmann's life was short but full of incident. He was born in what was then (and is now once again) St Petersburg, on 29 June 1888, and studied mathematics at the university there from 1906 to 1910. He became a member of the mathematics faculty at the university, specializing in theoretical meteorology, and served as a volunteer in the Russian air force as a technical expert making weather observations at the front during the First World War. This involved many flights over enemy territory as an observer, leading to at least one crash landing and the award of the George Cross to Friedmann for valour.

He lived through the turmoil of the revolution of 1917 to become a full professor at Perm University, but he was caught up in the civil war and had to flee when the city was overrun by the White Russians (Friedmann supported the revolution, and had been active in left-wing politics since his schooldays). After the Bolsheviks recaptured Perm, Friedmann helped to reorganize the university in difficult circumstances, before returning to Petrograd (as it then was) in 1920 to carry out research in meteorology at the Academy of Sciences and the Main Geophysical Observatory. He was soon in charge of weather observations throughout the Soviet Union. By the time he died, in 1925, the city had been renamed Leningrad.

According to some sources, the work for which Friedmann is now famous was carried out in 1917, during the siege of Petrograd, almost as soon as he learned of Einstein's general theory (the circumstances cannot have been that dissimilar to those under which Karl **Schwarzschild** responded to Einstein's new theory). But these ideas were not published until 1922, in a paper which introduced two features of key importance to modern *cosmology*. First, from the outset Friedmann realized that he was dealing with a *family* of solutions to Einstein's equations, a set of cosmological models. He understood that there was no unique solution to the equations, as Einstein had hoped. Second, he incorporated the idea of expansion into his models from the beginning.

If we think of the *spacetime* of the Universe as being curved, analogously to the curved surface of a soap bubble, Friedmann's calculations showed how this curvature might be changing with time.

In some variations on the theme, the bubble expands forever; in others, it expands to a certain size, then collapses back upon itself as gravity overcomes the expansion. And there are more complicated models. But in all of them there is a period during which the whole universe expands in such a way that it will produce a recession velocity proportional to distance. This is exactly what Edwin *Hubble* and his colleagues found from studies of the *redshifts* of galaxies in the late 1920s.

But Hubble and his colleagues seem to have known nothing of Friedmann's work at the time, although Einstein had seen it. According to the cosmologist George *Gamow*, who was one of Friedmann's students in the 1920s, Friedmann wrote to Einstein before his work was published, but received no reply until a colleague from Russia who was visiting Berlin mentioned the work to Einstein. This elicited what Gamow called a 'grumpy letter' (*My World Line*, Viking, 1970), confirming that Friedmann's work was correct; it was only after he received that letter that Friedmann published his results.

Friedmann died in Leningrad on 16 September 1925. According to the official biographies, he died of typhoid. According to Gamow, however, he died of pneumonia following a chill that he contracted while flying in a meteorological balloon. Since Gamow was a student of Friedmann, his version may be more reliable, and there is no doubt that Friedmann did take part in a balloon flight to an altitude of 7,400 m in July 1925, two months before he died. Either way, Friedmann did not live to see the observational verification of his calculations, and it was not until Georges *Lemaître* independently solved Einstein's equations in the same way that Friedmann had done, that these cosmological models began to be taken seriously.

Further reading: Edward Harrison, *Cosmology*; John Gribbin, *In Search of the Big Bang*; E. A. Tropp, V. Ya. Frenkel and A. D. Chermin, *Alexander A. Friedmann*, Cambridge University Press, 1994.

Friedmann universe models   The basic *cosmological models*, first developed by Aleksandr *Friedmann* in 1922, from Albert *Einstein*'s equations of the *general theory of relativity*. Essentially the same solutions to Einstein's equations were found five years later by Georges *Lemaître*.

fundamental forces   (= fundamental interactions) The four forces that operate between *elementary particles*. In order of strength, starting with the weakest, they are *gravity*, the weak nuclear force, the electromagnetic force, and the strong force.

In the 1980s, there were claims that a 'fifth force' had been discovered, and this was interpreted, for a time, as a possible modification of gravity (in effect, an 'antigravity' force) with a range of a few tens of metres. But careful experiments showed that all of the effects attributed to the fifth force could, in fact, be explained by gravity, once allowance had been made for variations in the *density* of geological layers in the regions where the experiments were carried out. There is no evidence that any forces other than the four known ones operate in the Universe, and one of the primary objectives of particle physicists is to explain the workings of all four forces in one mathematical package (see *grand unified theories*).

The relative strengths of the four forces cover an enormous range. In units in which the strength of the strong force is 1, the strength of the electromagnetic force is 10^{-2} (just 1 per cent of the strength of the strong force), the strength of the weak force is 10^{-6} (one-millionth the strength of the strong force) and the strength of gravity is a mere 10^{-40}. This means, for example, that the electromagnetic force of repulsion between two electrons is 10^{38} times greater than the gravitational force of attraction between the same two electrons. Gravity is so extraordinarily weak that it plays virtually no part in interactions involving pairs of fundamental particles or a few particles.

But gravity was the first of the four forces to be investigated scientifically, and to be described (by Isaac *Newton*) by a satisfactory mathematical theory. This is because gravity is additive – the more particles you put together in one lump, the stronger the gravity of the lump of matter becomes – and has a very long range, falling off only as 1 over the square of the distance from the lump of matter (it obeys an inverse square law). So the gravity of the Sun is very large, because it contains very many particles, and it has a very long range, holding the planets in their orbits.

Indeed, it is the difference in size between the electromagnetic force and gravity that makes stars so big. The electric forces operating between the nuclei inside a star like the Sun tend to keep them apart, because the nuclei all have positive charge, and like charges repel one other. In a similar way, the electrons in the outer parts of an atom all have negative charge, and if you try to push two atoms together, they will be kept apart by the repulsion between their respective electron clouds. But although the electromagnetic force also follows an inverse square law, and in principle it has a very long range, each atom has zero net charge, because the negative charge in its electron cloud is exactly balanced by the positive charge of its nucleus (the electrons do not fall into the nuclei because of quantum effects; see *quantum theory*). So if you put very many atoms together, the overall charge is zero, while the gravitational forces associated with the collection of atoms increase.

Once you have about 10^{38} atoms together in one lump, the force of gravity felt by the atoms in the middle of the lump (the weight of all the atoms above) will be so strong that the individual atoms are crushed together, so that nuclei can come into contact and the processes of *nuclear fusion* that keep stars hot can begin. So, simply by comparing the strengths of the electromagnetic force and the force of gravity, you could predict that all stars must contain at least 10^{38} nuclei.

It does not quite work like that, because the 10^{38} atoms are not all concentrated at a single point. Instead, they are spread out over the volume of the star. This gives gravity a handicap, reducing its effectiveness by one-third, because the volume of the lump of matter goes as the cube of its radius. So gravity can actually crush atoms together and cause nuclear fusion once the lump contains about 10^{57} atoms, because 38 is two-thirds of 57. A lump of matter containing 10^{57} hydrogen nuclei (a number you can remember as the Heinz soup parameter) is indeed the size of a star just a little bit smaller than our Sun, with a mass about 85 per cent of the Sun's mass.

Because of quantum effects, it is possible to make slightly lighter stars, but not anything lighter than about 10 per cent of the Sun's mass. All of that mass is needed to make the star heavy enough to crush two atoms together and make one pair of nuclei fuse – although, of course, once it is that heavy it will crush all the atoms in its interior, and cause many nuclei to fuse.

Another way of appreciating the weakness of gravity is to consider the fall of an apple from a tree. The stalk of the apple is held together by electromagnetic forces, operating between atoms and *molecules*. The stalk contains only a few molecules; but it takes the combined gravitational pull of all the particles in the Earth to pull on the apple strongly enough to break the stalk and make the apple fall to the ground.

The two other forces, the strong and weak nuclear forces (usually referred to just as the strong force and the weak force, with the 'nuclear' taken as read), do not obey an inverse square law, and only have very short range, with an influence that extends roughly across the size of an atomic nucleus. The strong force operates directly between *quarks*, binding them together to make *hadrons*, including the protons and neutrons of atomic nuclei (members of the *baryon* family). The force leaks out of individual nucleons to influence the particles next door, holding protons and neutrons together in nuclei even though the electromagnetic force of repulsion between the protons is trying to blow the nucleus apart.

Because the strong force is about 100 times stronger than the electromagnetic force, you would expect the electromagnetic force to dominate, making nuclei unstable, if they contained more than about 100 protons (in this case, since all the protons have the same charge, the electromagnetic force *is* additive, while, because of its very short range, the strong force is not and can only provide a link between adjacent *nucleons*). In fact, the situation is made slightly easier for the strong force by the presence of neutrons in the nucleus, but the heaviest stable nuclei still contain only just over 200 nucleons, and none of them contains as many as 100 protons (even plutonium only has 94 protons in its nucleus). Once again, a simple understanding of the balance between two of the fundamental forces explains what might otherwise be a puzzling feature of nature, in this case the limit on the number of stable elements.

One unique feature of the strong force is that, up to the limit of its range, it is *stronger* for quarks that are further apart. As long as the three quarks in a nucleon stay within about 10^{-15} m of each other, they do not feel the force much at all – it is as if they are joined by loose elastic bands that are about that long. But once a quark tries to move further than 10^{-15} m from its partners, the 'elastic bands' start to stretch, pulling it back into place. The more it tries to move away, the more the elastic stretches, and the stronger the force pulling it back becomes. It can only escape from the nucleon if so much energy is put in (perhaps by a collision with another particle) that the elastic band snaps, creating two new quarks out of pure energy, one on each side of the break (see *special theory of relativity*).

The escaping quark will be attached to one of the new quarks, forming a bound pair called a *meson*, while the second new quark takes its place in the nucleon.

The weak force behaves even less like a force in the everyday sense of the term, and is an interaction responsible for the process of *beta decay*. It operates between *leptons*, and during the decay of *hadrons*, in which leptons are produced. But both the weak interaction and the electromagnetic interaction can be combined in one mathematical description, known as the electroweak theory. This portrays them as different facets of a single force, the electroweak interaction; it is one of the triumphs of particle theorists, reducing the number of different forces to three, and (perhaps) pointing the way towards an even more comprehensive mathematical package in which the strong force is combined with the electroweak force.

In *classical mechanics*, forces between particles are described in terms of *field equations*, and a 'field of force' is envisaged surrounding a particle that exerts a force on other particles. In *quantum theory*, the forces (or interactions) are carried (or mediated) by particles. The electromagnetic interaction is mediated by *photons*, exchanged between charged particles; the weak interaction is mediated by particles known as intermediate vector bosons, exchanged between leptons (and, in some cases, between a lepton and a hadron); the strong force is mediated by gluons; and gravity is mediated by gravitons. There is direct evidence for the existence of all these force carriers except gravitons; it is almost certain that gravitons (which are also predicted by string theory) do exist, but the extreme weakness of gravity makes it impossible to monitor the way they are exchanged between particles.

Further reading: Frank Close, *The Cosmic Onion*; Paul Davies, *The Forces of Nature*.

fundamental particles   See *elementary particles*.

fusion   See *fission and fusion*.

Galactic centre   The centre of our own *Galaxy*, which may be the site of a moderately large *black hole*, several million times more massive than our Sun. The galactic centre is obscured by dust, and cannot be seen by *optical telescopes*, but it is a strong *radio source* (*Sagittarius A*) and is surrounded by a ring of gas and dust orbiting at a speed of about 110 km/sec. The central source is also visible in *X-rays* and in the *infrared*.

Galactic plane   The imaginary circle on the sky marked out by the densest concentration of stars in our *Milky Way Galaxy*. It is tilted at about 63° compared with the Earth's equator.

Galactic poles   The imaginary points 90° north and south of the *Galactic plane*. The north galactic pole is in the *constellation* Coma Berenices, and the south galactic pole is in the constellation Sculptor.

Galactic rotation   Everything in the *Galaxy* is moving in its own *orbit* around the *Galactic centre*. The orbital speed for a *star* like the Sun depends on the amount of mass inside the orbit, not on the total mass of the Galaxy, so it does not obey *Kepler's laws*. The Sun and nearby stars are moving at about 220 km/sec and take about 225 million years (a *cosmic year*) to complete one orbit.

galactic wind   An outflow of hot gas from any *galaxy*. It is often a result of a burst of star formation, possibly triggered by the merger of two galaxies (see *galaxy formation*), and is detected by its emission of *X-rays*.

Galactic year   See *cosmic year*.

galaxies   Huge collections of *stars*, held together by *gravity* to form an 'island' in space. The largest galaxies contain thousands of billions of stars and may be several hundred thousand *light years* in diameter. Even the smallest 'dwarf' galaxies contain millions of stars. Our *Milky Way* Galaxy contains a few hundred billion stars. In round terms, the size of a galaxy compared with the size of the *orbit* of the Earth around the Sun is in about the same proportions as the size of your body compared with an individual atom.

In spite of their great size, most galaxies are so far away from us that they can be seen only with the aid of telescopes. Only the nearest large galaxy, the *Andromeda galaxy*, and two small companions to the Milky Way, the *Magellanic Clouds*, can be seen as faint patches of light on the sky with the unaided human eye. An estimated 50 billion galaxies are visible to modern telescopes (including the *Hubble Space Telescope*), but only a few thousand have been studied systematically.

Galaxies are divided into two main classes, *elliptical* and *disc* galaxies, by their appearance. In addition to the visible bright stars we can see, galaxies are embedded in large amounts of *dark matter*, revealed by its gravitational influence on the way galaxies move. Most galaxies occur in

clusters; the most distant galaxy known (dubbed 8C 1435 + 635 and identified in 1994) has a *redshift* of 4.25, and is seen by light which left it when
the Universe was only 20 per cent of its present age (see *look back time*).

Galaxy   Our home in the cosmos, an island of *stars* (hundreds of billions
of stars broadly similar to our Sun), gas and dust held together by gravity
to form a *disc galaxy* some 30 kiloparsecs across surrounded by a *halo* of
visible *globular clusters*, and embedded in a much more extensive halo of
dark matter, detected only by its gravitational influence.

The most obvious visible feature of our Galaxy (given the capital letter
to distinguish it from all the other *galaxies* in the Universe) is the band of
light on the sky called the *Milky Way*; our Galaxy is sometimes referred to
as the Milky Way Galaxy, or just as the Milky Way (the word 'Galaxy'
actually comes from the Greek for 'Milky Way'). The Milky Way proper is
a faint band of light across the sky (brighter in the Southern Hemisphere
than in the north), which is shown by telescopes to be made up of a vast
number of individual stars, too faint (and too close together on the sky)
to be picked out as individuals by the unaided human eye. This band of
light is actually our view of the billions of stars that make up the disc of
our Galaxy, from a viewpoint embedded in the disc, about two-thirds of
the way out from the centre of the Galaxy to the edge of the disc. There is
nothing special about our place in the Galaxy.

In places, the Milky Way is obscured by dark clouds of interstellar
matter, so that it looks as if there are holes in the Milky Way. Observations
using radio telescopes have shown, however, that the material of the Milky
Way is genuinely distributed in a disc, with *spiral arms* of material trailing
through the disc. Our Galaxy is indeed an ordinary disc galaxy like the
millions of others we see, with the aid of telescopes, scattered across the
sky. There is nothing special about our place in the Universe.

Although it is 30 kiloparsecs across (with the Sun about 9 kiloparsecs
out from the centre), the disc of our Galaxy is only about 300 *parsecs* thick
in its outer regions.

As in other disc galaxies, at the centre of our Galaxy there is a bulge
of stars, resembling a small *elliptical galaxy*. This bulge is about 7 kiloparsecs
across and 1 kiloparsec thick; it lies in the direction of the constellation
Sagittarius, as seen from Earth. There may be a supermassive *black hole* at
the very centre of the Galaxy. The central region of the Galaxy, like
the halo, contains only old stars (about 15 billion years old), known as
Population II, and very little gas or dust; the disc contains stars of all ages,
including young *Population I* stars.

Population II stars are thought to be left over from the first burst of
star formation when the Galaxy itself formed. The youngest stars are
concentrated in a thin layer about 500 parsecs thick at the centre of the

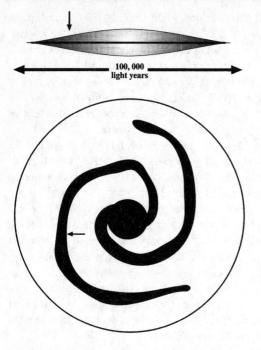

Schematic representations of the edge-on and plan views of our **Galaxy**; the arrows indicate the approximate location of the Sun in the disc of stars.

disc, where star formation is still going on. Slightly older stars (2 to 5 billion years old) are spread through the entire disc. The Sun is one of these intermediate-age stars.

In round numbers, the total number of stars in the Galaxy is roughly the same as the number of grains of rice that could be squeezed into a warehouse the size of a cathedral. But if each star were represented by a single rice grain, then the equivalent scale model of the disc of our Galaxy would have a diameter the same as the distance from the Earth to the Moon (about 400,000 km).

The stars of the disc, along with the clouds of gas and dust, orbit around the centre of the Galaxy in a similar way to the way in which the planets orbit around the Sun. The speed of each star in its orbit around the centre of the Galaxy depends on its distance from the centre – stars further out from the centre move more slowly than those close to the centre. The Sun is moving at about 250 km/sec in its orbit, and takes roughly 225 million years to travel once around the Galaxy (an interval sometimes referred to as the cosmic year). Studies of the way in which the stars move indicate the nature of the gravitational field of the Galaxy as a whole, and reveal its overall mass. This is about 1,000 billion times the mass of our Sun, roughly ten times the mass of all the stars in the Milky Way put together. This is strong evidence for the existence of *dark matter* in our Galaxy, extending far beyond the bright disc of stars in the Milky Way.

As new stars continue to form in the Galaxy, run through their life cycles and fade away, while everything moves at its own speed around the centre, the details of the overall appearance of the Galaxy are constantly changing on astronomical timescales, although the basic picture stays much the same. This has been likened to the life cycle of a living organism, and astronomers freely use terms like 'birth', 'evolution' and 'death' to describe the behaviour of stars and galaxies. The key to this picture of the Galaxy as an evolving 'living system' is the way clouds of material are repeatedly squeezed as they pass through the spiral arms.

Superficially, the spiral pattern made by the arms of bright stars in a galaxy like our own is rather like the swirling pattern made by cream as it is stirred into a cup of dark coffee. But just as the spiral pattern made by the cream soon gets blended into a smooth brown colour by the *differential rotation* of the liquid, so the spiral arms of our Galaxy should get smeared out in a few rotations of the whole Galaxy – within about 1 billion years. But unlike the cream being stirred into your coffee, the spiral arms are not 'wound up' by rotation. The pattern is actually caused by the presence of hot, young stars which shine brightly along the edges of the spiral arms, and they are produced by clouds of gas and dust moving into the spiral pattern and being squeezed by a shock wave there. The shock wave is the fundamental spiral feature.

The spiral pattern is the visible part of a density wave which is moving around the Galaxy in the same direction as the stars are moving, but more slowly. You can picture how this happens by imagining looking down on the Galaxy from a vantage point in space high above the disc. Each star orbits the centre of the Galaxy in an almost perfectly circular orbit. But it also weaves in and out a little in its orbit, because of the gravitational influence of other stars and dark matter. In effect, the star moves in a small circle, while the centre of that circle follows the circular orbit around the Galaxy – rather like the old picture of *epicycles*, but on a galactic scale. This weaving makes the stars bunch up in some places, as a density wave. The density wave moves around the Galaxy at a speed of about 30 km/sec, while the stars and clouds of material in the disc move at speeds of 200–300 km/sec, overtaking the density wave and passing through it.

This produces a shock wave along the edge of the spiral pattern, like a huge, spiralling sonic boom. Clouds of gas and dust pile up in a traffic jam just behind each spiral arm, along the inside of the curve, and are squeezed, triggering bursts of *star formation*.

The process is largely self-sustaining. The largest stars formed in this way run through their life cycles quickly, and explode as **supernovae**. Such an explosion, sending a blast wave through the nearby interstellar material, is just the thing to trigger the collapse of other clouds to make more stars. Once the process gets going it spreads, like a forest fire, across the affected region of the Galaxy. Smaller, longer-lived stars, and remaining clouds of interstellar material, continue on their way through the arm and around the disc. But as material passes through the spiral arm and on its way around the Galaxy, it is not long (by astronomical timescales) before it meets the next spiral arm, and is squeezed again. Everything in the disc is repeatedly subjected to this squeezing process as it orbits around the centre of the Galaxy.

Altogether, astronomers estimate that there may be 3 billion solar masses of material in the form of **molecular clouds** orbiting in the disc within the orbit of the Sun. This is 15 per cent of the total mass of stars in the same region of our Galaxy (excluding the central bulge). The clouds are made from material ejected by stars at the end of their lives; cloud stuff is constantly being turned into stars, and star stuff is constantly being recycled to produce new clouds.

Computer simulations show that the natural form of disturbance that will be produced in a rotating disc galaxy through this process is indeed a spiral; these **models** also show that the spiral pattern cannot persist for very long unless the whole disc is embedded in a large halo of dark matter. The gravitational influence of the dark matter stabilizes the spiral pattern and makes it last much longer, before eventually changing its shape to form a bar across the centre of the galaxy. Many such barred spirals are known, and it is likely that our Galaxy will evolve in this direction. There is some evidence that there is already a small bar of stars across the centre of our Galaxy, but it is very difficult to observe this region because it is obscured by gas and dust in the plane of the Milky Way.

The most striking feature of the whole process is the way in which massive stars (which each live for only 1–10 million years) form from clouds which collapse on timescales of 100,000–1 million years, and yet maintain a pattern of spiral arms which persists for at least 1 billion years. In the process, the rate at which dust and gas is converted into new stars is almost the same as the rate at which old stars return matter to the interstellar medium. In both cases, it amounts to only a couple of solar masses of material each year, throughout the disc of our Galaxy. It doesn't sound much, but it adds up dramatically over a few million years.

The whole process maintains a quasi-steady state, reminiscent of a living organism in which a collection of individual cells, many of which have a lifetime of only a few weeks, maintain the pattern of a human body for 70 years or more. Your body 'loses' cells all the time (chiefly from your skin and the lining of your gut), but you do not 'wear out' because those cells are constantly being replaced. Similarly, the stars in the spiral arms are constantly moving on and fading away as they age, but the spiral pattern does not 'wear out' because bright blue stars are constantly being born along the inner edges of the arms.

Incidentally, there are a thousand times more cells in your body than there are bright stars in the Galaxy, so the Galaxy is a much simpler 'organism' than a human being.

Further reading: John Gribbin, *In the Beginning*; Nigel Henbest and Heather Couper, *The Guide to the Galaxy*.

galaxy formation and evolution   Astronomers' ideas about the origins and evolution of *galaxies* changed dramatically in the early 1990s, thanks to the development of improved telescopes (including the **Hubble Space Telescope**) and instrumentation, which made it possible to look further back in time by studying fainter, more highly *redshifted* galaxies, seen as they were when the Universe was young. Until this breakthrough, it was widely accepted that the galaxies we see today all formed essentially at the same time, just after the **Big Bang**, and that they had each evolved internally as the Universe aged. The new picture is of a dynamic, changing Universe in which galaxies compete with one another for 'living space', merge, and absorb other galaxies. The most dramatic difference between the old and new pictures is that, whereas *elliptical galaxies* were once thought to be the oldest systems, they are now regarded as relatively recent products of interactions involving *disc galaxies* and other systems.

Disc galaxies (sometimes known as spirals) are much more common than ellipticals in photographs of the sky, but the biggest ellipticals are much bigger than any spiral. A typical disc galaxy, like our own Galaxy, may contain 100 billion stars, but the biggest ellipticals contain 100 times more stars. But there are also many dwarf ellipticals, some of which are no bigger than a *globular cluster*, containing about 1 million stars. Since many of these dwarf systems must be too faint to be seen, there must be many more ellipticals than we have yet identified.

Astronomers used to think that ellipticals are old because they contain mainly cool, red stars, and very little dust and gas. Since cool, red stars are old, it was argued that elliptical galaxies are old. But although disc galaxies contain many hot, young stars, and star formation is still going on amidst the clouds of gas and dust in these galaxies, they do also contain many old (**Population II**) stars, concentrated in a central bulge and scattered through a spherical halo around the disc. The new evidence shows that the old, red stars in elliptical galaxies have actually come from disc galaxies. Ellipticals form either when two spiral galaxies collide with one another, stripping away their discs, or from mergers in which an existing elliptical galaxy swallows up a disc galaxy. This is why the biggest ellipticals are so big.

The evidence comes from observations of galaxies in the act of merging, and from computer simulations of such events. When two disc galaxies collide, the thin discs are stripped off and the systems merge into a single 'starpile' shaped like an elliptical galaxy. The stars themselves do not collide with one another, but the interacting gravitational fields of the two systems pull all the stars into one ball. Clouds of gas and dust from the colliding galaxies really do collide with one another, sending shock waves rippling through the new system and triggering a wave of *star formation* (see *starburst galaxies*).

When a large elliptical galaxy absorbs a smaller disc galaxy, the elliptical grows and still looks like a single star system. But computer simulations show that many stars from the disc galaxy end up following similar orbits to one another inside the enlarged elliptical. Photographs show that this is indeed the case. Instead of being featureless starpiles, ellipticals are criss-crossed by stripes and arcs of brighter light, corresponding to the remains of disc galaxies that have been swallowed up but not yet fully digested.

The further out into the cosmos we look, the further back in time we see, because light takes a finite time to travel across space. In our part of the Universe, *clusters of galaxies* contain many ellipticals and have an overall reddish colour. At distances corresponding to a *look back time* of about 5 billion years, the clusters are much bluer (showing that active star formation was more common then), and the Hubble Space Telescope has shown that many of the objects in these distant clusters are pairs of disc galaxies in the act of merging. Ellipticals today contain so little gas and dust only because it was all turned into stars during those mergers.

The largest ellipticals sit at the centres of clusters today, continuing to grow by absorbing any other galaxy that comes too close, like a spider getting fat by sitting in its web and waiting for food to come its way. Only 1 per cent of all the galaxies we see today are actively involved in such mergers, but they happen so quickly (compared with the age of a galaxy) that astronomers calculate that half of all the galaxies we now see have been involved in mergers with galaxies roughly the same size in the past 7 or 8 billion years. Mergers must have been even more common when the Universe was younger and galaxies were closer together.

Disc galaxies themselves are now thought to have formed from mergers between smaller units early in the life of the Universe. The range of ages of the globular clusters, from about 14 billion years to about 7 billion years, suggests that our Galaxy formed over a period of several billion years from an amalgamation of about 1 million smaller gas clouds. When each 'new' gas cloud collided with the growing Galaxy, the shock wave would trigger a burst of star formation, forming a new globular cluster or making a contribution to the bulge of stars at the centre of the Galaxy. Left-over material would then settle down into the disc, eventually forming the *spiral arms*. Computer simulations show that this whole process of mergers is particularly effective in the context of *dark matter* models, with the gravity of the dark matter holding everything together – indeed, if there were no dark matter in the Universe, galaxies as we know them probably could not have formed at all.

There are still puzzles that have yet to be explained concerning the origin and evolution of galaxies. For example, in the 1980s researchers at the AT&T Bell Laboratories identified huge numbers of dwarf galaxies at *redshifts* corresponding to a look back time of 2–3 billion years. We see these dwarf galaxies as they were when the Earth was half its present age. At that time, life on Earth had not even emerged from the seas and on to the land; but if there had been astronomers alive on Earth then, their telescopes would have shown the night sky blazing with blue dwarf galaxies, each about one-hundredth of the size of our Galaxy.

There were so many of these galaxies that their appearance on astronomical photographs today has been described as like 'cosmic wallpaper'. But we do not see anything like them active today.

It may be that those dwarf galaxies were an intermediate stage in the formation of disc galaxies, and have been absorbed into the spirals we see closer to home. Or it may be that the dwarf galaxies that made up the cosmic wallpaper simply burned out. Because they were so small, with relatively weak gravitational fields, the *supernovae* resulting from the first wave of star formation in these tiny galaxies may have produced shock waves powerful enough to sweep all the remaining gas and dust in the dwarfs away into intergalactic space, leaving no raw materials for new stars to form from. The modern equivalents of those dwarfs (or at least, their fossil remains) may still be there, but now containing only old, fading stars, too faint to be seen from Earth.

Before the era of the blue dwarfs, the Universe contained galaxies much larger in size than those we see today, but no larger in mass. They were simply more spread out, because gravity had not had time since the Big Bang to pull all the pieces together into more compact form. Whole groups of galaxies evolved together like an ecosystem of living creatures, competing for raw materials (clouds of gas to be turned into new stars), absorbing one another, and adapting to the changing conditions as the Universe itself expanded and aged. Listening to astronomers talking about emerging populations, evolution and variations that become extinct (like the blue dwarfs), it is sometimes hard to remember that they are indeed astronomers talking about galaxies, not biologists talking about the evolution of life on Earth.

Galilean telescope Simple *refracting telescope* of the kind used (but not invented) by *Galileo*, with a convex objective lens and a concave eyepiece in a long tube.

Galileo A NASA spaceprobe launched in 1989 and scheduled to reach Jupiter in December 1995. The first spaceprobe intended to go into orbit around Jupiter, sending back data about the planet and its moons for about two years.

Galileo (Galilei, Galileo) (1564–1642) Italian mathematician and natural philosopher, and arguably the first modern astronomer, since he was the first person to make systematic observations of the heavens using a *telescope*.

Galileo Galilei, who is invariably referred to solely by his first name, was born in Pisa on 15 February 1564 (the same year that William Shakespeare was born), the son of a well-known musician and scholar, Vincenzio Galilei, who was interested in mathematics and the idea that the mathematical relationships between the positions of the *planets* reflected the mathematical relationships in music (the 'harmony of the spheres'). Galileo was educated privately until his family moved to Florence in 1575, then studied at a monastery until 1581. He returned to Pisa to study medicine, but turned his attention instead to mathematics, physics and astronomy, never completing his medical studies.

While studying in Pisa, probably in 1583, he realized while watching a lamp swinging to and fro in the cathedral that the time taken for each swing of the same pendulum was the same regardless of the amplitude of the swing – he made the discovery by timing the swinging of the lamp with his pulse. He used this discovery as the basis for the design of a clock, which was constructed, after Galileo's death, by his son.

Galileo left Pisa without taking his degree in 1585, and studied at home in Florence. In 1589 he became professor of mathematics at the University of Pisa, where he gained fame by refuting the old idea, dating back to the time of the Ancient Greeks, that a heavy object will fall faster than a light object. There is no evidence, though, that he proved that both objects fall at the same rate by dropping a cannon ball and a musket ball from the leaning tower in the city.

In order to study the effect of *gravity* on falling objects, Galileo actually carried out a series of experiments with balls rolling down inclined planes, so that their 'falling' motion was slowed down and could be timed accurately. These studies helped to establish the way in which gravity accelerates a falling object, and pointed the way for Isaac *Newton*'s study of

Galileo (1564–1642).

motion in the 17th century. Galileo understood that motion is relative.

Much of this work was carried out in Padua, where Galileo became professor of mathematics in 1592 and stayed for eighteen years. Although Galileo never married, in 1599 he began a liaison with Marina Gamba, and the couple had two daughters and a son.

In 1609 Galileo learned of the invention of the telescope by Dutch scientists (but see *Digges*), and made his own instrument, which he used to look at the Moon, Jupiter and its *moons* (the four largest of which he discovered), and the *stars*. He described his observations in a book, *The Starry Messenger*, which was published in 1610 and made him famous. The success of the book brought him to the attention of the Grand Duke of Tuscany, who became his patron while he continued his scientific studies (turning now to hydrodynamics) in Florence. When he moved to Florence, he left Marina behind in Padua, and she married someone else.

In 1613 Galileo publicly supported the Sun-centred *model* of the Solar System put forward by Nicolaus *Copernicus*. In 1616 the Catholic Church declared that this work was heretical, and Galileo was ordered to abandon his support for the theory. He refrained from public support for Copernicus' ideas until 1624, when Urban VIII became Pope. Thinking that the time was ripe for a change, he obtained permission to publish what was supposed to be a balanced discussion of the rival Earth-centred and Sun-centred models, but when the resulting book (*Dialogue Concerning the Two Chief World Systems*) appeared in 1632 it was clear which system the author favoured. The book was banned, and Galileo was charged with heresy. Threatened with torture and 69 years old, he was forced to reject officially the idea that the Earth moves around the Sun, but allegedly muttered 'eppur si muove' ('yet it does move') just after making his formal recantation.

Galileo was sentenced to life imprisonment, commuted to house arrest, and worked at his villa near Florence until he died eight years later, a month short of his 78th birthday. He had been blind for the last four years of his life, but completed a book summing up his life's work, the *Discourses*, with the aid of his disciples Vincenzo Vivani and Evangelista Toricelli. The manuscript was smuggled out of Italy and the book was first published in Holland in 1638; Galileo died at Arcetri on 8 January 1642. On 31 October 1992, after three centuries of careful deliberation by the Church, Pope John Paul II formally retracted the sentence passed on Galileo by the Inquisition.

GALLEX   Solar *neutrino* detector based on gallium (the name comes from GALLium EXperiment). Based in the underground laboratory at Gran Sasso, in northern Italy, this is a joint European/US/Israeli experiment which uses 30 tonnes of gallium to 'trap' passing neutrinos. See *solar neutrino problem*, *SAGE*.

gamma ray astronomy   Investigation of the nature of the Universe using the highest-energy forms of *electromagnetic radiation*. This radiation, which extends from *X-ray* energies up to energies a trillion times greater, can be detected only by instruments flown on satellites and rockets above the Earth's atmosphere. In wavelength terms, gamma rays are electromagnetic waves with wavelengths shorter than 0.1 nanometres.

Gamma rays from space have been studied by a series of satellites since the end of the 1960s; the most important of the pioneering gamma ray observatories were SAS II, launched by NASA in 1972, and *COS B*, a European satellite launched in 1975. These show that there is a band of radiation at these wavelengths coming from the plane of the *Milky Way*, as well as individual sources of gamma radiation associated with objects such as *supernova* remnants and *pulsars*. There is also a faint background of gamma radiation coming from all directions in space, which may be due to the activity of many distant galaxies.

One particularly intriguing source of gamma radiation lies at the centre of our Milky Way Galaxy, and produces radiation at an energy of 511,000 *electron Volts*, caused by *electrons* and *positrons* annihilating one another. This may be related to the activity of a *black hole* at the galactic centre.

The most baffling, and intriguing, gamma ray sources are known as 'bursters'. They were first detected in the late 1960s by satellites sent up by the US Air Force to monitor nuclear explosions on Earth linked with atomic weapons testing, but the discovery was not declassified and made public until 1973. These very powerful sources appear and fade away again in a matter of a few seconds, but during its brief lifetime a gamma ray burster can shine as brightly at gamma ray energies as all the other gamma ray objects in the sky put together. On average, one of these bursts is detected every day. Nobody knows what causes gamma ray bursts, or even whether the sources lie in our own Milky Way Galaxy or far away across the Universe.

gamma rays   *Electromagnetic radiation* with wavelengths in the range 10^{-10}–10^{-14}m, corresponding to energies of 10,000 *electron Volts* (10 keV) to 10 million electron volts (10 MeV) per *photon*. They are similar to *X-rays*, but with higher energy (that is, shorter wavelengths).

Gamow, George ('Joe') (1904–68)   Ukrainian-born American physicist who made the first calculations of conditions in the *Big Bang*, predicted the existence of the *background radiation* and played a part in cracking the genetic code of DNA, the molecule of life.

Gamow was a larger-than-life character whose leaps of imagination took him from nuclear physics to *cosmology* and on to molecular biology. His enormous sense of fun led him to perpetrate several classic practical jokes (see *alpha beta gamma theory*), and he also found time to write a stream of books popularizing science, some of which are still in print today. Born in Odessa on 4 March 1904, the son of a teacher, Gamow lived through the turmoil of war and revolution before entering Novorossysky University in 1922, when he was eighteen. He had become interested in astronomy at the age of thirteen, when his father gave him a telescope as a birthday present, and he soon transferred from Novorossysky to the University of Leningrad, where he studied optics and then cosmology, working as a student under Aleksandr *Friedmann*, from whom he learned first hand about *cosmological models*, and completing his PhD studies in 1928, three years after Friedmann died.

Fresh from his PhD work, Gamow travelled around Europe, working for a time at the University of Göttingen, then at the Institute of Theoretical Physics in Copenhagen, the Cavendish Laboratory in Cambridge, and back in Copenhagen, in the years from 1928 to 1931. These three centres were at the heart of the revolution then taking place in the development of *quantum theory*, and during his visit to Göttingen Gamow made his first major contribution to science, applying quantum theory to explain how an *alpha particle* can escape from an atomic *nucleus* (the same explanation was independently derived by the American Edward Condon (1902–74)). This explanation of *alpha decay* can also be turned around to give insights into the way alpha particles (nuclei of *helium*) combine with other nuclei (*fusion*; see also *nucleosynthesis*).

Gamow was called back to the USSR in 1931, and was appointed Master of Research at the Academy of Sciences in Leningrad, and professor of physics at Leningrad University. He was deeply unhappy under the Stalinist regime of the time (no place for an imaginative practical joker), and when he obtained permission to attend a conference in Brussels in 1933 he seized the opportunity to leave the Soviet Union permanently, taking up a post at the George Washington University in Washington, DC, in 1934. He stayed there until 1956, before moving on to the University of Colorado, where he worked until his death at the age of 64. In the same year that he moved to Colorado, Gamow was awarded the Kalinga Prize by UNESCO in recognition of his work in popularizing science, best represented by his 'Mr Tompkins' series.

When he settled in the USA, Gamow took to signing his letters to friends 'Geo.', an abbreviation that he was convinced was pronounced 'Joe'; so he became 'Joe' to his friends and colleagues from then on.

Although he also carried out investigations of *beta decay* and the evolution of *stars*, as well as working on the Manhattan Project in the Second World War, and later on the development of the hydrogen bomb, Gamow is best known for his contribution to the theory of the Big Bang. Starting in 1946, and working with his students Ralph *Alpher* and Robert *Herman*, Gamow showed how primordial helium would have been manufactured from hydrogen nuclei (protons) and neutrons in the Big Bang itself, and predicted that the Universe should be filled with a weak background of microwave radiation left over from the Big Bang. This prediction was forgotten until the accidental discovery of the radiation by Arno *Penzias* and Robert *Wilson* in the 1960s.

Perhaps because of the delay before the radiation was discovered, it is not always appreciated that there is a direct connection, through his student Gamow, between modern studies of the cosmic microwave background radiation and the cosmological models of Friedmann, originally developed as long ago as 1917.

In the 1950s, Gamow became fascinated by the problem of cracking the genetic code of DNA. Although he did not solve the problem himself, and enthusiastically followed up several false leads, he acted as a catalyst in encouraging other people working on the problem, and made several significant minor contributions to the effort. His most important contribution was the idea that the sequence of smaller molecules along a DNA molecule itself could indeed be 'read' as a code, a kind of four-letter alphabet. Gamow died in Boulder on 20 August 1968.

Further reading: George Gamow, *Mr Tompkins in Paperback*; John Gribbin, *In Search of the Big Bang*.

Ganymede, photographed here by the Voyager spaceprobe, is the largest
moon in the Solar System.

Ganymede   The largest *moon* of Jupiter, with a diameter of 5,262 km (the
largest moon in the Solar System). It has a very low density, less than twice
that of water. It is bright overall, with an *albedo* of 0.42, but with many
dark, heavily cratered surface features.

gauge theory   An approach to developing a *unified theory* of the *fundamental forces* which is based on the idea of symmetry. All of the successful
models of the particle world today are based on gauge theory.

Gauge theory gets its name from the way in which the point from
which measurements are made can be 'regauged' in these *models*. For
example, if you have a ball sitting on one step on a staircase, and then
move the ball down to the next step, the amount of gravitational energy
stored by the ball decreases by a definite amount. The change in energy
depends only on the difference in height of the two steps. You can measure
the height of each step from the bottom of the staircase, or regauge your
measurements to take the distance of each step from the centre of the
Earth, or from anywhere else, and this has no effect on the outcome of
the calculation. This is called gauge symmetry.

Exactly equivalent gauge symmetries apply for electromagnetic interactions, such as moving an electron around in a magnetic field. But it turns out that the mathematical description of these phenomena is gauge symmetric only if the *photon* has zero mass. This matched what physicists already knew about the photon. The equivalent descriptions of other forms of particle interactions are more complicated, but one of the triumphs of gauge theory is that it predicted the existence of three counterparts to the photon (dubbed the W^+, W^- and Z^0 *bosons*) which were later discovered by experimenters.

Gauge theory plays an important part in the theory of *inflation*, which describes the earliest stages in the expansion of the Universe. According to inflation, the driving force for the initial expansion came from a breaking of initial gauge symmetry involving the fundamental interactions.

Gauss, Karl Friedrich (1777–1855) German mathematician and astronomer who pioneered the development of *non-Euclidean geometry* (important in the *general theory of relativity*) and did important work in calculating orbits of planets, as well as extensive work in mathematics and physics. A magnetic unit is named after him.

Geminga A powerful source of *gamma rays* and *X-rays* in the *constellation* Gemini. It is probably a *binary system* in which two *neutron stars* are in orbit around one another. The distance to Geminga is not known accurately, but may be less than 200 *parsecs*.

general theory of relativity Theory of *gravity* developed by Albert *Einstein* in the early part of the 20th century, and presented to the Prussian Academy of Sciences in 1915. Because gravity is the dominant force in the Universe at large (thanks to its very long range), the theory is also a theory of *cosmology*, and underpins all modern *models* of how the Universe got to be the way it is.

Einstein's *special theory of relativity*, published in 1905, deals with the dynamical relationships between objects moving at constant speeds in straight lines. It does not deal with accelerations, or with gravity, which is why it is called the 'special' (meaning 'restricted') theory. Einstein always intended to generalize his theory to deal with accelerations and gravity, but it took him ten years (not all of the time devoted exclusively to the general theory) to find a satisfactory mathematical description of the dynamics of the Universe and everything in it. Fortunately, we do not need to go through the mathematics in order to understand Einstein's theory, because it can be described very clearly in terms of geometry and physical pictures.

Indeed, the whole point about Einstein's theory is that it gives us a physical picture of how gravity works; Isaac *Newton* discovered the inverse square law of gravity, but explicitly said that he offered no explanation of *why* gravity should follow an inverse square law ('hypotheses non fingo'). The general theory of relativity also says that gravity obeys an inverse square law (except in extremely strong **gravitational fields**), but it tells us *why* this should be so. That is why Einstein's theory is better than Newton's, even though it includes Newton's theory within itself, and gives the same 'answers' as Newton's theory everywhere except where the gravitational field is very strong.

There were two key physical insights which led Einstein to the general theory, and which help us to understand the physics behind it. First, he realized that if someone fell from the roof of a tall building, they would not feel the force of gravity at all until they hit the ground (ignore the wind resistance in this simple picture). They would be weightless – in *free fall*, as we would now put it. In other words, the acceleration associated with the fall exactly cancels out the force of gravity, or *acceleration and gravity are equivalent* (the **equivalence principle**, first formulated in this way by Einstein in 1907).

The second physical insight extends this equivalence to the effect of gravity on light. Instead of a person falling from a roof, Einstein now imagined a windowless elevator falling freely inside its shaft, the cable having snapped and all the safety devices having failed. According to the equivalence principle, there would be no way for a physicist inside the elevator, equipped with all the usual instruments of a physics lab, to tell whether the elevator was accelerating towards an unpleasant collision with the ground or was floating freely in the depths of space.

But what would happen to a beam of light shone across the falling elevator from one side to the other? In the weightless 'room', **Newton's laws of motion** must apply, and the light beam must travel in a straight line from one side of the elevator to the other. But now imagine how things will look to somebody outside the elevator. Suppose that the elevator has walls of glass, and that the path of the light beam is measured by sensitive instruments on each floor that the elevator falls past. Because the 'weightless' elevator and everything in it is actually being accelerated by gravity, in the time it takes the light beam to cross the elevator the falling laboratory has increased its speed. The only way in which the beam can hit the spot on the other wall exactly opposite the spot it started from is if it has followed a curved path, bending downwards to match the increase in speed of the elevator. And the only thing that could be doing the bending is gravity.

So Einstein inferred that, if gravity and acceleration are precisely equivalent, gravity must bend light, by a precise amount that could be calculated. This was not entirely a startling suggestion: Newton's theory, based on the idea of light as a stream of tiny particles, also suggested that

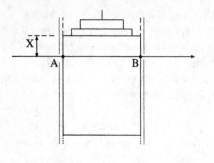

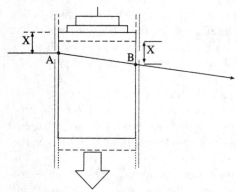

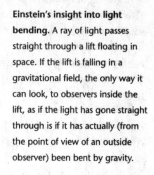

Einstein's insight into light bending. A ray of light passes straight through a lift floating in space. If the lift is falling in a gravitational field, the only way it can look, to observers inside the lift, as if the light has gone straight through is if it has actually (from the point of view of an outside observer) been bent by gravity.

a light beam would be deflected by gravity. But in Einstein's theory, the **deflection of light** is predicted to be exactly twice as great as it is according to Newton's theory. When the bending of starlight caused by the gravity of the Sun was measured during a **solar eclipse** in 1919, and found to match Einstein's prediction rather than Newton's, the general theory of relativity was hailed as a scientific triumph.

By then, Einstein had provided a physical picture of what goes on to bend light. Imagine empty space (strictly speaking, **spacetime**) to be represented by a stretched rubber sheet, like the surface of a trampoline. On such a surface, you can represent light rays by rolling marbles across the surface; they travel in straight lines, and the rules of **Euclidean geometry** apply. Now imagine placing a heavy object, such as a bowling ball, on the sheet, to represent the Sun. The sheet curves under the weight, and if you roll a marble across the sheet, near to the 'Sun', its trajectory will be bent as it follows the curve around the heavy weight. Such curved trajectories are the **geodesics** in the curved 'space' of the distorted rubber sheet, and on the curved surface the rules of geometry are now those of *non-Euclidean geometry*. The presence of matter, said Einstein, causes four-dimensional spacetime to curve in an equivalent way. The curvature of spacetime then affects the motion of everything passing through the region of curved spacetime, including light beams and planets. The situation is summed up in a neat aphorism: 'matter tells space how to bend; space tells matter how to move'.

There is one important point about this picture that sometimes causes confusion. We are dealing not just with curved space (in spite of the aphorism!), but with curved spacetime. The orbit of the Earth around the Sun, for example, forms a closed loop in space, and the Earth is held in its orbit by the Sun's gravity. If you imagine that this closed orbit represents the curvature of space caused by the Sun, you might leap to the conclusion that space itself is closed around the Sun – which is obviously not true, because the Sun is not a *black hole* and light (and other things) can escape from the Solar System.

In fact, the Sun and the Earth are both following their own *world lines* through four-dimensional spacetime, first described in those terms by Herman *Minkowski* in 1908. In that description, time and space are geo-metrically equivalent, but related by the speed of light, which is 300 million metres per second. So each second of time is equivalent to 300 million metres in the time direction. The Earth and the Sun are moving through spacetime in very nearly the same direction, from the past into the future, and instead of the orbit of the Earth around the Sun being a closed loop, in four dimensions it is a stretched-out helix, twisting around the world line of the Sun.

Look at it another way. It takes light about $8\frac{1}{3}$ minutes to get from the Sun to the Earth, so the diameter of the Earth's orbit is about 52 *light minutes*. But instead of taking 52 minutes to complete its orbit, the Earth actually takes a year (525,600 minutes) – and in that time it has moved along its world line in the time direction by a full 525,600 minutes, 10,000 times more than its equivalent journey through space, and 63,000 times the equivalent 'distance' from the Earth to the Sun. So the four-dimensional 'orbit' of the Earth around the Sun is a helix so tall and thin that the pitch is 63,000 times bigger than the radius.

The general theory of relativity makes many predictions which have been tested by experiment many times. These include the deflection of light, the *advance of the perihelion* of Mercury, the *gravitational redshift* and *gravitational time dilation*. It has passed every test with flying colours. Its greatest triumph, however, has been in explaining the observed behaviour of the *binary pulsar*. There is no doubt that the general theory is a good and accurate description of the behaviour of matter in the Universe, and of the relationship between space, time and matter. If it is ever improved upon, any better theory will have to incorporate the general theory within itself, just as the general theory incorporates Newton's theory of gravity within itself.

When Einstein applied the equations of the general theory to describe the behaviour of the Universe at large, he was surprised to find that in their pure form the equations did not allow the possibility of a 'static' Universe – they said that space must either be contracting or expanding as time passed. At that time, in 1917, the Universe was though to be static. This led Einstein to introduce an extra term, the **cosmological constant**, to hold the Universe still. But within a few years it was realized that we really do live in an **expanding Universe**, and that the cosmological constant was not needed. This discovery can also be regarded as confirmation of a prediction made by the general theory, even though Einstein himself did not realize its significance when he made the original calculations.

It is the general theory of relativity that tells us how the Universe has evolved away from an initial **singularity**, and which implies that the Universe was indeed born in a superdense state, in the **Big Bang**.

Further reading: John Gribbin, *In Search of the Edge of Time*; Kip Thorne, *Black Holes and Time Warps*; Clifford Will, *Was Einstein Right?*

geodesic   The equivalent in curved space of a straight line on a flat piece of paper – indeed, such a straight line is a special case of a geodesic, the shortest distance between two points. *Photons* always travel along geodesics.

geosynchronous orbit   An orbit of a *satellite* around a rotating *planet* in which the time taken for the satellite to make one orbit is the same as the time taken for the planet to rotate once. Because of this, a satellite in a geosynchronous orbit seems to hover above one point on the equator of the planet (in a geostationary orbit), or (if the orbit is slightly inclined to the equator) to trace out a figure-of-eight pattern on the sky.

The time taken for a satellite to complete one orbit depends only on its height and the mass of the planet it is orbiting. Satellites in lower orbits take less time to complete each orbit. For the Earth, the rotation period is 23 hours 56 minutes and 4.1 seconds, and the orbit which has the same rotation period lies 35,900 km above the equator. Powerful rockets are needed to place satellites in such a high orbit, but the cost and effort of doing this are well worthwhile for communications satellites, navigation satellites and some Earth observation satellites, including some weather satellites. With just three satellites evenly spaced around the Earth in geosynchronous orbits, it is possible to beam communications to any point around the world; it is because TV satellites are in geosynchronous orbit that your receiving dish does not have to track around the sky as the

Earth rotates and the satellite moves.

The importance of geosynchronous orbits for communications was first appreciated by Arthur C. Clarke (better known as a science fiction writer), who published papers outlining the possibilities in the 1940s, long before the technology existed to make this a practical reality.

German–Spanish Astronomical Centre Observatory in Almeria in southern Spain, at an altitude of 2,160 m on the Calar Alto mountain. It is associated with the Max Planck Institute for Astronomy, in Heidelberg. Its main *telescopes* are a 3.5 m *reflector*, a 1.5 m reflector, a 1.2 m reflector and a 0.8 m *Schmidt camera*. The 3.5 m telescope is identical to one at the *European Southern Observatory*.

Giacconi, Riccardo (1931–) Italian-born American physicist who pioneered the development of *X-ray astronomy*. He was head of the team that launched a rocket in 1962 intended to monitor X-rays from the Moon, but which found an intense source of X-rays from beyond the Solar System, now identified as *Scorpius X-1*. He was also involved in the *Uhuru* project and studies of *Cerenkov radiation*.

Giant Metre-wave Radio Telescope (GMRT) An *aperture synthesis* array north of Pune, in India. The array is made up of 30 fully steerable dish *antennas*, each with an *aperture* of 45 m, operating at wavelengths in the metre and decimetre bands. It is the largest radio telescope in the world operating at these wavelengths, with twelve dishes in a central array 1 km square, and the others distributed along three railway tracks each 14 km long in a 'Y' shape.

giant molecular clouds Large clouds of gas, mainly composed of *molecules* of *hydrogen*, in our *Galaxy*. Revealed by *radio astronomy* techniques in the 1970s, these clouds contain millions of times as much mass as our Sun, and provide the raw material from which new *stars* are made.

giant planets The four largest planets in our Solar System (Jupiter, Saturn, Uranus and Neptune), all of which are chiefly composed of gas and are much bigger than the rocky *terrestrial planets*.

giant star Any star which has a diameter roughly 10 to 100 times as big as our Sun's, and a *luminosity* in the range from 10 to 1,000 times the luminosity of the Sun. See *stellar evolution*.

Ginga A Japanese *X-ray astronomy satellite*, launched in 1987.

Giotto A spaceprobe launched by the *European Space Agency* in 1985, which carried out a close encounter with *Halley's Comet* in 1986. The mission provided important confirmation of the accuracy of the 'dirty snowball' model of *comets*.

glitch A sudden change in the period of a *pulsar*, probably caused by a 'starquake' rearranging the material the pulsar is made of.

On average, the period of a pulsar slowly increases, as the spinning *neutron star* loses energy and spins more slowly. The *Crab pulsar* (which is the youngest pulsar known) slows down, for example, by one part in a million every day. But occasionally the Crab and other young pulsars (most notably, the *Vela pulsar*) suddenly speed up. In the case of the Vela pulsar, such a glitch may shorten the pulsar's period by twenty times the amount it usually increases in a day, putting the rotation rate of the neutron star back to where it was three weeks previously. Such glitches occur every few years.

The most likely explanation is that these young pulsars are spinning so rapidly that their shape is distorted, so that they bulge outwards at the equator. As the rotation period slows down, the effect of the *centrifugal force* responsible for this bulge decreases. But for a time, the stiff outer crust of the neutron star (about 1 km thick on a neutron star 10 km in diameter) holds its old shape, even under the intense pull of the star's gravity. At some point, however, gravity overcomes the stiffness of the crust, and it cracks in a starquake, readjusting its shape to match the smaller equatorial bulge corresponding to the slower rotation speed of the pulsar. Like a spinning ice skater pulling in her arms, this enables the neutron star to spin faster for a while.

Global Oscillation Network Group   See *GONG*.

globular cluster   A densely packed ball of *stars*, containing hundreds of thousands (or even millions) of individual stars. The globular clusters in our *Galaxy* are spread through a spherical *halo* around the Galaxy, and contain some of the oldest stars in the Galaxy. Although this halo extends over a sphere, so that globular clusters are not concentrated in the plane of the *Milky Way*, most of the globular clusters are no further from the centre of the Galaxy than we are. Similar globular clusters distributed in similar spherical haloes have been identified in other *galaxies*. Some globular clusters seem to be slightly flattened, but most are almost perfectly spherical.

There are about 150 globular clusters known in our Galaxy. At the heart of a globular cluster, the stars are so close together that there may be as many as 1,000 stars in a single cubic *parsec* of space (for comparison, there is no other star within 1 parsec of our Sun!). The way these clusters are distributed, their low *metallicity*, and the ages of the stars they contain, suggest that they formed when the Galaxy was young, 15 billion years or more ago, not long after the *Big Bang* itself. They contain mostly *Population II* stars, many of which have evolved to become *red giants*.

Because all of the stars in an individual globular cluster must have formed together, and they are all at essentially the same distance from us, the exact way in which the stars are distributed in a *Hertzprung–Russell diagram* (in particular, the point at which the red giants turn off from the

The globular cluster M15 (also known as NGC 7078), which lies in the direction of the constellation Pegasus.

main sequence to form the *horizontal branch*) gives an indication of the age and distance of all the stars in the cluster. This technique was important in estimating the size of our Galaxy, and gives the oldest cluster ages as about 18 billion years; any satisfactory model of the Big Bang must therefore give an age for the Universe of about 20 billion years or more, so that the globular clusters had time to form.

gluon   An *elementary particle* which carries the strong interaction (one of the *fundamental forces*) between *quarks*, and holds particles such as *protons* and *neutrons* together. It is a member of the *boson* family, carrying out a role similar to the way *photons* carry the electromagnetic force between electrically charged particles.

glycine   An important *molecule* in living systems, an *amino acid* which is one of the building blocks of proteins. Glycine has been identified by *spectroscopy* in *interstellar material* in a region of *star* formation about 100 *parsecs* from the *galactic centre*. This is strong evidence that there may be life elsewhere in the *Galaxy*. See *Hoyle, Fred, life in the Universe* and *panspermia*.

Goddard Space Flight Center (GSFC)   American facility in Greenbelt, Maryland, which is run by NASA and operates both *satellites* and sounding rockets for scientific research, as well as a global network of tracking stations. It is named after Robert Goddard (1882–1945), who developed the first liquid-fuelled rocket in 1926.

Gödel, Kurt (1906–78) Austrian-born American mathematician who worked at Princeton alongside Albert *Einstein*, and discovered a solution to the equations of the *general theory of relativity* which describes a rotating universe *model* in which *time travel* would be possible.

Gold, Thomas ('Tommy') (1920–) Austrian-born American physicist who was one of the originators (in the 1940s) of the *Steady State Hypothesis*, and who was the first person to suggest in print (in 1968) that *pulsars* are rapidly rotating *neutron stars*.

Goldilocks effect See *anthropic principle*.

GONG Acronym for the Global Oscillation Network Group, a project which monitored the behaviour of the Sun from six observing stations spaced roughly evenly around the world during the 1990s. The particular aim of the project is to monitor the way in which the Sun oscillates, using *helioseismology* to probe the Sun's interior in a similar fashion to the way in which seismologists study earthquakes to find out the structure of our planet. With six observing stations, the Sun can be monitored continually, 24 hours a day, even if one of the stations is out of action for any reason; the sites were chosen to minimize cloud cover.

At each site an automated instrument makes simultaneous measurements of the way the surface of the Sun is moving, using the *Doppler effect*, at 65,000 points across its surface. The whole project, run by the US National Solar Observatory, involves more than 150 individual scientists from 61 different research centres in 15 different countries, and generates a gigabyte of data every day – for comparison, a 300-page book contains less than half a megabyte of information, so each day's data from GONG is equivalent to more than 2,000 such books. Hardly surprisingly, the data are still being analysed and will provide a resource for years to come, but the project has already helped to refine models of how the Sun works, and to confirm the accuracy, for example, of calculations of the temperature at the heart of the Sun.

Gould Belt A belt of stars and gas which is tilted at about 16° to the *galactic plane* and contains a large number of *giant stars*. First noted by Sir John *Herschel* in 1847, and later studied by the astronomer B. A. Gould, the belt is a band of young stars branching off from the nearest spiral arm of our *Galaxy*.

Graham Smith, Francis See *Smith, Francis Graham*.

GRANAT A Russian *satellite* launched in 1989 to make *X-ray* and *gamma ray* observations of the sky.

grand unified theories (GUTs)   General name for theories or *models* which attempt to describe the physical behaviour of all particles and forces (the *fundamental interactions*) in one set of mathematical equations. Such a theory, which has not yet been discovered, is sometimes referred to as a theory of everything, or TOE. Only slightly tongue in cheek, some physicists say that their Holy Grail is a single equation that defines a TOE and can be written on the front of a T-shirt.

This is not a completely ludicrous dream, because there has already been considerable success in unifying the physicists' description of the material world. As recently as the middle of the 19th century, electricity and magnetism seemed to be two separate forces, but the work of James Clerk *Maxwell* showed that they are really two facets of the same fundamental interaction, now known as electromagnetism, described by one set of equations. By the middle of the 20th century, this description had been improved to include the effects of *quantum mechanics*, and in the form of quantum electrodynamics (QED) is one of the most successful theories physicists have ever come up with, correctly predicting the behaviour of interactions involving charged particles such as electrons to very many decimal places.

QED is a *gauge theory*, and because of its success it has become the archetype for physicists trying to develop theories that describe the behaviour of the other fundamental interactions. The essence of QED is that charged particles such as electrons and protons interact with one another by exchanging *photons*, which are regarded as the *quanta* of the electromagnetic field. In an analogous way, the *weak interaction*, which is responsible for processes such as *beta decay* in *nuclei*, is described as being mediated by the exchange of particles which play the equivalent role to photons. They are known as intermediate vector *bosons*.

In the 1960s, physicists found a mathematical theory which combines QED and the weak interaction in one mathematical package. This became known as the electroweak theory, and it made specific predictions about the nature of the intermediate vector bosons. The theory requires three of these entities, dubbed W^+, W^- and Z^0, and predicts what their masses should be. These particles were discovered in the 1980s, with exactly the properties predicted by the theory.

The next step towards a TOE would be to include the *strong nuclear interaction*, which holds the particles in the nucleus together, in the package. This has not yet been achieved, but as an intermediate step physicists have developed a description of the strong interaction using a gauge theory based upon the success of quantum electrodynamics. In this picture, the strong interaction is seen as arising from the exchange of *gluons* (the equivalent in this theory of photons in QED) between *quarks*.

Because some of the properties of quarks (actually equivalent to a different kind of charge) have been somewhat whimsically given the names of colours, this theory is called quantum chromodynamics, or QCD, in a conscious mimicry of QED.

Unfortunately, whereas QED requires just one kind of photon, and the electroweak theory adds in only three intermediate vector bosons to the calculations, QCD requires eight distinct gluons, which makes the theory that much more complicated and difficult to work with. Even so, there are realistic prospects of finding a way to include QCD with the electroweak theory into one unified description of the particle world; but it is much harder to see how *gravity*, the fourth of the fundamental interactions, could be included in the package. Even without gravity, the hoped-for combination of electromagnetism, the weak interaction and the strong interaction into one package is itself often referred to as a GUT, reserving the name TOE for the final theory which, physicists hope, will include gravity as well.

The reason why it is so hard to include gravity in the picture is the extreme weakness of gravity compared with the other three forces of nature. In one way, though, gravity is as simple to deal with as electromagnetism, because it requires only one kind of mediating particle, the massless *graviton*.

The difficulty of including gravity in a TOE can be seen by looking at the way in which, physicists suggest, the four fundamental interactions 'split off' from a single unified interaction as the Universe emerged from the Big Bang. One of the essential differences between photons and the intermediate vector bosons and gluons is that photons have no *mass*, whereas the other particles do have mass. Because they are massless, photons can be created easily, and can travel (in principle) right across the Universe. The bosons that mediate the weak and strong forces cannot do this. During an interaction, the mass needed to 'create' the appropriate set of bosons is borrowed from the vacuum, in line with the uncertainty principle of quantum mechanics, which says that so-called 'virtual' particles can flicker in and out of existence, provided that they do not last long enough for the Universe to 'notice' their existence. The more mass such a particle has, the more energy it has to borrow during its brief existence, and the more quickly it must repay the debt. This limits the range over which the boson can move before it completes its task and disappears.

But when the Universe was very young, it seethed with the energy of the primordial fireball. Provided that the density of this energy was high enough, even gluons and intermediate vector bosons could extract enough energy from the fireball to become real particles, and travel everywhere in the fireball. They would have been *exactly* equivalent to photons, not merely analogous to them, and all the fundamental interactions would have been equally strong and equally long-ranged. But as the Universe expanded and cooled, they successively lost some of their powers, and became short-range particles restricted to the places we see them today, notably inside the nucleus.

Gravity itself is still the odd one out, on this picture, and according to the best theories we have it would have been the same strength as all the other forces when the Universe as a whole was at a temperature of 10^{32} K. That was just 10^{-43} seconds after the moment at which the Universe emerged from a *singularity*, when everything that we can see today was contained in a volume no bigger across than the *Planck length*. A more realistic way of looking at this is to say that the Universe was born with an age of 10^{-43} seconds, and that there was no 'before' during which gravity really was ever the equal of the other forces. It was just after this that *inflation* is thought to have occurred.

As the Universe began to expand smoothly and cool, the other three forces were still unified. But at a temperature of 10^{28} K, just 10^{-36} seconds after the beginning, it became too cool to maintain the carriers of the strong force, which became restricted to the distances we know today. At 10^{-12} seconds, with a temperature of 10^{15} K, the Universe became too cool to support the intermediate vector bosons, and the weak force also became restricted to short ranges. This occurred when the whole Universe was at a temperature equivalent to the highest energies yet achieved in particle accelerators on Earth – which is one reason why the electroweak theory has been much more firmly established (by comparison with experiments) than QCD has been.

On this picture, it is easy to see why it has proved so difficult to include gravity in unified theories. Curiously, though, gravity had already been included in a unified theory with electromagnetism before the strong and weak interactions were discovered! This approach to unification was largely forgotten (or ignored) for many years after the two 'extra' forces were identified, but it now looks like a front runner to produce the long-sought theory of everything.

The *general theory of relativity* describes gravity in terms of the curvature of four-dimensional *spacetime*. Soon after Albert *Einstein* had come up with this idea, it was discovered that writing out the equivalent of Einstein's equations of the general theory to represent curvature in five dimensions produced the familiar *field equations* of Einstein's theory, alongside another set of field equations, which are precisely Maxwell's equations of electromagnetism. Within a few years, in the 1920s, this five-dimensional unification of gravity and electromagnetism was even extended to include quantum effects, and became known as *Kaluza–Klein theory* after the two researchers who pioneered the work.

All theories that involve adding extra dimensions into the calculations are now known as Kaluza–Klein theories, but the approach was largely ignored for many years because it requires not just one but several 'extra' dimensions to include the effects of the more complicated weak and strong interactions, which were discovered just after the initial triumph of Kaluza–Klein theory. If a photon is a ripple in the fifth dimension, then (crudely speaking) the Z particle might be regarded as a ripple in the sixth dimension, and so on.

The revival of interest in this kind of theory happened in the 1980s, for two reasons. First, attempts at constructing grand unified theories were getting horribly complicated, and some of them seemed in any case to require extra dimensions in order to work. If you had to have many extra dimensions anyway, why not use the Kaluza–Klein approach? Second, mathematical physicists became interested in string theory, which describes entities that we used to think of as point-like particles as tiny pieces (much, much smaller than protons) of one-dimensional 'string'. String theory also only 'works' in many dimensions, but it produces an enormous bonus – gravity.

Theorists were playing with the equations that describe the interactions of these strings in many dimensions, when they discovered that closed loops of string described by some of the equations have exactly the properties required to provide a description of gravity – the loops of string are, in fact, gravitons.

Nobody was trying to use the theory to describe gravity, which was expected to be the hardest of the fundamental interactions to bring into the string net; but it just fell out of the equations of its own accord. Unfortunately, nobody knows why this should have happened – there is no physical insight into what the theory *means* – and string theory is still largely a mathematical toy, with no grounding in physics. It's rather as if some mathematician had discovered Maxwell's equations in a world where nobody knew about magnetism or electricity; the equations are very pretty, but what are they trying to tell us?

One of the experts on string theory, Michael Green (then of Queen Mary College in London), pointed out in an article in *Scientific American* in 1986 (volume 255, no. 3, page 44), that in string theory 'details have come first; we are still groping for a unifying insight into the logic of the theory. For example, the occurrence of the massless graviton and the gauge particles that emerge from superstring theories appears accidental and somewhat mysterious; one would like them to emerge naturally in a theory after the unifying principles are well established.'

That 'groping for a unifying insight' continues in the 1990s. Physicists still hope to find a theory of everything, and they strongly suspect that it must involve an understanding of a multidimensional Universe and probably of particles as tiny strings. But they are still a long way from being able to write the answer to 'life, the Universe and everything' on the front of a T-shirt.

See also **fundamental forces**.

granulation   Grainy pattern on the surface of the Sun, visible with the aid of solar **telescopes**, caused by the convection of hot gases in the surface layer. Individual 'grains' are 300–1500 km in diameter, and last for a few minutes.

gravitational collapse   Strictly speaking, the collapse of any object as a result of the gravitational pull of all of its constituent parts on each other. When astronomers use the expression without any qualification, however, they are usually referring to the stage at the end of the life of a massive star when it can no longer hold itself up against the inward tug of gravity by generating heat in its core through **fusion**. This happens when there is no more nuclear fuel left in the core for the star to burn.

With nothing left to hold the outer layers of the star up, they collapse inward very rapidly, in less than a second. This is the gravitational collapse. It releases gravitational energy, which blasts most of the star's mass out into space in a **supernova** explosion. The central core of the star continues the collapse, eventually forming a **white dwarf**, a **neutron star** or a **black hole**, depending on how much matter has been left behind.

gravitational constant   The universal constant G that is a measure of the strength of the **gravitational force**. Claims that G may actually vary slowly with time (as the Universe expands) or with distance have not been borne out by observations.

gravitational field   The idea that the gravitational influence of any object at any point in space can be represented by a number indicating the 'strength' of gravity at that point. Strictly speaking, the gravitational field of an object extends to fill the entire Universe, but in practice the influence of an object is significant only in its immediate vicinity (although in the case of a **quasar** or a **galaxy** the 'immediate vicinity' may extend over millions of **parsecs**.

Field theory was first developed by James Clerk **Maxwell** in the 19th century, to describe electromagnetism. Albert **Einstein** developed his field theory of gravity (the **general theory of relativity**) early in the 20th century. The important feature of both theories is that they describe the field in terms of a set of **field equations** which indicate the value of the field at any point, and that the value changes smoothly from one point to the next, so that the field has nearly the same strength for neighbouring points.

Before field theory was developed, the picture physicists had of inter-actions between particles was of something reaching out across the gap between the particles and affecting them directly – action at a distance. Field theory says that, rather, the action is a local phenomenon, with each particle interacting with the field at its own location, and the field interacting with each particle, although the overall structure of the field depends on the nature and distribution of all the particles.

See also *Mach's Principle*.

gravitational force   See *gravitational field*.

gravitational instability   The tendency for a small irregularity in a cloud of material to grow because of *gravity*. In a cloud of gas in space, any region slightly more dense than average will get more dense still as it attracts material from its surroundings, while any region with below average *density* will thin out more as it loses material to nearby denser regions.

gravitational lens   A cosmic magnifying glass produced when the *gravity* of an object bends light from a more distant object and makes the image of the distant object brighter to astronomers.

In some cases, the object acting as the gravitational lens is a *galaxy*, and the bending of light by the galaxy can produce multiple images of more distant objects such as *quasars* or other galaxies. Several systems in which this effect occurs are known; some astronomers suggest that, as well as these obvious cases of gravitational lensing, as many as two-thirds of all known quasars may have been brightened by the effect of gravitational lensing.

Gravitational lensing also occurs on a smaller scale, when a dark object

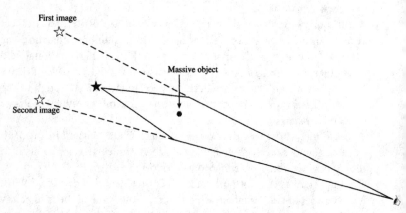

Gravitational lens. Greatly exaggerated representation of the way gravity can bend light to create multiple images of a distant object.

Gravitational Lens G2237+0305

Gravitational lens. A spectacular example of a gravitational lens. The four corner images in this Einstein Cross are all views of the same distant quasar, produced by light being bent around the central galaxy.

in our own *Galaxy* passes directly in front of a more distant star (for example, one of the stars in the *Magellanic Clouds*) and makes its image brighten briefly. Examples of this kind of gravitational lensing were first observed, by several teams of astronomers, in 1993, and confirmed the existence of compact dark objects (*MACHOs*) in our Galaxy. Lensing by individual stars in this way is sometimes called 'microlensing'.

Gravitational lensing is a prediction of Albert Einstein's **general theory of relativity**, and the bending of light involved is explained as a consequence of the distortion of **spacetime** near a massive object, so that the light rays are following **geodesics** through curved space. Observations of gravitational lensing provide evidence that Einstein's general theory is indeed a good description of the way gravity works.

Studies of the effect of gravitational lenses on light from distant quasars also contribute to the debate about the age of the Universe and the rate at which it is expanding. According to results published in 1995 by a team of astronomers from Tartu Observatory, in Estonia, and Hamburger

Sternwarte, in Germany, the time delay between light flickers going one way around the intervening galaxy and the equivalent flickers going the other way around the galaxy for two images of a quasar known as QSO 0957 + 561 implies that the value of the **Hubble constant** must be less than 70 km per second per Megaparsec.

The great attraction of the gravitational lensing technique is that it is a 'clean' measure of the Hubble constant, and that it deals with objects that are very far away. When the distant quasar flickers, all you have to do is note the flicker in one image and wait until the same flicker shows up in the other image. Since light travels at the speed of light, this tells you how much further the second path around the intervening galaxy is.

Because of the geometry of the situation, the actual distances to the intervening galaxy and the quasar itself cancel out of the calculation, and the time delay gives a direct measure of the Hubble constant. But you have to be patient – in the case of QSO 0957 + 561 the delay is 423 days, and you have to monitor several flickers to be sure you have got it right.

See also **Einstein cross**, **Einstein ring**.

gravitational mass   The amount of matter in an object, defined in terms of the **gravitational force** that it exerts. Gravitational mass is exactly equivalent to **inertial mass**, but there is no accepted theory of why this should be so. See **Mach's Principle**.

gravitational radiation   Ripples in the fabric of **spacetime** caused by the motion of objects having **mass** when they move in certain ways. Gravitational radiation, which is associated with acceleration and orbital motion, is a prediction of Albert Einstein's **general theory of relativity**, and travels at the **speed of light**. The theory says that this radiation is completely negligible except where there are strong **gravitational fields**; although the radiation has yet to be detected directly, its existence was spectacularly confirmed by observations of the **binary pulsar** in the 1980s.

The image of matter as solid lumps embedded in a stretched rubber sheet, representing spacetime, makes the origin of gravitational radiation clear. When one of the lumps vibrates, it sends out ripples through the sheet, and these ripples set other lumps of matter vibrating. This is analogous to the way in which a vibrating charged particle sends out electromagnetic radiation in the form of waves which shake other charged particles; but gravitational radiation is very hard to detect, because it is only 10^{-38} times as strong as electromagnetic radiation.

One way in which to detect gravitational radiation is to suspend a large bar of material, protected as far as possible from all other sources of vibration, and watch it with sensitive instruments to see if it is disturbed by the passage of gravitational waves. This was tried, using large bars of aluminium, in pioneering experiments in the 1960s and 1970s. Those experiments were so sensitive that they could monitor vibrations in the bar caused by vehicles driving by in the streets outside the laboratory, but they did not succeed in identifying the 'signature' of gravitational radiation. This was no surprise because, if Einstein's theory is correct, any such radiation in the vicinity of the Earth would be too weak to produce detectable shaking of the bars. Nevertheless, it was, of course, worth carrying out the experiments, both to make sure that there was nothing going on that Einstein's theory had not predicted, and to develop techniques that could be used in more sensitive gravitational radiation detectors. Such detectors are now being built, and if Einstein's theory is correct and they work as planned, they may detect gravitational waves around the beginning of the 21st century.

There are two sources of such radiation which ought to produce ripples in spacetime big enough to be detected by the next generation of instruments. The first is the collapse of the core of a large star to form a **neutron star** or a **black hole** when the outer layers of the star explode in a **supernova**. Such events are rare by human timescales, but a detectable supernova will occasionally occur in our Galaxy – on average, about once every 25 years. When they do occur, they should produce enormous amounts of gravitational radiation in a short time – the energy equivalent (mc^2) of the entire mass of our Sun in a burst of radiation lasting for just 5 microseconds (for comparison, the rate of gravitational radiation produced by the motion of the Earth in its orbit around the Sun is just 200 watts, equivalent in energy terms to the output of an ordinary light bulb).

Even if such an event occurred 10 kiloparsecs away, near the centre of our Galaxy, the amount of gravitational radiation reaching the Earth from it would be equivalent in energy terms to all the electromagnetic energy, across the entire **spectrum**, reaching us from the Sun in about 100 seconds. Such an outburst would be relatively easy to detect. But because such events are rare, the best bet for the first direct observation of gravitational radiation is to detect the kind of gravitational radiation produced in systems like the binary pulsar, where two very dense stars are in orbit around one another.

Such a system is like an extreme version of a weightlifter's barbell. Viewed in the plane of rotation, this produces gravitational waves which can be visualized in terms of their effects on a circular ring in the same plane. Physicists call this kind of radiation 'quadrupole radiation'.

Quadrupole radiation can be understood most simply in terms of radiation from electric charges. A pair of electrical charges, one positive and one negative, forms a dipole, and when these two charges move (vibrating in and out, or rotating about one another) they produce dipole electromagnetic radiation. A dipole is electrically neutral overall, even though it can radiate in this way.

A pair of dipoles together can make a quadrupole, with two positive charges and two negative charges. When the charges in such an array move in an appropriate way (for example, with one dipole rotating around the other), they produce quadrupole radiation. Unlike electricity, however, mass comes with only one 'sign', so there is no gravitational equivalent of electromagnetic dipole radiation. Two masses which rotate around one another actually behave like a pair of dipoles, producing the gravitational quadrupole radiation which is visualized in terms of its effect on that circular ring.

As the wave passes by, the ring is simultaneously squeezed in one direction and stretched in another direction at right angles, so that it becomes an ellipse. Then, the pattern reverses, first returning the circle to its original shape, then distorting it into an ellipse at right angles to the first ellipse. This pattern of alternate squeezing and stretching in two directions at right angles is the characteristic signature of quadrupole radiation. In order to detect such radiation, you need just three test masses, placed in a right-angled 'L' shape, to monitor the distortion of spacetime as the wave passes. And you need some *very* accurate measuring instruments.

The approach being used in the new generation of gravitational radiation detectors involves placing the three heavy masses in evacuated tubes several kilometres long, built underground. The test masses have mirror surfaces, and are monitored using laser beams which bounce to and fro off the mirrors in the evacuated tunnels. The laser beams from each arm of the detector are combined in an **interferometer**, which measures changes in the positions of the test masses in terms of the wavelength of light. A typical design involves tunnels 3 km long, and can measure changes in the separation of the masses at each end of one of these tunnels to an accuracy of 10^{-18}m – less than the size of an atomic nucleus. The whole experiment is very similar to the **Michelson–Morley experiment** used, at the end of the 19th century, in the fruitless attempt to detect the motion of the Earth relative to the ether.

Eventually, equivalent systems may be built on much larger scales in space or on the Moon. Meanwhile, just in case, astronomers keep tabs on the positions of distant spacecraft (such as the **Voyager probes**), using measurements of the **Doppler effect** on their radio signals to see if they are being shaken by gravitational waves. No such shaking has yet been observed.

It is just possible that all of these experimental efforts to detect gravitational radiation may be upstaged by a completely different approach. According to some versions of the theory of **inflation**, the interaction between gravitational waves and matter when the Universe was very young should have produced a characteristic pattern in the distribution of matter across the Universe at the time when matter last interacted directly with the *background radiation*, about 300,000 years after the *Big Bang*. If so, this pattern should be preserved as a fossilized remnant in the background radiation itself, and may be detectable within a few years as observations of the background radiation using instruments like those in the *COBE satellite* and ground-based detectors improve. The ripples in the background radiation may contain information about ripples in the fabric of spacetime.

Further reading: Kip Thorne, *Black Holes and Time Warps*.

gravitational radius See *Schwarzschild radius*.

gravitational redshift See *redshift*.

gravitational time dilation Slowing down of clocks caused by a *gravitational field*, as predicted by the *general theory of relativity*. See *redshift*.

graviton Hypothetical particle required by *quantum theory* to carry the force of *gravity* between two objects that have *mass*. It is a *boson*, carrying out the equivalent role for gravity that the *photon* does for the electromagnetic force between two charged particles. Gravitons occur naturally in some versions of *string theory*.

gravity The force of attraction that exists between any two objects that have *mass*. This is one of the four *fundamental interactions* known to physicists.

Isaac *Newton* realized, in the 17th century, that gravity acts in the same way for all objects anywhere in the *Universe* (that it follows a universal law), and that the attraction between two objects is proportional to the product of their masses and inversely proportional to the square of the distance between them – the famous 'inverse square law of gravitation'. This law explains both the fall of an apple from a tree and the nature of the orbits of the Moon around the Earth and the planets around the Sun.

Albert *Einstein* explained the inverse square law, early in the 20th century, as a result of the way *spacetime* is distorted by the presence of matter. His *general theory of relativity* therefore goes further than Newton's theory of gravity, but includes Newton's theory within itself.

Gravity is the weakest of the four forces of nature, but because the gravitational influence adds up for every single particle in a lump of matter, and because the force has a very long range (in principle, infinite range), the overall effect of a lot of particles pulling together can be very strong. The gravity of the Earth holds everything down on its surface, the gravity of the Sun holds planets in their orbits, and the gravity of everything in the *Galaxy* holds the stars themselves in their orbits. In extreme cases, gravity can cause the collapse of spacetime into a *black hole*. Except for in the first split-second after the beginning of time (during the era of *inflation*), gravity is the only force that has to be taken into account in describing the evolution of the Universe at large.

great attractor   A large concentration of *mass* in the direction of (but far beyond) the *constellations* Hydra and Centaurus.

Studies of the way *galaxies* move, using the *Doppler effect*, show that our own *Galaxy*, the *Local Group*, the *Local Supercluster* and other *clusters of galaxies* in our part of the Universe are all streaming in the direction of the great attractor. This motion is superimposed on their motion as part of the *expanding Universe*. The motion of our Galaxy in that direction, at a speed of about 600 km/sec, is also shown by measurements of the *background radiation*. This radiation is slightly hotter in the direction we are moving, and slightly cooler behind us, because we are running 'head on' into the radiation in front and running away from the radiation behind.

All of these observations can be explained if there is a concentration of mass equivalent to about a million *galaxies* like our own about 40 million *parsecs* away in the direction we are moving, tugging on us through its *gravity*. Although there does seem to be a large concentration of clusters of galaxies in the right place to form the great attractor, it is difficult to study this concentration of mass because it is largely obscured by dust in the *Milky Way*. If these observations are correct, however, they provide strong evidence that the overall density of the Universe is very close to the critical value needed to make it 'closed' (see *cosmological models*, *Omega point*).

Great Wall   See *large-scale structure*.

Green Bank Telescope   *Radio telescope* at Green Bank in West Virginia, which had an *aperture* of 91 m. It started operations in 1962, but collapsed in 1988. A replacement with an aperture of 100 m, the world's largest fully steerable dish *antenna*, was scheduled to begin operations in 1995.

greenhouse effect   The process by which the atmosphere of a planet traps heat near the surface of the planet, making the surface warmer than it would be if the planet did not have an atmosphere. The effect works because incoming radiation from the Sun (mainly in the form of visible light) passes through the atmosphere and warms the surface below. The warm surface radiates energy back out into space, but this outgoing radiation has much longer wavelengths than the incoming solar energy – in the infrared part of the spectrum rather than in the form of visible light. The outgoing infrared radiation is partly absorbed by gases in the atmosphere, and these warm gases in their turn re-radiate the energy, some going up through the atmosphere and out into space, some back down towards the ground to keep it warm.

The clearest example of the power of the greenhouse effect is shown by comparing the temperature on the surface of the Earth with the temperature on the surface of the Moon. On the airless Moon, averaging over the cold dark side and the hot sunlit side, the temperature is about –18 °C. The Earth is at essentially the same distance from the Sun, and if it too were an airless ball of rock, it would also have a temperature of –18 °C. In fact, the average temperature over the whole surface of the Earth is about 15 °C. Our planet is kept some 33 degrees warmer than it would otherwise be by the power of the greenhouse effect.

But the greenhouse effect is *not* the process that keeps the inside of a greenhouse warm! Inside a greenhouse, air is warmed by the ground, which has itself been warmed by the Sun. The hot air would like to rise by convection, but it cannot because the glass roof of the greenhouse prevents this. So the hot air stays inside the greenhouse and gets even hotter, while air just outside the greenhouse rises through convection and allows cooler air to come in and take its place.

Greenstein, Jesse Leonard (1909–)   American astronomer (born in New York City) who was one of the leading pioneers of *radio astronomy*, and was involved in the identification of the *redshift* of the first known *quasar*, 3C 48.

Greenwich Mean Time (GMT)   The time, defined in terms of the apparent motion of the Sun across the sky, on the 0° meridian, which passes through Greenwich, in London, the original site of the *Royal Greenwich Observatory*.

Gregorian calendar   The calendar widely used on Earth today, introduced by Pope Gregory XIII in 1582. It replaced the less accurate *Julian calendar*.

Because the time taken by the Earth to orbit once around the Sun (the year) is not an exact number of days, any calendar system which has the same whole number of days in each year will gradually get out of step with the seasons. The Gregorian system solves this problem by having an extra day in each year whose number in the Christian era exactly divides by four (a leap year), *except* for century years, *unless* the century number exactly divides by 400. So the year 2000 is a leap year, but 1900 was not and 2100 will not be. In 'ordinary' years there are 365 days, so there are 366 days in each leap year.

Over 400 years, the Gregorian calendar gives an average length for each year of 365.2425 days, which is close enough to the true average, 365.2422 days, for most purposes. The 'error' will only add up to 30 per cent of a day in a thousand years, but eventually it could be corrected by omitting a leap day, perhaps around the year 4400.

When the Gregorian calendar was introduced in Catholic countries in 1582, the old calendar had got so far out of step with the seasons that 10 days had to be omitted, so that the day after Thursday, 4 October on the old system became Friday, 15 October on the new system. In Britain and some other parts of the world, the calendar was not reformed until

September 1752, and by then 11 days had to be omitted to get in step with the new calendar, so that the day after 2 September became 14 September. In Russia, the old calendar stayed in use until after the revolution in 1917, which is why that revolution is sometimes referred to as the October Revolution. Even though it took place on 7 November according to the Gregorian calendar, that was 26 October on the old calendar.

On the Julian calendar, New Year's Day had originally been set as 25 March, to coincide with the vernal *equinox*. By 1582 the calendar had drifted so much that the equinox fell on 11 March; after the Gregorian reform of the calendar, the equinox fell on 21 March, but the Gregorian calendar stipulated that New Year's Day became 1 January. So for centuries, for part of the year different countries did not agree on what year it was, let alone what day or what month.

The Jewish era is counted from the supposed time of the Creation, 3761 BC on the Christian calendar, and the Muslim era counts from the hegira in AD 622.

Gregorian telescope Type of *reflecting telescope* proposed by James Gregory in 1663. The telescope uses two curved mirrors, one parabolic and the other ellipsoidal. Gregory could not obtain mirrors accurate enough to make the telescope work before the design was superseded by the simpler *Newtonian telescope*.

group sunspot number See *sunspot number*.

Gum nebula A huge *emission nebula* which spreads across the southern *constellations* Vela and Puppis. It is about 35 arc degrees across and 400 *parsecs* away from us, with a diameter of about 250 parsecs. It was caused by the effects of a *supernova* explosion that must have been visible from Earth about 1 million years ago.

Guth, Alan Harvey (1947–) American physicist who came up with the idea of *inflation*, and who described the Universe, famously, as 'the ultimate free lunch'.

Guth followed a conventional route to becoming a research physicist in the 1970s. He was born on 27 February 1947 in Brunswick, New Jersey, and attended high school in Highland Park, New Jersey, where his family had moved three years later. He entered MIT in 1964 and took first a Bachelor's and then a Master's degree in physics before going on to complete his PhD in 1972, specializing in theoretical particle physics. At that time, there were many more bright young physicists around than there were permanent jobs for them, and like many of his contemporaries Guth worked in a series of short-term posts, first at Princeton University, then at Columbia University and then at Cornell University. While he was at Cornell, Guth heard Robert *Dicke* give a talk on the *flatness problem*, in the spring of 1979.

In October 1979, Guth moved on again, to spend a year at the Stanford Linear Accelerator Center. Following Dicke's talk, he had been reading up on *cosmology*, and on Thursday, 6 December 1979, after a discussion with Harvard physicist Sidney Coleman, everything crystallized in his mind. Working late that night at home, and into the small hours of 7 December, Guth came up with the basic idea of inflation, the importance of which he immediately appreciated. On the relevant page of his notebook, written that same night, there is a short, five-line passage headed in block capitals SPECTACULAR REALIZATION and boxed around with a double line.

A couple of weeks later, Guth, still learning cosmology, learned of the *horizon problem*, and realized that his 'spectacular realization' solved that as well. The first version of his inflationary *model* was published in 1981, but even before then the idea had spread like wildfire through the astro-nomical community. In the wake of the success of his insight, Guth became first a visiting associate professor and then, in June 1981, full associate professor in the Department of Physics at MIT. There he has remained, and is now Victor F. Weisskopf Professor of Physics.

Guth continues to work on the implications of inflation, including the possibility of *baby universes*. He was luckier than he appreciated at the time to have made his spectacular realization when he did – cosmologists in the then Soviet Union were already working along very similar lines, and some of them, notably Andrei *Linde*, have been closely involved in developing and popularizing the idea of inflation since the early 1980s.

Further reading: John Gribbin, *In Search of the Big Bang*.

H_0 See *Hubble constant*.

HI region A region of space where there is a large amount of ordinary *hydrogen* gas, in the form of neutral *atoms*, perhaps with a small amount of molecular hydrogen present. Because the atoms are electrically neutral, the clouds are sometimes referred to as H^0 regions.

Hydrogen is the most common *element* in the Universe, and provides the bulk of the material from which *stars* are formed. HI regions are typically about 5 *parsecs* across, and may contain 50 times as much matter as our Sun, in a cloud at a temperature of about 70 *Kelvin* (the range of typical temperatures for these clouds is from about 25 K to about 250 K). The clouds are too cold to radiate visible *light*, but they are easily identified by *radio astronomy* techniques, because neutral hydrogen produces a characteristic emission at radio *frequencies*, at a wavelength of 21 cm. The *Doppler effect* of motion on this radio 'line' enables astronomers to work out how the clouds of hydrogen are moving.

Although the *density* of atoms in an HI region is very low by everyday standards – only about 50 atoms per cubic centimetre – overall these clouds contain about half of the mass in the space between the stars in the *spiral arms* of our own *Galaxy* and other *disc galaxies*. Overall, this interstellar material contains about a tenth as much matter as all the bright stars in a *galaxy* like our own.

HII region A region of space where there is a large amount of *hydrogen* gas that has been ionized – that is, where the neutral *atoms* have been broken apart into their constituent *nuclei* (in this case, individual *protons*) and *electrons*. Because the ionized atoms have a residual positive charge, these clouds are sometimes referred to as H$^+$ regions.

The *energy* which is responsible for ionizing the hydrogen in these clouds comes from hot, young *stars* embedded in the HII regions. These stars have formed from clouds of gas in space, and shine brightly in the *ultraviolet* part of the *spectrum*, where light has enough energy to knock electrons free from their individual atoms. The HII regions themselves are roughly spherical in shape, and up to about 200 *parsecs* in size. The density inside such a cloud is about 1,000 times greater than in an *HI region*. The temperatures in these clouds range from about 1,000 K to about 10,000 K.

The hot material in these clouds radiates in the *infrared*, *ultraviolet* and *optical* parts of the spectrum, producing colourful images in astronomical photographs – the *Orion nebula* is a classic example of such a glowing cloud of material surrounding a stellar nursery. The free electrons in the cloud also radiate, in the *radio* part of the spectrum, as they interact with magnetic fields.

Because the largest HII regions all seem to have the same size, measurements of the apparent size of HII regions in other *galaxies* have been used to estimate the distances to those galaxies – the smaller the HII regions seem to be, the further away the galaxy must be.

hadron Any of the fundamental particles that interact through the *strong force* (see *fundamental forces*). All hadrons are composed of *quarks*; *protons* and *neutrons* are members of the hadron family, but this also includes more massive unstable particles. See *baryons*.

hadron era The time during the early stages of the *Big Bang*, between 10^{-35} and 10^{-6} seconds after the outburst from a *singularity*, when the behaviour of the Universe was dominated by the *strong nuclear interaction*, the force which operates between *hadrons*. During this time, conditions were so extreme that individual hadrons could not exist, and most of the matter in the Universe was in the form of a soup of *quarks* and *gluons*. Also known as the quark era.

Hakucho The first Japanese *X-ray astronomy* satellite, launched in 1979. It studied X-ray *pulsars* and X-ray *bursters* during a 5-year lifetime.

Hale, George Ellery (1868–1938) American astronomer who was instrumental in building the large telescopes which transformed the study of the Universe in the 20th century.

Hale was born in Chicago on 29 June 1868, the son of a wealthy industrialist who was involved in the manufacture of elevators. His early interest in designing and building instruments, encouraged by his father, turned to astronomy under the influence of a neighbour, Sherburne W. Burnham, who was an enthusiastic amateur astronomer. Hale studied physics at MIT, graduating in 1890, and spent two years operating a solar observatory in Chicago that had been built under his direction, using his father's money. In 1892, Hale took up a post at the university of Chicago, and from 1895 to 1905 he was the director of the university's **Yerkes Observatory**. This centred around a 40-inch (1 m) **refracting telescope** (still the largest refractor in the world today) built with funds provided (thanks to Hale's charm) by Charles T. Yerkes.

Always searching for ways to build bigger and better telescopes, and skilful at getting money from individuals such as Yerkes, Hale was looking for a home for a 60-inch (1.5 m) mirror that his father had obtained in 1896. He persuaded the Carnegie Institution of Washington to provide financial backing for the **Mount Wilson Observatory** in California, where he became director in 1904 and the new mirror came into action in 1908. By 1918 it had been joined by the 100-inch (2.5 m) **Hooker telescope** (named after the benefactor who paid for it). This instrument was used by Edwin **Hubble** and his colleagues to discover the expansion of the Universe (see **redshift**), and it was the biggest instrument of its kind for 30 years.

In 1923, Hale resigned his directorship of the Mount Wilson Observatories, following a nervous breakdown brought on by overwork. But his idea of a quiet retirement, at the age of 55, involved building a small observatory near his home in Pasadena, and inventing a new kind of spectroscope to study the Sun. He then tried to raise money to build an observatory in the Southern Hemisphere, but failed, suffering a second nervous breakdown. Soon, though, he was back in action with another scheme, at work persuading the Rockefeller Foundation to provide funds for a 200-inch (5 m) **aperture** telescope, and in 1929 $6 million were donated for the purpose. It took nearly 20 years to complete the project, which became the centrepiece of the **Palomar Observatory**, initially run by the California Institute of Technology. Hale died in Pasadena on 21 February 1938, long before the work was completed, but the instrument, commissioned in 1948, was named the **Hale telescope** in his honour.

Although Hale will be remembered chiefly for his success in establishing three major observatories, he also made observations, and among many contributions to astronomy he discovered, in 1905, that **sunspots** are cooler than their surroundings on the solar surface, not hotter; in 1908 he discovered (using **spectroscopy**) the magnetic fields associated with **sunspots**, and in 1919 he discovered (in collaboration with W. S. Adams) the roughly 22-year long 'double sunspot' cycle of solar magnetic activity. He also founded the *Astrophysical Journal*, one of the most important astronomical research publications, and was a key figure in transforming the Throop Polytechnic Institute in Pasadena into Caltech.

Hale Observatories Name used during the 1970s for a group of observatories made up of the *Palomar Observatory*, the *Mount Wilson Observatory*, Big Bear Solar Observatory and *Las Campanas Observatory*.

Hale telescope The 5 m (200-inch) aperture *reflecting telescope* at the *Palomar Observatory*, which was the largest telescope in the world for three decades after its completion in 1947.

half life The time taken for exactly half of the *atoms* in a sample of radioactive material to undergo *radioactive decay*. It is one of the curious features of *quantum theory* that radioactive decay is an entirely probabilistic process, and works in this statistical way. Although the half life is different for different radioactive *isotopes*, if you start out with, say, 10,000 atoms of a particular radioactive isotope then after one half life 5,000 will have decayed, after the next half life a further 2,500 will have decayed, in the next half life a further 1,250 will decay, and so on. Any individual atom (strictly speaking, an individual *nucleus*) in the original 10,000 might decay immediately, or might sit around until all of the other 9,999 atoms have decayed before following suit. There is no way to tell in advance which atoms will decay quickly and which ones later, and, indeed, the atoms themselves do not 'know' their fate.

The term is also used to describe the decay of unstable particles – the *neutron*, for example, has a half life of just over 10 minutes for *beta decay*. The average lifetime of a radioactive particle is always rather more than the half life, because although half of the particles always decay in the first half life, some of the rest hang around for several half lives before finally giving up the ghost; but the half life can be thought of as a typical lifetime for the relevant type of particle or nucleus.

Halley, Edmond (1656–1742) English astronomer, mathematician and physicist who realized that the *comet* which now bears his name is in a regular *orbit* around the Sun under the influence of *gravity*, in accordance with the law of gravity discovered by Isaac *Newton*. He also compiled a *star catalogue*, detected the *proper motion* of some stars using historical records, and initiated a research programme that led to a good estimate of the distance from the Earth to the Sun (the *astronomical unit*). He was the second Astronomer Royal, succeeding John *Flamsteed* in 1720 and holding the post for the rest of his life.

Halley was born at Haggerton, near London, on 8 November 1656. He was the son of a wealthy businessman, and studied at the University of Oxford, where he published three scientific papers on astronomical topics and wrote a book about **Kepler's laws**, but left in 1676 without a taking a degree. This was no handicap to his career, because his book had already been noticed by the Astronomer Royal, Flamsteed, who encouraged him in a project to catalogue stars of the Southern Hemisphere from the island of St Helena, in the South Atlantic. The project took two years, and was not entirely successful, but on his return to England in 1678 Halley was promptly elected a Fellow of the Royal Society, at the age of 22.

His adventures over the next three decades included travelling around Europe to meet other scientists, two years as Deputy Comptroller of the Mint in Chester, the command of a Royal Navy warship (the *Paramour*), and diplomatic missions on behalf of the government. He also made pioneering studies in meteorology and magnetism, and was a key influence in persuading Isaac Newton to publish his epic book, the *Principia* – Halley even funded publication of the book, although this was not entirely an act of altruism since he made a small profit out of it. In 1703 he became professor of geometry at Oxford University, the same university he had left without a degree in 1676, and, now nearly 50 years old, settled down to a more conventional academic career.

Halley's great astronomical work was his study of comets. He calculated the orbits of 24 comets, and realized that the comets seen in 1456, 1531, 1607 and 1682 were separate visits of the same comet. He predicted it would return again in 1758, which it duly did (16 years after Halley's death), confirming both his own reputation and the power of Newton's law of gravity.

From 1710 onwards, Halley also made a careful study of the work of **Ptolemy**, which included a copy of a star catalogue dating from the second century BC, originally compiled by **Hipparcus**. He discovered that some of the star positions given in the catalogue differed considerably from those of the same stars in the 18th century, and realized that the stars had moved across the sky during the intervening centuries.

In 1679, Halley had suggested that observations of Venus as it passed across the face of the Sun (transits) could be used, through a variation on the *parallax* technique, to determine the distance to the Sun. In 1716 he set out detailed suggestions for observation of the transits due in 1761 and 1769. These were observed, long after his death, and used to determine the Sun–Earth distance just as he had anticipated. Sixty-two observing stations monitored the transit of 1761, and a similar number the second transit. When all the data had been analysed, they gave a distance from the Earth to the Sun of 153 million km, closely agreeing with the modern value of 149.6 million km for the astronomical unit. In this way, Halley made his last contribution to astronomy 27 years after he had died (at Greenwich on 14 January 1742) at the age of 85.

Halley's Comet, photographed in 1910 from the Royal Observatory at the Cape of Good Hope, South Africa.

Halley's Comet   A *comet* which orbits the Sun once in about 76 years, between 35 *astronomical units* (beyond the orbit of Neptune) and 0.6 AU (between the orbits of Mercury and Venus). It is visible from Earth during its close approach to the Sun, and was named after Edmond *Halley*, who realized that three comets seen in 1531, 1607 and 1682 were all the same object returning to the inner Solar System, and correctly predicted that it would be seen again in 1758.

halo   Originally a term used to describe the spherical region surrounding a *disc galaxy* (like our own *Milky Way*) through which the *globular clusters* and some individual stars move. More recently, the term has been extended to include the much larger spherical region, dominated by the gravitational influence of *dark matter*, in which *galaxies* are thought to be embedded.

   The bright halo of our own *Galaxy* has roughly the same diameter as the disc of the Galaxy (about 30 kiloparsecs) and contains hot gas as well as the stars of the globular clusters and other old, *Population II* stars. These stars are thought to have formed when the Galaxy was young, before the bulk of the primordial hydrogen and helium had settled down into the disc that now forms the *Milky Way*. Similar haloes are seen around other disc galaxies.

The presence of a much larger dark halo is revealed by studies of the way in which galaxies rotate. Overall, in an individual disc galaxy the amount of matter required to explain the observations is about ten times as much as the matter we can see in the form of bright stars, so 90 per cent of the mass of a galaxy like our own must be in the form of dark material in the extended halo. Some of this may be in the form of very faint stars, dubbed *brown dwarfs*, made of the same kind of material (*baryonic matter*) as the visible stars of the Milky Way. Such objects resemble the planet Jupiter, rather than the Sun. Evidence for the existence of at least some objects of this kind in our own Galaxy has come from studies of the *gravitational lens* effect they produce on the light of more distant stars. Some of the material in the dark halo may, however, be in a non-baryonic form, as particles known as *WIMPs* left over from the *Big Bang*.

In 1994 astronomers at the *Kitt Peak National Observatory* reported the first detection of a faint glow of starlight surrounding a distant galaxy and tracing out the dark halo. This confirmed the reality of dark haloes, but many more observations will be needed to determine whether all of the halo mass is in the form of faint stars, or whether these stars are embedded in a sea of WIMPs.

More prosaically, astronomers also use the term halo in the everyday sense to describe a ring of light around an object.

Hanbury-Brown, Robert (1916–)   British pioneer of *radio astronomy* who was born in Aruvankadu, in India, and worked on radar during the Second World War. He developed the technique of *interferometry* using radio telescopes.

Harvard College Observatory   See *Smithsonian Astrophysical Observatory.*

Harvard–Smithsonian Center for Astrophysics (CfA)   See *Smithsonian Astrophysical Observatory.*

Hawking, Stephen William (1942–)   British theoretical physicist who has made major contributions to our understanding of *black holes* and the origin of the Universe. Hawking, the son of a doctor specializing in tropical medicine, was born in Oxford, where his mother had gone to escape the Blitz while pregnant, on 8 January 1942, 300 years to the day (as Hawking is fond of pointing out) after *Galileo* died. When he was two weeks old, his mother took him back to the family home in Highgate; in 1950 the family moved to St Albans, where Hawking was educated at the local private school before going to Oxford to study physics. He graduated in 1962 and moved to Cambridge to work for a PhD in theoretical physics.

A brilliant student who had obtained a First Class degree with almost effortless ease (he later said that he had done something like 1,000 hours work during his three years at Oxford, an average of one hour a day), Hawking initially found the transition to postgraduate research difficult.

Stephen Hawking (born in 1942).

He became ill late in 1962, and when the illness was diagnosed as motor neuron disease (also known as ALS; a progressive deterioration of the nerves that control voluntary muscle action) his initial reaction was to give up. But then he decided to finish his PhD, and flung himself into his research, 'working hard,' as he later put it, 'for the first time in my life. To my surprise, I found I liked it' (S. W. Hawking, *A Short History*, privately produced pamphlet).

Although Hawking's illness has continued to run its course, he has continued to work hard and has produced a stream of important contributions to cosmology and astrophysics, as well as writing a best-selling book about his work.

Hawking's first important discovery concerned the nature of the Big Bang. The theorist Roger *Penrose*, then working in London, had proved in the mid-1960s that matter that falls into a black hole must be crushed into a mathematical point at its centre, known as a *singularity*. It could not orbit around within the *event horizon* of the black hole without being crushed into the singularity. Hawking turned these calculations around to show that in the expanding Universe, everything must have come *from* a singularity at the beginning of time. He later investigated the way in which black holes may leak energy out into the Universe, shrinking until they explode in an outburst of energy. This work involves mixing ideas from relativity theory, *quantum theory* and thermodynamics, and means that any black hole has a characteristic temperature.

Later in his career, Hawking tried to find a quantum theory of gravity, which would make it possible to unite the four forces of nature (the strong and weak *nuclear interactions*, electromagnetism and *gravity* itself) in one 'theory of everything'. This work has met with limited success, but has suggested that our Universe may exist as a self-contained 'bubble' in spacetime, with no 'edge' in either space or time (this *no boundary condition* is a way of avoiding the singularity at the Big Bang after all). A related possibility is that our Universe is one of many interconnected bubbles (see *baby universes*).

Further reading: Stephen Hawking, *A Brief History of Time*; Michael White and John Gribbin, *Stephen Hawking: A life in science*.

Hawking radiation *Radiation* emitted from near the surface of a *black hole*. This radiation corresponds to a loss of *energy* from the black hole, and for a small black hole the loss may make it shrink away until it disappears. The amount of energy radiated each second depends only on the mass of the black hole (it is greater for smaller black holes), and gives each black hole a characteristic temperature.

The simplest way to understand how the process works is to think in terms of the uncertainty principle of *quantum theory*. One form of the uncertainty principle says that over very short time intervals the Universe itself cannot be sure just how much energy there is in every tiny volume of space. Because of this uncertainty, pairs of particles (such as an *electron* and its *antimatter* counterpart the *positron*) can pop into existence out of nothing at all, provided they very quickly annihilate one another and disappear again.

This process is thought to be going on all the time, everywhere in 'empty' space. When it happens just outside the *event horizon* of a black hole, one of the particles may be captured by the hole, while the other one escapes into the Universe at large. The energy needed to promote the particles into reality has, in effect, come from the *gravitational field* of the black hole – but the hole has made two particles and only swallowed one, so overall it has given up one particle's worth of energy, which the escaped particle has carried off. So the *mass* of the black hole has been decreased by an equivalent amount.

For ordinary black holes made out of dead stars, the effect is of no importance because they will swallow up other particles from their surroundings, more than making up for the mass lost by Hawking radiation. But if tiny black holes were produced in the *Big Bang*, they could lose mass faster than they can swallow up new matter, evaporating and eventually disappearing in a blast of radiation. A black hole with the mass of our Sun would have a temperature of only one-ten-millionth of a Kelvin – but one with the mass of a proton would have a temperature of 120 billion Kelvin.

Any black holes that size formed in the Big Bang should be exploding today in a burst of X- and gamma-radiation. One of the wilder speculations in astronomy, by no means universally accepted, is that gamma ray 'bursters' (see *gamma ray astronomy*) may mark the death throes of such mini black holes.

Hayashi, Chushiro (1920–) Japanese astrophysicist who made contributions to the *Big Bang* theory in the 1950s and to the understanding of *stellar evolution* in the 1960s. The evolutionary track of a young star as it approaches the *main sequence* on the *Hertzprung–Russell diagram* is named after him.

Hayashi track See *Hayashi, Chushiro*.

Haystack Observatory A *radio telescope* with a 36.6 m dish, located in Massachusetts to the northwest of Boston. It is run by the Massachusetts Institute of Technology, mainly at very short radio wavelengths, 6–8 mm. It was built in the 1960s, and originally used for radar studies of the Moon and the *terrestrial planets*.

HDE 226868 See *Cygnus X-1*.

heat death of the Universe See *arrow of time*.

heavy elements In astronomy, any *element* except *hydrogen* and *helium*. Astronomers often refer to all of the heavy elements as 'metals', even though they include non-metallic elements such as *carbon* and *oxygen*.

All of the heavy elements have been built up by *fusion* reactions inside *stars*, from primordial hydrogen and helium (see *B^2FH*). They are scattered through space in stellar explosions (see *supernovae*), providing the raw material from which later generations of stars, richer in heavy elements, and planetary systems can form. The variation in the abundances of heavy elements in stars (their 'metallicity') indicates the relative ages of the stars, and provides clues to the evolution of stars and *galaxies*. The oldest stars in our *Galaxy*, which are found in the *globular clusters* of the *halo*, have much smaller relative abundances of heavy elements than a younger star like the Sun, some as little as 0.2 per cent of the Sun's metallicity.

See also *cosmic abundance*.

heavy hydrogen See *deuterium*.

Heinz soup parameter   See *fundamental forces*.

Heisenberg, Werner Karl (1901–76)   German physicist, born in Würzburg on 5 December 1901, who made many contributions to the development of *quantum theory*, but is best known for his formulation of the *uncertainty principle* in 1927.

helioseismology   The technique of investigating the interior structure of the Sun by studying the way in which its surface moves, analogous to the way in which seismologists find out about the interior structure of the Earth by studying the way the crust moves in earthquakes.

The discovery of the surface oscillations that make it possible to probe the Sun's interior in this way was made by accident in 1960, by researchers from CalTech planning to study random (or chaotic) movements of the hot gas at the surface of the Sun. But the discovery could not be exploited until the 1980s, when the technology needed to analyse these oscillations in detail had been developed.

All of these studies, including the discovery, depend on measurements of the *Doppler effect*, causing a tiny *redshift* in light from patches of the solar surface moving away from us, and a tiny *blueshift* in light from patches of the solar surface moving towards us. As a particular patch of the solar surface oscillates in and out, the Doppler effect on light from that part of the Sun changes in a rhythmic and regular way. You can get some idea of the precision of the instruments needed to measure this effect from the nature of the original discovery – the CalTech team found that patches of the Sun's surface oscillate intermittently, bouncing in and out five or six times in the space of roughly half an hour before the oscillation stops, with velocities of about 500 km/sec and an overall displacement of about 50 km. They move a patch of the Sun's surface in and out over a distance equivalent to just 2 per cent of the Sun's diameter.

At first, these oscillations seemed to be a purely localized phenomenon. But in the 1970s several astronomers independently came up with the insight that each of these short-lived local vibrations could be better explained as an effect caused by literally millions of much smaller vibrations, sound waves trapped inside the Sun and making the surface ring like a bell. What seemed to be a series of 5-minute oscillations was actually the superposition of hundreds of different *frequencies* of oscillation, with periods ranging from about 3 minutes to about 1 hour. The Sun was behaving like a gong in a sandstorm, being repeatedly struck by tiny particles of sand, with new vibrations starting up and old vibrations dying away all the time. The resulting mixture of tones (possibly caused by random vibrations in the first place) is rather like the sort of thing you might hear if the lid of a grand piano was being thumped repeatedly at random; mixed up with the random thumping, you would hear the gentle sound of each of the strings of the piano, vibrating at its natural frequency to produce a pure note, and combining with the pure notes from the other strings to make a great chord.

The waves inside the Sun really are sound waves, like the sound waves that vibrate inside the pipe of a church organ. They combine to disturb the surface of the Sun in a regular way because of the way the speed of sound varies at different depths inside the Sun.

Deeper layers of the Sun are hotter than the surface layer, and they are also more dense than the surface layer. As a result, the speed of sound is greater in the deeper layers. When a sound wave starts moving downwards from the surface of the Sun, the bottom of the wave therefore moves faster than the top of the wave, which brings the wave curving back up from the bottom of the region where convection is important (see *Sun*) and back up to the surface. The waves cannot escape from the surface, so they bounce back into the interior of the Sun, like light reflecting from a mirror. As the whole process repeats, the wave loops its way around the Sun, diving repeatedly into the zone of convection and back up to the surface.

The depth to which each wave penetrates, and therefore the length of each 'hop' around the surface, depends on the wavelength of the vibration. In most cases, the waves bounce around the Sun and eventually fade away without producing a noticeable effect on the surface. But in some cases the length of each hop is just right to make an exact number of hops fit into one circuit of the Sun. In that case, although the wave may bounce six, or twelve, or 600 times in each circuit of the Sun, it always ends up back where it started, and then repeats its journey, reflecting from the same patches on the surface for as long as the wave persists, giving each of them a rhythmic push as it passes. The pattern is called a 'standing wave', and it is exactly equivalent to the standing waves that make a plucked guitar string or a blown organ pipe produce a pure note.

By analysing the notes produced by an organ pipe, a competent physicist could tell you what the dimensions of the pipe were, without ever seeing it. In the same way, by analysing the surface 'notes' produced by the standing waves in their circuits of the Sun, astrophysicists can tell you about conditions in deeper layers, without ever having seen below the surface of the Sun. Although the calculations are made more difficult because the Sun is a fully three-dimensional object, not a linear tube like an organ pipe, the principles are exactly the same. One way of visualizing the overall effect of a particular mode of vibration is to picture a soccer ball made of alternating white and black hexagons (a real soccer ball is made of a mixture of pentagons and hexagons, but ignore that subtlety). Each black hexagon would represent a patch of the Sun moving inward (away from us), while each white hexagon would represent a patch of the Sun moving outward (towards us). Then about $2\frac{1}{2}$ minutes later the pattern would be reversed, with the regions that used to be moving outward now moving inward, and vice versa.

But first you need detailed observations of the way the surface of the Sun vibrates, and sophisticated mathematical techniques to unravel the patterns you can see into the pure notes that they are composed of. It turns out that each individual vibration moves the surface of the Sun in and out by only a few tens of metres (the diameter of the Sun is about 1 million kilometres), at speeds of only a few tens of centimetres per second. It is the combined effect of literally millions of these tiny vibrations that produces the larger, short-lived oscillations first seen in 1960.

The way in which the sound waves move through the convective zone depends on the temperature and depth of the convective zone, and on its composition. Helioseismology shows that the outer part of the Sun is made of 75 per cent hydrogen and 25 per cent helium, in line with what astrophysicists had expected (see *nucleosynthesis*), but that it is a little deeper than astrophysicists had previously thought (on the basis of their theoretical *models*), extending down to about 200,000 km from the surface and covering roughly 30 per cent of the distance from the surface of the Sun to the centre. Some of the sound waves that contribute to the oscillations of the solar surface pass right through the deep interior of the Sun, and these provide information about the conditions (especially the temperature) in the region near the centre, where nuclear *fusion* reactions are going on to keep the Sun hot, and where the solar *neutrinos* originate.

There are slight differences between the measured vibrations of the Sun and the predictions of the equivalent solar vibrations in standard solar models developed before the advent of helioseismology, and this is leading to a revision of the models and an improved understanding of how the Sun and *stars* work; one possibility, still being investigated, is that the temperature at the heart of the Sun may be slightly lower than astrophysicists used to think. This may have a bearing on, for example, the *solar neutrino problem*. Another important discovery from helioseismology is that the pattern of rotation we see on the surface of the Sun, in which the equator rotates more rapidly than regions at higher latitudes (closer to the poles), persists throughout the convective zone, but the inner region of the Sun rotates much more like a solid sphere. This has important implications, not yet fully worked out, for an understanding of the under-lying mechanisms that drive the *solar cycle* of activity.

Helioseismology has become such a valuable tool for probing the Sun that great efforts have been made to get the right kind of observations from instruments which can monitor the Sun continuously, without being affected by cloud cover or by the pattern of night and day. Some studies have been made from instruments on board artificial *satellites*; a joint French and United States team has carried out observations from the South Pole, where the Sun never sets in summer. But the most successful approach so far has been that of the Global Oscillation Network Group, which has set up instruments at sites around the world so that the Sun is always being monitored by at least one of them (see *GONG*).

helium   Second most common *element* in the Universe, and the second lightest, after *hydrogen*. There are two *isotopes* of helium. Each has two *protons* in its *nucleus*; in addition, the helium-3 nucleus contains one *neutron*, while the helium-4 nucleus (also known as an *alpha particle*) contains two neutrons. In helium *atoms*, each nucleus is accompanied by two *electrons*.

helium burning   A *nuclear fusion* reaction in which three *nuclei* of helium-4 combine to form one nucleus of *carbon-12*. The process takes place in *stars* that have used up all the *hydrogen* in their cores. See *B²FH*, *stellar evolution*, *triple alpha process*.

helium flash   When a star with one or two times as much mass as our Sun has exhausted all its *hydrogen* fuel, it contracts and the temperature in its core rises rapidly. When the temperature reaches about 100 million *Kelvin*, *helium burning* begins very suddenly, producing a flash of energy which halts the contraction. The onset of helium burning occurs more gradually in more massive stars.

Helmholtz, Hermann Ludwig Ferdinand von (1821–94)   German physiologist and physicist who was the first person to come up with a precise formulation of the law of conservation of *energy*, and made important investigations of the way in which the Sun generates its heat (see *Kelvin–Helmholtz timescale*).

Helmholtz was born in Potsdam on 31 August 1821. His father taught philosophy at the Potsdam Gymnasium, and his mother was a descendant of William Penn, who had founded the state of Pennsylvania in the USA. Although he was interested in physics and showed great academic ability, Helmholtz studied medicine at university, taking advantage of an arrangement that the army paid his fees in return for a promise that he would serve eight years as an army doctor after he qualified. During four years at the Friedrich Wilhelm Institute in Berlin, Helmhholtz completed his medical studies, but also found time to become a skilled pianist and to take courses in physics and mathematics.

Helmholtz obtained his MD in 1842, and by 1843 he was stationed in Potsdam as a regimental surgeon. The duties were sufficiently light that he had time to carry out experiments in a laboratory that he set up in the barracks. The official biographies report that his skill and reputation as a scientist became so great that he was officially released from his army duties in 1848; there are other reports, however, which suggest that after an official leave of absence to carry out scientific studies he simply refused to return to army life, and was dismissed somewhat ignominiously from the service.

In 1849, Helmholtz became professor of physiology in Königsberg, and went on to a series of other academic posts in a long and distinguished career. While at Königsberg he made his suggestion that the Sun might get its energy from a gradual *gravitational collapse*. This was his only important contribution to astronomy, but he was an inspiring teacher who pointed one of his students, Heinrich Hertz (1857–94) towards the discovery of radio waves. His other students included Albert *Michelson*. Helmholtz was one of the last great polymaths, such a dominant figure in German science in the second half of the 19th century that he was once described as 'the most illustrious man [in the country] next to Bismarck and the old Emperor'; he died in Berlin on 8 September 1894.

Henry Draper Catalog See *Draper, Henry.*

Henyey track The path followed by a *star* in the *Hertzprung–Russell diagram* during its evolution from the end of the *Hayashi track* to the *main sequence*. See *stellar evolution.*

Heraklides of Pontus (388–315 BC) Greek philosopher and astronomer who taught that the Earth turns on its axis once every 24 hours. The idea did not become widely accepted for 1,800 years. Heraklides also suggested that Mercury and Venus orbit the Sun, not the Earth, but thought that the Sun circled the Earth with the two planets moving in *epicycles* around the Sun.

Herbig–Haro Objects Small, faint *nebulae* which are thought to mark the sites of very young stars obscured from view by a cloud of *interstellar matter.*

Hercules X-1 An *X-ray pulsar* in a *binary system* in the *constellation* Hercules. The system consists of a visible star, HZ Herculis, with a *neutron star* companion that orbits the parent star once every 1.7 days. The pulsar has a period of 1.24 seconds, and is surrounded by an *accretion disc.*

Herman, Robert (1922–) Physicist who worked with George Gamow and Ralph Alpher in the 1940s on the model of the origin of the Universe in a hot *Big Bang*. The team made the first prediction that the Universe must be filled with a *background radiation* with a temperature of a few *Kelvin*, but the prediction was largely ignored until this radiation was discovered by accident in the 1960s.

Herschel family (Sir William, 1738–1822; Caroline, 1750–1848; Sir John, 1782–1871) William and Caroline Herschel were both born in Hanover (William on 15 November 1738, Caroline on 16 March 1750), the children of a member of the band of the Hanoverian footguards. After initially following in his father's footsteps, William moved to England in 1757, where he worked as a musician in various posts until appointed organist at a chapel in Bath in 1766. His sister Caroline, almost 12 years younger than him, joined William there in 1772 to act as his housekeeper.

William had become interested in astronomy and began to build his own *telescopes* and make observations; Caroline acted as his assistant, and moved on to her own studies of the heavens.

In 1781 the Herschels discovered a new planet, Uranus. This was a sensation at the time, and their work (largely attributed to William) came to the attention of the king, George III. He had a passionate interest in astronomy, and employed William as his private astronomer from 1782 onwards. The Herschels moved first to Datchet, near Windsor, and then to Slough (in 1786) where William stayed for the rest of his life. He built a series of large telescopes and made many pioneering observations over the next 30 years, and was knighted in 1816.

Herschel extended Charles *Messier*'s catalogue of 100 *nebulae*, eventually to more than 2,000 objects; he discovered two moons of Uranus (Titania and Oberon) and two moons of Saturn (Mimas and Enceladus); and he explained the appearance of the *Milky Way* as being our view of a *Galaxy* 'shaped like a grindstone' seen from the inside. The extent to which

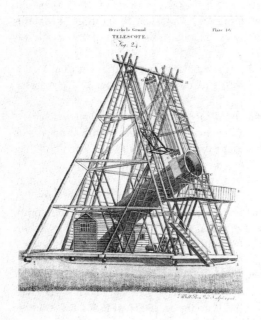

The great telescope used by William **Herschel** in his exploration of the heavens.

Caroline was involved in his work is not clear, but it is now accepted that she was more William's partner than his mere assistant. She discovered many nebulae in her own right, and after William's death in Slough on 25 August 1822 she returned to Hanover, where she prepared a catalogue of 2,500 nebulae and *clusters of stars*, for which she received the Gold Medal of the Royal Astronomical Society in 1828, at the age of 78. She died on 9 January 1848, just over two years short of her 100th birthday.

William had married a wealthy widow in 1788, and their son John was born at Slough four years later, on 7 March 1792. He graduated from Cambridge in 1813, and began to study law, but was drawn into working with his father, who had become ill in 1808, out of a sense of duty. He extended and revised his father's work, and went to South Africa in 1834, where, over a four-year period, he catalogued the nebulae and star clusters of the southern skies. He returned to London in 1838, and more or less gave up astronomy, although he spent years writing up the southern survey, which was not published until 1847.

John Herschel's other interests included chemistry and photography (he invented many of the techniques, and terms such as 'positive' and 'negative', still used by photographers), and was one of the most famous scientists of his time. He wrote a popular book, *Outlines of Astronomy*, which was published in 1849 and was the *Brief History of Time* of its day, and he was made a baronet at the coronation of Queen Victoria in 1837. He died at Hawkhurst, in Kent, on 11 May 1871, regarded as one of the greatest scientists of his time. But he had never had to take an academic post, and was always able to live from his private means and do what research he pleased. Between them, the Herschels opened up the observations of the heavens and set the scene for the developments of the 20th century.

Herschel Telescope   See *William Herschel Telescope*.

Hertz   The standard unit of *frequency* (symbol Hz), defined as one cycle per second. It is named after the German physicist Heinrich Hertz (1857–94), who discovered radio waves, confirming the predictions of James Clerk *Maxwell*. Visible light has frequencies in the range from about 750×10^{12} to $1,500 \times 10^{12}$ Hz.

Hertzprung, Ejnar (1873–1967)   Danish astronomer, born in Copenhagen, who worked as a chemical engineer before becoming an astronomer in 1902. He published the first presentation of what became known as the *Hertzprung–Russell diagram* in 1905, and later worked with Karl *Schwarzschild* at the University of Göttingen and in Potsdam.

Hertzprung–Russell diagram   A kind of graph in which the temperature (or *colour*) of each *star* is plotted against its *absolute magnitude* (a measure of the actual brightness of the star). The position of a star in the

Hertzprung–Russell diagram depends on its mass and its age, and studies of the way stars are distributed on the diagram help astrophysicists to work out how stars evolve.

The potential value of such a diagram in studying the nature of stars was first appreciated by Ejnar **Hertzprung**, working in Copenhagen in the first decade of the 20th century. His version was published in 1911, although it was derived from work published several years earlier. The same idea was arrived at independently by Henry Norris **Russell**, working in Princeton, who published his version in 1913. Hertzprung and Russell never worked together on the idea.

The essential feature of the H–R diagram is that it relates the colour of a star to its brightness. Brightness is measured 'up the page' (the y-axis of the graph), while temperature is measured 'across the page' (the x-axis), with the peculiarity that *cooler* stars are further to the *right* in the diagram. This way of measuring temperature is chosen because essentially it means that from left to right across the H–R diagram the colours of the stars correspond to the sequence O B A F G K M in the classification developed by Annie Jump **Cannon**.

Stars in the bottom right of the H–R diagram are faint, cool and red (with temperatures below 3,500 K), while stars in the top left of the diagram are bright, hot and blue-white (with temperatures above 25,000 K). Most visible stars lie on a band running from top left to bottom right in the diagram, which is called the **main sequence**. This corresponds to all the stars that, like the Sun, get their energy from the **fusion** of hydrogen **nuclei** into helium nuclei in their centres.

The colour of a star depends on the temperature of its surface, but its absolute magnitude depends on the total output of energy across the whole surface of the star. So a very big star can be relatively cool but still very bright, because there is a lot of cool surface giving out energy. Equally, although a star with a small volume and small surface area may be hot and white, it cannot be very bright because there is a limit to how much energy can escape across its surface each second without blowing the star apart. But on the main sequence all the stars are more or less the same size (they are all **dwarf stars**), even though they have different masses. This is the key to using the diagram to understand **stellar evolution**.

The position of a star on the main sequence depends only on its mass. A small star does not have to burn its hydrogen very quickly in order to generate enough heat to hold it up against the inward tug of gravity, so it sits at the cool end of the main sequence; but a massive star has to burn a lot of fuel every second to prevent itself from collapsing under its own weight, so it sits high up on the main sequence. One result of this is that the more massive stars at the top end of the main sequence burn out more

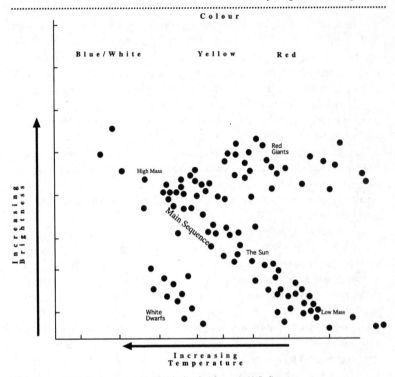

Hertzprung–Russell diagram. A highly idealised, schematic H–R diagram.

quickly than the cooler stars lower down the main sequence. As a star uses up its hydrogen fuel, it becomes slightly brighter and cooler, but still sits essentially on the main sequence. When all of the hydrogen at the centre of the star is used up, the core of the star shrinks, while the outer layers of the star expand (see ***stellar evolution***) as it becomes a ***red giant.***

For a star like the Sun, it takes about 10 billion years of main sequence life to use up the hydrogen fuel in its core. For an M star, with a mass less than about one-tenth of the mass of the Sun, it would take hundreds of billions of years. But for a star with five times as much mass as our Sun, it will take only 70 million years before the star has to become a red giant. The most massive stars on the main sequence are about 50 times as massive as the Sun and perhaps 20 times its diameter.

A red giant still produces a lot of energy, but this energy is now escaping from a much larger surface area, because the outer layers of the star have swollen. So the amount of energy crossing each square metre of the surface is less, and this is what determines the colour of the star. On the H–R diagram, as a star ages it leaves the main sequence and shifts to the right, where it lies in a band known as the *red giant branch*. Some stars end up in a short strip to the left of the red giant branch (but still to the right of the line of the main sequence), known as the *horizontal branch*; these are stars that have lost mass during their time as red giants. Many of them pass through a phase of activity as *RR-Lyrae* or *Cepheid variables*.

Eventually, perhaps after losing mass in a stellar explosion such as a *nova* or a *supernova*, the ageing star (if it has not become a *neutron star* or a *black hole*) runs out of fuel entirely and shrinks inward upon itself. Although the star is now cooling, because it has shrunk the escaping energy is passing through a smaller surface area, so the energy crossing each square metre increases, and the star becomes a hot but faint *white dwarf* in the bottom left of the H–R diagram, before it fades away entirely into a stellar cinder.

Of course, all of this takes far longer than any human timescale, and nobody has seen a star literally moving around the H–R diagram as it evolves. Astronomers have discovered the nature of stars in different parts of the diagram by studies of many stars at different stages in their life cycle, just as you could work out the life cycle of a tree by studying many trees at different stages in their life cycles in a wood over the course of a single year, rather than by watching a single tree for decades to see how it grew and aged. For stars, the observations are compared with computer *models* of how stars evolve, and used to refine those models.

When stars with different mass but the same age as each other are plotted in an H–R diagram, the exact pattern that is produced depends on their age. This shows up clearly when the stars of a *globular cluster* are plotted in this way, because all the stars in such a cluster did form together from the collapse of a single large cloud of gas. The brightest stars at the top left of the main sequence (the ones with most mass) burn their fuel first, because they need so much energy each second to stave off the eventual *gravitational collapse*. So they are the first to leave the main sequence and move across to the red giant branch.

If we could watch the same globular cluster for millions of years, plotting a new H–R diagram for the same stars in the cluster every 100 years or so, the main sequence would seem to shrink away from the top as the cluster aged, rather like a candle gradually burning down. Instead of a complete diagonal main sequence with a scattering of red giants on the branch to the right, the diagonal band at any time would only come up part of the way from the bottom right of the diagram, before turning off to the right. The exact point at which this turn-off occurs depends on the age of the cluster, and these turn-off ages for globular clusters (determined, once again, from a comparison between obser- vations of many real stars and the predictions of the computer models) provide one of the best indications of the ages of some of the oldest stars in our *Galaxy*.

The H–R diagram can also be used in determining the distances to *clusters of stars*, because the positions of the stars in the main sequence is related to their absolute magnitude. The further away a cluster is from us, the fainter the light from its stars will be, and the lower down the H–R diagram its main sequence will seem to lie. This enables astronomers to set the distance to a cluster by choosing the value which adjusts the *apparent magnitudes* of the stars by the right amount to make them match the standard main sequence.

Hewish, Antony (1924–) British radio astronomer who was involved in the discovery of *pulsars* in 1967.

Hewish was born on 11 May 1924 in Fowey, Cornwall, the son of a banker. His undergraduate studies in Cambridge were interrupted by war service at the Telecommunications Research Establishment in Malvern (working on airborne radar countermeasures), between 1943 and 1946. While at Malvern, he met Martin *Ryle*, and after completing his delayed degree in 1948 he joined Ryle's new radio astronomy team at the Cavendish Laboratory in Cambridge, where he received his PhD in 1952. He spent the rest of his career there, becoming professor of radio astronomy in 1971. He retired in 1989.

In the 1950s and into the 1960s, part of Hewish's work was as a member of the team responsible for several radio surveys of the sky, which resulted in a series of *catalogues* of *radio sources*. One of Hewish's special interests was a phenomenon known as *scintillation*, the radio equivalent of the twinkling of visible stars, caused by the effect of electrically charged material in space (ionized gas clouds) on the radio waves from a distant source as the waves pass through the clouds. This twinkling occurs only for objects that look small on the sky, and Hewish realized that it would provide a means for detecting *quasars*.

In the mid-1960s, Hewish was in charge of a project to construct a special *radio telescope* to search for quasars in this way. It was using this telescope that his student Jocelyn *Bell* discovered the first pulsar in 1967. Working with Bell, Hewish soon found several more pulsars, which were explained as rapidly spinning *neutron stars*. The entire project was funded out of a grant of £17,000, which has been described by Sir Bernard *Lovell* as 'one of the most cost-effective in scientific history' (Robert Weber, *Pioneers of Science*, Adam Hilger, 1988).

In 1974, Hewish shared the Nobel Prize in Physics with Ryle: Ryle's share was for his overall contributions to radio astronomy, Hewish's citation specifically mentioned his discovery of pulsars. The award was notable on two counts – it was the first time that any branch of astronomy had been recognized with the award of a Nobel Prize, and it failed to include Bell, who had actually made the discovery mentioned in Hewish's citation.

Hey, James Stanley (1909–)   British physicist who worked on radar during the Second World War and discovered that the Sun is a strong source of radio noise, which caused interference with radar equipment in the 4–8 m wavelength band. He also discovered the *radio source* Cygnus A, and pioneered the use of radar to study *meteors*.

hierarchical model of the Universe   The idea that matter in the Universe is grouped in successively larger clumps, like the way Russian dolls nest inside each other. Stars group together to form galaxies, galaxies are grouped in clusters, clusters of galaxies form superclusters, and so on.

The *model* was proposed by the Swedish astronomer Charles Charlier (1862–1934) at the beginning of the 20th century, and predates the *Big Bang* model of the Universe. If the Universe were perfectly hierarchical, it would be described as having fractal geometry, and would extend to infinity. In fact, although there is evidence of clustering up to the scale of superclusters, with a suggestion of structure on scales up to about 50 million *parsecs* across, the smoothness of the *background radiation* and other evidence suggests that on still larger scales the superclusters themselves, or clusters of superclusters, are distributed uniformly across space. Nevertheless, although the perfect hierarchical model with fractal geometry is not an accurate description of the entire Universe, the structure of the distribution of visible matter in the Universe does follow an enormous hierarchy ranging from fragments of *interstellar matter* at least up to superclusters of galaxies.

High Energy Astrophysical Observatory (HEAO)   General name given to three NASA *satellites* launched in the late 1970s, two of which observed the Universe at *X-ray* frequencies and one at *gamma ray* energies. HEAO-2 was also known as the *Einstein satellite*.

Hipparchus of Nicaea (second century BC, died around 127 BC)   Greek astronomer also known as Hipparchus of Rhodes, because he spent most of his later life there. He measured the length of the year to an accuracy of 6.5 minutes, and made the first realistic estimates of the distances to the Sun and Moon.

Hipparcos   *Satellite* launched by the *European Space Agency* in 1989 that carried out an extremely accurate survey of the positions of stars (see *catalogues*). The name is an acronym for HIgh Precision PARallax COllecting Satellite, a tortuous (and ridiculous) attempt to come up with something similar to *Hipparchus*.

Homestake Gold Mine   The site in Lead, South Dakota, where Ray *Davis* and his colleagues have been monitoring the flux of *neutrinos* from the

Hooker telescope. The 100-inch Hooker telsecope on Mount Wilson. This is the telescope used by Edwin Hubble to discover the expansion of the Universe.

Sun since the late 1960s. The detector is an Olympic-swimming-pool-sized tank of perchlorethylene 1.5 km below the surface of the ground. See *solar neutrino problem*.

homogeneity   The property of being the same everywhere. The Universe is thought to be homogeneous on the large scale, at least as far as we can see, with irregularities such as *galaxies* representing only minor deviations from homogeneity.

Hooker telescope   The main *reflecting telescope* at the *Mount Wilson Observatory*, named after its benefactor. The Hooker telescope has a main mirror with an *aperture* of 100 inches (2.5 m) and began operating in 1918. It was renovated in the early 1990s, and is still one of the most important ground-based telescopes. See *Hale, George Ellery*.

horizon problem   The puzzle that the Universe looks the same on opposite sides of the sky (opposite horizons) even though there has not been time since the *Big Bang* for light (or anything else) to travel across the Universe and back. So how do the opposite horizons 'know' how to keep in step with each other? The horizon problem is resolved by the *inflationary model*.

245

horizontal branch   A feature in the *Hertzsprung–Russell diagram* of a *globular cluster*, corresponding to low-mass stars that have lost mass during the course of their evolution, and now all have similar *absolute magnitudes*. See *stellar evolution*.

hot big bang   See *Big Bang*.

hot dark matter (HDM)   Generic name for any non-baryonic particles that emerged from the *Big Bang* moving at speeds close to that of *light*. In some *models* of the Universe, HDM particles contribute up to a third of the total *mass* (see *dark matter*). If *neutrinos* have mass, they could provide all of this HDM.

See also *mixed dark matter.*

Hoyle, Sir Fred (1915–)   British astronomer who explained how *elements* are manufactured inside *stars* by *nucleosynthesis*, developed the *Steady State* model of the Universe, made many other contributions to astronomy and *cosmology*, and wrote both popular science books and science fiction stories, some of which were turned into television series.

Hoyle was born at Bingley, in Yorkshire, on 24 June 1915, and after an extended confrontation with the school system (which was unable to cope with his precocious independence) he attended the local grammar school before going on to study at Emmanuel College, in Cambridge. He graduated in 1936, and after a period as a research student became a Fellow of St John's College in 1939. With typical Hoyle logic and disdain for convention, however, he never bothered to take a PhD (not an essential requirement in those days) because as a student he would pay less income tax. He did complete all the technical requirements for the degree, but did not formally carry through the bureaucratic procedure to receive it, although he did accept his MA in 1939. Hoyle's career was then interrupted by the Second World War, during which he worked on radar for the Admiralty. During the course of this work, he met Herman *Bondi* and Tommy *Gold*. In 1945 he became a lecturer in mathematics in Cambridge, and in 1958 he was appointed Plumian Professor of Astronomy and Experimental Philosophy. He founded the Institute of Theoretical Astronomy in Cambridge (now part of the Institute of Astronomy) in 1967, and served as its first director until 1973, the year after he had been knighted, when he resigned all his posts in Cambridge after a dispute about the future development of astronomical research. This resignation was the culmination of what had often been a stormy relationship between Hoyle and the authorities; Hoyle always meant what he said, and never made promises he could not keep, which left him frequently baffled by the workings of bureaucracy.

As well as working in Cambridge between 1945 and 1973, Hoyle was a frequent visitor to other research centres, such as the *Hale Observatory*

Sir Fred Hoyle (born in 1915).

and Caltech, and he maintained many of these connections, including an honorary post with University College, Cardiff, in the 1970s and 1980s, while in effect working as an independent scientist. He continued to publish scientific papers into the 1990s, although officially retired and living in Bournemouth.

Hoyle's contributions to *astrophysics* and *cosmology* are the most wide-ranging and important of any scientist of his generation, but because some of his ideas have not fitted in with the orthodoxy of the time he has not always received the recognition he deserves. The only comparable figure in the history of modern astronomy, in terms of the breadth and depth of his contributions, is Sir Arthur *Eddington*.

In the 1940s, Hoyle was one of the three co-founders of the *Steady State* model of the Universe (together with Bondi and Gold), but he developed the idea in an independent fashion, taking it in a different direction from Bondi and Gold. In spite of mounting evidence in the 1960s that was widely accepted as proof that the *Big Bang* model is a better description of the Universe, Hoyle never accepted this. Although he was the person who coined the term 'Big Bang', he intended it as a term of derision for a model he once described as being about 'as elegant as a party girl jumping out of a cake'. The problem with the Big Bang, in his view, is that it has what Hoyle has referred to as 'interesting and sinister properties', meaning the *singularity* at the beginning of time.

With his colleague Jayant Narlikar, from the 1960s onward Hoyle developed a version of the Steady State model which incorporates 'local' regions of rapid expansion; this model closely resembles some versions of *inflation*, but (as much through historical accident and fashion as anything else) inflation is usually regarded as part of the Big Bang description of the Universe.

In the 1940s and 1950s, Hoyle was the leading light in the team that explained how the *elements* are manufactured inside stars (see *B²FH*), and in the 1960s he played a major part in explaining how, if there had been a Big Bang, the lightest elements (such as *hydrogen*, *helium* and *deuterium*) would have been 'cooked' in the first few minutes of the life of the Universe. In a *tour de force* that is still recalled with awe by the astronomers who were present, Hoyle actually carried out the first step in this work, determining how much helium was manufactured in the Big Bang, during a course of undergraduate lectures he was giving on 'Extragalactic Astrophysics'. The former students who attended those lectures describe how the problem was solved before their very eyes, in 'real time', over a period of about three weeks. The success of this work, as it developed into a more complete theory of cosmological nucleosynthesis, is, ironically, one of the cornerstones of the Big Bang model today. Yaakov *Zel'dovich* carried out similar work independently.

Later in his career, and especially after he left Cambridge, Hoyle, in collaboration with Chandra Wickramasinghe, developed the idea that complex *molecules* of the kind essential to life exist in clouds of *interstellar matter*. These ideas were elaborated into a suggestion that life itself might have emerged initially in space, and that evolution on Earth might be influenced by the arrival of new kinds of life form ('viruses from space') carried into the inner Solar System by *comets*. The extreme development of these ideas alienated many people and may have contributed to the decision of the Nobel Committee not to give Hoyle a share of the Nobel Prize which Willy *Fowler* received for his contribution to the work of the B²FH team. But as increasingly complex molecules have been found in space by *spectroscopy* (including at least one *amino acid* and buckminsterfullerene), the basis of Hoyle and Wickramasinghe's work looks more solid. Even if the Earth is not being infected with viruses from space, it now seems highly likely that the prompt emergence of life on Earth soon after our *planet* was formed was in part because it was 'seeded' with organic material (amino acids and the like, even if not actual living cells) by cometary impacts.

As asides to his main lines of research (only a few of which have been mentioned here), Hoyle found time to explain how Stonehenge could have been used in ancient times as an *eclipse* predictor, contributed to the debate about the desirability of nuclear energy, came up with a new theory of ice ages, and wrote many popular and academic books about astronomy, as well as his science fiction and a highly entertaining and insightful autobiography. He was one of the most original scientific thinkers of the 20th century, and he has been proved right more often than he has been proved wrong.

Further reading: Fred Hoyle, *Home is Where the Wind Blows*; John Gribbin, *In the Beginning*.

H–R diagram See *Hertzsprung–Russell diagram*.

Hubble, Edwin Powell (1889–1953) American astronomer who proved that many objects classified as *nebulae* are other *galaxies* (beyond the *Milky Way*), discovered the relationship between *redshift* and distance, and inferred that the Universe is expanding.

Hubble was the fifth of seven sons of a lawyer in Marshfield, Missouri, and was born on 20 November 1889. He attended high school in Chicago, then studied law at the University of Chicago (graduating in 1911) and as a Rhodes Scholar at Queen's College in Oxford. But his interest in astronomy had been kindled by the work of George *Hale* in Chicago, and although Hubble was called to the Kentucky Bar in 1913 and practised law briefly, in 1914 he returned to Chicago to take up a post as a research assistant at the *Yerkes Observatory*, which had been founded by Hale.

Hubble was a natural athlete, who represented Oxford University during his time in England and fought an exhibition bout as an amateur boxer against the French champion Georges Carpentier; he was a good enough boxer to have been offered the chance of turning professional to fight the world heavyweight champion Jack Johnson, but was wise enough to turn down the invitation.

While working at the Yerkes Observatory, Hubble studied astronomy and completed his PhD in 1917, the year in which the United States entered the First World War. Hale offered him a post at the *Mount Wilson Observatory*, but first Hubble volunteered for the infantry and fought in France, where he was wounded by shell fragments in his right arm. He eventually took up the post at Mount Wilson in 1919, just when the new 100-inch (2.5 m) *Hooker telescope* there was coming into operation. He stayed there for the rest of his career.

Hubble was the right man in the right place at the right time to discover the true extent of the Universe. His PhD work had involved studies of the fuzzy patches of light on the sky known as nebulae, and he had concluded that, while some of these objects are gas clouds within our Galaxy, others were probably more distant objects, beyond the *Milky Way*. In the early 1920s, he was able to use the new telescope to show that the

Edwin Hubble (1889–1953).

Andromeda 'nebula' (now the ***Andromeda galaxy***) is in fact a collection of stars similar to our own Milky Way. He identified several of these stars as ***Cepheid variables***, which made it possible to calculate the distance to the Andromeda galaxy. Between 1925 and 1929 Hubble showed that the 'spiral nebulae' are all other ***galaxies***, and that the Milky Way is therefore just one ***spiral galaxy*** among many in the Universe. He later established the nature of ***elliptical galaxies*** as other 'islands' in the Universe.

In 1929, Hubble, building on the work of Vesto ***Slipher***, and with the assistance of Milton ***Humason***, established the relationship between redshift and distance for galaxies, now known as ***Hubble's Law***. This showed that the Universe is expanding, and suggested that it must have had a definite origin, at a certain moment in time. It was soon realized that these observations matched the predictions of Albert ***Einstein***'s ***general theory of relativity***, as developed in the ***cosmological models*** of Aleksandr ***Friedmann*** and Georges ***Lemaître***. Hubble's early measurements suggested that the Universe was only a few billion years old, but improvements in the technique by Walter ***Baade*** and others have revised this upwards substantially; the best estimate today is that the ***Big Bang*** occurred 15–20 billion years ago.

During the 1930s, Hubble studied the distribution of galaxies across the sky, and showed that it is isotropic – that the Universe is the same in all directions. In the 1940s, as one of the elder statesmen of astronomy, Hubble was involved in the completion of the 200-inch (5 m) *Hale telescope* on Mount Palomar, and he used the new telescope for studies of faint stars in the last few years of his life. He died in San Marino, California, on 28 September 1953.

Hubble constant   A number which gives the rate at which the Universe is expanding. This number is expected to be the same everywhere in the Universe, which is why it is called a 'constant', but it changes as time passes, because the rate at which the Universe is expanding is slowing down as gravity tugs on all the matter in the Universe. For this reason, many cosmologists prefer to use the term 'Hubble parameter'; in that terminology, the Hubble constant, usually denoted by the symbol H_0, is the value of the Hubble parameter at the present time.

The constant was first determined by Edwin *Hubble* at the end of the 1920s, from his studies of the *redshifts* and distances to relatively nearby galaxies. He found that the redshift, which is now explained as produced by the expansion of the Universe, is proportional to the distance to a galaxy; H_0 is the constant of proportionality that comes into the equation. An accurate determination of the Hubble constant depends on measuring distances to galaxies by means independent of the redshift (see *cosmic distance scale*). The best estimates have been revised considerably since Hubble's day, and are still uncertain, although most estimates lie in the range 40–80 kilometres per second per Megaparsec. If the correct value is indeed 40, for example, that would mean that the redshift of a galaxy 1 million parsecs away would be the same as that produced by a recession velocity of 40 km/sec, the redshift of a galaxy 2 Mpc away would correspond to a velocity of 80 km/sec, and so on.

The inverse of the Hubble constant (called the 'Hubble time') is a measure of the age of the Universe – the bigger the value of the constant, the younger the Universe must be.

Curiously, estimates of the value of the Hubble constant based on measuring the distances to relatively nearby objects (including studies of *Cepheid variables* in the Virgo cluster using the Hubble Space Telescope) tend to give larger values for the constant (implying a younger Universe) than estimates based on techniques which look further out into the Universe, notably the *Sunyaev–Zel'dovich effect* and *gravitational lensing*. Some astronomers wonder whether this means that the Hubble constant is not really a constant at all; others point out that both gravitational lensing and the Sunyaev–Zel'dovich effect are in principle 'clean' measures of the Hubble constant, involving no chain of inferences. By contrast, the traditional Cepheid technique involves several stepping stones into the Universe, and might have more scope for errors to creep in.

Gustav **Tammann**, who favours the lower end of the range for H_0, explains the disagreement as a consequence of the difference between optimism and pessimism among cosmologists. The optimists, he says, believe that their distance indicators are nearly perfect, and believe the distances worked out from some relation which is calibrated by local measurements, in our neighbourhood of the Universe. Taking the measurements at face value implies that the value of H_0 is relatively large, that it gets bigger still the further out into the Universe we look, and (from the inferred distances to other galaxies) our own Galaxy and the Andromeda Galaxy are unusually large specimens.

Pessimists like Tammann, on the other hand, accept that their observations of distant galaxies are intrinsically 'quite lousy', in Tammann's terminology. So they can't be sure that their measurements of brightnesses of distant galaxies are accurate (in fact, the pessimists are sure the measurements are not accurate). This problem is compounded because the further out we look into the Universe the harder it is to see faint galaxies. So catalogues of distant objects are dominated by the biggest galaxies, giving the impression that the average brightness increases with distance (a phenomenon known as the **Malmquist Effect**). Optimists who take the observations at face value in effect get smaller distances the further out they look into the Universe because of the scatter and the Malmquist Effect. So the galaxies they see are really bigger, brighter and further away than they infer. If this is corrected for in the way the pessimists prefer, it gives a lower value of H_0, which is the same everywhere in the Universe at the same time, and a revised distance scale which implies that our Galaxy and the Andromeda Galaxy are neither the biggest nor the brightest in the Universe, by a long way.

The jury is still out, but it is our view that the physical arguments favour the pessimists' position.

Hubble diagram Essentially a simple graph in which the **redshifts** of **galaxies** are plotted against their distances. This kind of diagram enabled Edwin **Hubble** to discover that the Universe is expanding, with redshift proportional to distance. See **cosmic distance scale**, **expanding Universe**.

Hubble radius From **Hubble's Law**, it follows that there must be a certain distance at which **galaxies** would be receding from us at the **speed of light**. We could never see anything from beyond this distance, known as the Hubble radius. The Hubble radius is equal to the distance that light could travel in the time since the **Big Bang**, so it is the same number in **light years** as the **age of the Universe** in years – roughly 20 billion light years, or 6,000 million **parsecs**.

Hubble's Law The recession velocity (indicated by **redshift**) of a distant **galaxy** (one outside the Local Group) is directly proportional to its distance

Hubble Space Telescope. Artist's impression of the Hubble Space Telescope in orbit, shortly after being launched from the Space Shuttle.

from us. See *cosmic distance scale, expanding Universe.*

Hubble Space Telescope (HST)   An orbiting observatory built jointly by NASA and the European Space Agency, launched in 1990. Flaws with the telescope were largely corrected by astronauts in 1993, and it is now producing superb images from above the obscuring layers of the Earth's atmosphere. Its main *telescope* is a 2.4 m reflector – almost as big as any ground-based telescope in operation before 1947.

Hubble time   See *age of the Universe, Hubble constant.*

Hulse, Rusell Alan (1950–)   American astronomer who received the Nobel Prize in Physics in 1993, for his discovery (jointly with Joseph *Taylor*) of the *binary pulsar.*

Hulse was born on 28 November 1950, in New York city. He was educated at the Bronx High School, and went on to study at The Cooper Union (a college in lower Manhattan), and then at the University of Amherst, where he received his PhD in 1975. It was while working as a research student at Amherst on his PhD project that Hulse carried out the observations which led to the discovery of the binary pulsar, using the *Arecibo radio telescope.*

After completing his PhD, Hulse worked at the US *National Radio Astronomy Observatory* from 1975 to 1977, but he was unable to find a secure post working in astronomy, and since 1977 he has worked at the Plasma Physics Laboratory of Princeton University, involved in research aimed at producing nuclear *fusion* in sustainable reactions that could be used to produce power on Earth from essentially the same process that keeps the Sun and stars hot.

Humason, Milton Lasell (1891–1972)   American astronomer who left school at fourteen and worked as a mule driver on the pack trains carrying materials up Mount Wilson, in California, for the construction of a new observatory. This led to a job as janitor at the observatory, then to one of night assistant, and finally in 1919 Humason was appointed to the staff in spite of his lack of formal training. He worked with Edwin *Hubble* on the *redshift* survey which led to the discovery that the Universe is expanding, measuring the redshifts of 620 galaxies between 1930 and 1957, when he retired.

Huygens, Christiaan (1629–95)   The greatest scientist between the time of *Galileo* and that of Isaac *Newton*, who (among other things) invented the first successful pendulum clock, designed and built improved astronomical telescopes, and developed a complete wave theory of light. He also discovered Titan (one of the moons of Saturn), and recognized the nature of Saturn's rings.

Huygens was born in the Hague on 14 April 1629, and was the son of a well-known poet and statesman, a member of a family with a tradition of public service to the House of Orange. René *Descartes* was a frequent visitor to the family home, and Huygens was educated to the highest standards of the day, studying mathematics and law at the University of Leiden from 1645 to 1647, then spending two further years studying law in Breda. But he never settled down to the career in diplomacy which was expected of him, and instead spent the next sixteen years as a gentleman scientist, living off an income provided by his father and free to study whatever he pleased. He was so successful, and achieved such a high reputation, that in 1666 he was invited to Paris to work under the auspices of the new Academie Royale des Sciences. He stayed there for fifteen years, but encountered religious prejudice to his Protestant views, and, partly as a result of this, he returned to Holland in 1681. Huygens also suffered from poor health, and only occasionally travelled to meet other scientists (including Isaac Newton, in London in 1687). In 1694 he fell ill again, but this time did not recover, and he died in the following year, on 8 July, aged 66.

Huygens first gained fame as the inventor of a pendulum clock which developed the ideas of Galileo into a practicable form. A Dutch clockmaker built a clock to Huygens' design in 1657, and it was so successful that similar clocks were soon installed in church towers across Holland,

Christiaan Huygens (1629–95).

and then throughout Europe. Huygens also studied mathematics, making early developments in probability theory, and dynamics, including *centrifugal force* and the nature of acceleration. He could not take this work as far as Newton did, because unlike Newton he did not develop the mathematical tool needed to do the job, differential calculus. Coincidentally, in 1672 he met the German mathematician Gottfried Leibnitz (1646–1716), and gave him some tuition in mathematics. When Leibnitz developed his version of calculus (independently of Newton, who had not published his work), it was Huygens who formally recommended the paper to the French Academie in 1674. Its publication led to a blazing row about priority between Newton and Leibnitz, and between their respective supporters.

In 1655, Huygens had begun working with his brother Constantijn on making telescopes for astronomical studies. He developed an improved eyepiece that provided clearer images (the design is still used today), and in the same year, using their new telescope, he discovered Titan. He also realized that the strange features known at the time as the 'arms' of Saturn were, in fact, rings, and in 1659 he published this discovery in a coded form, as a Latin anagram which could be decoded to read 'It is surrounded by a thin flat ring, nowhere touching and inclined to the ecliptic'. He later explained the ring as made up of a swarm of tiny particles like myriads of miniature moons in orbit around the planet. Also in 1659, Huygens made the first observations of dark markings on the surface of Mars.

Through his work with telescopes, Huygens became interested in the nature of light, and developed a complete theory describing the behaviour of light in terms of waves. This was first published in a communication to the French Academie in 1678, but in a complete form only in 1690, in his book *Traité de la Lumière*. The rival theory of light as a stream of tiny particles ('corpuscles') was, however, championed by Isaac Newton, and because of his fame and prestige the wave theory languished throughout the 18th century. Today, *quantum theory* requires that light be described both in terms of waves and in terms of particles.

Although his achievements in science in the second half of the 17th century were second only to those of Newton himself, Huygens worked largely alone, never taught at a university, and had no students or disciples. Largely as a result, his work did not receive full credit at the time, and had less influence than it might have done on the development of science in the century following his death in the Hague on 8 July 1695.

Further reading: John Gribbin, *Schrödinger's Kittens*.

Hydrocarbons Compounds of *hydrogen* and *carbon*. There are thousands of possible compounds that can be made from these two *elements*. Several of the simpler hydrocarbons, including methane (CH_4) and acetylene (C_2H_2) have been identified in clouds of *interstellar matter*. Not to be confused with *carbohydrates*.

hydrogen The simplest *element*; symbol H. Each *atom* of ordinary hydrogen consists of a single *proton* (the *nucleus*) and a single *electron*; atoms of a rare *isotope* called *deuterium* ('heavy hydrogen') each have a nucleus consisting of a proton and a single *neutron*, again associated with a single electron. The unstable isotope tritium, produced in nuclear reactions, has an additional neutron in its nucleus, but spontaneously undergoes *beta decay* to become an isotope of *helium* (helium-3).

Hydrogen is by far the most common element in the Universe, making up 75 per cent of the mass of all the visible matter in stars and galaxies. Like helium, it was produced in the *Big Bang* in which the Universe was born. As well as in the visible stars, there are large amounts of hydrogen in clouds of gas in interstellar space (see *HI region*, *HII region*). Deuterium was also produced in the Big Bang, but in minute quantities. It is very difficult to detect in stars and interstellar material, but the few observations that have been made suggest that there are only a couple of deuterium atoms for every 100,000 hydrogen atoms in the Universe, closely agreeing with the predictions of the Big Bang model.

Because the structure of the hydrogen atom is so simple, it produces an easily identified characteristic *spectrum* both in visible *light* (where it produces a distinctive line, known as hydrogen alpha, at a wavelength of 656.28 nanometres in the red part of the spectrum) and at *radio* wavelengths (most notably, at a wavelength of 21 cm). These very well-defined features are important in measuring the *redshifts* of distant *galaxies* and *quasars*. Hydrogen also occurs in molecular form (as H_2) in *molecular clouds* of gas in space.

hydroxyl A so-called radical (an incomplete *molecule*), consisting in this case of a single *hydrogen atom* and a single *oxygen* atom bound together as OH. Essentially a water molecule (H_2O) lacking one of its hydrogen atoms, the hydroxyl radical is a common constituent of interstellar matter, and was the first interstellar 'molecule' to be detected (in 1963), from its characteristic absorption in the radio part of the *electromagnetic spectrum*. It has also been detected as an emitter of radiation.

Hyperion One of the moons of Saturn, discovered by W. C. Bond in 1848. It is an irregularly shaped object, roughly 350 km by 200 km, probably a piece of debris from a larger body broken up in a collision.

Iapetus One of the moons of Saturn, detected by Giovanni *Cassini* in 1671. Unusually, one side of this moon is dark, covered with unknown material, while the other is bright, ice-covered and heavily cratered. Nobody knows why this difference has arisen. Iapetus orbits Saturn at an average distance of 3.5 million km; it has a radius of 720 km and a mass 0.025 per cent of that of our Moon.

Icarus An *asteroid* with a diameter of 1.4 km, discovered in 1949 by Walter *Baade*. Icarus has an unusual, highly elliptical orbit which takes it closer to the Sun than Mercury.

ICARUS Acronym for Imaging Cosmic And Rare Underground Signals, the name of a particle detector used (among other things) to monitor *neutrinos* from the Sun. It is based in the Gran Sasso Underground Laboratory in northern Italy. The detector uses 200 tonnes of liquid argon-40 to 'catch' the particles of interest. See *solar neutrino problem*.

ice dwarf Term coined in the 1990s to describe objects in the *Solar System* which seem too large to classify as normal *comets*, too icy to be called *asteroids*, and too profuse to be *planets*. They are best thought of as the nuclei of very large comets, hundreds of kilometres in diameter, which populate the *Kuiper Belt* and the *Oort Cloud* in very large numbers – many millions. Chiron is typical of this relatively newly discovered component of the Solar System (although not now in a typical ice dwarf orbit), and even Pluto is regarded by some astronomers as an ice dwarf rather than a true planet. It has been suggested that from time to time an ice dwarf is perturbed into an orbit taking it into the inner Solar System, where it breaks up and may cause havoc among the **terrestrial planets**. See **Doomsday asteroid**.

image intensifier (image tube)   Device that uses electronics to amplify a faint image. Incoming **photons** stimulate the release of **electrons** from a sensitive surface, and these electrons are accelerated and focused by magnetic fields before producing an image on a screen similar to a TV screen. The image is brighter but more blurred than it would be without the intensification.

IMB   **Neutrino** detector similar to the **Kamiokande** detector, located in a salt mine near Fairport, Ohio. Both IMB and Kamiokande were originally designed to investigate the **radioactive decay** of **protons**, but were triggered by neutrinos from **Supernova 1987A**.

inertia   A measure of the reluctance of an object to change the way it is moving (an indication of the incredible heaviness of being).

An object that is at rest stays at rest unless and until some force makes it move, and a moving object continues to move at the same speed in the same direction unless and until some force changes the way it is moving. This behaviour was given the status of a law by Isaac **Newton** (his first law of motion); it was Newton who pointed out that the **mass** of an object can be defined in terms of inertia. The acceleration produced in a body with a certain mass by the application of a force is equal to the force divided by the mass ($f = ma$), so the **inertial mass** of the object is given by the force divided by the acceleration it produces ($m = f/a$). For the same mass, this is the same for all forces and accelerations. It is one of the puzzles of science, still not fully explained, that this inertial mass has exactly the same value as the **gravitational mass** of an object, which determines the strength of the **gravitational field** of the object (see **Mach's Principle**).

In everyday life, the consequences of Newton's first law are not obvious, because we live on the surface of a **planet** where everything is pulled downwards by gravity. When objects move across the surface of a planet, they do not continue indefinitely, but are brought to a halt by friction. The full effects of inertia are much more obvious in conditions of **free fall** inside an orbiting spacecraft, where (as many TV broadcasts from space have demonstrated) objects really do keep moving in straight lines at constant speed until they hit something.

inertial frame   An inertial frame is a reference system in which Isaac **Newton**'s laws of motion apply, and objects move in straight lines at constant speeds unless a force changes their motion (see **inertia**). The surface of the Earth is not in an inertial frame, because objects are pulled downwards by the Earth's **gravity**, and accelerate until they hit the ground. Because of this, the behaviour of objects in inertial frames is not 'common sense' to us, and had to be worked out by scientific observations of the way things behave when the influence of gravity (and friction) can be eliminated. The behaviour of an object in an inertial frame is approximately demonstrated by the free movement of the puck across the surface of an air table in the amusement arcade hockey game; the puck is actually floating on a layer of air just above the surface of the table, with friction almost eliminated.

Newton thought that there was a special inertial frame against which the motion of everything in the Universe could be measured. Albert *Einstein* suggested, with his *special theory of relativity*, that all inertial frames are exactly equivalent to one another, and that anybody who is in a frame of reference where Newton's laws apply can regard themselves as at rest and measure the motion of everything in the Universe relative to their own frame. Even so, one frame of reference does seem to be more equal than others – the inertial frame in which the *background radiation* has exactly the same temperature in all directions and the most distant *galaxies* and *quasars* are distributed uniformly across the sky (see *Mach's Principle*).

inertial mass The amount of matter in an object, defined in terms of the amount of force needed to give it a certain acceleration. Inertial mass is exactly equivalent to *gravitational mass*, but there is no accepted theory of why this should be so. See *Mach's Principle*.

inflation General term for *models* of the very early Universe which involve a short period of extremely rapid (exponential) expansion, blowing the size of what is now the observable Universe up from a region far smaller than a *proton* to about the size of a grapefruit in a small fraction of a second. This process would smooth out *spacetime* to make the Universe flat (see *flatness problem*), and would also resolve the *horizon problem*.

Inflation became established as the standard model of the very early Universe in the 1980s. It achieved this success not only because it resolved many puzzles about the nature of the Universe, but because it did so using the *grand unified theories* (GUTs) and understanding of *quantum theory* developed by particle physicists completely independently of any cosmological studies. These theories of the particle world had been developed with no thought that they might be applied in *cosmology* (they were in no sense 'designed' to tackle all the problems they turned out to solve), and their success in this area suggested to many people that they must be telling us something of fundamental importance about the Universe.

The marriage of particle physics (the study of the very small) and cosmology (the study of the very large) seems to provide an explanation of how the Universe began, and how it got to be the way it is. Inflation is therefore regarded as the most important development in cosmological thinking since the discovery that the Universe is expanding first suggested that it began in a *Big Bang*.

Inflation. A curved surface, like the skin of a balloon, can be made to appear flat if it expands enormously. This is thought to be how spacetime was flattened by inflation when the Universe was young.

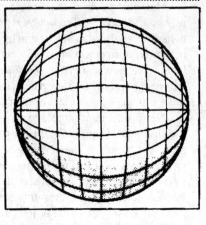

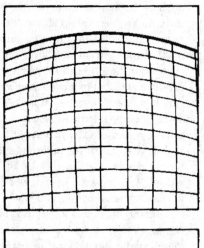

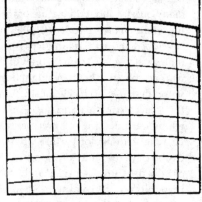

Taken at face value, the observed expansion of the Universe implies that it was born out of a *singularity*, a point of infinite *density*, 15–20 billion years ago. Quantum physics tells us that it is meaningless to talk in quite such extreme terms, and that instead we should consider the expansion as having started from a region no bigger across than the *Planck length* (10^{-35} m), when the density was not infinite but 'only' some 10^{94} grams per cubic centimetre. The first puzzle is how anything that dense could ever expand – it would have an enormously strong *gravitational field*, turning it into a *black hole* and snuffing it out of existence (back into the singularity) as soon as it was born (see *free lunch Universe*).

Other problems with the Big Bang theory before the development of inflationary models involve the extreme flatness of *spacetime* (which means that the expansion of the Universe is balanced against the tug of gravity, so that it sits precisely on the dividing line between expanding forever and one day recollapsing in a *Big Crunch*; see *density parameter*), and its appearance of extreme *homogeneity* and *isotropy*, most clearly indicated by the uniformity of the *background radiation*.

All of these problems would be resolved if something gave the Universe a violent outward push (in effect, acting like antigravity) when it was still about a Planck length in size. Such a small region of space would be too tiny, initially, to contain irregularities, so it would start off homogeneous and isotropic. There would have been plenty of time for signals travelling at the *speed of light* to have criss-crossed the ridiculously tiny volume, so there is no horizon problem – both sides of the embryonic *universe* are 'aware' of each other. And spacetime itself gets flattened by the expansion, in much the same way that the wrinkly surface of a prune becomes a smooth, flat surface when the prune is placed in water and swells up. As in the standard Big Bang model, we can still think of the Universe as like the skin of an expanding balloon, but now we have to think of it as an absolutely enormous balloon that was hugely inflated during the first split-second of its existence.

The reason why the GUTs created such a sensation when they were applied to cosmology is that they predict the existence of exactly the right kind of mechanisms to do this trick. They are called *scalar fields*, and they are associated with the splitting apart of the original grand unified force into the *fundamental forces* we know today, as the Universe began to expand and cool. Gravity itself would have split off at the *Planck time*, 10^{-43} seconds, and the strong force by about 10^{-35} seconds. Within about 10^{-32} seconds, the scalar fields would have done their work, doubling the size of the Universe at least once every 10^{-34} seconds (some versions of inflation suggest even more rapid expansion than this).

This may sound modest, but it would mean that in 10^{-32} seconds there were 100 doublings. This is enough to take a quantum fluctuation 10^{20} times smaller than a proton and inflate it to a sphere about 10 cm across in about 15×10^{-33} seconds. At that point, the scalar field has done its work of kick-starting the Universe, and is settling down, giving up its energy and leaving a hot fireball expanding so rapidly that, even though gravity can now begin to do its work of pulling everything back into a Big Crunch, it will take hundreds of billions of years first to halt the expansion and then to reverse it.

Curiously, this kind of exponential expansion of spacetime is exactly described by one of the first cosmological models developed using the *general theory of relativity*, by Willem *de Sitter* in 1917. For more than half a century, this de Sitter model seemed to be only a mathematical curiosity, of no relevance to the real Universe; but it is now one of the cornerstones of inflationary cosmology.

One of the peculiarities of inflation is that it seems to take place faster than the speed of light. Even light takes 3 billionths of a second (3×10^{-9} seconds) to cross a single metre, and yet inflation expands the Universe from a size much smaller than a proton to 10 cm across in only about 5×10^{-33} seconds. This is possible because it is spacetime itself that is expanding, carrying matter along for the ride; nothing is moving through spacetime faster than light, either during inflation or ever since. Indeed, it is just because the expansion takes place so quickly that matter has no time to move while it is going on, and the process 'freezes in' the original uniformity of the primordial quantum bubble that became our Universe.

The inflationary scenario has already gone through several stages of development during its short history. The first inflationary model was developed by Alexei Starobinsky, at the L. D. Landau Institute of Theoretical Physics in Moscow, at the end of the 1970s – but it was not then called 'inflation'. It was a very complicated model based on a quantum theory of gravity, but it caused a sensation among cosmologists in what was then the Soviet Union, becoming known as the 'Starobinsky model' of the Universe. Unfortunately, because of the difficulties Soviet scientists still had in travelling abroad or communicating with colleagues outside the Soviet sphere of influence at that time, the news did not spread outside their country.

In 1981, Alan *Guth*, then at MIT, published a different version of the inflationary scenario, not knowing anything of Starobinsky's work (*Physical Review*, volume D23, page 347, January 1981). This version was more accessible in both senses of the word – it was easier to understand, and Guth was based in the USA, able to discuss his ideas freely with colleagues around the world. And as a bonus, Guth came up with the catchy name 'inflation' for the process he was describing. Although there were obvious flaws with the specific details of Guth's original model (which he acknowledged at the time), it was this version of the idea that made every cosmologist aware of the power of inflation.

In October 1981 there was an international meeting in Moscow, where inflation was a major talking point. Stephen **Hawking** presented a paper claiming that inflation could not be made to work at all, but Andrei **Linde** presented an improved version, called 'new inflation', which got around the difficulties with Guth's model. Ironically, Linde was the official translator for Hawking's talk, and had the embarrassing task of offering the audience the counter-argument to his own work! But after the formal presentations Hawking was persuaded that Linde was right, and inflation might be made to work after all. Within a few months, the new inflationary scenario was also published by Andreas Albrecht and Paul Steinhardt, of the University of Pennsylvania, and by the end of 1982 inflation was well established.

Linde has been involved in most of the significant developments with the theory since then. The next step forward came with the realization that there need not be anything special about the Planck-sized region of spacetime that expanded to become our Universe. If that was part of some larger region of spacetime in which all kinds of scalar fields were at work, then only the regions in which those fields produced inflation could lead to the emergence of a large universe like our own. Linde called this 'chaotic inflation', because the scalar fields can have any value at different places in the early super-universe; it is the standard version of inflation today, and can be regarded as an example of the kind of reasoning associated with the **anthropic principle** (but note that this use of the term 'chaos' is like the everyday meaning implying a complicated mess, and has nothing to do with the mathematical subject known as 'chaos theory').

The idea of chaotic inflation led to what is (so far) the ultimate development of the inflationary scenario. The great unanswered question in standard Big Bang cosmology is what came 'before' the singularity. It is often said that the question is meaningless, since time itself began at the singularity. But chaotic inflation suggests that our Universe grew out of a quantum fluctuation in some pre-existing region of spacetime, and that exactly equivalent processes can create regions of inflation within our own Universe. In effect, new universes bud off from our Universe, and our Universe may itself have budded off from another universe, in a process which had no beginning and will have no end. A variation on this theme suggests that the 'budding' process takes place through black holes, and that every time a black hole collapses into a singularity it 'bounces' out into another set of spacetime dimensions, creating a new inflationary universe – this is known as the **baby universe** scenario.

There are similarities between the idea of eternal inflation and a self-reproducing universe and the version of the *Steady State hypothesis* developed by Fred *Hoyle* and Jayant Narlikar, with their *C-field* playing the part of the scalar field that drives inflation. As Hoyle wryly pointed out at a meeting of the *Royal Astronomical Society* in London in December 1994, the relevant equations in inflation theory are exactly the same as in his version of the Steady State idea, but with the letter 'C' replaced by the Greek Φ. 'This,' said Hoyle (tongue firmly in cheek), 'makes all the difference in the world'.

Modern proponents of the inflationary scenario arrived at these equations entirely independently of Hoyle's approach, and are reluctant to accept this analogy, having cut their cosmological teeth on the Big Bang model. Indeed, when Guth was asked, in 1980, how the then new idea of inflation related to the Steady State theory, he is reported as replying 'what is the Steady State theory?' But although inflation is generally regarded as a development of Big Bang cosmology, it is better seen as marrying the best features of both the Big Bang and the Steady State scenarios.

This might all seem like a philosophical debate as futile as the argument about how many angels can dance on the head of a pin, except for the fact that observations of the background radiation by *COBE* showed exactly the pattern of tiny irregularities that the inflationary scenario predicts. One of the first worries about the idea of inflation (long ago in 1981) was that it might be too good to be true. In particular, if the process was so efficient at smoothing out the Universe, how could irregularities as large as *galaxies, clusters of galaxies* and so on ever have arisen? But when the researchers looked more closely at the equations they realized that quantum fluctuations should still have been producing tiny ripples in the structure of the Universe even when our Universe was only something like 10^{-25} cm across – 100 million times bigger than the Planck length.

The theory said that inflation should have left behind an expanded version of these fluctuations, in the form of irregularities in the distribution of matter and energy in the Universe. These density perturbations would have left an imprint on the background radiation at the time when matter and radiation *decoupled* (about 300,000 years after the Big Bang), producing exactly the kind of non-uniformity in the background radiation that has now been seen, initially by COBE and later by other instruments. After decoupling, the density fluctuations grew to become the *large-scale structure* of the Universe revealed today by the distribution of galaxies. This means that the COBE observations are actually giving us information about what was happening in the Universe when it was less than 10^{-30} seconds old.

No other theory can explain why the Universe is so uniform overall, and yet contains exactly the kind of 'ripples' represented by the distribution of galaxies through space and by the variations in the background radiation. This does not prove that the inflationary scenario is correct, but it is worth remembering that had COBE found a different pattern of fluctuations (or no fluctuations at all) that would have proved the inflationary scenario wrong. In the best scientific tradition, the theory made a major and unambiguous prediction which did 'come true'. Inflation also predicts that the primordial perturbations may have left a trace in the form of *gravitational radiation* with particular characteristics, and it is hoped that detectors sensitive enough to identify this characteristic radiation may be developed within the next ten or twenty years.

Further reading: John Gribbin, *In the Beginning*.

Infrared Astronomical Satellite   See *IRAS*.

infrared astronomy   Studies of the Universe made using the infrared part of the *spectrum*, at wavelengths in the range from 1–1,000 micrometers. These wavelengths lie between the red (long-wavelength) end of the spectrum visible to our eyes and the part of the spectrum (at longer wavelengths than infrared) regarded as radio waves. Most of the infrared radiation from space is absorbed in the Earth's atmosphere, so infrared astronomy has to be carried out using telescopes on high mountains or instruments carried above the obscuring layers of the atmosphere on high-flying aircraft, balloons, rockets and satellites.

In space, however, infrared radiation is much more effective than visible light at penetrating clouds of *interstellar matter*. This is because longer wavelengths of radiation are less easily scattered by dust (the same effect is at work on sunlight in the atmosphere of the Earth, scattering short-wavelength blue light around the sky and leaving long-wavelength red light to produce spectacular sunsets). So infrared telescopes are very useful in probing regions obscured by dust, such as the centre of the *Milky Way*.

Infrared observations cover an enormous range of objects in the Universe, from planets, clouds of dust in space and the clouds in which new stars are forming (*HII regions*) to *active galaxies* and *quasars*. Infrared *spectroscopy* of cool clouds of gas and dust in our *Galaxy* has revealed the presence of ice and *hydrocarbons* in those clouds.

infrared background   Diffuse radiation from space, with wavelengths in the range 10–200 micrometres. It is probably produced by the combined faint emission from *stars*, *dust in space*, *infrared cirrus* and other *galaxies*.

infrared cirrus   Clouds of cool gas and dust above and below the plane of the *Milky Way* which radiate in the infrared part of the *spectrum* and show up as wispy structures resembling cirrus clouds in infrared maps of the sky.

infrared radiation   *Electromagnetic radiation* with wavelengths longer than

Interacting galaxies. This view of the Hercules cluster of galaxies, also known as Abell 2151, shows two disc galaxies interacting with one another at the centre of the picture.

those of visible *light* (beyond the red end of the *spectrum*), but shorter than those of *radio waves*. It is usually divided into three bands: near infrared, 0.8–8 micrometres wavelength; mid infrared, 8–30 micrometres; and far infrared, 30–300 micrometres. There is no clear distinction between long-wavelength infrared radiation and short-wavelength radio waves; the region just above 300 micrometres wavelength is usually referred to as submillimetre radiation.

Infrared is the radiation you feel as heat if you hold your hand near a fire or a hot radiator.

Infrared Telescope Facility   An infrared telescope with a 3 m *aperture* at the *Mauna Kea Observatory* in Hawaii; operated by *NASA*.

interacting galaxies   *Galaxies* that distort one another through the influence of their *gravitational fields*. This happens when a pair of galaxies, usually in a *cluster of galaxies*, pass close by one another. The resulting tidal distortions may produce wisps of *stars* that form luminous bridges of material between the two galaxies.

interferometry   A technique widely used in *radio astronomy* in which two or more separate *antennae* are linked electronically and study the same astronomical source. The antennae may be a few kilometres apart and physically linked by cables, or they may be on opposite sides of the world, making simultaneous observations which are recorded and later combined electronically by a computer. Either way, the combined observations enable astronomers to obtain information about details of the radio sources they are observing on a finer scale than would be possible with any of the single antennae involved, as if they were using a much larger antenna – this is a form of *aperture synthesis*.

An equivalent technique is also used in optical astronomy. The technique was pioneered by A. A. *Michelson* and colleagues at the *Mount Wilson Observatory* in 1920, using two mirrors mounted on a steel beam to deflect light from the same star on to the main mirror of the 100-inch (254 cm) *Hooker telescope*. Studies of the interference pattern made by combining the two beams of light made it possible to determine the angular size of the star *Betelgeuse* as 0.047 *arc seconds*.

Several variations on the optical interferometry theme have been developed and used successfully. Because light has much shorter wavelengths than radio, however, it is much harder to construct a full aperture synthesis system using optical telescopes than it is with radio telescopes. But such systems are now being tested, using telescopes a few tens of metres apart linked by laser beams (see *Keck telescope*).

intergalactic matter   Matter in the space between the *galaxies*. There is no evidence of dust in intergalactic space, but the shape of some *radio sources* (like a tadpole) shows that they are moving through a tenuous gas. Intergalactic *hydrogen* is also revealed by its characteristic radio emission at 21 cm wavelength. Much larger amounts of *dark matter* are also thought to lie between the galaxies.

International Atomic Time (IAT or, from the French, TAI)   Standard international time system based on *atomic time* and maintained by the Bureau International de l'Heure in Paris.

International Sunspot Number   See *sunspot number*.

International Ultraviolet Explorer (IUE)   *Satellite* launched by NASA in 1978 to make observations of the Universe at ultraviolet wavelengths (beyond the blue end of the visible *spectrum*) using a 45 cm *telescope*. It is a collaborative venture also involving the UK and the *European Space Agency*. IUE is in a *geosynchronous orbit*. It is the longest-lived astronomical satellite, and should be operational until the end of the 1990s.

interplanetary dust   A large cloud of material around the Sun, probably produced by material ejected from *comets* as they have broken up, and from collisions between *asteroids*. It extends at least 600 million km from the Sun.

interstellar absorption   See *cosmic dust*.

interstellar chemistry   Many *molecules* (more than 80 by the mid-1990s) have been discovered in clouds of gas and dust between the stars. Although the **nuclei** of different **elements** are built up inside stars by the processes of **nuclear fusion**, under the conditions of heat and pressure inside a star **atoms** cannot combine together to make molecules. So all of the variety of molecules seen in interstellar space must have been produced by chemical reactions going on in the clouds of gas and dust where we detect those molecules today. The complexity of the reactions involved in this interstellar chemistry is indicated by the complexity of some of the molecules identified – several contain ten or more atoms, and one is the **amino acid** called **glycine** (NH_2CH_2COOH), an essential building block of life on Earth.

Many of the molecules found in interstellar space are made up from carbon, oxygen and nitrogen (the most abundant elements manufactured from hydrogen and helium inside stars), together with hydrogen itself (see **CHON**). Simple compounds made up of carbon and hydrogen (CH) and carbon and nitrogen (CN) were discovered at the end of the 1930s, using optical **spectroscopy**. But the first real progress towards an understanding of interstellar chemistry came in the 1960s, when suitable **radio astronomy** techniques were developed to identify the characteristic radiation of polyatomic molecules in space. The hydroxyl compound (OH), water (H_2O), ammonia (NH_3) and formaldehyde (H_2CO) were soon identified.

In the 1970s, astronomers were surprised by the variety and complexity of organic molecules (that is, molecules that contain carbon atoms) found in space. These included ethyl alcohol (C_2H_5OH), which is present in one large complex of **molecular clouds** (known as Sgr B2) in sufficient quantities to make 10^{27} litres of vodka. As well as these complex molecules, interstellar clouds must also contain simple compounds such as oxygen (O_2), nitrogen (N_2), carbon dioxide (CO_2) and hydrogen (H_2), which are stable and form very easily from the basic atomic ingredients that are known to be present.

The key to interstellar chemistry is the presence of a large amount of carbon in the form of grains of graphite in these interstellar clouds. These show up from the way in which they absorb visible light from more distant stars, which can be explained by the presence of elongated grains about 0.1 millionths of a metre long, mostly made of carbon but with water ice and silicates present as well. It may seem odd to think of interstellar clouds as being laced with soot, but carbon is one of the most common products of **nucleosynthesis** inside stars, and there is a family of stars (known as carbon stars) which are shown by their spectra to have atmospheres relatively rich in carbon, and which vary regularly, puffing in and out with periods of a year or so, and ejecting material into interstellar space as they do so. The evidence suggests that many (if not all) stars go through a phase of such activity.

Most of the complex molecules are found in unusually dense clouds in space, where there is enough of the sooty dust to act as a shield, protecting the molecules from the strong *ultraviolet radiation* from nearby young stars, which would tend to break the molecules apart. These are exactly the clouds in which new young stars, and their associated planets, are forming. The molecules are probably built up by reactions that take place on the surfaces of dust grains, where atoms can 'stick' and have a chance to interact with one another. The molecules later evaporate from the surfaces of the grains. All of this makes it extremely likely that new planets are 'seeded' by quite complex molecules early in their existence; any molecules present in the interstellar clouds from which stars and planets form could easily be deposited on a planet by, for example, the impact of a large *comet*.

Interstellar chemistry involves not only interactions between material in gaseous form and the solid grains of dust in molecular clouds, but also interactions with the stars themselves. It is harder than you might think for such a cloud to collapse and fragment to form stars. When it starts to do so, gravitational energy is released in the form of heat, making the molecules in the cloud move faster and generating a pressure which resists further collapse. The cloud can collapse further only if this excess heat can be disposed of, in the form of electromagnetic radiation. This is produced by the molecules of compounds such as carbon dioxide and water vapour in the cloud. Then, when a young star begins to form, it produces copious amounts of ultraviolet radiation, which would tend to blow the cloud apart. Fortunately, though, the grains of carbon dust (or soot) in the cloud absorb the ultraviolet radiation and re-radiate it in the infrared part of the spectrum, at wavelengths which can escape much more easily into space. Carbon dust grains, and molecules produced by interstellar chemistry, are essential in the cooling processes without which the stars that edge the spiral arms of a galaxy like our own Milky Way would not form in such abundance.

In the early 1990s, astronomers found evidence of a complex molecule in the form of a ring in interstellar clouds. The evidence came from the NASA Ames Research Center, and concerned the detection of features interpreted as those of the spectrum of pyrene ($C_{12}H_{10}$), in which a dozen carbon atoms are joined together in a ring, with ten hydrogen atoms attached to it around the outside. This provided the first independent support for controversial claims made by Fred *Hoyle* and his colleague Chandra Wickramasinghe in the 1970s and 1980s.

Hoyle and Wickramasinghe have gone further than any other astron-omers in claiming that some of the features seen in the radiation from molecular clouds can be explained in terms of very large organic molecules called polymers. These form chains of repeating units. The basic com-

Interstellar chemistry. This typical cloud in space (the Lagoon nebula) is the kind of place where interstellar chemical reactions take place.

ponent of a polysaccharide chain, for example, is the so-called pyran ring, a hexagon made up of five carbon atoms and one oxygen atom (C_5O). These rings link together to make a chain when one of the carbon atoms joins on to another oxygen atom, which itself joins on to the next pyran ring in the chain, and so on. Once formed, a pyran ring shows one of the fundamental properties of life – it acts as a template, encouraging the formation of more identical rings which join up in a growing polysaccharide chain.

Hoyle and Wickramasinghe have suggested that even more complex polymers such as cellulose may already have been directly revealed by their spectral signatures in radiation from these clouds (other astronomers dispute this interpretation of the data), and that molecules of life itself may be present in the clouds but not yet detected. Once, these ideas were derided as so heretical that just voicing them may have cost Hoyle the Nobel Prize he deserved for his work on *nucleosynthesis*. In fact, the team have an impressive track record (they were, for example, the originators of the idea that interstellar clouds contain grains of soot, an idea established now beyond reasonable doubt), and their ideas look far less extreme now that *glycine* and pyrene have both been identified in space.

At the very least, it is now difficult to escape the conclusion that, when a planet like the Earth forms, its atmosphere and oceans are soon laced with complex organic molecules. Since interstellar chemistry seems to be the same in molecular clouds across the Milky Way, this suggests that the complex chemistry of other planets would be similar to that of the Earth, and that where life has evolved from that complex chemistry, it should be based on the same sort of compounds (including amino acids) that we are.

Further reading: John Gribbin, *In the Beginning*; Fred Hoyle, *Home is Where the Wind Blows*.

interstellar cloud   General name for *molecular clouds* and *HI regions*.

interstellar dust   See *dust in space*.

interstellar matter   Material between the stars of our *Galaxy*, making up about 10 per cent as much material as there is in the bright *stars* of the *Milky Way*. It includes *interstellar clouds* and *dust in space*. There are similar quantities of interstellar matter in other *disc galaxies*. See *galaxy formation and evolution*.

interstellar molecules   *Molecules* found in space, especially in *molecular clouds*. They are revealed by *spectroscopy*, using both *light* and *radio waves*. About 100 interstellar molecules have been identified in this way, mostly compounds of *carbon* (organic molecules); some contain more than ten *atoms*.

interstellar reddening   See *dust in space*.

inverse beta decay   The emission of a *positron* (equivalent to the absorption of an *electron*) from a nucleus during *radioactive decay*. See *beta decay*.

inverse square law   See *fundamental forces*.

Io   The innermost of the four large moons of Jupiter found by *Galileo*. It *orbits* 422,000 km from the *planet*, always keeping the same face towards Jupiter. The moon is yellowish-brown, and highly reflective, with an *albedo* of 0.61; it has a diameter of 3,630 km and a mass 20 per cent greater than that of our Moon, and it is the most volcanically active object in the Solar System, with the activity driven by internal heating caused by the strong tidal forces produced by Jupiter.

ion   An *atom* that has gained or (more commonly) lost one or more *electrons* so that it is no longer electrically neutral but carries an overall electric *charge*. An atom that has lost electrons becomes an ion with an overall positive charge equal to the number of electrons lost. Similarly, an atom that has gained electrons becomes a negative ion.

ionization   The process by which *electrons* are removed (usually) or added (occasionally) to *atoms* to turn them in to *ions*.

IRAS   Commonly used name for the Infrared Astronomical Satellite, launched by NASA on 25 January 1983. This was an international collaboration also involving Britain and Holland, which observed the Universe at infrared wavelengths (beyond the red end of the visible *spectrum*) for only ten months, but which mapped 95 per cent of the sky and made many important discoveries, including evidence from observations of *galaxies* that *spacetime* is flat (see *cosmological models*) and the discovery of *infrared cirrus*. The main *telescope* on IRAS had an *aperture* of 57 cm.

IRAS galaxies A type of *galaxy*, producing a large amount of *infrared radiation*, first identified by *IRAS*. Studies with *optical telescopes* show that many of these objects are interacting *disc galaxies* undergoing intense *star formation* activity as they merge with one another. See *galaxy formation and evolution*.

irregular galaxies Any *galaxy* that cannot be classified as either a *disc galaxy* or an *elliptical galaxy*. About 10 per cent of all known galaxies are irregulars. Many of them seem to be undergoing active *star formation*, and observations with *radio telescopes* often reveal the presence of a disc of gas within the galaxy; both these properties give irregulars a family resemblance to disc galaxies rather than ellipticals.

Isaac Newton Telescope (INT) A *reflecting telescope* with an *aperture* of 2.5 m, operated by the *Royal Greenwich Observatory* (RGO) but now based on La Palma in the Canary Islands. The INT was originally at Herstmonceux, in Sussex, where it began observing in 1967, but it was then moved to La Palma, where it was given a new mirror and began observing in 1984. It would probably have been cheaper to build a completely new instrument, but there were political reasons for the move, used to demonstrate the commitment of the RGO to the new *Roque de los Muchachos Observatory* on La Palma. The telescope is also used by Spain and Holland.

isotopes Atoms which have the same number of *protons* in their *nuclei*, but different numbers of *neutrons*. Isotopes of the same *element* have essentially identical chemical properties, because they have the same number of *electrons* (one for each proton) in each atom, but they have different masses. For example, *helium* comes in two varieties known as helium-3 (two protons plus one neutron in the nucleus) and helium-4 (two protons plus two neutrons).

isotropy The property of being the same in all directions. The Universe is isotropic – it looks the same (in broad terms) everywhere we look. The *background radiation* is extremely isotropic.

James Clerk Maxwell Telescope A *radio telescope* with a dish 15 m across designed to operate at submillimetre wavelengths (see *infrared astronomy*), at the *Mauna Kea Observatory* in Hawaii. It is operated by the *Royal Observatory, Edinburgh* as a joint venture involving the UK, the Netherlands and Canada. It began operating at the end of 1986.

Jansky A unit used in *radio astronomy* to describe the *energy* received from a source of radio noise. One Jansky, denoted by the symbol Jy, is 10^{-26}

watts per square metre per Hertz (1 hertz is one cycle per second). The ridiculously small size of the unit gives you some idea of how feeble the signals reaching *radio telescopes* on Earth are – all of the energy gathered in their telescopes from all the observations of every source ever studied by radio astronomers on Earth put together would not be enough to heat the water to make a single cup of coffee.

The unit is named in honour of Karl Jansky (1905–50), an American radio engineer who was the first person to detect radio emission from the *Milky Way*, and therefore was arguably the first radio astronomer. Jansky's discovery was made in 1931, while he was carrying out research for the Bell Telephone Laboratories into the origin of the background noise known as 'static' which affects radio communications.

Jansky, who was born at Norman, Oklahoma, on 22 October 1905, had joined Bell in 1928, after studying physics at the University of Wisconsin. He used a rotatable directional *antenna* to investigate the sources of static affecting shortwave radio (important at that time for ship-to-shore communications). As well as interference caused by such sources as electrical equipment and thunderstorms, he found a weak hiss of radio noise that appeared at almost the same time every day. Crucially, Jansky realized that the source 'gained' on the Sun by four minutes a day, meaning that it is fixed relative to the background stars (the time the Sun passes the zenith loses 4 minutes against the background stars each day because of the motion of the Earth in its *orbit*; see *sidereal time*). Jansky showed that the noise was coming from the direction of the centre of our *Galaxy*.

The discovery was published in December 1932. It made front-page news in the *New York Times*, but was not followed up by professional astronomers until after the Second World War; only the amateur Grote *Reber* showed much interest. Jansky died of heart disease on 14 February 1950; the unit was named after him in 1973.

Jeans, James Hapwood (1877–1946) British astrophysicist whose main contributions to the subject were rather technical and specialized, but who also wrote many popular works and made radio broadcasts about the Universe, and was largely responsible for publicizing the idea of the *heat death of the Universe*.

Jeans, the son of a journalist, was born at Ormskirk, in Lancashire, on 11 September 1877, but his family moved to Tulse Hill in London when he was three. He was educated at Cambridge University, where he became a Fellow of Trinity College in 1901. His book on the *Dynamical Theory of Gases* was published in 1904, a year before Jeans left Cambridge to work for a time at Princeton University. He returned to Cambridge in 1910, and after 1912 gave up his formal teaching post to concentrate on research as a private individual and on his writing. The only formal position in science that he held after 1912 was as an 'associate' of the **Mount Wilson Observatory** from 1923 to 1944.

Jeans investigated the behaviour of radiation in an enclosure, a key interest at the time in connection with the new **quantum theory**. He developed an equation to describe the behaviour of this radiation at long wavelengths, now known as the Rayleigh–Jeans law; it is important in the understanding of **black body radiation**. He also studied the dynamics of **stars**, and was one of the principal proponents of the 'tidal' theory of the origin of the Solar System, which suggested that the planets formed from a strand of material pulled from the Sun by a close encounter with a passing star; this theory is no longer regarded as a good description of the way the planets formed.

Although Jeans was interested in the way in which stars generate energy, and made the intriguing suggestion that it might come from the annihilation of electrons and positrons, at that time the processes of **nuclear fusion** were not understood and his suggestions later turned out to be wide of the mark. He also made an early suggestion (in 1928) that matter might be continuously created in the Universe, a forerunner of the **Steady State Hypothesis**, but he never developed the idea into a complete model of cosmology. Jeans was knighted in 1928; his many popular works include *The Universe Around Us* (1929) and *The New Background of Science* (1933).

Apart from his scientific work, Jeans was an expert on music, and became a director of the Royal Academy of Music in 1931. With his second wife, he wrote a book on *Science and Music*, which was published in 1938 and is still regarded as a definitive guide to the scientific aspects of music. He died at Dorking, in Surrey, on 16 September 1946.

Jeans' criterion A parameter which determines the size of regions in a cloud of gas with a certain **temperature** and **density** that are liable to **gravitational collapse**. The Jeans' criterion is only an approximate guide, but it predicts that, for example, the size of the objects formed by gravitational collapse at the time of the **decoupling era** in the early Universe would have been about that of a **globular cluster**.

jet Any narrow beam of material or **electromagnetic radiation** emerging from an **active galactic nucleus** or an **accretion disc**. Jets are found on all scales, from systems in our own **Galaxy** with a few times the mass of the Sun up to giant **elliptical galaxies** and **quasars**. They are particularly common at radio wavelengths, and are often thought to be associated with the activity around **black holes**.

The radio noise from a typical radio galaxy is concentrated in two huge lobes, one on either side of the galaxy. The lobes may be hundreds of kiloparsecs from the central galaxy, but connected to it by a faint trail (for a radio galaxy) or a prominent jet (for a quasar). The lobes are probably

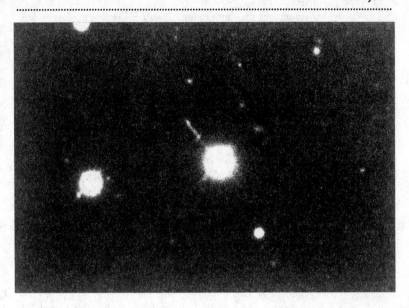

Jet. The quasar 3C 273, showing the jet of material produced by the activity of a supermassive black hole at the heart of the quasar. The jet extends nearly half a Megaparsec out into space.

regions activated by jets squirting out from the central region of the galaxy or quasar, perhaps including material expelled from the galaxy in a huge explosion long ago. If a jet of this kind is, say, 1 million *light years* long, that means that the central engine has been pouring out energy in the same direction for at least 1 million years, since nothing can travel faster than *light*.

What we actually 'see' (as light, radio waves or even X-rays) is *synchrotron radiation* produced by electrons moving in magnetic fields. The central powerhouse needed to produce all of this activity is a black hole of perhaps 100 million times the *mass* of our Sun, surrounded by a very fat torus of material forming a thick *accretion disc*. As matter funnels into the black hole, huge amounts of energy are released, and this follows the line of least resistance out from the active region, above each of the two poles (along the rotation axis of the torus), squirting both electromagnetic radiation and a *plasma* (essentially a mixture of electrons and protons) outward at nearly the *speed of light*. Imagine a ring doughnut, so fat that the hole in the middle is almost closed, with a knitting needle stuck through the hole to represent the jet, and you have a rough image of the powerhouse in a typical radio source.

Jet Propulsion Laboratory (JPL)   A centre which develops and operates unmanned deep-space probes such as *Voyager* and *Galileo* for NASA. It is based in Pasadena and run by the California Institute of Technology under contract from NASA.

Jodrell Bank   The site of the Nuffield Radio Astronomy Laboratories, near Macclesfield in Cheshire. The observatory is part of the University of Manchester, and is famous for the 250-foot (76.2 m) diameter fully steerable *radio telescope*, which began operating in 1957 and was named the Lovell Telescope, in honour of Sir Bernard *Lovell*, in 1987.

Jones, Sir Harold Spencer (1890–1960)   Tenth Astronomer Royal, succeeding Frank *Dyson* in 1933 and retiring from the post in 1955. He played a major part in arranging the move of the *Royal Greenwich Observatory* out of London, to Herstmonceux in Sussex.

Jovian planets   See *giant planets*.

Julian calendar   The calendar system brought into use in the Roman Empire by Julius Caesar in 46 BC, and named after him. It was actually devised by Sosigenes of Alexandria. It was modified slightly in the following decades,

Jodrell Bank. The 250-foot (76-metre) dish of the Mark 1A radio telescope at Jodrell Bank – the world's first giant radio astronomy dish.

and was established in its final form under Augustus in AD 8.

The Julian calendar institutionalized the division of the year into 12 months, and had 365 days in a year, except that every fourth year contained 366 days (this leap year arrangement was not properly established until the time of Augustus). This meant that on average the length of the year was exactly 365.25 days according to the Julian calendar; but because this is 11 minutes and 14 seconds longer than the actual length of a year, the calendar gradually got out of step with the seasons; the difference amounts to one day every 128 years. Nevertheless, the Julian calendar remained the standard calendar system until 1582, when it was supplanted in Roman Catholic countries by the *Gregorian calendar* (which makes more subtle use of leap years), and was still in use in Russia at the beginning of the 20th century.

It was not until the 6th century, by the modern calendar, that the Roman scholar Dionysius Exiguus suggested counting the years from the birth of Jesus. He fixed the Conception on 25 March and the Nativity on 25 December in the year AD 1, but in tying the new calendar to the old one he miscalculated slightly in his countback of the years and was a few years out, so that Jesus was actually born in about 5 BC. Use of the new calendar spread only slowly across Europe, being taken up by Bede in the 8th century and by Charlemagne in the 9th century.

Julian Date (JD) Confusingly named standard system for coordinating events recorded on different calendars, which has no direct connection with Julius Caeser or the *Julian calendar*. The system was devised in 1582 by the French scholar Joseph Scaliger (1540–1609), and named in honour of his father, Julius Caeser Scaliger. It arbitrarily sets 1 January 4713 BC as the starting date, from which the Julian Date is given by the number of days that have elapsed. So, for example, 19 March 1990 is JD 2,447,970. Julian days begin at noon (Greenwich Mean Time), whereas dates on the everyday calendar begin at midnight.

Jupiter The largest *planet* in the Solar System, fifth out from the Sun, orbiting once around the Sun every 11.86 years at an average distance of 5.2 *astronomical units*. Its diameter is about eleven times the diameter of the Earth, and it has a *mass* 0.1 per cent of that of the Sun (318 times the mass of the Earth), more than twice as big as all the Sun's other planets put together. Jupiter has at least sixteen *moons* and a faint ring system. It is chiefly composed of hydrogen and helium, and resembles a failed star rather than an Earth-like planet.

Kaluza–Klein models Strictly speaking, the Kaluza–Klein model is the unification of *gravity* and electromagnetism in one five-dimensional model, first proposed by the German Theodor Kaluza in 1919, and refined to take account of the requirements of *quantum theory* by the Swedish physicist

Oskar Klein in 1926 (Kaluza and Klein never worked together). This model, the equivalent in five dimensions of Albert Einstein's equations of the *general theory of relativity*, which describe gravity in terms of four-dimensional *spacetime*, yields not only Einstein's equations for gravity but also James Clerk Maxwell's equations that describe *electromagnetic radiation*. Electromagnetism is seen as a 'ripple in the fifth dimension' analogous to gravity being a 'ripple the fourth dimension'. But where is the fifth dimension?

The standard explanation is that the fifth dimension is hidden from us by compactification. You can make an analogy with a hosepipe. The pipe is made of a sheet of two-dimensional material rolled round in the third dimension; but from a distance it looks like a one-dimensional line. It is possible to describe every point in spacetime as being made up of a whorl of five-dimensional spacetime wrapped up so that it looks four-dimensional – provided that the 'wrapping up' happens on a scale much smaller than that of an atomic *nucleus*.

With the discovery of the two forces that operate within the nucleus (see *fundamental forces*) and the development of *grand unified theories* involving many dimensions, the term 'Kaluza–Klein model' has come to be applied to any version of grand unification which operates in many dimensions and requires compactification.

Kamiokande   Name of a Japanese *neutrino* detector located underground at the Kamioka mine (the name comes from Kamioka Neutrino Detector Experiment). It has been running since 1987, and has detected about half of the predicted number of neutrinos from the Sun (see *solar neutrino problem*).

Kant, Immanuel (1724–1804)   German philosopher who was also interested in *cosmology* and who, in an essay published in 1755, made one of the first suggestions that the *planets* formed from a cloud of material around the Sun, an idea later developed independently by Pierre *Laplace*.

Karl Schwarzschild Observatory   German observatory founded in 1960, 350 m above sea level at Tautenberg, near Jena, in what was then the German Democratic Republic. Its main instrument is a 2 m *reflecting telescope* that can be used either for *spectroscopy* or as a *Schmidt camera*.

Keck telescope   A 10 m (400-inch) diameter *reflecting telescope* regarded as the most powerful ground-based instrument for *optical astronomy*. To make it possible to operate a telescope of such a large size, the Keck telescope is constructed on pioneering principles from 36 separate hexagonal mirrors, each 1.8 m across, which fit together in a honeycomb-like array and are steered by a computer system to work as if they were one mirror.

The telescope is part of the *Mauna Kea Observatory* in Hawaii, and is operated jointly by the California Institute of Technology and the University of California. It was completed in 1992, and gets its name from the W. M. Keck Foundation, which provided the money for its construction.

The Keck telescope can be used either for optical astronomy or for *infrared astronomy*, and has made important observations of *quasars* and the nature of the material in intergalactic space, revealed by *spectroscopy*. A twin telescope, Keck II, is being constructed 85 m from the original Keck telescope. When it is complete, the two telescopes will be linked by laser beams to act as an optical *interferometer*.

Kelvin (Baron Kelvin of Largs, William Thomson, 1824–1907)   Scottish physicist and electrical engineer who made many contributions to thermo-dynamics and the theory of *electromagnetic radiation*, and was responsible for the success of the first transatlantic cable communications system. His main contribution to astronomy dealt with the ages of the Earth and the Sun.

Born in Belfast, in Ireland, on 26 June 1824, William Thomson was the son of James Thomson, who was then professor of mathematics at the Academical Institution in Belfast, and became professor of mathematics at the University of Glasgow in 1832, when William was eight. The boy and his elder brother James (1822–92), who also became a physicist, had been educated at home by their father, but William showed such precocious talent that he was formally enrolled at the University of Glasgow at the age of ten, to study science (then referred to as natural philosophy). He did not graduate from Glasgow, but went on to Cambridge, where he did graduate in 1845. He worked briefly in Paris, before becoming professor of natural philosophy in Glasgow in 1846, where he stayed for the rest of his life.

Thomson's early work on electricity and magnetism led to his involve-ment with the project to lay a transatlantic cable for communications between Britain and the United States. Two unsuccessful cables had been laid in the 1850s before a third one, based on principles spelled out by Thomson, was successful in 1866; it was for this practical contribution that Thomson was knighted, and his share of the income from the patents made him a wealthy man. When he was made a baron in 1892, choosing the name Kelvin from that of a stream that passed through the site of the University of Glasgow, it was as much for his contributions to the Victorian economy as for his prowess in science, although by then he had long been regarded as the leading British scientist of his generation.

Thomson made equally important contributions to the development of the science of heat and motion (thermodynamics), which was also of keen interest to an industrial nation where the steam engine was so important. In connection with this work, he proposed (in 1848) an 'absol-ute' scale of temperatures, which for all practical purposes is identical with

Lord Kelvin (1824–1907).

the *Celsius* scale except that the zero point corresponds to –273.16 °C; this is now known as the *Kelvin temperature scale*. It was Thomson who, in 1851, formulated the famous second law of thermodynamics, which says that heat cannot flow spontaneously from a cooler object to a hotter one, and which implies the *heat death of the Universe*.

Using his understanding of heat and thermodynamics, Thomson was able to calculate how long it would have taken for a ball of rock the size of the Earth to cool down to its present state from an initial molten state. His calculations were accurate, but did not take account of the heat still being generated inside the Earth by *radioactive decay*, which had not been discovered at that time. He also estimated the maximum age of the Sun, based on the most efficient *energy* source he could imagine, the slow release of gravitational energy by contraction (see *Kelvin–Helmholtz timescale*).

These age estimates, which said that the Sun and Earth could not be more than a few tens of millions of years old, brought Thomson into conflict with geologists and Darwinian evolutionists in the second half of the 19th century, who both said that the Earth must be much older than Thomson's calculations allowed. The conflict was resolved with the discovery of radioactivity and the development of the *special theory of relativity*, which allows the conversion of *mass* into energy.

As Lord Kelvin, Thomson died at Largs, in Ayrshire, on 17 December 1907. He is buried near Isaac **Newton** in Westminster Abbey.

Kelvin–Helmholtz timescale The length of time for which a *star* like the Sun could continue to radiate *energy* simply by contracting slowly under its own weight – about 20–30 million years.

In the middle of the 19th century, astronomers and physicists puzzled over how the Sun kept itself hot. They realized that if it were entirely made of coal, burning in an atmosphere of pure oxygen, it would be burnt out in less than about 100,000 years, and they suspected, from geological evidence, that the Earth had been warmed by the Sun for much longer than that. Hermann **Helmholtz**, in Germany, and William Thomson (later Lord **Kelvin**) in Britain, independently came up with the same solution to the problem. They showed that simply by shrinking slowly in upon itself, the Sun could shine as brightly as it does today for several tens of millions of years, as gravitational energy was converted into heat.

But even this Kelvin–Helmholtz timescale conflicted with the requirements of geology and of Charles Darwin's theory of evolution by natural selection, which suggested that life had evolved slowly on Earth over hundreds or thousands of millions of years. The confrontation between physics and biology was resolved only at the end of the 19th century, when the discovery of radioactivity and then the development of Albert **Einstein**'s special theory of relativity showed that there was another form of energy (the conversion of *mass* into energy) that could in principle keep the Sun hot for the timescale required by evolution. Nevertheless, the way in which a contracting cloud of gas in space does begin to heat up in the centre and form a young star, heating it to the point where *fusion* reactions begin, is very well described by the Kelvin–Helmholtz contraction process.

See also *proton-proton reaction*.

Kelvin temperature scale Scale named after Lord **Kelvin**, in which 0 °C corresponds to 273.15 K. See *absolute zero*, *Celsius*.

Kepler, Johannes (1571–1630) German astronomer who discovered the three laws of planetary motion that helped lead Isaac **Newton** to his universal law of gravity.

Kepler was born in Weil der Statt, near Stuttgart, on 27 December 1571, the son of a mercenary soldier, but grandson of the local burgomaster. He suffered smallpox at the age of three, and this affected both his eyesight and the use of his hands throughout his life. As a child, Kepler had been deeply impressed by seeing the great *comet* of 1577, and retained an interest in astronomy. He studied at the University of Tübingen, completing a course in philosophy, mathematics and astronomy with an MA in 1591. He then began (possibly at the insistence of his family) a three-year theological course, but left this in his final year (1594) when he was offered

Johannes Kepler (1571–1630).

˙a job teaching mathematics at the Protestant Seminary in Graz, Austria.

While working as a teacher, Kepler began to publish his ideas about mathematical harmony as an underpinning to the structure of the Universe, and his work was noticed by established astronomers including Tycho **Brahe** and **Galileo**. As a Lutheran, Kepler was caught up in the religious troubles of the time, and was forced to leave Graz in 1598, the year after he had married. After a year in Prague, he returned to Graz but was expelled again, returning to Prague, where he became Tycho's assistant in 1600. When Tycho died in 1601, Kepler became his successor as Imperial Mathematician to the Emperor, Rudolph II. He undertook the completion of Tycho's astronomical tables (a task that was to occupy him, off and on, until 1627), worked out the orbit of Mars and discovered the first two of his three laws while working in Prague – all against the background of the Thirty Years War. He published a major work on *Optics* in 1604, and observed and wrote about the comet seen in 1607 (now known as Halley's Comet).

But eventually the troubles of the time caught up with Kepler. In 1611 his wife and son both died; in the same year, there was civil war in Prague and Rudolph was deposed. A year later Kepler moved to Linz, where he worked as a mathematician and surveyor for the States of Upper Austria. He remarried in 1613, and lived there for fourteen years. During that time his mother was tried for witchcraft in Würtemburg, and exonerated only after the trial had dragged on for three years. In 1619, Kepler's third law was published as part of his *De Harmonices Mundi*, a largely mystical work in which the third law was buried in the last of five chapters. By then, Kepler's great astronomical work was over, but his *Epitome*, published in seven volumes over a period of four years, summed up his achievements, provided an introduction to the ideas of **Copernicus**, and had a major influence on astronomy in the middle of the 17th century.

Kepler moved to Sagan, in Silesia, in 1628, where he worked for the Duke of Friedland, General Wallentein. In 1630 he was once again forced to move by religious persecution. On his way to Regensburg, in Bavaria, he caught a fever, and he died on 15 November.

The year after Kepler's death, his story *Solemnium* (*A Dream*) was published. It had been written twenty years earlier, and tells the story of a man who travels to the Moon. Arguably, this was the first science fiction story.

Keplerian telescope A simple *refracting telescope* using two lenses, both of them convex, in a tube. This gives a large field of view and good magnification, but produces an upside-down image (which doesn't matter much in astronomy!).

Kepler's laws The three laws of planetary motion, discovered by Johannes **Kepler** early in the 17th century, using data gathered by Tycho **Brahe**. The first law states that the *planets* move in elliptical *orbits*, with the Sun at one focus of the ellipse; the second states that an imaginary line joining the centre of a planet to the centre of the Sun sweeps out equal areas in equal times (so a planet moves faster in its orbit at the end of the orbit closer to the Sun); and the third law says that the square of the true period of the orbit (measured relative to the *fixed stars*) is proportional to the cube of the planet's average distance from the Sun.

Isaac **Newton** explained these laws as due to *gravity* operating as an inverse square law of attraction between the Sun and each planet – that is, with the strength of the force on a planet at each point in its orbit proportional to 1 divided by the square of the distance from the Sun to the planet at that moment. The equivalent laws apply to any object that is in a closed orbit about any other object (including the Moon and artificial satellites orbiting the Earth), except in extreme cases where effects described by the *general theory of relativity* become important (for example, in the *binary pulsar*). (See diagram on p. 284.)

Kerr, Roy Patrick (1934–) New Zealand mathematician (professor of mathematics at the University of Canterbury, Christchurch), who has made important contributions to the understanding of *black holes*. He developed the *Kerr solution* of the equations of the *general theory of relativity*.

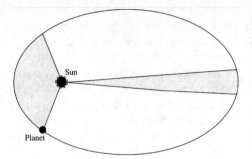

Kepler's laws. Kepler realised that the speed of a planet in its elliptical orbit changes in such a way that it always 'sweeps out' equal areas in equal times. This insight led to the discovery of the inverse square law of gravity.

Kerr black hole   See *Kerr solution*.

Kerr–Newman solution   A solution to Albert *Einstein*'s equations of the *general theory of relativity* that describes a rotating, electrically charged *black hole*. It was developed by Ezra Newman and colleagues at the University of Pittsburgh in 1965, starting from the *Kerr solution*. This is the most general mathematical description of a black hole. Setting rotation equal to zero in the Kerr–Newman solution gives the *Reissner–Nordstrøm solution*; setting charge equal to zero gives the Kerr solution, and setting both charge and rotation to zero gives the *Schwarzschild solution*.

Kerr solution   The mathematical description of a rotating, uncharged *black hole*, which is the kind most likely to exist in our Universe. This solution to Albert *Einstein*'s equations of the *general theory of relativity* was found by the New Zealander Roy *Kerr*, working at the University of Texas in 1963. Kerr actually discovered his solution to the equations purely from a mathematical standpoint – he was not looking for any physically meaningful result, but it turned out that the solution he found describes the way in which *spacetime* is dragged around by the rotating black hole. One of the consequences of this is that the *singularity* at the heart of the black hole is no longer a mathematical point, but a ring. According to the equations, it would be possible to dive through this ring and emerge in another region of spacetime (possibly another universe, or at a different place and/or a different time in our Universe). This has led to speculation about the use of black holes as *time machines*.

See also *wormholes*.

kiloparsec (kpc)   Unit of distance, equal to 1,000 *parsecs*.

kinetic energy   The *energy* an object possesses because of its motion. You have to put energy in to make an object go faster, and when the object is brought to a halt that energy is given up again as heat. That is why the brakes of a car get hot when it stops.

Kuiper, Gerard Peter (1905–73) Dutch-born American astronomer who played a major part in reviving the interest of astronomers in the study of the Solar System, and contributed to the development of *NASA*'s programme of unmanned exploration of the *planets*. Among many other achievements, he developed the theory of the origin of *comets*, and pioneered the use of aircraft to carry infrared *telescopes* above the absorbing layers of the atmosphere (see *Kuiper Airborne Observatory*).

Kuiper Airborne Observatory (KAO) A flying observatory for *infrared astronomy*, carried on board a Lockheed C141 Starlifter aircraft and run by *NASA* from the Ames Research Center at Moffett Field in California. It can operate at an altitude of 12 km, above 99 per cent of the obscuring water vapour in the atmosphere, and uses a 0.9 m diameter *telescope*. It has been operating since 1975, and is named after the astronomer Gerard *Kuiper*, who worked on NASA's scientific space programme. It was the KAO that discovered the rings of Uranus.

Kuiper belt A band of *comets*, thought to contain up to 1 billion individual objects, which lies between about 35 and 1,000 *astronomical units* from the Sun, in the same plane as the *planets* but beyond the orbit of Pluto. Comets from the Kuiper belt are occasionally disturbed by gravitational forces and fall into the inner Solar System. *Chiron* may have come from the Kuiper belt, which is named after the astronomer Gerard *Kuiper*. See also *Oort cloud*.

Lagrange, Joseph Louis (1736–1813) Italian-born French mathematician whose contributions to the development of the theory of dynamics are important in calculations of the *orbits* of *planets*, *satellites* and *moons*.

lagrangian points Points of stability in the orbital plane of a two-body system (such as the Earth–Moon system), where small objects may be trapped in the *gravitational field*. They were first identified theoretically by Joseph *Lagrange* in 1772. It has been suggested that the lagrangian points would be good places to locate large artificial colonies in space.

Landau, Lev Davidovich (1908–68) Azerbaijani-born Russian theoretical physicist (his father was a petroleum engineer, his mother a doctor), awarded the Nobel Prize in 1962 for his work on the nature of liquid *helium*. He entered Baku University at the age of fourteen, and moved on to Leningrad University when he was sixteen, graduating in 1927 and being awarded the equivalent of a PhD by Leningrad University (where he was a contemporary of George *Gamow*) when he was 21.

In 1932, Landau independently discovered the *Chandrasekhar limit* for the mass of a stable *white dwarf*, and went on to speculate about the possible existence of *neutron* cores inside large stars. This work led directly to Robert *Oppenheimer*'s investigation of the theory of *neutron stars*, and to the determination of the **Oppenheimer–Volkoff limit** for the mass of a stable neutron star, an important step towards the theory of *black holes*. Landau was also a great teacher, who created the strong theoretical physics group in Moscow, and was co-author (with E.M. Lifshitz, 1917–69) of a series of influential textbooks.

Laplace, Pierre Simon, Marquis de (1749–1827)   French mathematician and astronomer whose contributions to science ranked in his lifetime second only to those of Isaac *Newton*.

Laplace was born at Beaumont-en-Auge, in Normandy, on 28 March 1749. Some accounts refer to him as coming from a poor farming family; but there is evidence that, although he was not rich, Laplace's father was far from the poverty line, a local magistrate who made some of his money from the cider trade. Between the ages of seven and sixteen, Laplace was educated as a day-pupil at a local college run by Benedictines, which his father intended would lead to a career in the Church. He showed sufficient academic promise, however, to be allowed to go on to study at Caen, and then, in 1768, he moved on to Paris, where he obtained a post at the Ecole Militaire, where he became professor of mathematics.

Throughout the turmoil of the following decades, which saw the French Revolution and its aftermath of war followed by the restoration of the monarchy, Laplace thrived. He became one of the leading figures in the French Academy of Sciences, served in government posts before and after the revolution (including being a member of the Commission of Weights and Measures that introduced the metric system), discussed science with Napoleon, served in the Senate, and was even, briefly, Minister for the Interior, in 1803. Having voted in 1814, in the Senate, for the restoration of the monarchy, his public career briefly went into decline, but he was rewarded in 1817 when the restored king, Louis XVIII, made him a marquis.

Laplace made his scientific reputation with a great work updating Newton's investigations of planetary movements. Even Newton had been baffled by one aspect of this problem. He knew that one planet on its own, orbiting the Sun in accordance with *Kepler's laws*, would move forever in a perfectly elliptical orbit. But with two or more planets orbiting the Sun, it seemed that the extra gravitational influence of the other planets might upset the equilibrium, eventually sending the planets tumbling out of their orbits. Newton had no answer to the question of why this did not happen, and fell back on suggesting that the hand of God might be needed from time to time, to nudge the planets back into their proper orbits.

Laplace, in the mid-1780s, showed that in fact these perturbations are largely self-correcting. At the time, the orbit of Jupiter was known to be shrinking slowly, and the orbit of Saturn was seen to be expanding. Laplace showed that this is part of a long cycle, with a period of 929 years, in which these two *giant planets* wobble about the strict Keplerian orbits. This discovery led to possibly the most famous remark ever made by Laplace. When this breakthrough in celestial mechanics was published in book form, Napoleon commented to Laplace that he had noticed that there was no mention of God in the book. Laplace replied that 'I have no need of that hypothesis.'

Laplace was also one of the first people to consider the possibility of *black holes* (independently of John *Michell*), and suggested that the planets had formed from a primordial nebula of material around the Sun. Just as Newton had developed calculus in order to carry out his astronomical calculations, Laplace developed new mathematical techniques, which are still used today, in his astronomical studies. He also laid the foundations of probability theory. He died in Paris on 5 March 1827, 100 years, to the month, after the death of Newton.

Large Magellanic Cloud   See *Magellanic Clouds*.

large-scale structure   As far as the observable Universe is concerned, 'large-scale' means larger than 100 *Megaparsecs*, referring to the overall distribution of visible matter in the Universe as revealed by studies of the *redshifts* of very many *galaxies*. The most distinctive feature of the Universe on these scales is that the visible matter is distributed in thin sheets and filaments that wrap around huge voids, giving the appearance of a mass of intersecting bubbles – the Universe looks frothy on the largest scales.

Although the sheets and filaments take up only about 2 per cent of the volume of the visible Universe, they contain essentially all of the visible matter. The largest of these structures, called the Great Wall, is just under 100 Mpc away from us and is built up from a sheet of *clusters of galaxies* and *superclusters* that covers a region 225 Mpc long and 80 Mpc wide, but is only 10 Mpc thick.

This frothy appearance of the Universe, like many other factors, is an indication that the overall *density* of the Universe is close to the critical value needed to make *spacetime* flat (see *cosmological models*). Computer simulations show that, if there is a large amount of *dark matter* in the Universe and bright galaxies form only where the overall density reaches peak levels, the distribution of bright galaxies is very similar to that in the real Universe.

Las Campanas Observatory   Observatory on the Cerro Las Campanas, 2,300 m above sea level, near La Serena, in Chile. Owned by the Carnegie Institution of Washington, the main instruments at the observatory are a 2.5 m *reflecting telescope* and a 1 m reflector. A 6.5 m reflector is planned for the site.

Leavitt, Henrietta Swan (1868–1921)   American astronomer who worked at the Harvard College Observatory with Edward *Pickering*, and discovered the *period–luminosity relation* of *Cepheid variables*.

Lemaître, Georges Édouard (1894–1966) Belgian astrophysicist, cosmologist and priest, who developed the first version of what became the *Big Bang* description of the Universe.

Lemaître was born on 17 July 1894, at Charleroi, in Belgium. He trained as a civil engineer at the University of Louvain, then volunteered for the Belgian army, in which he served as an artillery officer during the First World War and was awarded the Belgian Croix de Guerre.

After the war, Lemaître returned to academic life, completing a PhD in physics at Louvain in 1920, and then entered a seminary, becoming an ordained priest in 1923. But he did not practise as a priest, instead travelling on a scholarship from the Belgian government which took him to Cambridge, in England, and Harvard and MIT in the United States. This was just at the time when Edwin *Hubble* and his colleagues were beginning to find evidence for the *expanding Universe*, and Lemaître was well up to date with the new developments when he returned to Belgium in 1927 to take up the post of professor of astrophysics at the University of Louvain, where he stayed for the rest of his career.

In 1927, Lemaître published a solution to Albert *Einstein*'s equations of the *general theory of relativity* that described the expanding Universe. This was essentially the same discovery made by Aleksandr *Friedmann* a few years earlier, but Lemaître was unaware of Friedmann's work at the time. Unlike Friedmann, however, Lemaître specifically suggested that *galaxies* might provide the 'test particles' which would show the expansion of the Universe.

This paper attracted little attention at the time (it was published in Belgium in a rather obscure journal), but it made more of a splash when Arthur *Eddington* learned of it and arranged for a translation to be published in the *Monthly Notices of the Royal Astronomical Society* in 1931, after the announcement of the discovery of the relationship between *redshift* and distance for galaxies (*Hubble's law*). Over the following years, culminating with his *Hypothesis of the Primal Atom* (published in 1946), Lemaître developed the idea that the Universe had originated in a 'primeval atom', a sphere only about 30 times bigger than the Sun but packed with all the matter we see in the Universe today, which had exploded (some time between 20 billion and 60 billion years ago), like the *fission* of an unstable *nucleus*, to create the expanding Universe.

This was the first scientific attempt at describing a creation event associated with the beginning of the expansion of the Universe, and although Lemaître's ideas did not have a major impact on the thinking of most astronomers in the 1930s, he succeeded in popularizing the idea of the primeval atom (or 'cosmic egg') for a wide audience. When George **Gamow** and his colleagues began to develop the Big Bang idea in the 1940s, they were working more in the tradition of Friedmann (who had been one of Gamow's teachers) than Lemaître, but Lemaître remained a staunch supporter of the Big Bang idea for the rest of his life, and lived to see it becoming accepted as the standard model of cosmology. He died at Louvain on 20 June 1966, the year after the announcement of the discovery of the **background radiation** that is interpreted as a remnant of the Big Bang fireball.

length contraction   See *Fitzgerald contraction*.

lens galaxy   See *elliptical galaxy*.

lenticular galaxy   See *elliptical galaxy*.

leptons   One of the two families of elementary particles from which all atomic matter is made. The family has six members – the **electron**, **muon** and **tau** particles and their associated **neutrinos**. See also **quarks**.

Le Verrier, Urbain Jean Joseph (1811–77)   French astronomer who in 1846 predicted the existence of a planet beyond Uranus by studying perturbations in the orbit of Uranus. He passed his calculations on to the Berlin Observatory, where the planet Neptune was discovered on 23 September 1846. The English astronomer **Adams** had made the same prediction independently.

Lick Observatory   The first observatory in the United States to be built on a mountain top. It was completed in 1888 on Mount Hamilton in California, 1,283 m above sea level. It was funded by an eccentric millionaire, James Lick, who is buried beneath the 91 cm **refracting telescope**, which was originally the observatory's main instrument and is still the second largest refractor in the world. The main telescope at Lick now is a 3 m reflector, completed in 1959. The observatory is run by the University of California, Santa Cruz.

life cycle of a galaxy   See *origin and evolution of galaxies*.

life cycle of a star   See *stellar evolution*.

life in the Universe   The simplest working definition of a living thing is that it can reproduce itself by taking raw materials from the surrounding environment and rearranging them to make new copies of the living thing. Unlike a crystal, which is clearly non-living even though it is able to 'grow' if provided with the right raw materials, the processes of growth, replication and self-repair that are associated with life go on inside living things, and can dramatically change the raw materials, so that in extreme cases a single cell (a fertilized egg) can transform the raw materials it takes in into a complex organism containing billions of cells; a crystal can only make more identical crystal.

A key feature of the way that life operates is that it feeds off a flow of *energy*, using the energy to build up complexity and order in an apparent contradiction of the second law of thermodynamics, which says that the amount of disorder in the Universe (entropy) always increases (see *arrow of time*). But the local decrease in entropy represented by life is always more than compensated for by an increase elsewhere. In the case of life on Earth, our bubble of decreased entropy is paid for by the huge increase in entropy associated with the changes going on inside the Sun that produce the energy on which we depend.

Some scientists and science fiction writers (often individuals doubling up in both roles) have speculated that life could exist, in forms that would seem exotic to us, anywhere that there is a flow of energy that can be tapped. In extreme cases, this could happen inside a cloud of interstellar material, making the cloud itself sentient (see, for example, Fred *Hoyle*'s novel *The Black Cloud*), or under the conditions of high gravity and intense magnetic fields at the surface of a neutron star (see Robert Forward's *Dragon's Egg*).

But although such speculations are appealing, one of the most dramatic discoveries of recent years has been the amount of material in the clouds of gas and dust between the stars that is involved in the same kind of chemical processes, based on *carbon*, that are important for life on Earth (see *CHON, interstellar chemistry*). This material is present not just in the *Milky Way*, but in other *galaxies* – the compound formaldehyde, for example, has been detected by its characteristic radio emission in the galaxy NGC 253. The inference is that the complex carbon chemistry which led to the development of life on Earth was a natural development from the only slightly less complex chemistry of the clouds of material from which the Solar System formed, and that where other Earth-like *planets* exist, complex chemistry may have developed towards life in a similar way. Leaving aside any considerations of whether life might develop on different kinds of planet, such as the *giant planets*, this implies that there could be millions of civilizations involving life forms based on carbon chemistry in our Galaxy today (see *Drake equation*). It also suggests that there may well be more than one (to say the least!) civilization comparable to our own in the galaxy NGC 253, where carbon-based life forms may have developed *radio astronomy* and may be discussing the possible significance of the discovery of formaldehyde in the collection of stars which we like to call 'our' Galaxy.

Traces of single-celled life forms (algae similar to those alive today) have been found in rocks much more than 3 billion years old, which were laid down less than 1 billion years after the formation of the Earth. It seems that life arose almost as soon as the Earth had cooled to the point where it could exist, which strengthens the case that the young planet was seeded with complex organic material from space, perhaps including *amino acids* such as glycine.

Some researchers have gone further, suggesting that the young Earth was seeded with actual living cells from space. In one variation on this theme, Fred Hoyle and his colleague Chandra Wickramasinghe propose that life evolved in interstellar clouds over a very long period of time before the Solar System had even formed. Francis Crick (who shared a Nobel Prize with James Watson for the discovery of the structure of the life molecule, DNA) has espoused an alternative proposal, that the Galaxy has been deliberately seeded with life by a civilization which arose billions of years ago when the Galaxy itself was young. This idea is known as *panspermia*; it builds on earlier speculations by other researchers (notably the Swede Svante *Arrhenius*), but still leaves unanswered the question of where the original civilization came from, which perhaps makes it less plausible than the alternative proposal. Either proposal, though, extends the timescale available for evolution from the roughly $4\frac{1}{2}$ billion years of Earth history to three or four times as long, depending on exactly how old the Universe is. The one inescapable conclusion seems to be that life did not start from scratch (that is, from simple atoms and molecules of carbon, oxygen, nitrogen and hydrogen) on the surface of the Earth.

Some scientists have looked at life in a different way in recent years, and have come up with ideas which offer a new perspective on the Universe at large. In the 1970s, James Lovelock developed the hypothesis that the entire Earth could be regarded as a single living organism (Gaia). The idea has proved controversial, but has encouraged new ways of thinking which have provided insights about the workings of the various feedbacks that maintain conditions suitable for life forms like us to exist on Earth. Extending the analogy outwards, Lee Smolin, of Syracuse University in New York, has suggested that entire galaxies might be regarded as living systems:

> there seems to be a kind of ecology in the physics of spiral galaxies by means of which the structures responsible for star formation – the spiral arms and the associated clouds of dust and gas – are maintained for time scales much longer than the relevant dynamical time scales ... These must involve self-organizing cycles of materials and energy of the kind that one sees in diverse non-equilibrium states as well as in biological systems.

Extending the analogy still further, Smolin has suggested that the entire Universe might be regarded as alive and having evolved (in the strict Darwinian sense) from earlier generations of universes (see *baby universes*). Andrei *Linde* has also speculated about the Darwinian evolution of universes.

The suggestion that galaxies and the Universe can literally be alive is far from being widely accepted, but just as the Gaia hypothesis encouraged Earth scientists to think about our planet in a different way, so this 'biological paradigm' at the very least encourages astronomers to think about the Universe and its contents in new and insightful ways. Astronomers are beginning to use concepts like evolution and population dynamics, taken on board from biology, even if they are careful to stress that these are 'only metaphors'. The idea of evolving populations has, for example, transformed our understanding of *galaxy formation and evolution*.

Further reading: Francis Crick, *Life Itself*; John Gribbin, *In the Beginning*.

light   In astronomy, light is usually regarded simply as the range of *electromagnetic radiation* which human eyes are sensitive to, with *wavelengths* in the range of about 380–750 nanometres. This part of the electromagnetic *spectrum* is bounded by *ultraviolet radiation* (at shorter wavelengths) and by *infrared radiation* (at longer wavelengths). It corresponds, in terms of colours, to the rainbow range from red (at the long-wavelength end) through orange, yellow, green, blue and indigo to violet (at the short-wavelength end).

Our eyes have evolved and adapted to be sensitive to light because it is there – radiation in this part of the spectrum is produced copiously by the Sun, and is not absorbed by the atmosphere of the Earth, so it penetrates to the ground. Astronomy first developed by observing light from the Sun, planets and stars, initially with the unaided human eye and later with the aid of telescopes; with modern detectors and photographic plates sensitive to a slightly wider range of wavelengths than the human eye, in *optical astronomy* the range of the spectrum used is 300–900 nm, but is still called light. There is no essential difference between light and other forms of electromagnetic radiation such as *X-rays* and *radio waves*; they simply have different wavelengths.

Since the development of *quantum theory* in the first quarter of the 20th century, light has also been regarded as having particle-like properties. In appropriate circumstances, the energy in light can be thought of as being carried by a stream of particles, called photons. Many modern detectors, such as *charge coupled devices*, are essentially photon counters that record the arrival of individual photons from faint astronomical sources and build them up into images over long periods of time.

light bending   See *deflection of light*.

light cone   A way of representing the movement of a flash of *light* through *spacetime*. In three-dimensional space, light spreads out spherically from a source. In order to represent four-dimensional spacetime on paper, relativists use a diagram in which the flow of time is represented up the page, while all of the dimensions of space are represented by a line across the

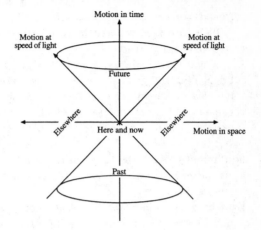

Light cone. The Minkowski diagram represents all of space and time on one diagram. From 'here and now', the only events we can have knowledge of are within the past light cone, and the only events we can influence are in the future light cone.

page (this representation develops from the work of Hermann **Minkowski**). The spread of a flash of light out from a single point on the page is represented by two diagonal lines moving up the page, each at 45 degrees to the vertical. If you now imagine rotating the paper around the vertical axis, these lines will sweep out a conical surface – the light cone. In this representation, any signal travelling slower than light traces out a track within this future light cone.

The light cone is important because Albert **Einstein** showed, with his *special theory of relativity*, that no signals carrying information can travel faster than light. This means that any event occurring at a point in spacetime can influence the region of spacetime only within its own future light cone. Extending the light lines on the diagram down the page traces out a mirror image of the future light cone, called the past light cone. Again, because no signal can travel faster than light, any point in spacetime can only be influenced by, and have knowledge about, events that occurred within its past light cone. The puzzle of the *horizon problem* arises because regions on opposite sides of the sky look exactly the same, even though they lie outside each other's light cones.

light curve A graph showing the way in which the brightness of an object, such as a *variable star*, changes as time passes. The light curve may be presented in terms of either *absolute magnitude* or *apparent magnitude* for the source. This distinction is often less important than the shape of the curve, which can indicate what kind of object is varying. The variations in brightness represented by a light curve may be due to intrinsic changes in the source (for example, the changes in brightness of *Cepheid variables*), or they may occur when one member of a *binary system* eclipses the light from its companion. Variations in the brightness of sources in other parts of the *electromagnetic spectrum* (such as those in the *X-ray* band) are also referred to as light curves.

Light curves do not have to show periodic variations – the sudden brightening and gradual fading of a *supernova*, for example, is also represented by a light curve. But where the light curves do show periodic variations they can provide a wealth of additional information – for example, about the *orbits* and *masses* of *stars* or *black holes* in binary systems.

light minute   The distance travelled by *light* in one minute. See *light year.*

light second   The distance travelled by *light* in 1 second – that is, 299,792.5 km. See *light year.*

light time   The time taken for *light* or other *electromagnetic radiation* to travel between two points. The light time along the average distance between the Sun and the Earth is 499 seconds, another way of saying that the average distance to the Sun is 499 *light seconds.*

light year   A unit of distance often used by science fiction writers and occasionally by astronomers. One light year is the distance that *light*, travelling at a speed of 299,792,458 metres per second, can travel in 1 year – that is, 9.46×10^{12} km (9.46 million million km), or 0.3066 *parsecs*. A light year is a measure of *distance*, not of time. Light months, light days, light hours and *light seconds* are derived in a similar way from the distance travelled by light in the appropriate time interval.

Linde, Andrei Dmitrivitch (1948–)   Russian-born cosmologist who played a major role in the development of the theory of the very early Universe known as *inflation*.

Linde was born in Moscow on 2 March 1948. He studied physics at the Moscow State University from 1966 to 1971, and then worked for a PhD in particle physics and cosmology at the Lebedev Physical Institute, also in Moscow. This work involved a study of cosmological phase transitions (analogous to the conversion of water into ice), which later became a basic ingredient of inflationary cosmology; he received his PhD in 1974, and remained at the Lebedev Institute, where he developed this work further and also made important contributions to the understanding of *fermions* in the context of *gauge theories*.

In the late 1970s, there was considerable interest among the cosmologists in Moscow in *cosmological models* based on what is now called inflation, but it was the independent work of Alan *Guth* (who gave inflationary cosmology its name) in the United States that brought the possibilities of inflation to the attention of a worldwide audience of astronomers at the beginning of the 1980s. In 1982, Linde suggested (in a paper that became the most often cited physics paper of the year) an improved version of inflation that became known as 'the new inflationary model', in 1983 he developed the chaotic model of inflation, and in 1986 he suggested that the Universe might be part of a self-reproducing system of *baby universes*.

Linde moved to the United States at the end of the 1980s, first visiting Harvard University and then taking up a permanent post at Stanford University, where he became a professor of physics. In the mid-1990s, he was still developing the idea of the Universe as a self-reproducing system that sprouts other inflationary universes and is itself a sprout from another universe. See also *monopole universe*.

LMC = *Large Magellanic Cloud.*

LMC X-3 An *X-ray source* in the *Large Magellanic Cloud*. LMC X-3 is a *binary system* with an orbital period of 1.7 days; the X-rays almost certainly come from an *accretion disc* around a *black hole*.

Local Group The name for the small group of *galaxies* to which our own *Galaxy* belongs. The Local Group also includes the *Andromeda galaxy*, the *Magellanic Clouds* and several *dwarf galaxies*. See *clusters of galaxies*.

local standard of rest (LSR) A *frame of reference*, centred on the Sun, in which the motions of all the other nearby *stars* would average out to be zero. The Sun is moving relative to the LSR at a speed of about 20 km/sec. The LSR itself is moving at about 250 km/sec around the *Galaxy* (see *cosmic year*).

local supercluster A large association of *galaxies* and *clusters of galaxies* which includes the *Local Group* and the *Virgo* and *Coma* clusters. This huge entity, first identified by Gerard *de Vaucouleurs* in 1955, has a radius of about 30 million *parsecs*; the Local Group is near the edge of the local supercluster.

Lockyer, (Joseph) Norman (1836–1920) British astronomer who discovered *helium* from its spectroscopic fingerprint in sunlight, was the first scientist to notice the link between solar activity and weather cycles, and founded the journal *Nature*.

Lockyer came into science by the back door, from an interest in astronomy as a serious amateur. He had been born in Rugby on 17 May 1836, the son of a surgeon-apothecary, and went through a conventional education to become a clerk at the War Office in London. He settled in Wimbledon, married in 1861, and became an enthusiastic amateur observer of the Sun.

Lockyer was lucky to be around at just the right time to take advantage of the new technique of *spectroscopy*, which had been developed at the end of the 1850s by Gustav Kirchoff (1824–87) and Robert Bunsen (1811–99). He grasped the opportunity with both hands. Using spectroscopy, Lockyer showed that the darkness of *sunspots* is caused by the presence of relatively cool gas absorbing light, and by the end of the 1860s, though still working at the War Office, he was using the still relatively new concept of the *Doppler effect* to study the movement of gases in the atmosphere of the Sun.

Through his spectroscopic studies of the Sun's atmosphere during an *eclipse* in 1868, and observations of the same eclipse made by Pierre Jansen (1824–1907), Lockyer found lines in the solar spectrum that could not be explained as being produced by any of the known *elements*. He inferred that there must be an element present in the Sun that had not been detected on Earth, and named it helium from the Greek word for the Sun, helios. Helium was only identified on Earth in 1895, by Sir William Ramsay (1852–1916) and Sir William Crookes (1832–1919), and Lockyer was knighted two years later, in 1897.

In 1869, Lockyer was one of the founders of the scientific journal *Nature*, which he edited for the next 50 years. His reputation as a scientist enabled him to leave the War Office, and, after working in a couple of other positions, in 1890 he became director of the Solar Physics Observatory in South Kensington. He only resigned from this post in 1911, when he was in his mid-seventies, because the observatory was moving to Cambridge and he wanted to stay in the south of England.

Lockyer was also interested in the megalithic monuments found in Britain and northern France, and suggested that the stone circles and avenues at Stonehenge must have been aligned precisely to mark the rising of the Sun at the summer solstice at the time it was built. Because of tiny changes in the position of the Sun as seen from Earth, by Lockyer's day the stones seemed to be out of alignment by 1 degree and 12 *arc seconds*. From the size of this discrepancy, he concluded that the stones were erected around 1840 BC, with an error of ±200 years. Modern dating techniques using radioactive carbon (carbon-14) show that it was built around 1848 BC, with an error of ±250 years.

Lockyer travelled on several eclipse expeditions, suggested (long before the discovery of the *electron*) that unusual spectral lines in the Sun could be explained by *atoms* being broken down into simpler constituents, and after the transfer of the Solar Physics Observatory to Cambridge remained an active observer using his private observatory in Sidmouth until just before he died – at Salcombe Regis, in Devon, on 16 August 1920, when he was 84.

look back time   The time it has taken *light* from the particular astronomical object that is being studied to reach us. Because the speed of light is finite, the further away from us an object is, the longer it takes light from that object to reach us. Light from a star 10 *light years* away takes 10 years to reach us; light from a *quasar* 5 billion light years away takes 5 billion years to reach us. Even light from the Sun takes just over 8 minutes to reach us, so the look back time to the Sun is just over 8 minutes.

This means that astronomers are literally looking back further into the past when they study the light (or other *electromagnetic radiation*) from more distant objects. If the output of light from the Sun were turned off, by magic, as you read these words, then it would be more than 8 minutes before the sky here on Earth went dark. Although the *supernova* explosion that created the *Crab nebula* was seen on Earth by Chinese astronomers in the year 1054, the site of this stellar explosion is about 6,500 light years away from us. So the light from the explosion took all that time to reach the Earth, and the events seen by the Chinese astronomers in 1054 had actually occurred in about the year 5446 BC by terrestrial calendars.

Some distant astronomical sources are studied by light which left them when dinosaurs roamed the Earth; others by electromagnetic radiation which left them before the Solar System had even formed. By studying objects such as *galaxies* at different distances across the Universe, astronomers can find out how the Universe has changed as it has evolved out of the *Big Bang*.

Lorentz, Hendrik Antoon (1853–1928)   Dutch physicist best known for developing the equations that describe how the properties of a moving body (such as its length) are affected by its motion. See *special theory of relativity* and *Fitzgerald, George Francis*.

Lorentz–Fitzgerald contraction   See *Fitzgerald contraction*.

Lorentz transformation equations   A set of equations developed by Hendrik *Lorentz*, which describe the changes that have to be made to measurements of lengths and intervals of time when dealing with different *inertial frames*.

George *Fitzgerald* was the first person (at the end of the 19th century) to point out that the failure of experiments such as the *Michelson–Morley experiment* to detect evidence of the Earth's motion through the ether (thought at the time to be the medium through which light travelled, like waves moving through water) could be explained if all measuring rods and the apparatus used in these experiments shrank by a certain amount in the direction of the Earth's motion (see *Fitzgerald contraction*). In 1904, Lorentz took this idea further by developing the equations now known as the Lorentz transformations, which describe how not only the length of a moving object but its other properties must 'transform' when viewed by observers moving at different velocities, in order for the ether to be undetectable.

Lorentz originally developed the equations to describe how electromagnetic fields would look to observers moving at different velocities, putting the relative velocities of observers into the equations of electromagnetism that had been developed by James Clerk **Maxwell**. A year later, Albert **Einstein** showed that, although there is no need for an ether, these transformation equations are a natural consequence of his **special theory of relativity**, and apply to mechanical systems, not only to electromagnetic fields. It is the special theory of relativity that tells us that moving objects shrink and gain mass, while moving clocks run slow; it is the Lorentz transformation equations that tell us just how much the moving objects shrink and gain mass, and just how slow moving clocks run.

Lovell, Sir (Alfred) (Charles) Bernard (1913–) British astrophysicist who worked on radar during the Second World War and used the expertise he developed there (and surplus radar equipment) to help pioneer the development of **radio astronomy** in Britain after the war. He founded the observatory at **Jodrell Bank**, where he was the first director.

Lovell Telescope See **Jodrell Bank**.

Lowell, Percival (1855–1916) American astronomer who turned to the subject full time only at the age of 39, after a wide-ranging career in business and as a diplomat and author. Famous for his ideas about life on Mars, he also founded the **Lowell Observatory** and was instrumental in encouraging Vesto **Slipher** to carry out the work that led to the discovery of the **redshifts** of **galaxies**.

Lowell Observatory Observatory in Flagstaff, Arizona, founded by Percival **Lowell** in 1894 at an altitude of 2,210 m. It is still run as a private observatory, with four **telescopes** at the original site and four more (the largest a 1.8 m reflector) at the Anderson Mesa, 24 km away.

luminosity The luminosity of a **star** or other object is the amount of energy that it radiates every second. Sometimes the luminosity is given in terms of the total amount of **electromagnetic radiation** emitted at all wavelengths, when it is called the bolometric luminosity; sometimes the luminosity is given for a particular band of wavelengths. The luminosity of a star depends both on the temperature of the surface of the star and the area of the surface – a bigger star will radiate more energy at the same temperature than a smaller star. So two stars which have the same surface temperature (and therefore the same colour) may have very different luminosities, while two stars which have the same luminosity may have quite different surface temperatures (and colours).

But the total amount of energy generated by a star is essentially determined by the amount of pressure it has to generate in its core to hold itself up against collapsing under its own weight; so there is a clear relationship between mass and luminosity (see **mass–luminosity relation**) which holds for all stars that are holding themselves up by generating heat through **nuclear fusion** (it does not hold for **white dwarfs** and **neutron stars**, which are held up by pressure generated by quantum effects).

Luminosities of stars and other objects are often given as multiples of the luminosity of the Sun, which is 3.83×10^{26} watts. Some faint stars generate only about one-ten-thousandth of the Sun's luminosity, while some bright, short-lived stars have 1 million times the Sun's luminosity. The luminosity of a star is directly related to its *absolute magnitude.*

luminosity function   A measure of the number of objects (usually stars or galaxies) with different *luminosities* (or *absolute magnitudes*) in a standard volume of space, such as a cubic *parsec* or a cubic *Megaparsec*. The luminosities being measured may be in visible light, or radio waves, or any other chosen region of the electromagnetic spectrum. The luminosity function tells you what the relative numbers of objects with different brightnesses are, and provides information about how populations vary from place to place in our *Galaxy* (or across the Universe).

For example, within about 10 parsecs of the Sun, there is a peak in the luminosity function for stars with an absolute magnitude of about 15. The absolute magnitude of the Sun is 4.79, and the way this measure works, higher numbers correspond to *fainter* stars. So the local luminosity function tells us that our neighbourhood is dominated by stars considerably fainter than the Sun – M-type dwarfs each with about one-ten-thousandth of the Sun's luminosity. The situation is quite different in a *globular cluster*, where brighter stars predominate, and so on.

In the Universe at large, the luminosity function for galaxies and *quasars* shows how the average brightness of these objects changes with *redshift*, and therefore how it has changed as the Universe has evolved (see *look back time*). This tells astronomers that quasars were both brighter and more common when the Universe was young.

lunar eclipse   A lunar eclipse occurs when the Moon passes through the shadow of the Earth – that is, when the Moon is exactly on the other side of the Earth to the Sun. There are two parts to the Earth's shadow. In the umbra, the Sun is completely obscured by the Earth, and it is completely dark; in the penumbra (surrounding the umbra), the Sun is partially obscured by the Earth. As the Moon enters the Earth's shadow, it first passes through the penumbra, and its brightness fades gradually (scarcely noticeably, as far as the human eye is concerned). But when the Moon enters the umbra, a dark shadow spreads across its surface until (if it is a total eclipse) the Moon is completely dark. The process then mirrors itself as the Moon emerges from the Earth's shadow on the other side. If only part of the Moon passes through the umbra, it is called a partial eclipse.

Lunar eclipses can only occur at the time of Full Moon, when the Sun, Earth and Moon are in a line and the Moon is fully illuminated, so they are guaranteed to be highly visible, spectacular events (weather permitting). But they do not occur at every Full Moon, because the **orbit** of the Moon around the Earth is slightly tilted (at an angle of about 5°) relative to the **ecliptic** (the plane of the Earth's orbit around the Sun). Total lunar eclipses occur only when the time of Full Moon corresponds to the point in its orbit where the Moon is crossing the plane of the ecliptic; this means that such eclipses recur with a regular period of just over 18 years. Because lunar eclipses can be timed with such precision, records of eclipses in ancient times can sometimes be used to date historical events.

Lyman alpha cloud   See *Lyman alpha forest.*

Lyman alpha forest   A mass of lines seen in the **spectra** of quasars with high **redshift**, produced by the effect of clouds of cold **hydrogen** in inter-galactic space (Lyman alpha clouds) on the light from the quasar as it passes through the clouds. It is sometimes referred to simply as the Lyman forest.

Hydrogen is both the most common **element** and the simplest, and its spectrum is very well understood. The nature of the spectral features in hydrogen now known as the Lyman lines (or Lyman series) was studied in detail by the American physicist Theodore Lyman (1874–1954) in the first two decades of the 20th century. The brightest of these lines, known as Lyman alpha, occurs in the ultraviolet, at a wavelength of 121.6 nanometres in the laboratory.

For quasars with high redshift, this strong emission line of hydrogen is shifted so far towards the red end of the spectrum that it appears in the visible region. Because it starts out in the ultraviolet, at even shorter wavelengths than blue light, the redshift moves it first into the *blue* part of the visible spectrum, even though the shift is still 'towards' the red end of the spectrum. Ordinary galaxies can only be studied in detail out to a redshift of about 0.3, but the Lyman forest provides a probe out to much greater distances. For a redshift of 1.7, the Lyman alpha line is seen at 330 nm, and for this and higher redshifts it stands out as a high peak, like a tall mountain, in a graph plotting the energy emitted at different wavelengths in a typical quasar spectrum. But in very many cases, just to the blue side of this tall peak (corresponding to slightly smaller redshifts), there are very many dips in the spectrum, like a series of steep, narrow valleys. These are absorption lines produced where cold gas takes energy out of the light from the quasar as it passes through; they are caused by the presence of clouds of cool gas in space at different redshifts between us and the quasar.

The range of redshifts covered by the Lyman forest corresponds to hundreds or thousands of millions of **light years** across the expanding Universe, and the light from a single quasar may contain dozens of Lyman alpha lines at different redshifts. Statistical studies of the details of these lines show that they originate in dark clouds about as big as a small *galaxy*, some 10,000 *parsecs* across, each containing only between 10 million and 100 million times as much mass as our Sun. Such clouds cannot exist in isolation in space – they would disperse quickly, in a sense evaporating. They must be held together by the gravitational pull of *dark matter* in which they are embedded. So the Lyman forest provides support for the idea that the Universe is nearly flat (see *cosmological models*) and contains a lot of dark matter.

Observations made by the *Hubble Space Telescope* have shown that identical Lyman alpha lines appear in the spectra of two quasars which lie close together in the sky (Q0107–025 A and Q0107–025 B). This means that the cloud must be very large, obscuring part of the light from each quasar. It may be more than 300 kiloparsecs across (more than ten times as large as the visible part of our Galaxy); the observations suggest that it may be in the form of a flattened disc or sheet of gaseous material.

Lyman alpha line   See *Lyman series*.

Lyman series   A series of lines in the ultraviolet *spectrum* of *hydrogen* discovered by the American physicist Theodore Lyman (1874–1954) in 1914. They provide a characteristic 'fingerprint' that is especially useful in determining *redshifts*. The most important of these, the Lyman alpha line, occurs at 121.6 nanometres in the unredshifted spectrum. For high redshift objects, this line is shifted into the visible part of the spectrum.

Mach, Ernst (1838–1916)   Austrian philosopher and physicist whose ideas influenced Albert *Einstein*'s thinking at the time he was developing the *general theory of relativity*.

Mach was born at Turas, in Moravia, in what is now the Czech Republic but was then part of the Austro-Hungarian Empire, on 18 February 1838; he moved with his family to Unter Siebenbrunn, near Vienna, two years later. His father, himself well educated in both arts and sciences, had retired to a farm there, where young Ernst was taught both academic subjects and practical aspects of farming and skills such as carpentry. He continued to be educated at home until the age of fifteen, then studied at the local high school (gymnasium) before going on to university in Vienna in 1855. He obtained a doctorate in physics in 1860, and taught physics and mathematics in Vienna until in 1864 he became a professor of mathematics at the University of Graz. In 1866 he switched to the chair of physics at Graz, but he soon moved on, in 1867, to become professor of experimental physics at the University of Prague, where he stayed until 1895, when he was appointed professor of philosophy (the exact title was Professor of History and Theory of Inductive Sciences) at the University of Vienna. He retired from this post in 1901, having suffered a severe stroke in 1897, but remained active and served for twelve years as a member of the upper house in the Austrian parliament.

As this outline of his career shows, Mach did not stick to one narrow line of research. In his experimental work, he investigated vision and hearing, but he was also interested in the behaviour of waves, and in 1887 he published photographs showing the shock waves associated with projectiles moving through the air. In recognition of his work in this area, the Mach number was named after him in 1929; a speed of Mach 1 is equal to the speed of sound, Mach 2 twice the speed of sound, and so on. He was also one of the first people to carry out a detailed investigation of the *Doppler effect*.

Mach's main influence on science was, however, more philosophical. He said that science should reject anything that could not be directly observed, and that all our information about the world comes through our senses, so anything which cannot be sensed has no meaning. This line of thought was influential in the development of the standard interpretation of *quantum theory* (the Copenhagen Interpretation) in the 1920s and 1930s. But this extreme view of what could be regarded as 'reality' also led Mach to reject the concept of atoms. Mach also thought that the order in which discoveries had been made had a big influence on how they were interpreted and given priority; some of these ideas are echoed in modern investigations by historians of science into the way concepts such as *quarks* became established in physics (see John Gribbin, *Schrödinger's Kittens and the Search for Reality*).

Isaac *Newton*'s idea of an absolute space and time against which everything else could be measured was not something that could be sensed directly, and was rejected by Mach. He argued that properties such as *inertia* depend on the relationship of an object to everything else in the Universe (see *Mach's Principle*). These ideas, especially as presented in Mach's book *Mechanics*, published in 1863, were a strong influence on Einstein in developing the general theory of relativity. But Mach did not approve of where Einstein had taken his ideas, and at the time of his death (in Munich, on 19 February 1916) he was preparing a rebuttal of Einstein's theory.

McCrea, Sir William Hunter (1904–) Irish cosmologist, mathematician and astrophysicist who has made fundamental contributions to many areas of 20th-century science.

McCrea was born in Dublin on 13 December 1904, before Ireland became independent, and he was therefore a British citizen. His father was a teacher, and the family moved to England three years later, first to Kent and then to Derbyshire, where they settled. McCrea was educated at Chesterfield Grammar School in Derbyshire before taking his degree at Cambridge University as a member of Trinity College. He graduated in

Sir William McCrea (born in 1904).

1926 and worked at the University of Göttingen for part of the research studies that led to the award of a PhD from Cambridge in 1929. He worked as a lecturer in mathematics at Edinburgh University from 1930 to 1932, then as assistant professor of mathematics at Imperial College in London, before becoming professor of mathematics at Queen's University, Belfast, in 1936. He officially held this post until 1944, but from 1943 to 1945 he was actually involved with wartime work for the Admiralty in London. He then became professor of mathematics at Royal Holloway College in London, where he stayed for twenty years, until he left in 1966 to establish the astronomy research group at the University of Sussex, where he was appointed research professor of theoretical astronomy. Although he officially retired in 1972, he became an emeritus professor at Sussex, and continued to be active in research into the 1990s. He was knighted in 1985.

McCrea's first important contribution to astrophysics came with his Ph.D. studies, which used *spectroscopy* to show that there is vastly more hydrogen than any other element in the Sun. Astonishingly, this was a new discovery at the end of the 1920s. Previously, astronomers had thought that the composition of the Sun roughly matched that of the Earth, being rich in iron and *heavy elements*; it took another twenty years or so for the true composition of the Sun to be understood and linked with the processes of *fusion* that keep it hot, but the first step in the right direction was taken by McCrea (and independently by the German Albrecht Unsöld) at the end of the 1920s.

McCrea also studied the evolution of galaxies, the formation of stars and planets, and the formation of molecules in *interstellar clouds*. He worked with Edward *Milne* on the development of a cosmological model of the Universe which uses *Newtonian mechanics*, but which reproduces essentially all of the features of models based on the *general theory of relativity*. This Newtonian cosmology is imperfect because it cannot deal with what happens at the 'edge of the Universe' – the boundary conditions – but it still provides intriguing insights into how the Universe works. It also means that any mathematician since Isaac *Newton* might have discovered the equations that describe the expanding Universe, and if the expansion had been discovered before the general theory of relativity had been developed, the Milne–McCrea model could have provided a reasonable explanation of the discovery.

McCrea was also one of the principal investigators of the implications of the *Steady State hypothesis*, suggesting how the *C-field* could be linked to sites of matter creation. This mathematical description of the behaviour of the C-field is essentially the same as the mathematical description of what is now known as *inflation*. McCrea suggested that stars form from the accumulation of many small objects, not from the collapse of single clouds of gas, contributed to the debate about the *solar neutrino problem*, and made contributions to the development of quantum physics. In the 1970s he proposed that comets form in the spiral arms of the Galaxy and are collected by the Solar System as it travels around the Galaxy in its orbit, and linked this idea with a new theory of Ice Ages.

McCrea has also made many 'non-scientific' contributions to science in the 20th century. For example, he served as a governor of the Dublin Institute of Advanced Study for ten years from its foundation in 1940, and in 1983 (at the age of 78) he was sent by the Royal Society to Argentina to restore friendly scientific relations after the Falklands War, a task carried out with his usual tact and charm. A quiet man with a strong sense of humour and a stubborn streak, he was an excellent teacher and wrote several clear books on topics such as relativity theory; many of his former students have gone on to become influential figures in astronomy in their own right (and one to become the principal author of this book). He once commented, during his Presidential Address to the *Royal Astronomical Society*, in 1963, that 'I am always surprised when a young man tells me he wants to work at cosmology; I think of cosmology as something that happens to one, not something one can choose.' It certainly happened to him.

MACHO Project Collaboration between astronomers based in California and in Australia to search for *MACHOs*, using a specially-built *charge coupled device* camera at the 1.3 m *reflecting telescope* of the *Mount Stromlo and Siding Spring Observatories*. It is headed by Charles Alcock at the Lawrence Livermore Laboratory in California.

MACHOs Acronym for Massive Astronomical Compact Halo Objects, a name deliberately chosen to contrast with *WIMPs* (many astronomers have a childish sense of humour and delight in bad puns). MACHOs represent one possible form of *dark matter* in the Universe, and (if they exist) are composed of *baryons*, essentially the same sort of stuff that the Sun, the stars and ourselves are made of.

The standard model of *nucleosynthesis* in the *Big Bang* allows for the possibility that there may be as much as ten times more baryonic material than we can see in the form of bright stars and galaxies. This would not be enough to make the Universe flat (see *cosmological models*), but it could be enough to provide all of the dark matter required to explain the dynamics of galaxies in clusters, and the way individual galaxies, including our own Milky Way, rotate.

A very small amount of this dark baryonic matter could be in the form of gas and dust, but since we do not see the obscuring effects of huge amounts of gas and dust in space (except in localized regions such as the plane of the Milky Way), most of it must be locked up in compact objects, roughly equivalent to dark stars. There are two ways in which this could happen. Either the dark baryonic matter is locked up in large numbers of planet-sized objects (roughly the size of Jupiter) that are too small to burn nuclear fuel (failed stars often known as *brown dwarfs*), or it may have been processed through very large stars which exploded when the Universe was young and left matter behind in the form of massive *black holes*.

Evidence from *gravitational lensing* experiments, such as *OGLE*, suggested in the early 1990s that the extended *halo* of our *Galaxy* does contain MACHOs, most probably in the form of brown dwarfs, although the black hole option has not yet been ruled out.

Mach's Principle The idea that *inertia* is caused by the interaction of an object with all of the other matter in the Universe. *Galileo* seems to have been the first person to realize that it is not the velocity with which an object moves but its acceleration that reveals whether or not forces are acting upon it. On Earth, there always are external forces (such as friction) at work, and so we have to keep pushing an object just to keep it going at a constant velocity. But the natural tendency is to keep moving in the same direction at constant speed unless an external force does act (see *Newton's laws*). But what do you measure velocities and accelerations against?

Isaac **Newton** thought that there was a preferred *frame of reference* in the Universe, defined by absolute space. Space is a tricky thing to pin down – you can't hammer a nail into it and measure your velocity relative to the nail. But Newton thought that the existence of the preferred frame of reference could be demonstrated by experiments on rotating objects – specifically, a bucket of water. He described the experiment in his great book, the *Principia*, published in 1686:

> The effects which distinguish absolute motion from relative motion are, the forces of receding from the axis of circular motion ... if a vessel, hung by a long cord, is so often turned about that the cord is strongly twisted, then filled with water, and held at rest together with the water; thereupon, by the sudden action of another force, it is whirled about the contrary way, and while the cord is untwisting itself ... the surface of the water will at first be plain, as before the vessel began to move; but after that, the vessel, by gradually communicating its motion to the water, will make it begin sensibly to revolve, and recede by little and little from the middle, and ascend to the sides of the vessel, forming itself into a concave figure (as I have experienced), and the swifter the motion becomes, the higher the water will rise.

Newton is talking about what we now call **centrifugal force**, and his comment 'as I have experienced' is pertinent, because unlike many of his predecessors and contemporaries Newton actually did experiments – he didn't just imagine how things 'ought' to work in an ideal world. You can experience the same thing, on a smaller scale, by stirring your cup of coffee. The liquid is pushed to (and up) the sides by centrifugal force, leaving a dent in the middle. But what Newton pointed out is that it is not motion relative to the container that matters, but, in some sense, the absolute motion of the liquid.

At the start of the experiment, the bucket begins to move, but the surface of the liquid stays flat, even though there is relative motion between the liquid and the bucket. Then, as friction makes the liquid rotate, the concave depression builds up (or down), even though there is now no motion of the liquid relative to the bucket. Finally, you can grab the bucket to stop it rotating; now, the liquid inside keeps on rotating, with a concave depression, just like the coffee stirred around in your cup. Somehow, the liquid 'knows' it is rotating, and behaves accordingly. But rotating relative to what?

Newton said it was rotating relative to fixed (or absolute) space. But 30 years later the Irish philosopher and mathematician (and bishop) George Berkeley (1685–1753) argued that all motion is relative, and must be measured against something. Since 'absolute space' cannot be perceived,

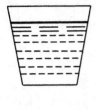

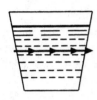

Mach's Principle. Newton's bucket experiment. *Top left:* With both bucket and water stationary, the surface of the water is flat. *Bottom left:* With the bucket rotating and the water stationary, the surface of the water is flat. *Bottom right:* With both bucket and water rotating, the surface of the water is curved. *Top right:* With the bucket stationary and the water rotating the surface of the water is curved.

that would not do as a reference point, he said. If there were nothing in the Universe but a single globe, he continued, it would be meaningless to talk about any movement of that globe. Even if there were two perfectly smooth globes in orbit around one another, there would be no way to measure that motion. But 'suppose that the heaven of fixed stars was suddenly created and we shall be in a position to imagine the motions of the globes by their relative position to the different parts of the Universe'. In effect, Berkeley argued that it is because the coffee in your cup knows that it is rotating relative to the distant *stars* that it rises up the sides of the cup in protest.

The same argument applies to accelerations in straight lines; Berkeley's reasoning would say that the push in the back you feel when you are in a car that accelerates away from a standing start is because your body knows that it is being accelerated relative to the distant stars and **galaxies**. But Berkeley was 150 years ahead of his time. Although there was some discussion of his ideas in the 18th century, they were largely ignored. Interest in the idea only really developed in the 1860s, when it was taken up by Ernst **Mach**.

Mach added very little to the ideas put forward by Berkeley, although he did make the intriguing suggestion that, if we want to explain the equatorial bulge of the Earth as due to centrifugal forces, 'it does not matter if we think of the Earth as turning round on its axis, or at rest while the fixed stars revolve around it'. It is the *relative* motion that is responsible for the bulge.

Albert **Einstein** learned of the idea that acceleration has to be measured relative to the **fixed stars** from Mach's work, and it was Einstein who gave the idea the name 'Mach's Principle'. When Einstein set out to develop his **general theory of relativity**, he intended to come up with a theory that would incorporate Mach's Principle as a natural consequence. He was only partially successful – the equations of the general theory incorporate this feedback between distant objects and accelerated motion only if the Universe is closed (see **cosmological models**), and perhaps not even then. But since **inflation** suggests that the Universe is indeed closed, this is not such a drawback as it seemed in Einstein's day.

If the local reference frame, the standard of rest, really is determined by some averaged-out effect of all the matter in the Universe, there ought to be some way to test this. One approach would be to place a test object inside a (very massive!) spherical shell of material, with the thick shell made to rotate rapidly relative to the distant galaxies. If Mach's Principle is correct, there should be a small dragging effect from the rotating shell, trying to tug the test object around with it. More modest effects may be detectable from studies of the behaviour of gyroscopes floating in **free fall** in orbit around the Earth; experiments along these lines have been devised by a team at Stanford University, but have not yet flown in space.

There is, though, another line of attack on the problem. What Berkeley and Mach called the 'fixed stars' are actually part of a system that is itself rotating – the Milky Way. Even before other galaxies were firmly identified (indeed, before Mach was born), William **Herschel** and other astronomers had provided good evidence that the Milky Way is a flattened disc of stars, its shape clearly determined by rotation and centrifugal force. At the end of the 19th century, Mach (or some other philosopher) might well have argued that there were only two ways in which the whole **Galaxy** could be seen to be under the influence of centrifugal force. Either Newton was right, and the whole system of 'fixed stars' is rotating relative to absolute, empty space; or Berkeley and Mach were right, and there must be some distribution of matter, far away across the Universe, that establishes a frame of reference against which the rotation of our Galaxy is measured. The existence of vast numbers of far-distant galaxies could actually have been predicted, on the basis of Mach's Principle and the knowledge that our Galaxy is rotating, decades before the work of Edwin **Hubble** which established the scale of the Universe!

Further reading: John Gribbin, *In Search of the Big Bang*.

Maffei galaxies Two **galaxies** only discovered in 1968 (by the Italian astronomer Paolo Maffei) because they lie almost in the plane of the **Milky Way** and are almost hidden by **interstellar dust**, except at red and infrared wavelengths (see **interstellar reddening**). Although they lie in almost the same direction on the sky, the galaxies are unrelated. Maffei I is a member of the **Local Group**, a giant **elliptical galaxy** about 1,000 **kiloparsecs** away;

Magellanic Clouds. The Small Magellanic Cloud, an irregular galaxy which is a companion to our Milky Way.

Maffei II is a *disk galaxy* about 5,000 kiloparsecs away, and probably not a member of the Local Group.

Magellanic Clouds   Two small irregular *galaxies* that are *satellites* of our *Milky Way*. They are about one-tenth as far from us as the *Andromeda galaxy*, the nearest large galaxy to our own. This means that they are close enough for individual *stars* in the Magellanic Clouds to be studied in detail, but so far away that all the stars in either one of the two Clouds can be regarded as at roughly the same distance from us. This has made the Clouds crucial stepping stones in determining the *cosmic distance scale*.

The Clouds are easily visible to the naked eye in the sky of the southern hemisphere, and are named after the Portuguese explorer Ferdinand Magellan (about 1480–1521), who was the first European to describe them in detail, in 1519. The Large Magellanic Cloud (LMC) has a diameter of about 10,000 *parsecs* and is about 50,000 parsecs away; the Small Magellanic Cloud (SMC) has a diameter of about 6,000 parsecs and is about 60,000 parsecs from us, but seems to be very extended in the line of sight. They both contain large numbers of *Population I* stars, and proportionally more gas than our *Galaxy.* The LMC seems to contain young, blue *globular clusters*, which are not seen in our Galaxy. Both of the Clouds are wrapped in a cloud of cold *hydrogen* gas, called the Magellanic Stream, which was pulled out of one or (probably) both of them by tidal forces when they passed close to our Galaxy some 200 million years ago. The stream contains about 1 billion times as much mass as our Sun.

Because the Clouds provide astronomers with samples of stars all at roughly the same distance from us, they are especially useful in studies of *stellar evolution.* If one star in, say, the LMC is twice as bright as another star in the LMC, we can be sure that it really is twice as bright, and not that it looks brighter just because it is much closer to us. For the same reason, studies of *Cepheid variables* in the Clouds provide the key to the *period–luminosity relation* for Cepheids, which is the single most important number in calibrating distances across the Universe.

magnetic field   The region around a magnet or a flowing electric current in which a force is exerted on a magnetic object (see *field theory*). The magnetic field is one manifestation of electromagnetism, one of the four *fundamental forces.* Because, like *gravity*, magnetism has a long range, magnetic fields associated with astronomical objects such as *planets*, *stars*, *galaxies* and *quasars* play an important part in many astrophysical processes.

magnitude scale (also known as Pogson scale)   Scale used by astronomers as a measure of the brightnesses of astronomical objects. The original magnitude scale was based on how bright objects look to the human eye; the Greek astronomer *Hipparchus* ranked *stars* in a scale from 'first magnitude', for the brightest stars he knew, to 'sixth magnitude', for those that can just be seen by the unaided human eye. In the middle of the 19th century, however, it became appreciated that the way the human eye works is not linear, but follows a logarithmic rule. So a star of the first magnitude is much more than six times brighter than a star of the sixth magnitude.

In 1856 the English astronomer Norman Pogson (1829–91) proposed that, in order to achieve a precise scale that matches the traditional scale based on human vision, in absolute terms a difference of 5 magnitudes should correspond to a factor of 100 difference in brightness. In other words, a difference of 1 magnitude corresponds to a difference in brightness of 2.512 times (because $2.512^5 = 100$). A star that is 2 magnitudes brighter than another is 2.512^2 times brighter, and so on.

This is the scale used by astronomers today, with the actual brightnesses measured by light-detecting machines, no longer estimated by eye. Because of the way Hipparchus defined the original magnitude scale, the *dimmer* a star is, the *greater* is its magnitude on the Pogson scale. And because brighter stars than Hipparchus considered have to be accounted for, negative numbers have to be used as well. Magnitudes are measured in different *wavelength* bands (in different *colours*) or over the entire electromagnetic *spectrum* (the bolometric magnitude). See also *apparent magnitude*, *absolute magnitude*, *luminosity*.

main sequence The band in the *Hertzprung–Russell diagram* occupied by *stars* which, like our Sun, are shining as a result of converting *hydrogen* into *helium* in their cores by the process of *nuclear fusion*. About 90 per cent of all bright stars sit on the main sequence. The lifetime (L) of a star on the main sequence, in years, is approximately given by taking the cube root of its mass (M) (in units where the mass of the Sun is 1) and dividing this into 10 billion. That gives $L = 10^{10}/M^{1/3}$. The cube root of 8, for example, is 2, so a star 8 times more massive than the Sun will live for about 5 billion years, roughly half as long as the Sun.

Maksutov, Dmitri Dmitrievich (1896–1964) Soviet (as it then was) expert in astronomical optics who designed and built several large *reflecting telescopes*, and invented the *Maksutov telescope*. He worked in Odessa in the 1920s, then at the State Optical Institute in Moscow, and, after 1952, at Pulkovo, near Leningrad.

Maksutov telescope An improved version of the *Schmidt camera*, developed by Dmitri *Maksutov* in the 1940s. The system gives very good-quality images over a wide field of view, making it ideal for astronomical photography. Although such telescopes are strong and compact, ideal for amateur astronomers, they use correcting lenses which are difficult to manufacture for large *aperture* instruments, which limits their professional applications.

Malmquist effect An illusion named after the Swedish astronomer Gunnar Malmquist (1893–1982), by which the average brightness of a group of distant objects (stars or galaxies) seems greater the further away they are, because the fainter members of the group are too faint to be seen and taken into account. See *Hubble constant*.

Mariner probes A series of spaceprobes sent by NASA to investigate the inner planets of the Solar System. Among the highlights of the series, Mariner 2 reached Venus in 1962, Mariner 4 sent back pictures of Mars in 1967, Mariner 9 was the first spaceprobe to go into orbit around Mars (in 1971), and Mariner 10 sent back pictures from Mercury in 1974. The two *Voyager probes* were originally named Mariner 11 and Mariner 12.

Mars Fourth *planet* out from the Sun, orbiting once every 686.98 days at a distance varying between 1.38 and 1.67 *astronomical units*. Mars turns on its axis once every 24 hours 37 minutes and 23 seconds. It has a diameter

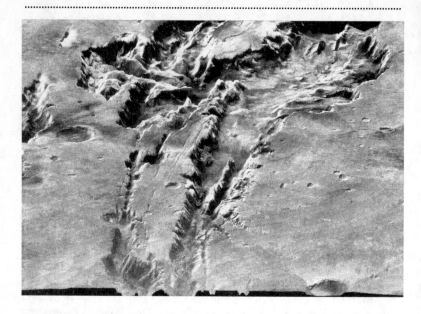

Mars. The giant Valles Marineris canyon on Mars, shown in a computer-processed image developed from data from the Viking spacecraft.

of 6,795 km (roughly half that of the Earth) and a mass a little more than one-tenth that of the Earth. It is a desert planet with a thin atmosphere (mainly carbon dioxide), where temperatures range from 26 °C to –111 °C.

maser   Name derived from the acronym for Microwave Amplification by Stimulated Emission of Radiation. The term 'laser' was later derived by replacing the word 'microwave' by 'light'; several naturally occurring masers have been identified in astronomical systems.

Both masers and lasers result from interactions between electro-magnetic radiation and atoms. The difference is the *wavelength* of the radiation involved, microwaves having much longer wavelengths than light. In both cases, if an atom or molecule is in a suitable energetic state (an excited state), then the passage of an electromagnetic wave with a certain wavelength past it can trigger it to give up energy in the form of more electromagnetic radiation with exactly the same wavelength. This reinforces the passing wave, which can then interact with more excited atoms to build up an intense pulse of radiation with one very precisely defined *frequency*. This stimulated emission effect was predicted by Albert *Einstein* in the 1920s, using *quantum theory*.

In some astronomical clouds, molecules are energized to excited states by radiation from nearby stars. If some of these molecules spontaneously give up their energy as microwaves, the microwaves can trigger other excited molecules of the same kind to give up their energy as well, in a cascade that produces a bright beam of radiation with a well-defined wavelength. Maser action associated with the presence of water (H_2O), the hydroxyl radical (OH), silicon oxide (SiO) and methanol (CH_3OH) has been detected in *molecular clouds* and in the atmospheres of old stars. In the case of hydroxyl masers, hundreds of such clouds are now known; the first OH maser was identified in the *Orion nebula* in 1965.

Maskelyne, Nevil (1732–1811) Fifth Astronomer Royal, succeeding Nathanial *Bliss* in 1765 and serving in the post until he died. Born in London on 6 October 1732, Maskelyne was educated at Trinity College in Cambridge, where he became a Bachelor of Divinity in 1754. Although ordained a year later, he went to work as an assistant to James *Bradley* at the Royal Greenwich Observatory. Even after he became Astronomer Royal, however, he continued to carry out clerical duties in a variety of parishes. One of his main interests was in techniques for determining longitude at sea, and in 1767 he founded the *Nautical Almanac*. He died in Greenwich ·
on 9 February 1811.

mass A measure of the amount of stuff there is in an object. Mass can be measured in either of two ways – by the extent to which the object resists being accelerated by a force (its *inertia*; see also *Newton's laws of motion*) or by the strength of the *gravitational field* associated with the object. It is a curious and still not fully explained phenomenon that these two masses (the inertial mass and the gravitational mass) are always exactly the same for the same object (see *Mach's Principle*).

Provided they are in the same gravitational field, equal masses feel an equal gravitational force, which is called their weight. So the mass of an object can be specified in terms of how much it would weigh on the surface of the Earth – the standard unit is 1 kilogram (kg). But an object with a mass of 1 kg would not weigh 1 kg if it were moved to, say, the Moon, where the gravitational force is weaker and each kilo of mass would weigh only 160 grams. Your weight depends on where you are in the Universe, but your mass (strictly speaking, your rest mass; see *special theory of relativity*) stays the same wherever you are.

Astronomers usually measure mass in terms of the mass of the Sun, which is 2×10^{30} kg.

mass–luminosity relation An approximate correlation between *mass* and *luminosity* that holds for *stars* on the *main sequence* of the *Hertzsprung–Russell diagram*. Roughly speaking, the more massive a star is, the hotter it must be inside in order to generate enough pressure in its interior to hold itself up against *gravitational collapse*. This was first properly appreciated by Arthur *Eddington*, who discussed the implications in 1924.

The mass–luminosity relation is only approximate, because other factors, such as its chemical composition, also affect the brightness of a star, and three slightly different mass–luminosity relations are found to apply to stars on different parts of the main sequence. For stars like the Sun (with masses from about 0.3 to 7 times the mass of the Sun), the luminosity is proportional to the fourth power of the mass (M^4), while for brighter, more massive stars, luminosity is proportional to the cube of the mass (M^3), and for dim *red dwarf* stars, the luminosity goes as the mass to the power of 2.5 ($M^{2.5}$). *White dwarfs* and *red giants* do not follow these simple rules.

mass number   Number denoting the combined total of *protons* and *neutrons* in the *nucleus* of an *atom* of a particular *isotope* of an *element*, such as *helium*. This distinguishes the particular isotope being referred to – for example, helium-3 or helium-4.

Mauna Kea Observatory   A set of telescopes near the summit of Mauna Kea, Hawaii. Telescopes operated by several different nations, including the USA, UK, France and Canada, share the observatory site 4,200 m above sea level. These include conventional optical telescopes and telescopes sensitive to *infrared* radiation. One of the telescopes on Mauna Kea, the *Keck telescope*, is the largest optical telescope in the world.

Maunder, Edward Walter (1851–1928)   British astronomer who worked (initially as an assistant) at the *Royal Greenwich Observatory* and carried out a detailed study of *sunspots*. He found, from historical records, that there was hardly any sunspot activity between 1645 and 1715 (now known as the *Maunder Minimum*), but the discovery roused little interest among astronomers in his lifetime.

Maunder minimum   The 70 years following 1645, during which very few *sunspots* were observed. Named after E. W. *Maunder*, who discovered the minimum in 1890 while studying old records of solar activity at the *Royal Greenwich Observatory*. Although for many decades some astronomers remained sceptical about the reality of the Maunder minimum, arguing that it might reflect a lack of observations of the Sun rather than a lack of sunspots between 1645 and 1715, in the 1970s these doubts were laid to rest by new studies of the historical data and by analyses of the amount of radioactive carbon (carbon-14) in tree rings. Carbon-14 is produced in the atmosphere (in carbon dioxide, which is taken up by plants during photosynthesis) by interactions involving solar *cosmic rays*, and these studies showed that the Sun was indeed quiet during the second half of the 17th century.

The Maunder minimum coincided almost exactly with the reign of Louis XIV, the French 'Sun King'. It also coincided with a period of extreme cold on Earth known as the Little Ice Age. It is hard to escape the conclusion that the lack of sunspots was somehow linked with the cold here on Earth, but no entirely satisfactory explanation of how the link worked has yet been found.

Max Planck Institute for Astronomy German institute that runs an observatory at Calar Alto, in Spain, from its headquarters in Heidelberg.

Max Planck Institute for Radio Astronomy German institute that runs an observatory at Effelsberg, near Bonn, from its headquarters in Bonn. The main instrument is a 100 m steerable dish known as the Effelsberg Telescope.

Maxwell, James Clerk (1831–79) Scottish physicist whose many achievements included developing the three-colour theory of colour vision and the equations which describe the behaviour of *electromagnetic radiation*.

Maxwell was born in Edinburgh on 13 June 1831. He came from a wealthy family (his father was a lawyer) and studied at the University of Edinburgh from 1847 to 1850, then went on to Cambridge University, where he graduated in 1854. He worked in Aberdeen and at King's College in London, but when his father died in 1865 he returned to the family estates in Scotland and carried out research privately. In 1871 he was persuaded to come back into the mainstream of academic life, and became the first professor of experimental physics in Cambridge, where he was instrumental in setting up the Cavendish Laboratory, which opened in 1874. He died of cancer, in Cambridge, on 5 November 1879, at the age of 48.

In one of his earliest papers, published in 1859, Maxwell proved that the rings of Saturn must be made up of many small particles each in their own *orbit* around the *planet*, and could not be solid objects. He later explained (and demonstrated) how all of the colours perceived by human eyes can be made up of combinations of red, green and blue light – a technique used by space scientists in the 1980s to (among other things) create full colour images of the rings of Saturn from data sent back to Earth by the *Voyager probes*.

But Maxwell's greatest achievement was to discover the equations which describe the behaviour of electromagnetic radiation. These equations contain a constant, c, which gives the velocity of electromagnetic waves. It turned out that this constant was exactly equal to the *speed of light*, so Maxwell realized that he had proved that light is a form of electromagnetic wave. He also predicted, on the basis of these calculations, the existence of what are now called radio waves, which move at the same speed, c.

Curiously, Maxwell's equations do not specify the *frame of reference* in which c is measured. Although Maxwell and his contemporaries thought that electromagnetic waves must move through an *ether*, relative to which the speed was measured, Albert *Einstein* realized that c is an absolute constant, and that the speed of light is the same in all *inertial frames*. This was one of the key insights that led Einstein to develop the *special theory of relativity*.

mean density of matter   The average *density* of matter across the entire Universe. The actual density is not very well known, although many theorists favour the idea that the Universe contains exactly enough matter for its *gravity* to make *spacetime* flat (see *cosmological models*, *density parameter*).

   If the Universe does contain precisely this critical density of matter, then the mean density of matter in the Universe is 5×10^{-27} kg per cubic metre, assuming a value of 55 km per second per *Megaparsec* for the *Hubble constant*. The amount of visible bright matter in the Universe corresponds to a few per cent of this critical density, but the way members of *clusters of galaxies* move, and the way individual galaxies rotate, shows the presence of a large amount of *dark matter*, sufficient to bring the mean density of matter up to about a third of the critical value. This matter is definitely present, and may well be all in the form of *baryons*. The extra matter needed to make spacetime flat would (if it exists) have to be non-baryonic.

   If all of the known matter in the Universe were in the form of *atoms* of *hydrogen*, and it were spread uniformly throughout all of space, then there would be just one hydrogen atom in each cubic metre of the Universe. To make spacetime flat would require the equivalent of just over three hydrogen atoms in every cubic metre of space.

Megaparsec (Mpc)   A unit of distance, equal to 1 million *parsecs*.

Mercury   Closest planet to the Sun, orbiting once every 87.97 days at an average distance of 0.39 *astronomical units*. Mercury rotates once every 58.64 days, so that three 'days' on Mercury last for two of the planet's 'years'. The planet is very heavily cratered, with essentially no atmosphere; temperatures range from 190 °C to –180 °C. Mercury has a diameter of 4,880 km (intermediate in size between the Moon and Mars) and a mass about 5 per cent of that of the Earth.

MERLIN   Acronym for the Multi-Element Radio-Linked Interferometer Network, a system of *radio telescopes* in Britain linked by *microwave* communications to *Jodrell Bank*. The telescopes are used for *interferometry* and *aperture synthesis* work, building up detailed maps of *radio sources* that can be processed to resemble photographs.

meson   See *elementary particles*.

Messier, Charles (1730–1817)   French astronomer who was chiefly interested in *comets*. In order to avoid wasting his time by misidentifying other fuzzy blobs on the sky as comets, he compiled an important *catalogue* of

Mercury. The heavily cratered surface of Mercury, photographed by the spaceprobe Mariner 10 on 29 March 1974, from a distance of 400,000 km.

faint astronomical objects, many of which are now known to be *galaxies*.

Messier Catalogue   *Catalogue* of faint astronomical objects compiled by Charles *Messier* in the second half of the 18th century as an adjunct to his interest in *comets*. The Messier Catalogue is now regarded as his chief scientific legacy. The final version of the catalogue lists 110 objects, many now known to be *galaxies* (such as M31, the *Andromeda galaxy*), but there are five mistakes in the catalogue (numbers M40, M47, M48, M91 and M102), so the actual number of objects in it is 105. (See photo on p. 318.)

metallicity   See *heavy elements*.

metals   Astronomers somewhat cavalierly define every *element* except for *hydrogen* and *helium* as a 'metal', so to an astronomer the most common 'metals' in the Universe are *carbon*, *oxygen* and *nitrogen*, none of which would be regarded as a metal by a chemist. See *heavy elements*.

meteor   Trail of light across the night sky produced when a fragment of material from space burns up in the atmosphere of the Earth. It is often known as a shooting star. Because particles of dust and fragments of rock *orbit* the Sun in streams (probably left by the break-up of *comets*), many

Messier Catalogue. One of the 'nebulae' in the Messier Catalogue, M74 (also known as NGC 628), which is shown by this modern Charge Coupled Device image to be a spiral galaxy.

meteors occur in showers at certain times of the year, when the Earth crosses one of these streams.

meteorite   Piece of rocky or metallic material from space that has survived the passage through the Earth's atmosphere without burning up completely as a *meteor*, and has hit the surface of the Earth. Meteorites may be fragments of *comets* or *asteroids*, and occasional large impacts produce large *craters*. The largest of these have been linked with ecological disasters causing mass extinctions of life on Earth, such as the death of the dinosaurs.

meteoroid   General name for the kind of objects which may turn into *meteors* or *meteorites* if they strike the Earth. So a lump of rock falling on Mars, for example, is a meteoroid, not a meteorite. While the lump of rock is floating about in space, it may be called a meteoroid or an *asteroid*.

metre   Standard unit of length, originally defined (by the National Assembly in revolutionary France in the 1790s) as one-ten-millionth of the distance from the North Pole to the equator. Since 1983 the definition of the metre has been as the distance travelled by *light* in 1/299,792,458 seconds. Several people have suggested redefining the metre so that the speed of light is exactly 300,000,000 metres per second, but unfortunately

---

the suggestions do not seem to have been taken seriously by the people who decide such things.

metric A way of measuring the relationships between the positions of events in space, or in *spacetime*. The word 'metric' comes from the same root as the word 'geometry'; the key point about metric measurements is that a suitable choice of metric provides a measure of distances that is independent of the *frame of reference* from which the measurements are made.

This can be understood in terms of the famous theorem of **Pythagoras**, about the lengths of the sides of a right-angled triangle. If the length of the hypotenuse of the triangle is s and the lengths of the other two sides are x and y, then Pythagoras' theorem tells us that $s^2 = x^2 + y^2$. This defines the shortest distance between the two points at either end of the hypotenuse. In three dimensions, if you measure the distances between two points in three directions at right angles to one another, the shortest distance between the two points is given by the equation $s^2 = x^2 + y^2 + z^2$. It doesn't matter where you choose to measure the x's, y's and z's from (where the origin of your measuring system is), the value of s is always the same (for the same two points).

Hermann **Minkowski** realized that you can do the same thing in four dimensions, three of space and one of time, using the *special theory of relativity* developed by Albert **Einstein**, and the *Lorentz transformation equations*. Now, $s^2 = x^2 + y^2 + z^2 - c^2t^2$.

The factor of c, the speed of light, comes in to make the time part of the equation have the same units of length as the other parameters, and Einstein's theory tells us that in a sense time behaves like 'negative distance', which is where the minus sign comes from. Since c is equal to 30 billion centimetres per second, this means that in a sense 1 second is equivalent to −30 billion centimetres; since there is a minus sign in front of the c^2t^2 term, time is also in some sense associated with the square root of a negative number (a so-called complex number), which makes visualizing *spacetime* harder but doesn't raise any problems with the calculations. In spite of these complications, the important point in metric terms is that the separation s between two points in four-dimensional spacetime is always the same, wherever you choose to measure the x's, y's, z's and t's from, and *whichever way you are moving*, provided you are in an *inertial frame*.

This four-dimensional separation is called the interval between two events, and it is the same for observers in all inertial frames. Different observers may measure each of the parameters x, y, z and t differently, but the metric combination of the four parameters will give the same interval s.

This particular metric (the Minkowski metric) describes the relationship between events occurring in flat spacetime, and is a good approximation to the way things are related in the Universe at large. More complicated metrics have to be used when dealing with curved spacetime – for example, in the vicinity of **black holes** – where the **general theory of relativity** provides the appropriate description of what is going on.

Meudon Observatory   The **astrophysics** section of the **Paris Observatory**, founded as an independent observatory in 1876, but merged with the Paris Observatory in 1926.

Michell, John (1724–93)   English geologist, astronomer and clergyman who was the first person to come up with the notion of **black holes**.

Some books give the impression that the Reverend John Michell was a country parson who had some kind of dilettante interest in science, and who stumbled across the idea of black holes almost by chance. This is far from the truth. Michell was a serious scientist before he became a country rector, and he was elected as a Fellow of the Royal Society in 1760, before he entered the Church. His reputation had been made by his study of the disastrous earthquake that had struck Lisbon in 1755. Michell established that the damage had been caused by a disturbance in the Earth's crust actually centred under the Atlantic Ocean, and he is regarded today as the father of the science of seismology.

The exact date and place of birth of Michell are not known, but he came from Nottinghamshire, where he was probably born in 1724, and studied at the University of Cambridge, where he graduated in 1752 and became first a Fellow of Queen's College, and then, in 1762, Woodward Professor of Geology. At the time, it was normal for academics at the university (which had ecclesiastical origins) to take Holy Orders, not necessarily with any intention of working in the Church; but in 1764, a year after becoming a Bachelor of Divinity, Michell left the university to become the rector of a parish at Thornhill, in Yorkshire.

This did not stop him from continuing an active interest in science, and William **Herschel** was a regular visitor to the rectory. In 1767, Michell published a paper pointing out that there are too many **binary stars** visible in the sky to have arisen just by the chance juxtaposition of stars scattered at random, and that some of these binaries must be physically associated with one another. He was also the first person to make a realistic estimate of the distance to a star. Using an argument based on the apparent brightness of **Vega**, he calculated that the star is about 460,000 times further away from us than the Sun is, which is about 25 per cent of the accepted modern value of the distance.

Michell also made important investigations of magnetism (before he left Cambridge), and while at Thornhill invented an instrument known as a torsion balance which could be used to measure the exact strength of very small forces. He intended to use this to measure the strength of *gravity* from the force exerted on a delicate rod by a larger *mass*, but died before he could carry out the experiment. It was carried out, using Michell's technique, by his friend Henry Cavendish (1731–1810), a brilliant English polymath who made many contributions to science, and the results were published in 1798. The Cavendish Laboratory in Cambridge, founded in 1871, is named after Henry Cavendish.

But what now seems Michell's most prescient work is often not mentioned at all in the reference books. The first mention of what he called 'dark stars' was made in a paper read to the Royal Society on Michell's behalf by Cavendish in 1783. This was an impressively detailed discussion of ways to work out the properties of stars, including their distances, sizes and masses, by measuring the gravitational effect on *light* emitted from their surfaces. Everything was based on Isaac *Newton*'s idea of light as being made up of tiny particles (sometimes known as corpuscles), and the supposition that 'the particles of light' are 'attracted in the same manner as all other bodies with which we are acquainted' by gravity.

Michell realized that if a star were big enough then what we now call the *escape velocity* from its surface would exceed the speed of light. Among the many detailed arguments in his long-forgotten, but now famous paper, he pointed out that as a result:

> If there should really exist in nature any bodies whose density is not less than that of the sun, and whose diameters are more than 500 times the diameter of the sun, since their light could not arrive at us ... we could have no information from sight; yet, if any other luminiferous bodies should happen to revolve around them we might still perhaps from the motions of these revolving bodies infer the existence of the central ones with some degree of probability, as this might afford a clue to some of the apparent irregularities of the revolving bodies, which would not be easily explicable on any other hypothesis.

A sphere 500 times bigger than the Sun would be about as big across as the Solar System, so the kind of dark star envisaged by Michell is very similar to the kind of black hole now thought to lie at the hearts of *quasars*. His idea of locating black holes in binary systems from their gravitational influence on the orbits of their companion stars exactly describes the way in which the first black hole known in our *Galaxy*, *Cygnus X-1*, was identified in 1972, 189 years after Michell's paper had been read to the Royal Society.

Michell died at Thornhill on 9 April 1793.

Michelson, Albert Abraham (1852–1931) German-born American physicist who made many determinations of the *speed of light* and carried out an experiment with Edward *Morley* which famously failed to find any evidence for an 'ether' through which *light* was transmitted. See *special theory of relativity*.

Michelson–Morley experiment An experiment carried out in the 1880s, initially by Albert *Michelson* and later in collaboration with Edward *Morley*, in an attempt to detect the motion of the Earth through the *ether*, by measuring differences in the speed of light determined along the line of the Earth's motion and at right angles to that line. Equivalent experiments have been carried out many times since by other experimenters, and all come up with the same result – no effects attributable to the motion of the Earth can be seen in any measurements of the speed of light. The experiments show that there is no ether, and that the measured speed of light does not depend on how the measuring apparatus is moving.

The Michelson–Morley experiment is often regarded as having provided the impetus for Albert *Einstein* to develop the *special theory of relativity*, which says that the speed of light is an absolute constant, c, and will be measured to have the same value by all observers, in any *inertial frame*, regardless of how they are moving through space, or at what speed the source of the light is moving through space. But Einstein always told enquirers that he had been unaware of the Michelson–Morley experiment at the time he developed the special theory, in 1905. The impetus for his work at that time came in part from James Clerk *Maxwell*'s equations describing the motion of electromagnetic waves (including light waves); those equations include the speed of light as an absolute constant.

micrometeorite A particle of *interplanetary dust* so small (weighing less than 1 millionth of a gram) that it does not burn up in the Earth's atmosphere as a *meteor*, but falls gently down to the surface of the Earth. The 'rain' of micrometeorites adds about 4,000 tonnes of matter to the Earth each year.

microwave background radiation See *background radiation*.

microwaves *Radio waves* with *wavelengths* in the range 1–30 cm. The term is sometimes used to include the millimetre wave band, which fills the gap between the far *infrared* part of the *spectrum* at 1 mm and the start of the true microwave band at 1 cm. Microwaves are used in *radio astronomy*, most famously in investigations of the *background radiation*, but also in the study of *interstellar molecules*. On Earth, microwaves are used in radar and telecommunications, and as the source of heat used to cook food in a microwave oven. In these terms, the Universe can be thought of as a microwave oven operating at a temperature of –270.3 °C.

Milankovitch, Milutin (1879–1958) Yugoslavian climatologist who developed the theory that the cycle of Ice Ages on Earth is caused by changes in the Earth's orbit which alter the balance of heat between the seasons.

Milankovitch theory (Milankovitch model)   An explanation of recent Ice Age cycles on Earth in terms of changes in the orientation of the Earth in space and in the shape of its *orbit* around the Sun.

Over periods of tens of thousands of years, the Earth wobbles like a spinning top (precesses) as it moves around the Sun, and the shape of the orbit itself changes from almost circular to slightly more elliptical and back again. These changes combine to alter the balance of heat between the seasons. Although the total amount of heat arriving at the Earth from the Sun in one year is always the same (assuming the output of heat from the Sun is always the same), sometimes there is a large difference between summer and winter, and sometimes there is a smaller contrast between the seasons. The effects combine to produce rhythms roughly 26,000 years long, 40,000 years long and 90,000–100,000 years long in climate change on Earth.

Although he was not the first person to investigate the links between climatic changes and astronomical rhythms, Milutin *Milankovitch* carried out immensely detailed calculations which established the plausibility of the connection. It has since been borne out by improved calculations using electronic computers and by studies of past climates using data from deep-sea cores. It now seems that, with the present-day geography of the globe, the 'natural' state of our planet is in an Ice Age, and that it is only pulled out of this frozen state into an interglacial (like the conditions that exist today) when the astronomical rhythms provide strong summer heating in the Northern Hemisphere, melting the snow from the continents that surround the north polar region.

In round terms, the pattern of recent climate changes explained by the astronomical rhythms (over the past few million years) is one of about 90,000 years of Ice Age conditions followed by 10,000–15,000 years of interglacial, and then another Ice Age, and so on. We are living at the end of an interglacial that began about 15,000 years ago.

Further reading: Mary and John Gribbin, *Being Human*; John Imbrie and Katherine Palmer Imbrie, *Ice Ages*.

Milky Way   Band of *stars* across the sky, marking the plane of our *Galaxy*. This is also used as an alternative name for the Galaxy.

millimetre astronomy   Study of the Universe using *electromagnetic radiation* with wavelengths in the range 1–10 mm. It is a branch of *radio astronomy* that is particularly important for the study of *interstellar molecules* using *spectroscopy*.

millisecond pulsars   Very rapidly spinning *neutron stars* that radiate pulses of *radio waves*. Unlike 'normal' *pulsars*, millisecond pulsars are thought to be relatively old stars that have been spun up by the *accretion* of material, from a companion in a *binary system* or (just possibly) from *interstellar material*. There is no pulsar known which literally spins 1,000 times a second, but the name 'millisecond pulsar' is a justifiable slight exaggeration for objects which actually have periods of a few milliseconds.

The first known millisecond pulsar, PSR 1937 + 211, was identified in 1982, by astronomers from Princeton University using the *Arecibo radio telescope*. It is still the fastest known, with a period of 1.56 milliseconds, corresponding to a spin rate of 642 rotations every second – for an object with about the same mass as the Sun, but only a few kilometres in diameter. The second object of this kind to be discovered, PSR 1953 + 29, which has a period of 6.1 milliseconds, is known to be in a binary system, orbiting an unseen companion once every 120 days.

Although these discoveries came as a surprise, radio astronomers soon found that there are many millisecond pulsars in *globular clusters*, and that more than half of them are in binary systems. There are 11 millisecond pulsars in the single cluster 47 Tuc, all with periods of less than 6 milliseconds, and more than 30 millisecond pulsars known in globular clusters altogether. About the same number had been found in the plane of our Galaxy by the middle of 1994. It seems likely that all of these objects were originally in binary systems, and that the isolated millisecond pulsars that we see today have lost their companions, either because of close encounters with other stars (which are not infrequent events in the crowded conditions inside globular clusters), or because the radiation from the pulsar has destroyed its companion (see *black widow pulsar*).

Because *supernova* explosions of the kind which create ordinary pulsars would disrupt binary systems, it is extremely likely that at least some millisecond pulsars started out as *white dwarf* stars in binary systems (perhaps associated with the smaller *nova* explosions) and gained enough mass by accretion to exceed the *Chandrasekhar limit*, making them collapse into the neutron star state, spinning them up and boosting their *magnetic fields* as they did so.

Mills cross   A type of *radio telescope*, using the principle of *interferometry*, made up of two lines of *antennas* at right angles to one another. First developed in Australia in the 1950s, it was named after the astronomer B. Y. Mills.

Milne, Edward Arthur (1896–1950)   British astrophysicist and cosmologist, who was born in Hull (on 14 February 1896) and educated at Trinity College, Cambridge, where his studies were interrupted by war work on ballistics. He graduated in 1919, and worked in Cambridge from 1920 to 1925, as professor of mathematics at the University of Manchester from 1925 to 1928, and as Rouse Ball Professor of Mathematics in Oxford (the post now held by Roger *Penrose*) from 1928 onwards (again, with an interruption for war work). He died while attending a scientific conference in Dublin, on 21 September 1950.

In the 1920s, working initially in collaboration with Ralph Fowler (1889–1944) in Cambridge, Milne made pioneering theoretical investigations into the structure of *stars*. He later became interested in *cosmology*, and introduced the idea of the *cosmological principle*, that the Universe should look the same from everywhere within it. With William *McCrea*, in the early 1930s, Milne showed that the expanding Universe, predicted by the equations of the *general theory of relativity*, could actually be described very well in the context of *classical mechanics* and Isaac *Newton's* theory of *gravity*. He also espoused the idea that the 'constants' of physics, such as the gravitational constant, G, actually change with time as the Universe ages. This is now largely discredited, but it provoked considerable interest at the time.

Mimas   One of the moons of Saturn, discovered by William *Herschel* in 1789. It has a diameter of only 390 km, and is marked by a *crater* 130 km across. Largely made of ice, Mimas orbits Saturn at a distance of 185,500 km.

mini black holes   Tiny *black holes* (much smaller than an *atom*) which might have been left over from the *Big Bang*. Such small black holes could not be manufactured today, but in principle any amount of matter, no matter how small, will become a black hole if it is squeezed into a sufficiently small volume. Density fluctuations in the very early Universe could conceivably have created mini black holes even before the era of Big Bang *nucleosynthesis*. This is intriguing for several reasons, one of which is that such black holes would not have been formed from *baryons*, and therefore could provide the non-baryonic *dark matter* which many theorists would like to be around to make *spacetime* flat (see *cosmological models*).

The idea of mini black holes was popularized by Stephen *Hawking* in the early 1970s, who showed that such objects must evaporate and eventually explode as they emit *energy* in the form of *Hawking radiation*. The smaller the black hole is, the sooner the explosion will occur. Miniholes that each have a mass of about 10^{13} grams (the mass of a cubic kilometre of rock, but squeezed within a volume roughly the same as that of a *proton*), created in the Big Bang, should be exploding today, 15 billion years or so after the Big Bang.

Unfortunately, the observed background of *gamma rays* from the sky is so low that it seems that very few such miniholes are now exploding, and that comparably few of their lighter counterparts have exploded in the past.

Most theorists doubt that mini black holes exist. They could account for gamma ray bursts from space but there would be too few of them to make spacetime flat.

Minkowski, Hermann (1864–1909)   German mathematician who was one of Albert *Einstein*'s teachers, and who came up with the idea of four-dimensional *spacetime* as the stage on which physical interactions take place. He was an uncle of the observational astronomer Rudolph Minkowski (1895–1976), who was involved in the identification of the first known extragalactic radio source, *Cygnus A*, in 1956.

Hermann Minkowski was born at Alexotas, in Lithuania (then under Russian rule), on 22 June 1864. His family moved to Königsberg, then in Germany (now Kaliningrad, a Russian enclave) in 1872, and he studied at the University of Königsberg, receiving his PhD in 1885, and lectured at the University of Bonn before returning to Königsberg as associate professor of mathematics in 1894. Just two years later, he moved on to become a professor of mathematics at the Zurich Federal Institute of Technology from 1896 to 1902, where Einstein was one of his pupils. The young Einstein, although clever, did not make a favourable impression on his teachers because of his casual approach to formal courses; it was Minkowski who described him as a 'lazy dog', who 'never bothered about mathematics at all'.

In 1902 Minkowski became professor of mathematics at Göttingen University, where he stayed for the rest of his life. He was essentially involved in pure mathematics, including number theory and geometry. It was through his understanding of the more abstract side of mathematics and geometry in more than three dimensions that he developed the idea of four-dimensional spacetime, presented in a lecture in Cologne in 1908. In that lecture, his opening words set the scene for a revolutionary new insight into the significance of the *special theory of relativity*, an insight which (although Einstein was not initially impressed by it) would shortly provide the key to the development of the *general theory of relativity*:

> The views of space and time which I wish to lay before you have sprung from the soil of experimental physics, and therein lies their strength. They are radical. Henceforth space by itself, and time by itself, are doomed to fade into mere shadows, and only a kind of union of the two will preserve an independent reality.

The work was published in 1909, after Minkowski's death on 12 January that year from complications following an attack of appendicitis.

Minkowski diagram   See *light cone*.

minor planets   Another term for *asteroids* – lumps of rock much smaller than the main planets. Most of the minor planets orbit around the Sun in a belt between the orbits of Mars and Jupiter. This asteroid belt is between 1.7 and 4.0 *astronomical units* from the Sun, where the objects orbit with periods of between 3 and 6 years. It was once thought to be the debris

Minor planets. The asteroid 243 Ida, photographed by the Galileo spaceprobe, en route to its encounter with Jupiter in December 1995. The spacecraft passed just over 2,000 km from Ida on 28 August 1993.

from a planet that had broken up, but it now seems more likely that the minor planets are cosmic rubble of the kind from which planets formed, so that the asteroid belt represents a failed planet rather than an exploded planet.

Possibly, the material originally formed into a few objects each about the size of Mars, which collided with one another and broke up; if so, most of the original material in the asteroid belt has since been lost, since the total mass of all the minor planets put together is today only about 15 per cent of the mass of the Moon. Intriguingly, the main belt of minor planets is at just the right distance from the Sun to fit **Bode's Law.**

Some minor planets (about 5 per cent) have quite different orbits, and some of these are in elliptical orbits which take them closer to the Sun than the Earth's orbit, crossing the Earth's orbit as they do so. Such objects have collided with the Earth in the past (helping to explain where material 'lost' from the asteroid belt has gone), and may do so in the future (see **doomsday asteroid**).

It is estimated that more than 100,000 minor planets can be photographed, but fewer than 6,000 have actually been studied and given numbers and (in most cases) names. The largest of the minor planets is

Ceres, with a diameter of 933 km; the smallest known asteroids have diameters of a few hundred metres, but objects down to the size of pebbles and dust grains are probably present in the asteroid belt. Only ten of the minor planets are bigger than 250 km, and only 120 are larger than half this size.

Mira The first *variable star* to be discovered, by the Dutch astronomer David Fabricius in a series of observations in the 1590s and 1600s. Mira is a *red giant* and varies with an average period of 332 days, brightening and fading as the star pulses in and out. The star is about 60 *parsecs* away and is orbited by a faint companion, VZ Ceti, that is thought to be a *white dwarf* with an *accretion disk*.

Miranda One of the *moons* of Uranus, discovered by Gerard *Kuiper* in 1948. Pictures from *Voyager* 2 show that the surface of Miranda is a jumble of different types of 'terrain', suggesting that the moon was once shattered by an impact and has been gently stuck back together by *gravity*. It is 470 km in diameter.

missing mass See *dark matter*.

Mitchell, Maria (1818–99) American astronomer who worked as a librarian and then as a computer with the US Nautical Almanac Office before being appointed professor of astronomy and director of the observatory at the new Vassar College in 1865. She was the first American woman astronomer of note, and discovered a *comet* in 1847.

mixed dark matter See *dark matter*.

models In *cosmology* and *astrophysics*, mathematical descriptions of the Universe or objects such as stars, which can be used to test theories about how they work.

Designers of everyday items such as cars or aircraft can make physical models of the objects out of wood or metal, and can study the behaviour of these models (for example, in a wind tunnel) to find out how the real object will behave. It is not possible to build a physical model of the Universe or a star and analyse it in the same way, but it is possible to set up systems of equations that describe the behaviour of these objects allowed by the laws of physics. Such equations may, for example, relate the pressure, density and temperature at the heart of a star to its total mass and luminosity. By adjusting the parameters that go into the mathematical model (perhaps by specifying a different luminosity), it is possible to calculate how all of the other properties must change to match that adjustment. And by comparing the predictions of the models with observations of equivalent objects in the real Universe, theorists can find out how good their theories and models are as descriptions of the real world.

Today, such modelling almost always involves the use of high-speed electronic computers to deal with the drudgery of the calculations, and this has made it easier to calculate the behaviour of the models in detail. But the actual model is the system of mathematical equations, not the computer, and its behaviour can in principle still be worked out the way the behaviour of the first such models was investigated, using pencil and paper.

molecular cloud   An *interstellar cloud* in which most of the material present is gas in the form of *molecules*. Both small molecular clouds and giant molecular clouds are found in the *interstellar matter* in the plane of the *Milky Way*.

The small molecular clouds are a few *light years* across, and have *densities* corresponding to between about 1,000 and 10,000 molecules per cubic centimetre. Their temperatures are very low, in the range from 10–20 *Kelvin*, and they are largely composed of hydrogen gas in the form of H_2 molecules. They stay this cold because there are no *stars* inside the clouds to warm them up; some of them contain even colder cores, where the density of hydrogen is at least ten times greater.

The giant molecular clouds also contain large amounts of hydrogen, but in addition they contain large amounts of carbon monoxide (CO) and smaller quantities of other molecules (more than 60 types of molecule in some of the largest clouds). They each contain up to 10 million times as much mass as our Sun, and are between about 150 light years and 250 light years across. This makes them the most massive single entities in our Galaxy.

Unlike small molecular clouds, giant molecular clouds are associated with regions of active star formation, and they are heated by the *radiation* from the young stars they contain. One such cloud may contain several denser cores, each containing up to 1,000 times as much mass as our Sun and with a density of 100,000 molecules per cubic centimetre, from which *infrared radiation* shows the early stages of star formation going on. There is a giant molecular cloud, for example, associated with the *Orion nebula*. Some giant molecular clouds contain regions of *maser* activity. Several thousand of these giant molecular clouds are scattered across the Milky Way.

molecule   The smallest part of a chemical compound or *element* that can exist independently. Molecules are made of two or more *atoms* held together by electromagnetic forces. In some cases the atoms are identical, and the molecule is a unit of an element (for example, the *hydrogen* molecule, H_2). In most cases there is a variety of atoms, and the molecule is a unit of a compound (for example, the water molecule, H_2O).

Molonglo Radio Observatory   Australian radio observatory located near Canberra and operated by the University of Sydney. Site of a large *Mills cross* type of *radio telescope* completed in 1966 and later modified for *aperture synthesis* work.

monopole Hypothetical particle possessing a single 'flavour' of magnetism – an isolated north or south magnetic pole. Some *grand unified theories* predict the existence of monopoles, but they have never been found. One possible explanation is that during *inflation* the entire observable Universe expanded from a region containing just one monopole, which we have not been lucky enough to see.

monopole universe One of Andrei *Linde's* most intriguing recent speculations – that the entire Universe exists inside a single magnetic *monopole* produced by *inflation*. Many of the *grand unified theories* imply the existence of large numbers of magnetic monopoles in our Universe, but no such monopole has ever been detected. Standard models of inflation solve this 'monopole problem' by arguing that the seed from which our entire visible Universe grew was a quantum fluctuation so small that it contained only one monopole. That monopole is still out there, somewhere in the Universe, but it is highly unlikely that it will ever pass our way.

But Linde has discovered that, according to theory, the conditions that create inflation persist *inside* a magnetic monopole even after inflation has halted in the Universe at large. Such a monopole would look like a magnetically charged *black hole*, connecting our Universe through a *wormhole* in *spacetime* to another region of inflating spacetime. Within this region of inflation, quantum processes can produce monopole–antimonopole pairs, which then separate exponentially rapidly as a result of the inflation. Inflation then stops, leaving an expanding Universe rather like our own which may contain one or two monopoles, within each of which there are more regions of inflating spacetime.

The result is a never-ending fractal structure, with inflating universes embedded inside each other and connected through the magnetic monopole wormholes. Our Universe may be inside a monopole which is inside another universe which is inside another monopole, and so on indefinitely. What Linde calls 'the continuous creation of exponentially expanding space' means that 'monopoles by themselves can solve the monopole problem'. Although it seems bizarre, the idea is, he stresses, 'so simple that it certainly deserves further investigation'.

In a delicious touch of irony, Linde, who works at Stanford University, made this outrageous claim in a lecture at a workshop on the Birth of the Universe held in Rome, where the view of Creation is usually rather different. See also *free lunch universe*.

Moon The only natural *satellite* of our planet, Earth. The closest other astronomical object to our planet, and the most intensively studied, it is also the only other astronomical object that has yet been visited by people.

The Moon formed, with the Earth, about 4,600 million years ago. It contains a small, iron-rich core, but is mostly composed of rock and has a heavily cratered surface, caused by a bombardment of *asteroids* when the Solar System was young, chiefly between about 500 million and 700 million years after it formed. Volcanic activity continued up until about 2,000 million years ago, flooding large areas of the surface (known as 'seas', although they contain no water) with basalt that cooled and solidified into level plains. The Moon has essentially no atmosphere, because the pull of gravity at the surface of the Moon is only about one-sixth of that at the surface of the Earth, too weak to retain a blanket of gas.

Tidal forces have slowed the rotation of the Moon until it has become locked with its orbit around the Earth, so that the Moon always keeps the same side of its surface facing the Earth. It has a diameter of 3,476 km, an average distance from the Earth of 384,400 km (about 1.3 *light seconds*), and takes 27.322 days to complete one orbit (the sidereal month). It is by far the largest natural satellite, in comparison to the size of the parent body, in the Solar System, with a mass 0.0123 times that of the Earth (the Pluto–Charon double system does not seem to represent a true planet–moon combination). In many ways, the Earth–Moon system resembles a double planet.

moons Strictly speaking, the natural *satellites* of *planets*, although the term is occasionally stretched a little to include artificial satellites. Apart from Mercury and Venus, the two planets closest to the Sun, every planet in the Solar System has at least one moon (although it is stretching the definition more than a little to describe Charon, the companion of Pluto, as a moon, it is also stretching the definition of a planet to dignify Pluto by that term).

The Moon. A close-up view of our Moon, photographed from Apollo 12 in 1969.

Altogether, there are more than 65 known moons in the Solar System, and the number keeps going up as more and more small objects orbiting the *giant planets* are discovered. This total does not include the ring systems, such as the rings of Saturn, which contain innumerable small pieces of rock and ice that could properly each be regarded as an individual moon. Three moons (Titan, which orbits Saturn; Io, which orbits Jupiter; and Triton, which belongs to Neptune) have atmospheres. Titan is the largest moon in the Solar System, with a diameter of 5,150 km; the smallest objects dignified with the name moons are irregular rocky lumps in orbit around Jupiter and Saturn, each a few tens of kilometres across, which may be captured *asteroids*.

Morley, Edward Williams (1838–1923)   American chemist and physicist who had a passion for precise measurement and carried out an experiment with Albert *Michelson* which famously failed to find any evidence for an 'ether' through which *light* was transmitted. See *special theory of relativity*.

morning star   See *Venus*.

Mount Hopkins Observatory   Original name (until 1982) of the *Fred L. Whipple Observatory*.

Mount Palomar Observatory   Site of the famous 200-inch (5.08 m) *Hale telescope* completed in 1946. Run by the California Institute of Technology, the observatory includes two *Schmidt cameras* and a 1.5 m *reflecting telescope*, at an altitude of 1,710 m on a mountain near Los Angeles.

Mount Stromlo and Siding Spring Observatories   Observatory for *optical astronomy* run by the Australian National Observatories and located on Mount Stromlo, near Canberra, at an altitude of 770 m, and on Siding Spring Mountain, in New South Wales, at an altitude of 1,150 m. The main instruments are a 1.9 m *reflecting telescope* (at Mount Stromlo), and a 2.3 metre reflector (at Siding Spring). The Siding Spring site is shared by the *Anglo-Australian Observatory*.

Mount Wilson Observatory   Home of the 100-inch (2.54 m) diameter *Hooker telescope*, near Pasadena in California. It was established by George *Hale* in the early 20th century, using funds provided by Andrew Carnegie. Located at an altitude of 1,750 m, the observatory is now run jointly by Harvard University, the University of Southern California and the University College of Los Angeles. It includes solar telescopes and a 1.5 m reflector, as well as the Hooker telescope, which was refurbished in the early 1990s and is still in use.

moving cluster method   A rather neat trick for determining the distances to *stars* in the neighbourhood of the Sun (also known as moving cluster parallax).

The trick depends on finding a group of stars close enough for their motion across the sky to be measurable in a reasonable time (over a few years or decades). If the stars are part of an *open cluster* moving together through space, their individual tracks across the sky will seem to diverge from (or converge towards) a single point on the sky. This is an optical illusion exactly equivalent to the way the parallel lines of a long, straight railway track seem to converge on a point on the horizon, and it shows in which direction the group of stars is really moving through space.

By measuring the *redshifts* in the *light* from individual stars in the cluster, caused by the *Doppler effect*, astronomers can calculate the component of the velocity of the star towards or away from us in the line of sight. Combining this with the information about which direction the stars are moving in tells them how fast the stars are moving through space at right angles to the line of sight, in kilometres per second, or light hours per decade, or whatever units you like. And by combining this with the measurements of the angular distance each star moves across the sky in a year or a decade, they can work out how far away the star is.

The trick only works for clusters of stars that are within a few tens of *parsecs* of the Solar System, but it enabled astronomers to work out the distance to one cluster in particular, the Hyades cluster, which is about 40 parsecs away. This cluster contains many different types of star, with different *colours*, whose *luminosities* could be calibrated against their *spectral classification*.

MSW effect   See *solar neutrino problem*.

Mullard Radio Astronomy Observatory (MRAO)   The radio astronomy observatory of the University of Cambridge, site of the *Ryle telescope*. The technique of *aperture synthesis* was largely developed at this observatory. The MRAO is also the site from which *pulsars* were discovered, and it is part of the *MERLIN* group.

Multiple Mirror Telescope (MMT)   A *reflecting telescope* at the Fred L. Whipple Observatory, on Mount Hopkins (in Arizona) at an altitude of 2,600 m. The telescope used six mirrors, each 1.8 m across, working together to mimic the properties of a single mirror with an *aperture* of 4.5 m. It operated successfully throughout the 1980s, but in the mid-1990s the six small mirrors were replaced by a single mirror 6.5 m across. (See p. 334.)

multiple stars   Systems of *stars* which are gravitationally bound to one another, and which contain at least three members (systems with only two members are called *binary stars*). There is no clear distinction between a complex multiple star system and a small *cluster of stars*, but systems with six components (such as Castor) are generally regarded as multiples. In Castor, there are two binary pairs which each orbit around the *centre of mass* of the system (dancing around each other with a period of about 500 years), while another pair of cool *red dwarf stars* orbits much further out from the centre of mass. This hierarchical structure is typical of multiple star systems.

The Multiple Mirror Telescope (MMT) on Mount Hopkins, in Arizona. Six separate mirrors worked together in this telescope to mimic the power of a single mirror 4.5 metres across.

About a third of all 'binary' stars are known actually to be triples, with one component of the binary itself a binary. Similarly, about a third of all triples turn out to be quadruple star systems. Systems like Castor may represent about 0.1 per cent of all binary stars, and about 60 per cent of all stars have at least one companion. Multiple star systems are probably made up of stars that form together from the collapse of a single cloud of gas and dust in space.

muon See *elementary particles*.

naked singularity A *singularity* that is not concealed from view behind the *event horizon* of a *black hole*. The Universe may have been born out of a naked singularity. See *cosmic censorship*.

nanometre 1 billionth (10^{-9}) of a metre.

NASA Acronym for the National Aeronautics and Space Agency, the American agency created in 1958 (in response to the launch of Sputnik 1) to handle all non-military aspects of the US space programme. This includes development of scientific probes such as the *Voyager* and *Mariner* series, and the *Hubble Space Telescope*.

Nebula. The Horsehead nebula, in the constellation Orion, photographed using the 4-metre telescope at Kitt Peak National Observatory.

National Radio Astronomy Observatory (NRAO)   American radio observatory run from Charlottesville, in Virginia. The *radio telescopes* operated by the NRAO include one at Green Bank and the *Very Large Array*.

nebulae   Old name for any fuzzy patches of light on the sky. Many of these are now known to be other *galaxies*, beyond the *Milky Way*, and are sometimes referred to by the old name of external nebulae. The *Andromeda galaxy*, for example, is sometimes called the Andromeda nebula. Other nebulae are now known to be glowing clouds of gas within our own *Galaxy*, and they are often the sites of star formation. The *Orion nebula* is a classic example of this kind of nebula. The word 'nebula' is simply the Latin for 'cloud'.

Many nebulae are visible to the naked eye, but the invention of the telescope not only revealed many more nebulae than had been seen by the unaided eye, but also showed that many of the clouds are made up of *stars* too faint and close together to be distinguished by eye. In the first half of the 19th century, many astronomers, notably the *Herschels*, believed that all nebulae were made up of stars. The development of *spectroscopy* in the 1860s showed, however, that some nebulae are in fact clouds of gas. At that time it was still not clear whether the nebulae that are composed of stars lie within the Milky Way or beyond it; the question was not finally resolved until the work of Edwin *Hubble* and his colleagues gave the first good estimates of the distances to several external nebulae in the 1920s.

Within our Galaxy, bright *emission nebulae* are kept warm by the energy radiated by nearby stars, and show up red in astronomical photographs because of characteristic *radiation* from *hydrogen* in the red part of the *spectrum*. Other *reflection nebulae* show up in reflected starlight, and look blue, because of the way the starlight is scattered from dust particles in the nebula (this is exactly equivalent to the scattering that makes the sky look blue). Some dark *absorption nebulae* are visible only because they block out the light from more distant stars – they look like dark holes in the bright backdrop of the stars.

nebular hypothesis   The idea that the Solar System formed from a cloud of material (a 'nebulosity') in space. This basic concept dates back to 1755, when Immanuel *Kant* suggested, in his book *Universal Natural History and Theory of the Heavens*, that the *planets* had condensed out of a cloud of primordial matter. Pierre *Laplace* hit upon the same idea independently a little later, and published it in 1796 in his book *Exposition du système du monde*.

In its original form, the idea was that the planets had condensed out from irregularities in the Sun's atmosphere, which the proponents of the hypothesis envisaged as extending far out into space. This was developed into a *model* in which the whole Solar System was seen as forming from a great cloud of gas, most of which collapsed to form the Sun while a small amount of the material settled into a disc around the Sun, breaking up to form rings at different distances from the Sun, which then condensed to form the planets. Although the original ideas have been considerably modified over the past 200 years, and there are still difficulties in explaining why the Sun is spinning so slowly today if it formed in this way (like a spinning ice skater pulling in her arms, the primordial gas cloud should have spun faster as it contracted), this is still regarded as the most likely scenario for the formation of the planets.

Neptune   One of the *giant planets*, usually the eighth planet out from the Sun, although sometimes the orbit of Pluto takes it inside the orbit of Neptune (this is happening between 1979 and 1999). Neptune was discovered in 1846 on the basis of predictions made by John *Adams* and Urbain *Le Verrier*, but may have been seen and sketched by *Galileo* in 1613, without him realizing it was a planet. It orbits the Sun once every 164.79 years, at a distance of 30.06 *astronomical units*, in an almost circular orbit. Neptune, which contains 17.2 times as much mass as the Earth, has a diameter of 48,600 km, and eight known moons. It is very similar to Uranus.

Nereid Small *moon* of Neptune, discovered by Gerard *Kuiper* in 1949. It has a diameter of about 170 km and *orbits* Neptune at an average distance of 5.5 million km. Its mass is not known.

neutrino Fundamental particle, member of the *lepton* family, which has zero electric *charge*, very small (possibly zero) *mass*, and interacts with other particles only through the weak interaction (see *fundamental forces*) and (if it does have mass) *gravity*.

The need for neutrinos was suggested at the beginning of the 1930s, to explain anomalous behaviour observed during *beta decay*. Some of the *energy* involved in the process seemed to be 'lost', and Wolfgang *Pauli*, one of the pioneering quantum physicists, suggested the need for a 'new' particle, which he dubbed the neutron, to carry off the excess energy. The suggestion made so little impact that the name *neutron* was actually given to a quite different particle, discovered in 1932. But Enrico Fermi (1901–54) took up the idea proposed by Pauli, and when asked if Pauli's 'neutron' was the same as the newly discovered neutral particle he joked, 'No, Pauli's particle is only a neutrino'; the name stuck.

Neutrinos are extremely difficult to detect, but they are produced copiously in *nuclear fission* reactions, and were finally proven to exist in experiments in 1956 that monitored the flood of neutrinos emitted by a nuclear reactor at Savannah River in the United States. Twenty-five years earlier, Pauli had offered a case of champagne to any experimenter who could prove that his little neutral particles existed; he happily paid up on his longstanding promise.

Within a few years, John *Bahcall* and Ray *Davis* were making plans to detect neutrinos produced by the Sun, and the new science of *neutrino astronomy* was born. Neutrinos have since been detected from *Supernova 1987A*, and if they do have mass, they contribute some of the *dark matter* in the Universe.

Early in 1995, great interest was stirred by claims that an experiment at the Los Alamos National Laboratory, in New Mexico, had determined that neutrinos do have mass, and that it lies in the range 0.5–5 *electron Volts*. If these claims are borne out by further experiments, the implications are literally of cosmic importance. Although each neutrino has such a tiny mass, there are so many neutrinos in the Universe (about a billion in every cubic metre of space) that together they would provide more mass than all of the bright *stars* and *galaxies* put together. They would contribute about 20 per cent of the total mass required to make the Universe flat (see *cosmological models*), as predicted by the favoured theory of *inflation*.

This is just what cosmologists were hoping to hear. If all of the dark matter required to flatten the Universe were in the form of neutrinos, then galaxies would never have formed at all. The influence of these particles hurtling through space at close to the speed of light when the Universe was young would have smoothed out any irregularities in primordial gas clouds.

In fact, the distribution of galaxies in chains, clusters and filaments across the sky strongly suggests that they grew up under the influence of the gravitational pull of large amounts of so-called *cold dark matter*. This is made up of hypothetical particles which might be tens of times as massive as an *electron* (which itself weighs in at 500,000 eV), but which, unlike neutrinos (dubbed *hot dark matter*, by contrast) move only slowly through space.

The computer simulations of galaxy formation in the expanding Universe which most closely match the observed pattern of galaxies invoke the presence of 80 per cent cold dark matter and 20 per cent hot dark matter (the bright galaxies then make up about 1 per cent of the stuff of the Universe).

The technique used by the Los Alamos team involves monitoring the so-called oscillation of one kind of neutrino into another. There are three flavours of neutrino, one each associated with the electron, the muon and the tau particle. The Los Alamos experiment uses a beam of pure muon neutrinos which passes through a tank filled with 200 tonnes of mineral oil. Muon neutrinos cannot interact with the oil, but electron neutrinos can. If some of the muon neutrinos are converted into electron neutrinos, a few of them will interact with *protons* in the atoms of the oil to produce flashes of light, which are picked up by detectors around the tank of oil.

That is exactly what seems to have happened. And the rate at which it happens tells the experimenters that the neutrinos involved have masses between 0.5 eV and 5 eV. Although delighted by the news, in the spring of 1995 the attitude of most cosmologists was a cautious 'wait and see', at least until other experiments have confirmed these findings. If right, however, the discovery lends strong support to the standard theory of how the Universe and galaxies formed.

See also *solar neutrino problem*.

neutrino astronomy   The study of the Universe using detectors sensitive to the arrival of *neutrinos* from space. Neutrino astronomy began in the 1960s, thanks to the efforts of theorist John *Bahcall* and experimenter Ray *Davis* in developing a detector using chlorine (in the form of perchlorethylene) to monitor the arrival of neutrinos from the Sun. The Davis detector, which began operating in 1967, found only about one-third of the expected number of solar neutrinos (see *solar neutrino problem*); discussion of the significance of this discovery sometimes overlooks the fact that it was a great triumph to detect any neutrinos from space at all, and that the success of the Davis detector established neutrino astronomy as a new branch of the investigation of the Universe.

Most of the neutrino astronomy studies to date have dealt with solar neutrinos, which are produced in copious quantities inside the Sun and flood past and through the Earth. Neutrinos scarcely notice the existence of everyday matter, and if a beam of neutrinos like those produced inside the Sun were to travel through solid lead for a distance of 1,000 *parsecs*, only half of them would be captured by the nuclei of the lead atoms along the way.

Because neutrinos are so very reluctant to interact with other forms of matter (including the detectors), the instruments have to be very sensitive, and because they are very sensitive, the detectors have to be shielded from the influence of other particles, such as *cosmic rays*. So neutrino 'telescopes' are buried down deep mines, or in tunnels underneath mountains, where the layers of solid rock above screen out the unwanted particles. The neutrino telescopes also differ from the traditional image of a telescope in other ways.

The important piece of an instrument like the 120-inch reflecting telescope at the Lick Observatory is not the 50 tonnes or so of steel in the girders that support the mirror; what matters is the bit that actually interacts with photons of light – the thin film of shiny aluminium spread over the surface of the mirror. That amounts to just 1 cubic centimetre of aluminium, which, arranged in the right way on top of a mountain, is all you need if you want to study light from space. But whereas the surface of the 120-inch mirror is coated with just 1 cubic centimetre of aluminium, the interior 'surface' of the impressive steel box that holds the working part of the Davis detector (as big as an Olympic swimming pool) is 'coated' with more than 400,000 litres of perchlorethylene – and you need all that 'working surface', arranged in the right way underneath a mountain or down a mine, if you want to study neutrinos.

The Davis detector has actually recorded, on average, about one 'event' associated with the interaction of a neutrino from the Sun with a chlorine atom in the tank every two days since the end of the 1960s. This counting rate is typical of the level at which such experiments must operate; it's as if optical astronomers studying a particular *quasar* were restricted to seeing one photon every other day.

In the wake of the success of the Davis detector, other dedicated detectors for neutrino astronomy have been built, some (such as *SAGE* and *Gallex*) using gallium as the working material. In addition, particle detectors that were originally designed for other studies have proved capable of recording the arrival of neutrinos from space. Detectors such as the

Japanese *Kamiokande* experiment were designed to detect neutrinos produced by the *radioactive decay* of *protons*; this decay has never been observed, but the sensitivity of the detector (which actually consists of a cylindrical tank 15.6 m across and 16 m high, containing 3,000 tonnes of water) has enabled it to detect neutrinos from space as they (very occasionally) interact with the electrons associated with the water *molecules* in the tank. Other detectors based on the electron 'scattering' effect have been built in Australia and in the Gran Sasso Laboratory in northern Italy, and new kinds of detector are being planned and built around the world.

As far as neutrinos from the Sun are concerned, the picture initially became slightly more confused with the development of different detectors operating in different *energy* ranges. At first, it was not clear whether the different detectors agreed with one another, but as they have been run for several years, there has been some convergence on a consensus about the number and variety of neutrinos reaching us from the Sun.

This still does not agree completely with the predictions of the theorists, although the disagreement is not as large as the results from the Davis detector originally suggested. What is often not stressed sufficiently, however, is that even allowing for the existing uncertainties, and for the possible minor disagreement between observations and theory, the observations of solar neutrinos do broadly match the theoretical predictions. We are not talking about observations which require a complete reworking of the theoretical *models*, but of subtle differences which may suggest minor reworking – fine tuning – of the theories. The basic picture is consistent with *nuclear fusion* reactions going on at a temperature of about 15 million K in the heart of the Sun to provide the energy which keeps it shining. This is a great triumph for both theory and experiment.

Neutrino astronomy really came of age, however, when several detectors around the world recorded a burst of neutrinos associated with the explosion of *Supernova 1987A*. This supernova was seen in visible light on Earth on 23 February 1987, but that light had been travelling towards us from the *Large Magellanic Cloud* for more than 160,000 years, since the star that created the supernova exploded. When the records of the various neutrino detection experiments that were running around the world at the time were examined, they showed that a pulse of neutrinos had arrived at the Earth just before the light from the supernova. These were explained as neutrinos from the dying star, produced at the moment its core collapsed, some 3 hours before the energy released by that collapse was able to blast away the outer layers of the star and produce the glare of visible light. Had the neutrino detectors been monitoring in real time, they could have provided advance warning that a supernova was about to become visible – although, since the detectors are not directional, they could not have told the optical astronomers in which direction to point their telescopes.

In fact, because of their low counting rates and the need to make long observing runs in order to detect anything at all, neutrino telescopes are usually left to run for several days at a time before the associated data are analysed. Immediately after the supernova was seen, this regular pattern was changed and the data for the previous day were analysed straight away. Just two detectors were at the right stage of their operating cycle, and had the sensitivity, to pick up the neutrinos from the supernova. The Kamiokande team found a cluster of eleven neutrino events spread over an interval of 13 seconds (but most of them arriving in the first second of the burst), and a similar detector near Cleveland, Ohio (operated by the University of California-Irvine and the University of Michigan), found eight events spread over an interval of 6 seconds (remember that the counting rate for solar neutrino events in the Davis detector is one every two days).

The fact that the neutrinos all arrived within a few seconds sets tight limits on neutrino *mass*. If neutrinos have zero mass, like photons, then they all travel at the *speed of light*, and would arrive together (assuming they set out together) even after a journey as long as 160,000 years. But if neutrinos have mass – even if they all have the same mass – then the speed with which they travel depends on their individual energies. Just as a baseball hit more strongly will fly faster through the air, so the neutrinos given the biggest boost in the supernova explosion would travel faster and arrive first. The effect is more pronounced if the neutrinos have more mass; the fact that several neutrinos arrived within a second of one another at the beginning of the burst from the supernova, after a journey of 160,000 years, shows that the mass of each neutrino must be less than about 15 *electron Volts*. This has important implications for the possible contribution of neutrinos to the overall *density* of the Universe (see *dark matter*).

It is also still possible, of course, on the basis of these observations, that neutrinos have precisely zero mass and all travel at exactly the speed of light, but that they set out from the supernova at slightly different times. The important point is that they cannot have masses greater than 15 eV.

Astrophysicists calculate that, in a supernova explosion like the one seen in the Large Magellanic Cloud on 23 February 1987, about 10^{58} neutrinos are released when the core of the star collapses. These spread out in a spherical shell travelling outwards through space at (or very near) the speed of light. By the time the spherical shell reached Earth (when it had a diameter of more than 100 *kiloparsecs*), it had thinned out so much that 'only' about 300,000 billion (3×10^{14}) neutrinos passed through the Kamiokande detector, where just eleven collided with electrons and left a detectable trace. Well over 10 billion neutrinos from Supernova 1987A passed through the body of each fully grown adult on Earth at the time, and very probably just a few of those neutrinos interacted with electrons in the fluid inside the eyes of some of those people, producing a brief flash of light. One or two of us may have actually seen the influence of a neutrino from the supernova directly.

Some indication of the potential of neutrino astronomy can be seen from the way particle physicists used the observations of neutrinos from SN 1987A. The nineteen neutrinos detected from SN 1987A provide enough information to rule out the possible existence of one of the particles cherished by theorists, and to rule in another as the most likely candidate for the dark matter in the Universe.

If neutrinos have zero mass and all travel at the speed of light, the spread of about 12 seconds in their arrival time exactly matches the time required for the core of the star which formed the supernova to collapse into a ball of *neutrons* about 10 km across. Later arrivals carried slightly less energy than the first neutrinos to arrive, showing how the forming *neutron star* cooled, and closely matching theoretical predictions.

The outburst of neutrinos from the core of the collapsing star was driven by the gravitational energy released as the core collapsed. Under those energetic conditions, other particles could also have been produced, but only if they had the right combination of properties, one of which is their mass and another the strength with which they interact (couple) to other particles and radiation.

Because the neutrinos carried away most of the energy from the collapsing stellar core, there is a limit on how much energy could have been carried by these other particles. Two particles predicted by theory, but never yet detected, are of particular interest to physicists today – the photino, which is required by the theory known as *supersymmetry* (SUSY), and the *axion*, which is required to explain differences between left-handed and right-handed particles – their 'axial' symmetry.

Theory predicts all of the properties of these particles except their mass. Astonishingly, the information provided by studies of SN 1987A shows that if the photino does exist it must be very heavy, much more massive than the proton. If light photinos existed, they would have been formed in copious quantities in the explosion and carried away the energy that actually went into neutrinos. No other observation or experiment has been able to rule out the existence of a light photino.

But the supernova data match up very well with the possible production of a flood of axions in the explosion, carrying away the surplus energy not carried by neutrinos – provided that the axions each have a mass of about one-thousandth of an electron Volt (for comparison, an electron has a mass of 500,000 eV). Tiny though this is, theory suggests that there ought to be so many axions in the Universe that if they each have this much mass they can provide all of the dark matter which many cosmologists would like to have in order to 'close' the Universe and ensure that its present expansion will one day be reversed.

It is a breathtaking amount of information to extract from measuring the arrival times and energies of just nineteen neutrinos – so breathtaking that even the particle physicists seem dumbfounded. If that's what you can do with nineteen neutrinos, they can hardly wait for the next supernova, now that additional, more sensitive, neutrino detectors are up and running.

Since February 1987, the neutrino shell from that supernova has continued to expand outwards through the Universe, where it may be monitored by the neutrino detectors of other civilizations as it races on its way. Meanwhile, more (and more sensitive) neutrino detectors than ever before are ready and waiting here on Earth to detect the next neutrino burst, which may even now be approaching us at nearly the speed of light from a supernova explosion that has already happened, but whose light has not yet reached us.

neutrino oscillations See *solar neutrino problem*.

neutron An *elementary particle* found in all atomic nuclei except that of *hydrogen*. The neutron has no electric *charge* (hence its name) and its *mass* is very slightly greater than that of the *proton*. Outside the nucleus, a neutron will decay into a proton, an *electron* and an antineutrino. Each neutron is made up of three *quarks* and is a member of the *baryon* family.

neutron star A *star* made almost entirely of *neutrons*, with the density of an atomic *nucleus*. Such a star contains roughly the same amount of matter as there is in our Sun, but packed into a sphere about 10 km across.

At the beginning of the 1930s, Subrahmanyan *Chandrasekhar* had discovered that there is no way for a *white dwarf* star with more than about 1.4 times the mass of the Sun to hold itself up against *gravitational collapse* once its nuclear fuel is exhausted. The implication – extremely controversial at the time – was that any star left with more mass than this *Chandrasekhar limit* at the end of its lifetime would collapse indefinitely, forming what is now called a *black hole*. When the neutron was discovered, in 1932, some physicists and astronomers immediately began to speculate about the possible existence of stars made entirely of neutrons, intermediate in density between white dwarfs and stellar-mass black holes, and to wonder whether there was an upper limit on the mass of such stars. The Soviet physicist Lev *Landau* suggested that all stars might contain neutron 'cores', but calculations soon showed that, if 'neutronization' of a stellar core did begin to happen, it would be a runaway process in which the whole inner part of the star suddenly collapsed, releasing a vast amount of gravitational energy in an explosion. This tied in with a suggestion by Fritz **Zwicky** that neutron stars might be formed in **supernovae**.

By the end of the 1930s, all of these ideas were in print, together with calculations by Robert **Oppenheimer** and his student George Volkoff which showed that there is indeed an upper mass limit for neutron stars, now known as the **Oppenheimer–Volkoff limit**. Any star which ends its life with more than three times the mass of our Sun must collapse indefinitely. But it was a quarter of a century before the idea of neutron stars was taken seriously by most astronomers, and only the accidental discovery of **pulsars** in the mid-1960s convinced them that neutron stars really did exist.

It is now accepted that these superdense stars really do form in supernova explosions, where the intense pressure can create neutron stars with as little as one-tenth of the mass of our Sun (any lighter neutron star that tried to form would turn into a small white dwarf, as some of the neutrons converted themselves into protons by **beta decay**). Some neutron stars may form from white dwarfs with masses close to the Chandrasekhar limit, if they accrete enough extra material (perhaps from a companion in a **binary system**) to push their masses over the limit.

A neutron star has a solid crust of iron and similar **elements** (see **nucleosynthesis**), overlaying a region of 'normal' neutrons and a fluid inner part, mainly composed of superfluid neutrons and possibly with a central inner core of **quarks**. The density of matter in a neutron star is 1 million times greater than in a white dwarf and 1 million billion times greater than water, so that each cubic centimetre of the star would weigh about 100 million tonnes.

Newton, Isaac (1642–1727)   British physicist and mathematician, one of the greatest scientists of all time. He discovered the law of **gravity** and the laws of motion that bear his name, invented the mathematical technique of calculus, and carried out important studies in optics, including the design of a new kind of **reflecting telescope**. He also essentially invented the modern techniques of scientific investigation, in which ideas are tested and refined by comparison with experiments, instead of being plucked out of the air as more or less wild flights of fancy.

Newton was born at Woolsthorpe, in Lincolnshire, on Christmas Day 1642, according to the **Julian calendar** still in use in Britain at the time. This was 4 January 1643 on the **Gregorian calendar** already in use in Catholic countries at that time, which is the calendar we use today. So although it is sometimes said that Newton was born in the year Galileo died (on 8 January 1642 on the Gregorian calendar), this 'coincidence'

Isaac Newton (1642–1727).

depends upon comparing dates from two different calendar systems.

Newton's father died before the baby was born, and when young Isaac was three years old his mother remarried and he was left in the care of his grandmother. His mother was widowed again in 1658, and took him back into her home at Woolsthorpe, intending him to take over the farming of her land. Although very practical and good at making things with his hands, Newton had no interest in farming, and through the intervention of an uncle he was able to go back to school (in nearby Grantham) to prepare for a university education. In 1661 he went up to Trinity College in Cambridge, graduating in 1665.

Although he intended to take up a Fellowship in Cambridge, the university closed temporarily because of the plague, and for the next eighteen months Newton lived in Woolsthorpe, thinking about diverse scientific matters and carrying out experiments on *light* using prisms.

He returned to Cambridge in 1669, became a Fellow of Trinity in 1667, and was appointed Lucasian Professor of Mathematics in 1668, at the age of 26 (he held the post until he resigned from all his Cambridge appointments in 1701). He stayed in Cambridge for the next three decades, largely working alone, both at the scientific work for which he is now remembered and on alchemy experiments which seem bizarre to modern scientific eyes, but were entirely respectable at the time.

Newton was reclusive and highly secretive about his work, developing his ideas in isolation and often getting involved in bitter arguments with other scientists about priority; this may have been a legacy from his childhood, when after three years as an indulged fatherless child he was removed from his mother's everyday care and left to cope with life with his grandmother. He first came to the attention of the scientific community through his invention of the *Newtonian telescope*, on the strength of which he was elected a Fellow of the still relatively new Royal Society. He soon had a violent argument with Robert Hooke (1635–1703) concerning priority in developing the theory of light and colours, and withdrew back into his shell in Cambridge. He waited patiently until Hooke had died before publishing his *Opticks* in 1704, when he could have the last word.

By then, Newton was a household name, the President of the Royal Society, and about to be knighted (in 1705, by Queen Anne). The transformation came about through his work on gravity and mechanics, but this only saw the light of day because of the influence of Edmund *Halley*.

In the 1680s, there was considerable interest in scientific circles in the puzzle of why the *planets* should move in elliptical *orbits*, in line with *Kepler's laws*. On a visit to Newton in 1684, Halley was astonished to be told that Newton had long since solved that problem (in fact, the insight came to him during the plague year in Woolsthorpe); the reason for the elliptical orbits is that gravity obeys an *inverse square law*. Halley persuaded Newton to publish his explanation, and once Newton started writing, what had been intended as a short paper grew into his magnum opus, *Philosophiae Naturalis Principia Mathematica* (*The Mathematical Principles of Natural Philosophy*, known to all scientists simply as the *Principia*). This was published in 1687 – Halley even paid for the book to be published, but this was not entirely an act of altruism and he made a small profit from the success of the book.

The *Principia* is the most important scientific book ever published (even more influential than *A Brief History of Time*), and influenced the course of science over the next three centuries. As well as setting out the law of gravity and the laws of motion, it laid the foundations of the modern scientific method. After it was published, Newton made only minor contributions to science. He became involved in politics, serving on two occasions as a Member of Parliament, and was for a time deeply embroiled in his alchemy experiments. In 1692 he suffered a severe bout of mental illness and depression, which may have been linked to the toxic chemicals he worked with, but he recovered fully.

In 1696, Newton was made Warden of the Royal Mint and moved to London; he was promoted to Master of the Mint in 1698, and carried out the job with ruthless efficiency, pushing through a much needed reform of the currency. It was for this that he was knighted in 1705. Although his Presidency of the Royal Society, which lasted from 1703 until his death, was initially equally beneficial to the Society and helped to establish its reputation, in his declining years Newton lost his grip and the Society declined in influence, while he continued to wrangle bitterly with other scientists, including John *Flamsteed* and the German mathematician Gottfried Wilhelm Leibnitz (1646–1716) about his priority in making certain scientific discoveries. He died, after a painful illness, on 20 March 1727, and was buried in Westminster Abbey, prompting Voltaire to comment that England honoured a mathematician as other nations honoured a king.

In a foreword to a 20th-century edition of Newton's *Optics*, Albert *Einstein* wrote:

> Nature was to him an open book, whose letters he could read without effort. The conceptions which he used to reduce the material of experience to order seemed to flow spontaneously from experience itself, from the beautiful experiments which he ranged in order like playthings and describes with an affectionate wealth of detail. In one person, he combined the experimenter, the theorist, the mechanic and, not least, the artist in exposition. He stands before us strong, certain, and alone; his joy in creation and his minute precision are evident in every word and every figure.

Further reading: R. S. Westfall, *The Life of Isaac Newton*.

newton (N)   Unit of force, named after Isaac *Newton*. 1 newton is the force needed to give a *mass* of 1 kilogram an acceleration of 1 metre per second.

Newtonian mechanics = *classical mechanics*.

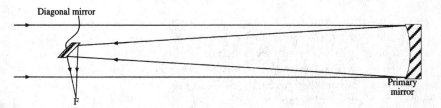

Newtonian telescope. Schematic representation of a Newtonian telescope.

Newtonian telescope   A type of *reflecting telescope* designed by Isaac *Newton* – it was this invention that first drew the attention of the Royal Society to Newton's work. In this system, light passes down the tube of the telescope on to a parabolic mirror, and is reflected up on to a smaller plane mirror which then reflects the light out through a hole in the side of the tube to a focus. The design is still widely used in small telescopes by amateur astronomers, but is unsuitable for large instruments because the focus would be inconveniently located.

Newton's law of gravity   The universal law, discovered by Isaac **Newton**, which says that the force of attraction between any two **masses**, anywhere in the Universe, is proportional to the product of the two masses (m_1 and m_2), divided by the square of the distance (r) between them. With the constant of proportionality written as G, the **gravitational constant**, this becomes $f = Gm_1m_2/r^2$.

Newton's laws of motion   Three fundamental laws describing the dynamical behaviour of objects, published by Isaac **Newton** in his great work, the *Principia*, in 1687. They state:

1. Every object continues in a state of rest or uniform motion in a straight line unless it is acted upon by a force.
2. When a force is applied to an object, this changes the momentum of the object (essentially, its speed) at a rate proportional to the force, and in the direction that the force is applied. In mathematical terms, the force F is related to the mass m of the object and the acceleration a that is produced by the equation $F = ma$.
3. Whenever a force (or, as Newton put it, an action) is applied to an object, the object pushes back with an equal and opposite reaction. So, for example, **gravity** pulls me downwards with a force equal to my **weight**, and the chair I am sitting on pushes back with an equal and opposite reaction, leaving me sitting still, not accelerating downwards (as I would if there were no intervening chair or floor) to the centre of the Earth. And while gravity pulls me towards the centre of the Earth, the **mass** of my body is pulling the Earth towards me with an equal force.

   On the surface of the Earth, these laws do not seem to tie in with common sense, because friction is constantly at work, bringing moving objects to rest. In fact, even friction is operating in accordance with Newton's laws. The laws can be seen at work more clearly inside a spacecraft in **free fall**, and they govern the behaviour of all astronomical objects.
   See **classical mechanics**.

N galaxy   A **galaxy** with a bright central **nucleus**, one of the many kinds of **active galaxy** intermediate between quiet galaxies and **quasars**. They are often **radio sources**.

NGC   Short for New General Catalogue. See **catalogues**.

nitrogen   One of the most common kinds of atomic material found in the Universe – fifth most common in terms of numbers of atoms, and sixth in terms of **mass**. It has an **atomic number** of 7, and the most common **isotope**

has seven **protons** and seven **neutrons** in each **nucleus**. Nitrogen plays an important part in the **carbon cycle**, which provides the **energy** inside massive **stars** on the **main sequence**, and is one of the key **elements** used by life (see **CHON**).

no boundary condition   A way to remove the **singularity** in the **Big Bang** from **models** of the very early Universe. This singularity is like an edge, or boundary, to the Universe – not in space, but in time. The existence of such a boundary naturally raises the question of what lies on the other side of the boundary – what came 'before' the Big Bang. In the late 1970s, Stephen **Hawking** suggested that such questions may be meaningless, because there is not really a boundary at 'time zero', and that the appearance of a singularity in the equations that describe the Big Bang is an artefact of the equations, not a real, physical edge to the Universe.

The model developed by Hawking and his colleagues depends on trying to apply the rules of **quantum theory** to the entire Universe, and is highly mathematical. Its basic features can, however, be understood in terms of a simple analogy.

Think of the entire Universe as being represented by a one-dimensional line, a circle drawn on the surface of a sphere. The Big Bang was a time when the Universe was very small, which we can represent by a tiny circle drawn around one of the poles of the sphere (the pole itself is 'time zero'). As time passes, the Universe expands, and this is represented by successive circles drawn further away from the pole and closer to the equator. Once these circles pass the equator, they get smaller and smaller, and vanish at the other pole in a **Big Crunch** (the **Omega Point**).

But although there is a 'time zero' at the pole where the expanding Universe is born, there is no 'edge' there, any more than there is an edge of the Earth at the North Pole. If you start walking due north near the North Pole, and keep going in a straight line, you will pass over the pole and find yourself heading due south. Similarly, if you could travel backwards in time to 'time zero', and kept going, you would pass right through it and find yourself going forwards in time. You would *not* end up 'before the Big Bang'.

Hawking's model of the Universe is a completely self-contained bubble of **spacetime**, with no edges – or, as he puts it, 'the boundary condition of the Universe may be that it has no boundary'.

non-Euclidean geometry   The rules of geometry that apply to curved surfaces, or to curved **spacetime**. **Euclidean geometry** applies only on flat surfaces and in flat spacetime.

There are two kinds of non-Euclidean geometry, either of which may apply to the spacetime of the Universe, and both of which have counterparts in two-dimensional form. Geometry with positive curvature is represented by the closed surface of a sphere, on which it is possible to travel in a straight line right round the surface and back to the starting point. On such a surface, the angles of a triangle always add up to *more* than 180°, and the exact amount they add up to depends on the size of the triangle.

Geometry with negative curvature corresponds to an open 'saddle surface', shaped like a mountain pass. It extends forever, and the angles of a triangle drawn in the surface add up to *less* than 180° – again, the exact amount they add up to depends on the size of the triangle. In our Universe of three spatial dimensions (plus one of time), positive curvature corresponds to a closed *model* and negative curvature to an open model (see *cosmological models*). If the Universe is non-Euclidean, the volumes of spheres are different from the volumes calculated in accordance with Euclidean geometry; but our Universe is so nearly flat that no such effect is measurable, even by counting the numbers of *galaxies* at different distances from us. See also Bernhard *Riemann*.

nova The explosive outburst of a faint *star* to become, temporarily, a brightly visible object, or 'new' star. In ancient times, the faint stars associated with novae were seldom visible to the naked eye, which is why they were thought to be completely new stars. But with photographic techniques much more sensitive than human eyes, modern astronomers discovered that there is often a faint star visible at the site of a nova in old photographs of that part of the sky. This has made it possible to study precursors of novae, as well as their aftermath, and to develop a good *model* of how they occur.

Most (almost certainly, all) novae are outbursts associated with *white dwarf* stars in *binary systems*, where the companion is a *red giant* in a close *orbit*. They are associated with the *accretion* of material from the companion (via an *accretion disc*), which builds up in a layer on the surface of the white dwarf. The flow of matter on to the white dwarf amounts to about 1 billionth of the mass of the Sun each year. When enough of this layer has built up, the pressure at its base causes an explosive outburst of *nuclear fusion* reactions, blasting the material out into space and causing the star to flare up brightly. The process then repeats – many novae have been seen to flare up repeatedly (such as the star T Coronae Borealis, in 1866 and 1946), and the rest are thought to be repeaters with timescales too long to have been monitored yet by human observers.

During a nova, the star brightens by about 10 *magnitudes* (increasing its brightness 100,000 times in a few days, then fading over a few months) and its surface temperature rises to about 100 million *Kelvin*. The material ejected in each outburst amounts to only roughly one-ten-thousandth of the *mass* of the Sun, but this is an important source of *heavy elements* which enriches the *interstellar medium* – there are about 25 novae each year in an ordinary *disc galaxy* like the *Milky Way*. The energy released in a nova is, however, only 1 millionth of that released in a *supernova*.

Novikov, Igor Dmitrievich (1935–)   Soviet physicist who, among other things, predicted (in the early 1960s) the existence of the *background radiation*, and wrote a paper suggesting that the ideal instrument to use in a search for this *radiation* would be the horn *antenna* at the Bell Laboratories in the United States. This was at the height of the Cold War, and the suggestion was not noticed by western astronomers at the time; when the predicted radiation was discovered using the very system suggested by Novikov, this came as a complete surprise to the discoverers, Arno *Penzias* and Robert *Wilson*.

Novikov studied at Moscow State University from 1959 to 1962, obtaining a first degree in astronomy and (in 1963) his PhD. He later worked at the Institute of Applied Mathematics in Moscow, headed the Department of Relatividic Astrophysics at the Space Research Institute in Moscow in the second half of the 1970s, held two professorships in Moscow in the 1980s and became head of the Department of Theoretical Astrophysics at the Lebedev Institute in Moscow in 1990. Since 1991, he has also been professor of astrophysics at Copenhagen University and at the NORDITA research centre in Copenhagen. As well as undertaking his technical work, he has written several popular books and many articles for a general audience.

Together with Kip *Thorne*, Novikov was instrumental in making the idea of *wormholes* and the investigation of *time travel* respectable at the end of the 1980s.

Further reading: John Gribbin, *In Search of the Edge of Time*.

nuclear fission   The process whereby the *nucleus* of a heavy *atom* splits into two or more parts, releasing energy and two or three free *neutrons* as it does so.

Fission releases energy because the most efficient way to store mass-energy in a nucleus is in the form of nuclei of iron and related *elements* (see *nucleosynthesis*). There is an essentially continuous gradient of energy storage efficiency from the heaviest elements down to iron, so that any process in which heavy nuclei can split to form lighter nuclei (down to iron) is favoured energetically. If nuclei of heavier elements can be split, forming lighter nuclei, this will release energy. In spite of this, many of these heavier nuclei are stable, once formed inside stars, even though it requires an input of energy (from a *supernova*) to make them.

Unstable heavy nuclei, such as those of uranium-235, may fission spontaneously. Fission can also be triggered if a fast-moving neutron strikes an unstable nucleus. Since fission itself releases neutrons from the nuclei that are splitting, if enough of a *radioactive* material such as uranium-235 is put together in one place, the spontaneous fission of one nucleus will trigger the fission of two or more nearby nuclei, each of which triggers the fission of at least two more nuclei, and so on in a cascade called a chain reaction. This is the process that releases energy in a so-called atomic bomb (really a nuclear bomb), and (in a controlled, slow process) in nuclear reactors used to generate electricity. In a bomb, the chain reaction runs away explosively because each fission triggers the fission of several more nuclei. In a nuclear reactor, the rate at which the reaction goes on is controlled by inserting material that absorbs some of the neutrons into the pile of uranium (or other radioactive material), so that on average each fission leads to the fission of just one more nucleus.

The fission of every nucleus in 1 kg of uranium-235 would yield 20,000 megawatt hours of energy (literally enough energy to keep a 20 megawatt power station running for 1,000 years), the same as would be released by burning 3 million tonnes of coal. See also *fission and fusion*.

nuclear fusion   The process whereby light *nuclei* fuse together to make one heavier nucleus, releasing *energy* as they do so.

Fusion releases energy because the most efficient way to store mass-energy in a nucleus is in the form of nuclei of iron and related *elements* (see *nucleosynthesis*). There is an essentially continuous gradient of energy storage efficiency from the lightest elements up to iron, so that any process in which light nuclei can fuse to form heavier nuclei (up to iron) is favoured energetically. Although nuclei of lighter elements can exist in stable forms, if the nuclei are squeezed tightly enough together (for example, by the extreme pressures that exist inside stars), they will over-come the repulsion caused by the positive electric charge on each nucleus and combine with one another, releasing energy as they do so.

Inside stars, fusion releases energy through two main processes, known as the *carbon cycle* and the *proton-proton reaction*. The extreme conditions (for example, temperatures in excess of 10 million *Kelvin*) required to force nuclei to fuse make it extremely difficult to encourage fusion under controlled conditions on Earth, but this is the process which provides the energy in a so-called hydrogen bomb, where the fusion is triggered by the explosion of an atomic bomb (see *nuclear fission*). There are hopes of achieving controlled nuclear fusion in reactors that will provide electricity.

The fusion of a pair of *deuterium* nuclei releases about 2 per cent as much energy as the fission of a single nucleus of uranium-235. Because each uranium-235 nucleus weighs about 100 times as much as a nucleus of deuterium, however, there are 100 times as many deuterium nuclei in 1 kg of deuterium as there are uranium-235 nuclei in 1 kg of uranium-235. So if all those deuterium nuclei fused in pairs, they would generate roughly the same amount of energy as the complete fission of 1 kg of uranium-235, the equivalent of burning 3 million tonnes of coal. See also *fission and fusion*.

nuclear interactions   See *fundamental forces*.

nucleon   Overall name for *protons* and *neutrons*, the two kinds of *elementary particle* found in the *nucleus* of an *atom*.

nucleosynthesis   The process by which *nuclei* of heavier *elements* are built up from nuclei of *hydrogen*, the simplest element. The first steps in this process of *nuclear fusion* took place in the *Big Bang*, converting a large amount of hydrogen into *helium*; but all except the lightest elements were manufactured from hydrogen and helium inside *stars*, in a process known as stellar nucleosynthesis.

Big Bang nucleosynthesis occurred after the Universe had expanded and cooled to the point where the temperature was about 100 billion *Kelvin*. This was about one-hundredth of a second after the outburst from a *singularity*, long after the era of *inflation*. At that time, the Universe was expanding and cooling quickly, but as it did so, primordial *protons* and *neutrons* were processed into one another by *beta decay* and *inverse beta decay*. This process ended when the Universe was about 3 minutes and 46 seconds old, when it had cooled to a temperature of 900 million K. With the proportions of protons and neutrons now 'frozen' as the Universe continued to cool and expand, some 25 per cent of the mass present in the form of *baryons* ended up as helium *nuclei*, with almost 75 per cent as hydrogen nuclei, together with tiny traces of a few other light elements such as *deuterium* and lithium.

This was the raw material for stellar nucleosynthesis, which was described in detail by Fred *Hoyle* and his colleagues in the 1950s (see *B^2FH*; Hoyle was also involved in developing the understanding of Big Bang nucleosynthesis in the 1960s, work so important that, in the words of Sir William *McCrea*, 'it was this paper that caused many physicists to accept hot big bang cosmology as serious quantitative science'). The essential process by which elements are built up inside stars involves adding helium-4 nuclei (also known as *alpha particles*) to existing nuclei, so the elements build up in steps of 4 *atomic mass units*. Helium itself is also manufactured out of hydrogen in *main sequence* stars (see *proton–proton chain* and *carbon cycle*), so there is plenty of helium around, later in the course of *stellar evolution*, to act as the raw material. But there is a bottleneck at the first step in the chain of nucleosynthesis, because the nucleus made by sticking two helium-4 nuclei together, beryllium-8, is extremely unstable, and breaks apart within 10 millionths of a billionth of a second after it forms. It was Hoyle's insight which showed how this bottleneck could be overcome in the *triple alpha process*, which fuses three helium-4 nuclei essentially instantaneously to make one nucleus of **carbon**-12.

Just as a helium-4 nucleus has slightly less mass than two individual protons and two individual neutrons added together, and so represents a more stable form of matter, so each carbon-12 nucleus contains slightly less mass than the three alpha particles which went to make it up. The 'extra' mass is given up as energy, in line with the famous equation $E = mc^2$, every time a carbon nucleus is formed in this way.

Once a star contains carbon-12, the rest is more or less plain sailing. Further fusion reactions take place, as long as the fusion processes continue to release energy. Adding another alpha particle makes oxygen-16, and successive fusion processes manufacture neon-20, magnesium-24 and eventually silicon-28. Along the way, **radioactive decay** processes that eject electrons or (more rarely) positrons from the nuclei form the other elements and **isotopes** – but the elements with nuclei that contain the equivalent of whole numbers of alpha particles remain the most common. The ultimate step in this chain occurs when pairs of silicon-28 nuclei combine to form iron-56 and related elements such as nickel-56 and cobalt-56. These 'iron peak' elements are the most stable of all, and to make heavier elements still it is necessary to put energy in to force the nuclei to fuse together.

One way to think of this is in terms of a valley representing the energies locked up in each nucleus, and therefore the stability of each nucleus. At the bottom of the valley, with the least energy per **nucleon**, are the iron peak elements ('peak', even though they are at the bottom of the valley, because they are relatively common, and stand out as a high point in the list of **cosmic abundances** of the elements). Going up one side of the valley, there are successively lighter elements, with hydrogen at the top. Think of each element as sitting on a little ledge on the side of the valley; it will stay on its ledge if it is not disturbed, but if it is given a small push, it will hop down to the next level, releasing more energy than there was in the original push. Up the other side of the valley, there are elements successively heavier than iron, such as lead and uranium, each sitting on its own ledge. But to make an element on a lower ledge move up to a higher ledge (converting it into a heavier element) requires a lot of energy. Given a chance, many of the very heavy elements will prefer to split apart, in the process known as **nuclear fission**, giving up excess energy and jumping back down towards the valley floor.

Elements heavier than iron can be built up inside stars when energetic neutrons penetrate into the nuclei and stay there. There are always some neutrons around, by-products of the various fusion reactions going on inside the star, and this build-up of heavier elements can go on slowly and steadily, through the **s-process**, as a result. Once again, the 'new' nucleus

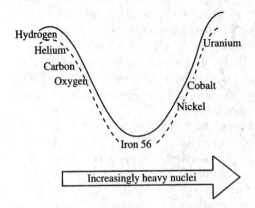

Nucleosynthesis. The 'valley of stability'. The nuclei of iron-group elements store energy more effectively than the nuclei of either lighter elements or heavier elements. So energy is released either when lighter nuclei fuse or when heavier nuclei fission.

produced may itself be stable, or it may eject a positron to become a different stable nucleus. This slow process of neutron capture can build up elements from iron-56 to bismuth-209, but in a process reminiscent of the beryllium bottleneck, if bismuth-209 captures a single neutron, it breaks up through *alpha decay*. All of this activity is going on inside *red giants* with masses less than about nine times the mass of our Sun.

Heavier elements still, as well as many neutron-rich nuclei between iron-56 and bismuth-209, are manufactured by a much more rapid process of neutron capture (the *r-process*) when there is a flood of high-energy neutrons available. This happens during *supernova* explosions, when gravitational energy from the collapsing core of the star drives the fusion reactions 'uphill', with two or more neutrons being captured by a nucleus in quick succession, followed by chains of beta decay. Many isotopes are produced by both processes. A handful of stable, slightly neutron-rich nuclei are produced only by the r-process and subsequent beta decays. And just 28 isotopes can be produced only by the s-process. Both processes end for very massive elements, where if the nuclei do form inside a supernova, they are quickly split apart either by alpha decay (emitting a helium-4 nucleus) or by nuclear fission (producing two roughly equal 'daughter' nuclei, each with about half the mass of the one that splits).

It may seem incredible that astrophysicists can describe these processes that go on inside stars in such detail, but the 'cross-sections' for the individual nuclear reactions involved are determined from studies using particle accelerators here on Earth, and the predictions of the *models* based on these cross-sections closely match the observed cosmic abundances of the elements. The implication is that we have a very good understanding of both what went on between the first hundredth of a second and 3 minutes 46 seconds after the birth of the Universe, and what goes on inside main sequence stars, red giants and supernovae today.

nucleus The central part of an *atom*, made of *protons* and *neutrons*, in which almost all of the *mass* of the atom is concentrated. In an atom, the nucleus is surrounded by a cloud of *electrons*. The size of the nucleus in relation to the electron cloud is roughly that of a grain of sand compared with the Albert Hall. Nuclei can also exist independently under extreme conditions such as those at the heart of a star.

The term nucleus was deliberately chosen by physicists in imitation of the use of the same word in biology to describe the central part of a living cell. In a further extension of the term, astronomers also use 'nucleus' to refer to the central part of a galaxy, and to the central part of a comet.

Nuffield Radio Astronomy Observatory See *Jodrell Bank*.

Oberon One of the larger *moons* of Uranus, discovered by William *Herschel* in 1787. It has a highly cratered surface and a diameter of 1,524 km, and orbits Uranus at an average distance of 582,600 km. Its mass is about 4 per cent of that of our Moon.

occultation Event in which one astronomical body passes in front of another. A *solar eclipse* is an example of an occultation, when the Sun is temporarily obscured by the Moon. Occultations of stars and other more distant objects by the Moon and planets also occur.

Ockham's razor A precept developed by the English logician and philosopher William of Ockham (about 1285 to 1349). He said that 'entities ought not to be multiplied except of necessity'. In modern language, if there are two possible explanations for something, and one explanation is simpler than the other, then the simpler explanation should be preferred. This has proved a useful (though not infallible) guide in the development of science.

OGLE Acronym for Optical Gravitational Lensing Experiment, a search for *MACHOs* by a joint Polish and American team of astronomers, using a *charge coupled device* detector on the 1 m *reflecting telescope* at the *Las Campanas Observatory*. Headed by Bohdan Paczynski, of Princeton University, who was the person who first suggested (in 1985) the possibility of observing gravitational microlensing by *dark matter* in our own *Galaxy*.

Olbers, Heinrich Wilhelm Matthäus (1758–1840) German amateur astronomer and mathematician (he worked as a doctor), who publicized the puzzle of why the sky is dark at night, developed an improved method of calculating the *orbits* of *comets*, and discovered two of the *minor planets*.

Olbers was born at Ardbergen, near Bremen, on 11 October 1758, the son of a Lutheran minister. He was educated locally, and showed a keen interest in mathematics and astronomy, but went to the University of Göttingen in 1777 to study medicine. There, he also attended lectures in mathematics and astronomy, and while still a medical student in 1779 he used his own observations of a comet discovered by Titius **Bode** to calculate the orbit of the comet. After further studies in Vienna, he qualified as a doctor in 1781, and settled in Bremen where he practised successfully until he retired in 1823. But while working as a doctor, he had built an observatory incorporating several **telescopes** on the upper floor of his house, where he carried out regular observations.

In 1796, Olbers discovered a comet of his own, and used a new method that he had devised to calculate its orbit. The success of this technique, which became the standard method used throughout the 19th century, established Olbers as a well-known figure in astronomy.

At that time there was great interest in the 'gap' in the Solar System between the orbits of Mars and Jupiter, where **Bode's Law** said that there ought to be a **planet**. The first minor planet (or **asteroid**) was found in this gap by the Italian Giuseppe Piazzi in 1801, and named Ceres. But Piazzi lost track of the object, and when the German Karl Gauss (1777–1855) calculated its orbit from Piazzi's observations, Olbers was the first astronomer to rediscover it, in January 1802.

While watching Ceres, Olbers discovered a second minor planet, Pallas, later that year; in 1807 he found another, Vesta. He also discovered four more comets, and calculated the orbits of eighteen comets found by other people. He was the first person to suggest that the cloud of material that forms the tail of a comet is produced from the cometary nucleus and pushed away by the Sun. He was also interested in the influence of the Moon on the weather, and the origin of **meteorites**.

But Olbers is most famous for his discussion of a puzzle that now bears his name, *Olbers' Paradox*, although he was not, in fact, the first person to think of it. In 1823, the year that he retired from medical practice at the age of 64, Olbers published his discussion of the puzzle of the dark night sky, and suggested that it could be resolved if there were a thin gruel of matter in the form of gas and dust spread between the stars. He argued that such clouds would block light and heat from distant stars, not realizing that in doing so eventually the clouds would get so hot that they would themselves radiate as brightly as any star.

Olbers died in Bremen on 2 March 1840.

Olbers' Paradox   The puzzle of why the sky is dark at night. This is only a puzzle if the Universe is eternal and unchanging, which was widely assumed to be the case until the 1920s. The real mystery is why nobody before then turned the 'paradox' on its head, and used the darkness of the night sky as a reason to argue that the Universe must have been born a finite time ago.

The term 'Olbers' Paradox' was popularized as a name for this puzzle by the cosmologist Hermann **Bondi** in the 1950s, in honour of Heinrich

Olbers, a 19th-century German astronomer who wrote a landmark paper discussing the puzzle. But just as it is not really a paradox, so the puzzle did not originate with Olbers. Like *Mach's Principle*, discussion of the problem goes back much further than the contribution of the person whose name it bears.

The puzzle can be stated very simply. It rests upon three assumptions: that the Universe is infinitely big, that it is eternal and unchanging, and that it is filled with *stars* (in the modern version of the puzzle, filled with *galaxies*) more or less the same as the stars of the *Milky Way* (or galaxies like the Milky Way itself). In that case, when we look out into the Universe we ought to see stars (galaxies) in every direction. Every single 'line of sight' ought to end on the surface of a star; so why do we see darkness in the gaps between the stars? The whole sky should be a blaze of light.

You can see why this is a puzzle by imagining yourself standing inside an infinitely large forest. It doesn't matter how far apart, or close together, the trees in the forest are; if it is really infinitely big, then wherever you look in a gap between the trunks of the nearer trees you will see the trunk of a more distant tree.

This is a simple version of the puzzle, the kind astronomers refer to as a 'hand-waving' argument. But the puzzle can be brought on to a secure mathematical footing, when it strikes with even greater force. Imagine that the Earth is at the centre of a large sphere encompassing many stars (galaxies). We can envisage a thin shell around this sphere, like the skin of an orange, that contains a certain number of stars (galaxies), all at the same distance from us. If the numbers involved are big enough (easy to arrange if the Universe is infinite), then we can say that, in round terms, each star (galaxy) contributes the same average brightness to the appearance of the night sky here on Earth. But the apparent brightness of each star (galaxy) is inversely proportional to the square of its distance from us. So the whole shell of stars (galaxies) contributes a brightness equal to the number of stars (galaxies) in the shell, multiplied by the average brightness and divided by the square of the distance to the shell.

If individual stars (galaxies) all have about the same absolute brightness, but they each look fainter the further away they are, you might guess that shells of this kind, dotted with stars (galaxies), will appear fainter if they have a larger radius. But a bigger shell will also contain more stars (galaxies). In fact, if the stars (galaxies) are distributed uniformly throughout the Universe, the number of stars (galaxies) in each shell turns out to be proportional to the square of the distance to the shell, because the surface area of a sphere goes as the square of its radius. This *exactly* compensates for the diminished apparent brightness of each star (galaxy) in the shell. So every spherical shell contributes the same brightness to the night sky.

The accurate calculation is actually much stronger than the hand-waving argument. With an infinite number of shells, each contributing the same amount of light, the sky should be infinitely bright. The best you can hope for is to argue that light from nearby stars blocks out some of the light from more distant stars – which still leaves the prediction that the sky should be as bright all over as the *surface* of a star like the Sun. The puzzle becomes not so much why the sky is dark at night, but why it is so dark even in the daytime.

The extent to which you find this paradoxical depends on how strongly you hold to the three basic assumptions – which tells us more about the culture of our recent ancestors than about the structure of the Universe. It was the Englishman Thomas **Digges**, writing in 1576, who discarded the Ptolemaic idea of the stars being attached to a single crystal sphere surrounding the Earth, and distributed the stars, in his imagination, into an endless infinity of space. Digges introduced the concept of infinity into the modern picture of the Universe (although in ancient times **Democritus** had considered the possibility of infinite space), and he also realized the need to explain why, in an infinite universe, the sky should be dark at night. He believed that the sky was dark because the more distant stars were simply too faint to be seen, an acceptable notion in the 16th century, but one which doesn't stand up once you work out how the light from spherical shells actually adds up (the accurate calculation uses Isaac **Newton**'s calculus, and Newton wasn't even born until 1642).

In 1610 the puzzle was investigated by Johannes **Kepler**, who seems to have been the first person to realize (even without the aid of calculus!) that the darkness of the night sky directly conflicts with the idea of an infinite universe filled with bright stars. He saw the darkness of the night sky simply as evidence that the Universe is finite in extent – he said, in effect, that when we look through the gaps between the stars we see a dark wall that surrounds the Universe. On this picture, instead of standing in an infinite forest you are standing in a small copse, and when you look through the gaps between the nearby tree trunks you see the world outside the copse. A century later, Edmund **Halley** also investigated the mystery, but he went back to the incorrect notion that more distant stars are simply too faint to be seen.

The first person to formulate the puzzle of the dark night sky in more or less the form outlined here was the Swiss astronomer Jean-Phillippe Loÿs **de Chésaux**, later in the 18th century. After a careful step-by-step calculation based on the sizes of stars and their separation, he estimated that there would be a star visible in every direction we look into space, provided that the Universe is (in modern terms) 10^{15} *light years* (1 million billion light years) across. Unlike Halley and Digges, de Chésaux realized that the geometry of the situation ensures that the faintness of distant stars is exactly compensated for by their increased numbers. But his 'explanation' of the puzzle was no better than theirs – he suggested that empty space simply absorbs the *energy* in the light from distant stars, so that the light gets fainter and fainter as it travels through the Universe.

And then along came Olbers, who discussed the problem in the 19th century, and came to a similar conclusion, arguing that the light from distant stars is absorbed in a thin gruel of material between the stars. What he failed to appreciate was that this would heat the gruel up, until it was radiating as much energy as it received – until, in fact, it was as bright as the stars themselves. So – what *is* the solution to the puzzle?

Clearly, at least one of the three basic assumptions must be wrong. We still cannot say definitely whether or not the Universe is infinite in extent; it probably is full of galaxies like the ones we can see; but we do know that it is *not* eternal and unchanging. The Universe as we know it began in a *Big Bang* some 15 billion years ago, and it is changing as *spacetime* expands, carrying the galaxies ever further apart from one another. A partial explanation of why the sky doesn't blaze with light is that the Universe is expanding and evolving as time passes. This produces the *redshift* in the light from distant galaxies, which in a very modest way does cause a weakening of the light on its journey superficially like the dimming proposed by de Chésaux.

Alas, if the Universe were infinitely old and had been expanding forever, with new galaxies being created to fill the gaps between the old ones as they move apart (as required by early versions of the *Steady State hypothesis*), the redshift alone would not save us from the implications of Olbers' Paradox. But what do we see when we look (with radio 'eyes') into the gaps between the stars and galaxies? What we detect there is the faint hiss of the cosmic *background radiation*, the equivalent of 'light' with a temperature of 2.7 K. That is the highly redshifted electromagnetic radiation from the time when the Universe was about 300,000 years old and was filled with radiation as hot as the surface of the Sun is today. If the Universe had not expanded since then, all of space would still be as hot as that, and would blaze as brightly as the surface of a star – so the redshift associated with the expanding Universe is one reason why the sky is dark at night, even though Olbers and his predecessors had no inkling of the existence of the Big Bang fireball.

But the reason why starlight has not been able to fill the Universe with energy (the original Olbers' Paradox) is simply that there hasn't been enough time for it to do the job. Light from a galaxy 50,000 light years away, say, takes 50,000 years to reach us, and in a Universe 15 billion years old we can only 'see' galaxies out to a distance of 15 billion light years (even if they formed immediately the Universe was born). Even if the Universe is infinite, light from more distant galaxies has not had time to reach us. In round terms, de Chésaux's calculation says that the Universe would have to be at least 10^{15} light years across *and* be at least 10^{15} light years old in order for every line of sight to end on a star (the numbers are bigger still if you take account of the way stars are clumped together in galaxies). That is roughly 1 million times older than the time that has elapsed since the Big Bang. If there was a definite moment of creation, and it was recent enough, then there is no puzzle about the darkness of the night sky.

The puzzle is why Newton and his contemporaries didn't hit on this resolution of the 'paradox'. The finite speed of light had been determined by the Dane Ole Rømer in 1676, and was well known to Newton, who mentioned it in his *Opticks*, published in 1704. When Halley read two papers on the puzzle of the dark night sky to the Royal Society in 1721, Newton was in the chair. Yet neither he nor anyone else pointed out that the puzzle could be resolved by assuming that no stars existed until relatively recently. This oversight is all the more baffling because at the time the Church taught that the Creation had taken place in 4004 BC. Any astronomer of Newton's day could have immediately calculated that no light from stars more than (1721 + 4,004) light years away had had time to reach the Earth – and a sphere of space with a radius of less than 6,000 light years is far too small (as de Chésaux's calculation shows) to hold enough stars to make the night sky bright. Perhaps the failure of Newton, Halley and their contemporaries to point this out actually indicates how little faith they had in the official date for the Creation.

So who was the great astronomer who first realized that by looking out into space and back in time we see the darkness that existed before the stars were born? None other than Edgar Allan Poe (1809–49), best known today for his short stories, who was also a keen amateur scientist and who delivered a lecture setting out the resolution to Olbers' Paradox in February 1848, just a year before he died at the age of 40. The lecture was published, later in 1848, as an essay entitled *Eureka*, in which he wrote of the dark gaps between the stars (which he called voids): 'The only mode, therefore, in which ... we could comprehend the voids which our telescopes find in innumerable directions, would be by supposing the distance of the invisible background so immense that no ray from it has yet been able to reach us at all.'

Nobody took much notice of these amateur speculations from a poet who died before he had time to promote his argument. Nor did anyone take much notice when the problem of the dark night sky was discussed in print by the Irish scientist Fournier d'Albe in 1907, who specifically said, 'if the world was created 100,000 years ago, then no light from bodies more than 100,000 light years away could possibly have reached us up to the present'. D'Albe himself had drawn on the ideas not of Poe but of Lord **Kelvin**, who had published his thoughts on the darkness of the night sky in 1904 in a volume of lectures, where they laid forgotten until they were dug out by Edward Harrison, of the University of Massachusetts, in the 1980s.

Even after the discovery that the Universe is expanding and must have had an origin at a definite moment in the past, nobody took full account of the importance of the insight provided by Olbers' Paradox until Harrison became intrigued by the history of the idea, and dug out the full story of it over a period of many years.·But it is still of more than just historical interest. Anybody from Newton to Edwin **Hubble** could have used the evidence of their own eyes to tell them that the Universe was born a finite time ago. And it is still true that one of the clearest pieces of evidence that there really was a Big Bang can be seen (literally!) with your own eyes, by looking up at the dark night sky.

Further reading: Edward Harrison, *Darkness at Night*.

Omega Point   Another name for the **Big Crunch**, in which a closed universe (see **cosmological models**, **density parameter**) collapses into a **singularity**, mirroring the way in which our Universe was born out of a **Big Bang**. The Omega Point would mark the end of time as we know it, but not necessarily an 'edge' to the Universe (see **no boundary condition**).

In the collapse towards the Omega Point, **galaxies** begin to overlap about a year before the Big Crunch. At about this time, the **background radiation** becomes hotter than the inside of a **star**, so stars break up into a hot soup of particles. An hour away from the Big Crunch (based on the rate at which **spacetime** is shrinking in the collapsing universe), the supermassive **black holes** at the hearts of galaxies would begin to merge. But at that moment, according to **models** developed by Werner Israel and colleagues at the University of Alberta, in Canada, the intense **gravitational fields** associated with the singularities in the interiors of the black holes would make the whole of the rest of the collapse take place in the **Planck time**, 10^{-43} seconds, instead of taking an hour. This is a collapse even more dramatic than the rapid expansion of the Universe which took place during **inflation**.

The most probable result of this catastrophic final collapse is that the superdense universe will 'bounce', beginning a new phase of expansion. The Omega Point becomes a new Big Bang, and our Big Bang may have been the Omega Point of another universe (or, if you like, of a previous cycle of the same universe; see *oscillating universe model*).

Further reading: John Gribbin, *In Search of the Edge of Time*.

Oort, Jan Hendrik (1900–92) Dutch astrophysicist whose main interest was in investigating the structure of *galaxies*, including our own. He was one of the pioneers of the use of *radio astronomy* to study the *Galaxy*, using the characteristic emission of *hydrogen* at a wavelength of 21 cm, and in the early 1950s he proposed the existence of what is now known as the *Oort cloud* of *comets*, developing this from an idea put forward in the 1930s by the Estonian Ernst Öpik (1893–1985).

Oort cloud A cloud of *comets*, surrounding the Sun in a spherical shell far beyond the orbit of the outermost planet. The comet cloud (also known as the Öpik-Oort cloud) lies between 30,000 and 100,000 Astronomical Units from the Sun, reaching roughly halfway to the nearest star. It may contain as many as a trillion (thousand billion) comets, with orbits oriented at random, not all in the plane of the orbits of the planets. The comets we see in the inner part of the Solar System are occasional visitors from the cloud, disturbed by the gravitational influence of nearby stars, falling in past the Sun on orbits that take millions of years to complete. Some of these visitors are then captured into short-period orbits and pass repeatedly through the inner Solar System – Halley's Comet is the classic example.

The total amount of matter in the Oort Cloud is probably only a few times the mass of the Earth. It is thought that the comets represent debris left over from the formation of the Solar System, and formed near the positions of the giant planets before being ejected into their present orbits by the gravitational influence of Jupiter.

Öpik-Oort cloud See *Oort cloud*.

open cluster A type of *cluster of stars* in which the stars are more spread out than in a *globular cluster*. They may contain a few dozen or up to several thousand stars in a region 1 *parsec* or so across. Open clusters contain young, hot *Population I* stars that have recently formed in the disk of the *Galaxy*. Sometimes known as galactic clusters. (See p. 364.)

open universe See *cosmological models*.

Oppenheimer, (Julius) Robert (1904–67) American physicist (born in New York on 22 April 1904) who worked on *quantum theory* in the 1920s and 1930s, and who is best known for the part he played in the production of the atomic bomb – he was the first director of the Los Alamos National Laboratories, where the bomb was developed.

Oppenheimer, who had long been known to have left-wing sympathies, was opposed to the development of the hydrogen bomb, and as a result his political loyalty to the United States was questioned during the McCarthy era; he lost his security clearance and could no longer work on secret projects. Despite this stigma, he served as director of the Institute

Open cluster. An open cluster of stars, NGC 3293 in the constellation Carina. The 'spikes' on the star images are produced by the imaging apparatus.

for Advanced Study in Princeton from 1947 to 1966, staying on there as a professor until his death from cancer on 18 February 1967. The award of the Enrico Fermi Prize on the recommendation of the US Atomic Energy Commission in 1963 was a belated attempt to atone for his victimization during the witchhunts of the previous decade.

It is much less well known that Oppenheimer contributed significantly to the development of a theoretical understanding of *degenerate matter*. Intrigued by the idea of 'neutron cores' proposed by Lev *Landau*, Oppenheimer and his student Robert Serber carried out an improved version of Landau's calculations. Then, working with another student, George Volkoff, Oppenheimer showed (in a paper published in 1939) that there is an upper mass limit (the *Oppenheimer–Volkoff limit*) above which no stable *neutron star* can exist. In a paper published in September 1939 in collaboration with yet another student, Hartland Snyder, Oppenheimer gave what is now regarded as the first clear description of the *astrophysics* (as distinct from the purely mathematical approach of researchers such as Karl *Schwarzschild*) of *black holes*.

There is still no more concise, clear way of expressing the understanding of the ultimate fate of a massive star than their words (*Physical Review*, volume 56, pages 455–9):

When all the thermonuclear sources of energy are exhausted, a sufficiently heavy star will collapse. Unless fission due to rotation, the radiation of mass, or the blowing off of mass by radiation, reduce the star's mass to the order of that of the sun, this contraction will continue indefinitely ... light from the star is progressively reddened ... an external observer sees the star asymptotically shrinking to its gravitational radius.

Partly because of the outbreak of the Second World War in the month this paper was published, it was a quarter of a century before these ideas were followed up, when the discovery of *pulsars* and their interpretation as spinning neutron stars persuaded astronomers that neutron stars really do exist, and that therefore black holes might exist as well.

Oppenheimer–Volkoff limit Limit on the maximum possible *mass* of a stable *neutron star*, determined by Robert *Oppenheimer* and his student George Volkoff in 1939. They showed that stable neutron stars could exist only if they had masses in the range 10–70 per cent of that of our Sun. Stars lighter than this range can only be *white dwarfs* or *brown dwarfs*. For masses greater than the upper limit of this range, as Oppenheimer and Volkoff wrote at the time (*Physical Review*, volume 55, pages 374–81, 'the star will continue to contract indefinitely, never reaching equilibrium'. In other words, it will become a *black hole*. More recent calculations suggest that the upper limit for a stable neutron star, still called the Oppenheimer–Volkoff limit, may be two or three times the mass of the Sun.

opposition The position in its *orbit* around the Sun when one of the *planets* further out from the Sun than the Earth is lies exactly opposite the Sun as seen from Earth (so it is exactly overhead at midnight). See *conjunction*.

optical astronomy Astronomy based on observations made using visible light, essentially the same part of the electromagnetic *spectrum* that our eyes are sensitive to.

optical pulsar Any *pulsar* that can be seen flashing in visible *light*, as well as through its emission of *radio waves*. The best known of these are the *Crab pulsar* and the *Vela pulsar*.

orbit The trajectory followed by any object, including *moons*, artificial *satellites* and *planets*, that is moving in the *gravitational field* of another object. For example, the Moon orbits the Earth, the Earth orbits the Sun, and the Sun orbits around the *Galaxy*. An object orbiting under the influence of a single mass must move in a circle, an ellipse, a parabola or a hyperbola (one of the 'conic' family of curves); the orbits of the planets around the Sun are elliptical, with the Sun at one focus of the ellipse. See *Kepler's laws*.

origin of the elements See *nucleosynthesis*.

origin of the Universe   See *Big Bang*.

Orion molecular cloud   Dense cloud mainly composed of hydrogen (but also containing other *molecules* including carbon monoxide) just behind, but touching, the *Orion nebula*. One of many *molecular clouds* in that part of the *Galaxy*, and probably a stellar nursery. It contains about 500 times as much mass as our Sun, at temperatures up to 100 K.

Orion nebula   A glowing cloud of gas in the *constellation* Orion, just visible to the unaided human eye in the middle of Orion's sword. The cloud is associated with a region of *star formation*, and is lit up by the young stars it contains, making a spectacular sight on astronomical photographs.

The *nebula* is about 400 *parsecs* away (almost on our doorstep by astronomical standards), and is part of a giant *molecular cloud* which covers almost all of the region of the sky outlined by the Orion constellation. The densest parts of the cloud absorb visible *light*, and can only be probed using *infrared* and *radio* techniques; they include hot spots associated with the birth of *stars*. Some of the youngest stars in the nebula, called the Trapezium stars, are only about 1 million years old, and shine very brightly in the *ultraviolet* part of the *spectrum*; it is their *radiation*, absorbed by the gas in the cloud and re-radiated as visible light, which makes the nebula glow. The glowing part of the cloud is an *HII region*.

The Orion nebula is a source of *X-rays*, and contains *Herbig–Haro objects*, a *maser* source and *T Tauri stars*. All of this activity going on so close to home makes the cloud one of the most intensively studied of all astronomical objects.

oscillating universe   See *cosmological models*.

oxygen   One of the most common types of atomic material found in the Universe – the third most abundant, after *hydrogen* and *helium*. It has an *atomic number* of 8, and the most common *isotope* has eight *protons* and eight *neutrons* in each *nucleus*. Oxygen plays an important part in the *carbon cycle*, which provides the *energy* source in massive *stars* on the *main sequence*, and is one of the key *elements* for life (see *CHON*).

Palomar Observatory   See *Mount Palomar Observatory*.

Palomar Sky Survey   A photographic *catalogue* of all of the northern sky and part of the southern sky (as much as can be seen from Los Angeles), made at the *Mount Palomar Observatory* in the 1950s, using a 48-inch *Schmidt camera*.

panspermia hypothesis   The idea that life did not originate on Earth, but was brought here in the form of single-celled organisms riding on a *meteorite*, perhaps produced by the break-up of another *planet*. It was first proposed by the Swedish chemist Svante *Arrhenius* in 1906.

The hypothesis has always roused a great deal of popular interest, although astronomers have tended to remain sceptical about it. The pan-

The Orion nebula, also known as NGC 1976, a great cloud of gas and dust in space where new stars are being born.

spermia idea was revived in a considerably different form by Fred *Hoyle* and his colleague Chandra Wickramasinghe in the 1970s; they suggested that life evolved in *molecular clouds* in space, and was carried to Earth on board a *comet* (or comets) rather than on a meteorite. Another variation on the theme has been espoused recently by Francis Crick, who shared the Nobel Prize with James Watson for their discovery of the structure of DNA. In this version, life originated on another planet somewhere in the *Milky Way* when the Universe was young, and gave rise to an intelligent life form which has deliberately seeded other planets with micro-organisms. See also *life in the Universe*.

parallax The apparent movement of an object across the sky when seen from two different points. Used to calculate the distance to the object by triangulation.

It is easy to see parallax at work. Hold a finger up at arm's length, and close one eye. Now close the open eye and open the one that was closed. Your finger seems to jump sideways compared with the more distant background. In principle, you could work out how long your arm is by measuring the angle across which your finger seems to jump, although there wouldn't be much point in this.

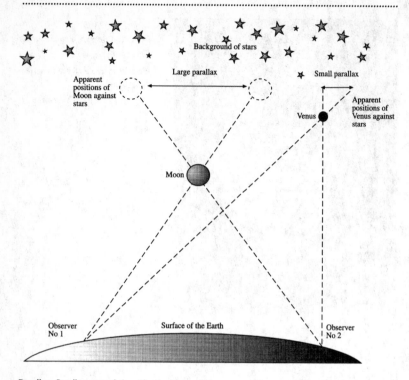

Parallax. Parallax at work (greatly exaggerated!).

There is a great deal of point, though, in doing the same kind of thing for astronomical objects such as the Moon and the ***planets*** of the Solar System. If one observer views the Moon directly overhead, for example, while another observer in a distant observatory sees it on the horizon at the same time, the parallax shift of the Moon between the two observations is 57 ***arc minutes*** relative to the ***fixed stars***, nearly twice the angular diameter of the Moon. By measuring this angle and the distance between the two observers, it is straightforward to work out the distance to the Moon, about 400,000 km, from the geometry of triangles.

Parallax measurements can just be pushed far enough to measure the distance to a few nearby stars, by making observations 6 months apart, when the Earth is on opposite sides of its orbit around the Sun. This baseline is nearly 300 million km long, twice the distance from the Earth to the Sun, and is long enough to produce a measurable shift in the positions of a handful of stars. In the 1830s the technique yielded the distance to 61 Cygni, the first published stellar distance to be determined in this way, as 3.4 *parsecs* (the actual parallax angle measured for this star is just 0.3136 *arc seconds*, which gives you some idea of how difficult the technique is to apply to stars). Together with similar measurements of the distances to *Alpha Centauri* (1.3 parsecs) and *Vega* (8.3 parsecs) at about the same time, this technique provided astronomers with their first direct measurements of the distances to stars, and the first indication of the true *cosmic distance scale*.

Paris Observatory (Observatoire de Paris) French national astronomy research centre, still based at the site in Paris where it was founded in 1667 – now the oldest astronomical observatory still in use. Observations are also carried out from Meudon, near Paris, and the observatory runs a *radio astronomy* site at Nançay.

Parkes Observatory Australia's national *radio astronomy* observatory, in New South Wales. The main instrument is a 64 m dish *antenna*, completed in 1961, now part of the *Australia telescope*.

parsec A measure of distance used by astronomers, equal to 3.2616 *light years* (3.0857×10^{16} m). A parsec is the distance from which the Earth and the Sun would appear to be separated by an angle of 1 *arc second*.

participatory universe See *Wheeler, John*.

particles See *elementary particles*.

Pauli, Wolfgang (1900–58) Austrian-born (in Vienna on 25 April 1900) Swiss physicist who made many contributions to the development of *quantum theory*, including the *Pauli exclusion principle* (which is crucial to an understanding of *white dwarfs* and *neutron stars*), and who first proposed the existence of the *neutrino*.

Pauli exclusion principle An expression of a law of nature which prevents any two *electrons* (or other *fermions*) from existing in exactly the same quantum state.

The principle was formulated by Wolfgang *Pauli* in 1925, specifically to explain the arrangement of electrons in atoms. By then, it had been well established that successively heavier *elements* (starting with hydrogen, the lightest element) have their electrons arranged around the central nucleus in a very well-ordered way. Hydrogen has just one electron, close to the nucleus. Helium has two electrons, each at the same distance from the nucleus. Lithium has three electrons, the first two at the same distance from the nucleus as the helium electrons (in the same 'electron shell', in the jargon of *quantum theory*), with the third slightly further out from the nucleus. The first electron shell can contain only two electrons, the second up to eight (corresponding to neon, with ten electrons in all), and there are similar limitations for still heavier elements, with electron shells wrapped like onion skins around the nuclei.

Pauli found that the number of electrons in each shell exactly matches the number of different combinations of quantum properties allowed for an electron in that shell. For example, in the innermost shell each electron has the same energy, but the two electrons spin in opposite senses, so they are (in principle) distinguishable from one another. The quantum rules are more complicated for outer shells, but in each case every electron has a unique set of quantum 'labels'.

Without the exclusion principle, all atoms would collapse to the size of a hydrogen atom, and chemistry would become far too simple to allow the existence of complex *molecules*, including the molecules of life. At a deeper level, it is the exclusion principle that prevents *white dwarfs* from collapsing, because the electrons in these *stars* cannot be squeezed into the same state (see *degenerate matter*). Only when the electrons physically combine with *protons* to make *neutrons* can the star collapse further, becoming a *neutron star*, which is itself held up because the same exclusion principle prevents all the neutrons from settling into the same state.

But the Pauli exclusion principle is obeyed only by fermions. It turns out that *bosons* (such as *photons*) *can* all crowd into the same quantum state. This important distinction between fermions and bosons can be likened to two different theatre audiences. One, full of well-behaved opera-goers, consists of people each sitting in their own numbered seat; the other, a collection of enthusiastic rock fans, ignores the seats and crowds to the front of the stage, jammed together in the same 'state'. Fermions are opera-lovers, bosons are rock fans.

Further reading: John Gribbin, *In Search of Schrödinger's Cat*.

Payne-Goposchkin, Cecilia Helena (1900–79)   British-born (in Wendover, Buckinghamshire, on 10 May 1900) American astronomer who worked with Harlow *Shapley* at Harvard and became, in 1956, the first female professor of astronomy at Harvard University. Her main interests were in stellar astronomy, and her observations in the 1920s helped to establish the dominance of hydrogen and helium in the composition of *stars*.

Peebles, (Phillip) James Edwin (1935–)   Canadian-born American cosmologist who has been especially interested in the physics of the *Big Bang*, the large-scale structure of the Universe, and the *background radiation*.

Peebles was born in Winnipeg, Manitoba, on 25 April 1935, and graduated from the University of Manitoba in 1958. He then studied under Robert *Dicke* at Princeton University for his PhD, which he received in 1962. He has stayed at Princeton for the rest of his career, and became professor of physics there in 1965 and Einstein Professor of Science in 1985. His first important piece of research was as a member of Dicke's team at Princeton in the mid-1960s, working on the project to detect the background radiation that was pipped at the post by Arno *Penzias* and Robert *Wilson* at the Bell Laboratories. He went on to make calculations of the amount of *helium* and *deuterium* that should have been produced in the Big Bang.

While continuing to investigate the physics of the very early Universe, Peebles has also studied the distribution of *galaxies* in order to investigate the way in which irregularities have grown as the Universe has expanded. He has written two major books that are now regarded as standard texts in cosmology – *Physical Cosmology* (1971) and *The Large Scale Structure of the Universe* (1979).

Early in the 1970s, working with Jeremiah Ostriker, Peebles found that the visible disc of stars in a galaxy like our *Milky Way* is extremely unstable, and would break up in less time than it takes for such a *disc galaxy* to rotate once. They found that the only way in which such discs could be stabilized to last for as long as the known age of our *Galaxy* would be if each disc is embedded in a spherical *halo* of *dark matter*, holding the bright stars of the disc in its gravitational grip. The evidence suggests that the total mass of this dark matter is about ten times that of the bright stars, in a typical disc galaxy.

In 1979, Peebles and Dicke drew the attention of the astronomical community to the *flatness problem*, the extraordinary coincidence that the rate at which the Universe is expanding sits exactly on the dividing line between allowing the expansion to continue forever (an *open universe*) and allowing gravity eventually to halt the expansion and bring about a collapse into a *Big Crunch* (a *closed universe*). This is like finding a perfectly sharpened pencil balancing on its point for millions of years. It was the work of Dicke and Peebles drawing attention to this peculiarity that lead Alan *Guth* to develop the theory of *inflation*.

Penrose, Roger (1931–) English mathematical physicist who has made major contributions to the development of the theory of *black holes*.

Penrose was born in Colchester, in Essex, on 8 August 1931. His father was an eminent geneticist, Lionel Penrose. Roger studied initially at University College, London, and received his PhD from the University of Cambridge in 1957. He held several short-term posts in London, Cambridge, Princeton, Syracuse and Texas universities, before settling as professor of applied mathematics at Birkbeck College, in London, in 1966. In 1973 he was appointed Rouse Ball Professor of Mathematics at the University of Oxford, where he has spent the rest of his career.

His first major contribution to the study of black holes, made in 1965, was to prove that anything which falls into a non-rotating black hole must be crushed into the *singularity* at the centre of the black hole – there is no way for material to travel in an *orbit* that sends it looping past and around the singularity. With Stephen *Hawking*, he then showed that the collapse of matter into such a black hole singularity exactly mirrors the expansion of the Universe out of the **Big Bang**, and that therefore there must have been a singularity at the birth of the Universe.

Since the laws of physics break down at a singularity, Penrose later developed the idea of **cosmic censorship**, which says that every singularity must be surrounded by an **event horizon** so that it cannot interact with the Universe at large, and the breakdown of physics is tastefully concealed from our sight. Other researchers, however, believe that **naked singularities** can exist, and point out that the singularity at the birth of the Universe represents just such a phenomenon.

At the end of the 1960s, Penrose described a process whereby energy could be extracted from a rotating black hole (a **Kerr black hole**). The process of energy extraction involves an object falling in on an orbit close to the black hole and then splitting in two; one part of the object falls into the hole, while the other escapes with more energy (including mass energy) than the original object. This was one of the ideas that led Hawking to the discovery that even non-rotating black holes emit energy in the form of **Hawking radiation**. Penrose also devised a method of representing events going on in and around black holes, using maps of **spacetime** which are now known as **Penrose diagrams**.

He has also been involved in attempts to find a mathematical way to unite **gravity** and **quantum theory**, using his own mathematical approach known as twistor theory. In 1990 his book *The Emperor's New Mind* became a best seller, in spite of containing complex technical arguments and many equations. In it, Penrose argues that consciousness is in part a quantum phenomenon, and that no artificial intelligence will ever be able to duplicate the phenomenon of human consciousness. He has also had a long interest in geometry in two and three dimensions. In the 1950s, together with his father, he devised a series of shapes which could be represented in two dimensions but could not be constructed in three dimensional form; some of these designs were used in drawings by the Dutch artist M. C. Escher. He later (in the 1970s) developed a tiling system, known as Penrose tiles, which completely covers a flat surface with just two kinds of rhombus-shaped tile, leaving no gaps but with no regularly repeating pattern. This later turned out to provide important insights into the nature of crystals.

Penrose diagram   A kind of **spacetime** diagram, particularly useful in studying what goes on inside **black holes**, developed by Roger **Penrose** from the **Minkowski diagram**.

As with the standard Minkowski diagram, time is represented 'up the page' and space is represented 'across the page', with the paths of **light** rays represented by lines at 45 degrees to the vertical. But in a Penrose diagram, distant regions of space and time are all mapped into a single diamond shape on a Minkowski-type diagram, so that the entire past and future

Penrose diagram. In a Penrose diagram, the entire Universe ouside a black hole is represented by the diamond on the left; the interior of the black hole is represented by the inverted triangle on the right. There is no way out of the black hole.

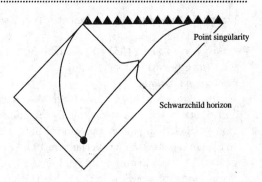

Point singularity

Schwarzchild horizon

Penrose diagram. The full equations actually tell us that another entire universe may have been born out of the Big Bang if it was indeed a singularity in the past. Observers from both our Universe and the other universe could meet inside a black hole – but in order to travel from one Universe to the other (A to B) you would have to travel into the black hole and out again, faster than the speed of light.

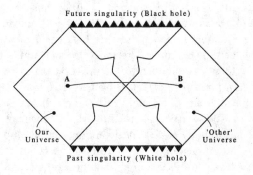

Future singularity (Black hole)

A   B

Our Universe

'Other' Universe

Past singularity (White hole)

history of the Universe can be represented on a single sheet of paper. This is purely a mathematical ploy, equivalent to the way in which the spherical surface of the Earth can be mapped on to a flat sheet of paper using, for example, Mercator's projection. Although the process does (like Mercator's projection) distort the picture somewhat (so that, for example, the region of spacetime inside the black hole that is being mapped gets half as much space in the diagram as the entire outside Universe), the Penrose diagram accurately shows how different regions of spacetime are connected to one another, and which regions can be reached from any chosen point without travelling faster than the *speed of light*.

Penrose diagrams can be used to illustrate the way in which spacetime is distorted in the region of rotating or electrically charged black holes, and show graphically how it may be possible in principle to travel through *wormholes* to other universes, to experience *time travel*, and for *white holes* to exist in the Universe(s).

Penzias, Arno Allan (1933–)   German-born American radio astronomer who (together with Robert *Wilson*) was the accidental discoverer of the *background radiation*.

Penzias was born in Munich on 26 April 1933, but was taken to the United States when he was a child by his family, who were Jewish refugees from Nazi Germany. They were one of the last families to escape (initially to England) before war was declared in 1939. He studied at the City College of New York, graduating in physics in 1954. After two years in the Army Signal Corps he went on to Columbia University in New York, where he received his master's degree in 1958 and his PhD in 1962. The year before the award of his doctorate, Penzias took up a post at the Radio Research Laboratories of the Bell Telephone Company, at Crawford Hill, near Holmdel in New Jersey. He has remained with the company, rising to become Executive Director of Research and Communication Sciences and then Vice President of Research. At the same time, he has held a series of posts at Princeton University, Harvard College Observatory, the State University of New York and Trenton State College, in New Jersey. Most of his work has dealt with the development of instrumentation for radio astronomy and *satellite* communications, and with related areas such as atmospheric physics. But he gained fame, and won the Nobel Prize (in 1978), for his first important piece of research.

At Crawford Hill, Penzias shared the single radio astronomy post offered by the Bell Labs with Robert Wilson, each of them spending half their time on other projects. They were allowed to use an *antenna* designed for work with early communications satellites, and found a source of cosmic radio noise which they could not explain, but which Robert *Dicke* and colleagues at nearby Princeton University interpreted as the echo of the *Big Bang*.

Further reading: Jeremy Bernstein, *Three Degrees Above Zero*.

perfect cosmological principle   See *cosmological principle*.

perihelion   The point in its *orbit* at which a *planet* or other body is at its closest to the Sun. The Earth is at perihelion on 3 January.

period–luminosity relation   The relationship between the period and *absolute magnitude* of a *Cepheid variable*, originally discovered by Henrietta Swan *Leavitt*. There are actually two period–luminosity relations, one for *Population I* Cepheids (also known as classical Cepheids) and one for *Population II* Cepheids (also known as W Virginis stars). See also *cosmic distance scale*.

phase transition   The change of a physical system from one state to another at the same temperature. For example, ice changes to liquid water at 0 °C, and water changes to vapour at 100 °C. Ice, liquid water and water vapour are the same chemical substance (all made of *molecules* of H_2O) in three different phases.

In the very early Universe, the change from a state in which *quarks* roamed freely to a state in which quarks are bound up inside *hadrons* (such as protons and neutrons) was a phase transition (see *quark–hadron phase transition*). In a sense, hadrons are 'frozen' quarks, the icy equivalent of liquid quarks. Earlier still, there were phase transitions associated with the splitting off of the *fundamental forces* from the single original force predicted by the *grand unified theories*.

When water vapour condenses to make liquid water, or when liquid water freezes to make ice, it gives up *energy* in the form of latent heat, because the state it is changing into stores less energy than the state it is changing from. In an analogous way, phase transitions in the very early Universe provided the energy which made the fireball of the Big Bang very hot during the era of Big Bang *nucleosynthesis*, and which ultimately left its trace in the form of the *background radiation*.

Such phase transitions can also produce defects in the structure of *spacetime*, analogous to the cracks that can form in a block of ice as it freezes. These defects may include *monopoles* and *cosmic string*, which may still exist in the Universe and could conceivably have provided the seeds on which *clusters of galaxies* have grown. The *scalar* fields that drove *inflation* were also associated with these phase transitions, but have long since faded away.

Phobos   The innermost of the two *moons* of Mars, discovered by Asaph Hall in 1898. Phobos is an irregular object roughly 27 km by 21 km by 19 km, orbiting Mars at a distance of 9,380 km once every 0.319 days. This is slower than the rate at which Mars rotates, so an observer on the surface of Mars would see the moon rising in the west and setting in the east. Like *Deimos*, Phobos is probably a captured *asteroid*. (See photo on p. 376.)

photoelectric effect   The process by which *light* falling on to the surface of a metal produces an electric current. At low levels of light intensity, the process can be described in terms of individual *photons* of light striking individual *atoms* of the surface and releasing individual *electrons*, and this also happens for other substances, not just metals.

Albert *Einstein* explained the photoelectric effect in these terms in 1905, making the then revolutionary suggestion that light (previously regarded purely as an electromagnetic wave) could be described in terms of particle-like quanta, what are now called photons. This was a key step in the development of *quantum theory*, and it was for this work, rather than for either of his two theories of relativity, that Einstein later received the Nobel Prize.

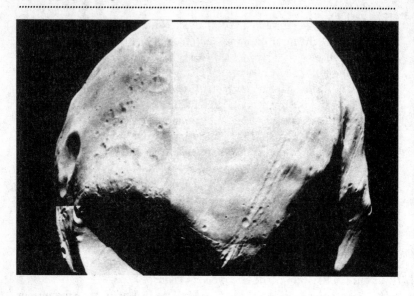

Phobos. A close-up view of Phobos, the larger of the two moons of Mars.

The photoelectric effect is important in astronomy because it enables astronomers to monitor the arrival of individual photons from a faint source, using *charge coupled devices* and similar detectors.

photographic magnitude   The *magnitude* of an astronomical object measured using a standard photographic emulsion, which is more sensitive in the blue and violet part of the spectrum than human eyes are.

photon   The quantum unit of *electromagnetic radiation*, which can be regarded as the particle of *light*. A photon has no rest *mass* or electric *charge*, and always travels at the speed of light if it is in a vacuum. It is a *boson*, and carries the electromagnetic force (see *fundamental forces*) between charged or magnetized objects.

Pickering, Edward Charles (1846–1919)   American astronomer, born in Boston on 19 July 1846, who was director of the Harvard College Observatory from 1876 to 1918. He dominated American astronomy in the last quarter of the 19th century, encouraging many young scientists to take up astronomy, including (unusually for the time) many women. He devoted an enormous effort to cataloguing, and pioneered the development of the *colour index* technique for classifying *stars*.

Pioneer spaceprobes   Series of unmanned spacecraft launched by *NASA* and used for a variety of investigations within the Solar System, including solar and lunar observations, Jupiter flyby missions, and both orbiting and lander missions to Venus. (See illustration on p. 378.)

pixel Word derived from the term 'picture element' to refer to the individual 'spots' on a digital image built up by, for example, a *charge coupled device*. A pixel is the smallest part of the image that can record a variation in brightness.

Planck, Max Ernst Karl Ludwig (1858–1947) German physicist, born in Kiel on 23 April 1858, who proposed, in 1900, that *electromagnetic radiation* can be emitted or absorbed only in definite units, which he called quanta. This explained the nature of *radiation* from a *black body*, and laid the foundations for *quantum theory*. Planck received the Nobel Prize for his work in 1918. It was, however, Albert *Einstein*, not Planck, who first suggested in modern times that electromagnetic radiation, including light, can actually be regarded as made up of individual particles, now called *photons* (see *photoelectric effect*).

Planck density The *density* of matter where 1 *Planck mass* occupies a volume 1 *Planck length* across. This corresponds to the density during the *Planck era*, effectively the state in which the Universe was born. The Planck density is 10^{94} grams per cubic centimetre.

Planck era The time when the entire Universe was at the *Planck density*, within 1 *Planck time* of the *singularity* at the birth of the Universe.

Planck length The length scale at which classical ideas about *gravity* and *spacetime* cease to be valid, and quantum effects dominate. This is the 'quantum of length', the smallest measurement of length that has any meaning. The Planck length is determined by the relative sizes of the constant of gravity, the speed of light and *Planck's constant*. It is roughly equal to 10^{-33} cm, which is 10^{-20} times the size of a *proton*.

Planck mass The *mass* of a hypothetical particle which would have an equivalent wavelength, according to *quantum theory*, equal to 1 *Planck length*. The Planck mass is about 10^{-5} grams, which sounds small until you compare it with the mass of a *proton*. It is actually 10^{19} times the mass of a proton, and 10^{16} times greater than the energies routinely attained in particle accelerators on Earth. This means that, in order to recreate the conditions that existed at the birth of the Universe, particle accelerators would have to be improved by a factor of 10,000 trillion.

Planck particle A hypothetical particle which is 1 *Planck length* across and contains one *Planck mass* at the *Planck density*. The known Universe may have originated in such a particle. See *inflation*.

Planck's constant A fundamental constant, denoted by h, which relates the energy of a *photon* to its *frequency* – that is, it relates the particle nature of a quantum entity to its wave nature. The value of h is 6.626×10^{-34} Joule seconds. Planck's constant turns up in many calculations in *quantum*

Pioneer spaceprobes. Illustration showing the main features of the identical Pioneer 10 and Pioneer 11 spacecraft. The large dish antenna dominates the various instrument packages.

theory – for example, in the ***uncertainty relation*** discovered by Werner *Heisenberg*.

Planck time   The time it would take *light*, moving at a speed of a few times 10^{10} cm per second, to cross a distance equal to the ***Planck length***. This is the 'quantum of time', the smallest measurement of time that has any meaning, and is equal to 10^{-43} seconds. No smaller time has any meaning, so within the framework of the laws of physics as understood today, we can only say that the Universe came into existence when it already had an age of 10^{-43} seconds, and when it had a *density* equal to the ***Planck density***.

planet   An astronomical object which is in *orbit* around a *star*, but does not have enough *mass* to become a star itself, and shines only by reflected *light*. This sets the upper limit for the mass of a planet as about one-twentieth of the mass of the Sun, perhaps 50 times as massive as Jupiter, the largest planet in our Solar System. A few 'super Jupiters' at the upper limit of the possible mass range for planets have been detected from their gravitational influence on the motion of the star they orbit. But studies of planets are essentially confined to the ones in our Solar System.

At the lower end of the mass range, it is hard to say where the definition of a planet should end. There are, for example, thousands of ***minor planets*** orbiting in a belt between the orbits of Mars and Jupiter. By tradition, though, the name planet is reserved for the nine major bodies (apart from the Sun) in the Solar System – Mercury, Venus, Earth, Mars, Jupiter, Saturn, Uranus, Neptune and Pluto. Even this standard definition is not entirely uncontroversial, since Pluto is a relatively small object in a peculiar orbit, and may well be an escaped *moon* or an unusual, large *asteroid* (it is only just over twice as big as Ceres, the largest of the minor planets, and less than half the size of Mercury, which has a radius of 2,439 km and is the smallest of the other eight planets).

Planets are more easily distinguished from *comets*, the other main kind of object found in the Solar System, which are more ephemeral entities largely made of ice and with nuclei only a few kilometres across.

planetary nebula A *nebula* associated with (surrounding) a star and looking like a disc (similar to the appearance of a planet, hence the name) through a telescope or in astronomical photographs. The term was first used by William *Herschel* in 1785.

A planetary nebula shines because light from the star it is associated with (especially *ultraviolet radiation*) is absorbed by the atoms in the nebula and re-radiated. In spite of the origin of their name, only about a tenth of all the planetary nebulae known (about 1,500 have been catalogued, but there are estimated to be ten times as many in the *Milky Way*) are really disc-shaped. Others look like rings of material, or have a double-lobed 'dumbbell' shape, or even a more complicated structure.

A typical planetary nebula has a temperature of about 12,000 *Kelvin*, contains about 20 per cent as much mass as our Sun, and is expanding away from the central star at speeds of about 20 km per second. Most expansion rates are determined from the *Doppler effect*. In one case, however, the growth of a nearby planetary nebula has been measured directly as 0.068 *arc seconds* per year. This is equivalent to measuring the increase in angular diameter of a tree that expands by 1 mm per year as it grows, from a distance of 100 km.

The stars at the centres of planetary nebulae are hot and blue, and fainter than ordinary *main sequence* stars. They are relatively old stars which are puffing material away into space (enriching interstellar matter with heavy elements) and starting to settle down to become *white dwarfs*, the cores of *red giant stars* that have lost their outer layers. On this picture, the visible nebula is only the inner part of a much larger expanding cloud of material, most of which cannot be seen because it is too far from the central star for the atoms in it to be lit up by the effect of the ultraviolet radiation.

The original red giant may have started out with up to eight times as much mass as our Sun, while the white dwarf that is left will have only a little more mass than the Sun. A planetary nebula is a short-lived phenomenon, each one perhaps lasting for 50,000 years, associated with the transition of a star from the red giant to the white dwarf states.

planet formation   Planets are thought to form together with their parent stars from the collapse of clouds of gas and dust in space. A system like the Solar System will have formed from a cloud large enough to fragment into several hundred stars like the Sun, which initially formed an *open cluster* of stars that have since spread out and gone their separate ways through the *Galaxy*. As an individual star forms, the hot ball of gas at the centre of the collapsing cloud fragment is surrounded by a relatively small amount of extra material, which settles into a disc around the young star, and eventually forms the planets.

It used to be thought that there were considerable problems in such a *model* with getting rid of the angular momentum of the collapsing cloud. As such a cloud shrinks inward upon itself, it will spin faster, like a spinning ice skater drawing in her arms. In order to shrink down to the size of the Sun, it seemed from some calculations that the proto-star would end up spinning impossibly fast, at close to the speed of light. However, because so many stars form from one collapsing cloud, they can all be spinning in different ways (some 'clockwise' and some 'anticlockwise'), sharing out the overall angular momentum. In addition, in the 1950s Fred *Hoyle* showed how magnetic fields associated with a young star can encourage processes which carry angular momentum away. What is regarded as clinching evidence for the standard model of planet formation comes from studies of young stars which are surrounded by dusty discs of exactly the kind invoked by the model.

Within those discs, the region closest to the young star is largely swept clear of the lightest *elements* (hydrogen and helium) by the heat of the star. The left-over material is in the form of grains which stick together and eventually clump into balls large enough to attract one another by gravity, building up to form planets like the Earth. Further out, where there is still plenty of hydrogen and helium, giant planets, like Saturn and Jupiter, form.

A minority view, espoused in particular by William *McCrea*, holds that proto-planets and *comet*-like objects form in *interstellar clouds* even before stars form, and that these objects are then captured by stars as they form, with many small, cold objects being gathered into orbit around the young star, colliding and sticking together to form planets like those we see in the Solar System. Extreme versions of this scenario suggest that the stars themselves form from the accumulation of these smaller objects. Either way, this would give planet formation a head start, by providing the young star with a retinue of ready-made proto-planets instead of it having to start building them up from scratch from tiny dust grains.

Pluto

Planet X   Originally, the name given by Percival *Lowell* to the *planet* he expected to find beyond the orbit of Neptune. Extensive efforts failed to find the planet in Lowell's lifetime. The discovery of Pluto in 1930 did not end the search for Planet X, because Pluto is too small to be the object Lowell was looking for. The existence of Planet X was suggested by discrepancies in the observed orbits of the outer planets, which, it was thought, could be explained by the gravitational influence of another planet. But extended observations of these orbits (remember that Neptune was only discovered in 1846 and takes 164 years to go round the Sun once, so we have not yet observed it for one complete orbit) now suggest that there are no discrepancies, and that Planet X does not exist.

plasma   A hot state of matter in which *electrons* have been stripped from *atoms* to leave positively charged *ions*, which mingle freely with the electrons. All of the matter inside a star is in the form of a plasma.

Plato (about 427–347 BC)   Greek philosopher who based much of his thinking on the idea of perfection, arguing that the Earth must be a perfect sphere and that all other objects in the Universe moved in perfect circles around the Earth. His influence extended until the work of Nicolaus *Copernicus*, nineteen centuries later, put astronomers on the right track. But at least Plato did teach that astronomy is a fundamental part of education.

Pluto   The ninth *planet* in the Solar System to be discovered, and usually the most distant from the Sun. However, Pluto's elliptical orbit sometimes takes it within the orbit of Neptune, and this is happening between 1979 and 1999. Pluto was discovered by the American astronomer Clyde Tombaugh (1906–) in 1930. Its average distance from the Sun is 39.44

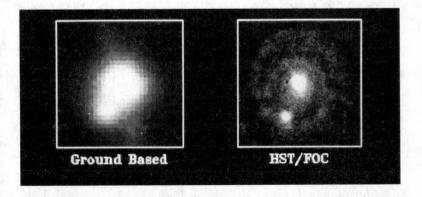

Pluto and its moon Charon, as seen from the ground and from the Hubble Space Telescope.

astronomical units, but the distance varies between about 30 and 50 AU at different parts of its orbit. At *aphelion*, light from the Sun takes nearly 7 hours to reach Pluto. It is separated from its moon Charon by only 19,500 km, and the average density for Pluto and Charon put together is 1.88 grams per cubic centimetre, less than twice the density of water.

According to observations made by the *Hubble Space Telescope* in 1994, Pluto is just 2,320 km across (two-thirds the size of our Moon) and it has a mass 0.3 per cent of that of the Earth. The temperature at the surface of Pluto is only 50 *Kelvin*; it orbits the Sun once every 248 years, locked by gravity so that it makes two orbits for every three by Neptune.

Pogson scale See *magnitude scale*.

Pond, John (1767–1836) The sixth Astronomer Royal, who succeeded Nevil *Maskelyne* in 1811 and retired from the post because of ill health in 1835. He made many much needed improvements to both the equipment and the observing techniques used at the Royal Greenwich Observatory, and introduced (in 1833) the practice of lowering a 'time ball' on a pole on top of the Observatory at 1 p.m. each day – the first public time signal.

Population I Term used to refer to the young *stars* that are found chiefly in the plane of our *Galaxy*, especially in the *spiral arms*. The distribution of these stars shows up clearly from the pattern of bright, hot blue stars that typify Population I, but cooler stars, including the Sun, are also part of Population I. The term was introduced in 1944 by Walter *Baade*; by extension, the name is also used to refer to similar stars in other *galaxies*.

Population I stars are relatively rich in *heavy elements*, because they have formed from clouds of gas and dust which had already been enriched by material processed by *nucleosynthesis* in previous generations of stars. Among other things, the presence of heavy elements in such protostellar clouds provides the raw material from which *planets* form, so planetary systems (like our Solar System) are expected to be associated only with Population I stars.

Population I stars are typically hot *main sequence* stars, often found in *open clusters* or other loose groups, and associated with *interstellar matter*.

Population II Term used to refer to the old, red *stars* that are found in the halo of our *Galaxy*, especially in the *globular clusters*, near the galactic centre, and in a thick layer much fatter than (and sandwiching) the plane defined by the *spiral arms*. By extension, the term is also used to refer to similar stars in other *galaxies*; *elliptical galaxies* are largely composed of Population II stars. The term was introduced in 1944 by Walter *Baade*.

Population II stars are relatively deficient in *heavy elements* because they formed long ago when the Galaxy was young, before the primordial hydrogen and helium had been processed by stellar *nucleosynthesis* to make the heavier elements. It is extremely unlikely that Population II stars have planets associated with them, because planets are made of heavy elements. The distinction with *Population I* is an astronomical convenience, and really there is a continuous gradation of stars from one extreme to the other.

Population III   Hypothetical generation of supermassive *stars* that may have existed before *galaxies* formed. The existence of Population III seems to be required because even the oldest stars around today, *Population II*, contain some traces of *heavy elements*, which can only have been produced in a previous generation of stars (see *nucleosynthesis*). Population III stars must have been very massive, running through their life cycles quickly and exploding, scattering a trace of *metals* through the material from which galaxies formed. They may have left behind remnants in the form of stellar-mass *black holes*.

positron   Positively charged *antimatter* counterpart to the *electron*.

potential energy   *Energy* possessed by an object as a result of its position – for example, in a *gravitational field*. An object on top of a tall building has more gravitational potential energy than if it were on the street below. If it is pushed off the building, that gravitational potential energy is first converted into *kinetic energy* as the object accelerates downwards, and then the kinetic energy is converted into heat when the object hits the ground. The potential energy comes from the work you have to do carrying the object up to the top of the building from street level in the first place. Potential energy always has to be measured relative to some standard state – in this case, relative to street level – although there may be a minimum value of the potential associated with a particular field, below which it is not possible to drop.

There are also potential energies associated with electric *charge* and with *magnetic fields*. A compressed spring stores potential energy, which is released when the spring is allowed to expand; this is an example of electrical potential energy, because the forces that operate between the atoms in the spring are electric forces. An electric battery stores potential energy in a chemical form (again, this really means in an electrical form, corresponding to the arrangements of electrons in atoms and molecules inside the battery) and releases it in the form of a flowing electrical current when the terminals of the battery are connected by a conductor.

In the very early Universe, potentials associated with the *scalar fields* that drove *inflation* provided energy during the *phase transitions* that occurred when the *fundamental forces* broke apart from the original single force predicted by *grand unified theories*. One of the curious features about gravitational potential energy is that it is always negative, compared with the energy stored in *mass* (which can itself be considered a form of potential energy, since $E = mc^2$). This is connected with the way energy has to be put

383

in, in order to take matter outwards (upwards) in a gravitational field, such as the field associated with the Earth. A cloud of particles infinitely far apart from one another that fell together to make a mass *m* at a central point would start with zero energy and release exactly *mc²* in kinetic energy as they fell in, ending up with *minus mc²*. This raises the intriguing possibility that the Universe could have appeared out of nothing at all, at a point in *spacetime* (see *free lunch universe*).

precession of Mercury   See *advance of the perihelion.*

precession of the equinoxes   Slow drift of the *equinoxes* westward across the sky, caused by the precession (wobble) of the Earth as it moves through *space*. It takes about 25,800 years for the equinoxes to move right around the sky. Coordinate systems, such as *right ascension*, that are based on measurements from the equinox must therefore be specified for a particular year; currently, the standard is set as the position corresponding to the start of the year 2000.

pressure   The amount of 'push' exerted by a compressed fluid (gas or liquid). Pressure is measured in terms of the force per unit area exerted by the fluid, and 1 pascal (Pa) of pressure is equal to a force of 1 *newton* per square centimetre. A pascal of pressure is about the same 'push' that the weight of a layer of butter exerts on a slice of bread. The average atmospheric pressure at sea level on Earth is 101,325 Pa.

primeval fireball   The state of the very early Universe when it was dominated by *radiation*. See *Big Bang, radiation era.*

primordial atom   See *Lemaître, Georges.*

primordial black hole   See *mini black hole.*

principle of equivalence   See *equivalence principle.*

principle of terrestrial mediocrity   See *cosmological principle.*

proper motion   The apparent angular movement of a star across the sky in a year. Relatively few stars, out of the 100 billion or so in our *Galaxy*, are close enough to us for their proper motions to be measurable, but the *satellite Hipparcos* measured the positions of more than 100,000 stars (still only 1 millionth of the number in the *Milky Way*) to an accuracy of 0.002 *arc seconds*, leading to the potential for measurements of many proper motions that equal or exceed 0.002 arc seconds per year. A second satellite which made a similar survey in a few years' time would be needed to pin these down precisely.

The first person to realize that stars move across the sky, and are not really 'fixed', was Edmund *Halley*, in 1718. He noticed that the positions of three bright stars recorded in old *catalogues* by the Greek astronomers *Hipparchus* and *Ptolemy* were not in the places where the Greeks had said they were. Since hundreds of other stars in the old catalogues were in the 'right' places, Halley inferred that this was not because the Greek astronomers had made a few mistakes, but because the stars in question – *Sirius*, Procyon and Arcturus – had moved across the sky since ancient times (Arcturus by a full arc degree, twice the angular size of the Moon as seen from Earth). It is no coincidence that these three stars are among the eight brightest in the sky; they look bright because they are close, and their motion through space shows up clearly on a timescale of a few centuries for the same reason.

The largest observed proper motion is 10.3 arc seconds per year, for *Barnard's star*. This is just over half of 1 per cent of the angular size of the Moon as viewed from Earth. The average proper motion for all the stars visible to the unaided human eye is about 0.1 arc seconds per year.

proper time   Time measured by a clock that is in the same *frame of reference* as the observer. Clocks in different frames of reference (either moving relative to the observer or in a different *gravitational field*) will measure a different flow of time. All are equally 'correct', but the proper time for you to use is the time in your frame of reference. See *special theory of relativity*.

protogalaxy   A huge cloud of gas in the process of collapsing to form a *galaxy*. Objects which roughly fit the bill have been seen in the visible Universe, but it is not certain that any of them are galaxies in formation. Existing galaxies must have formed from protogalaxies at extremely high *redshifts*.

proton   One of the particles from which *atoms* are made. The proton was the second atomic particle to be identified, by the British physicist J. J. Thomson (who had also discovered the *electron*) at the beginning of the 20th century. It is a member of the *baryon* family. The proton has 1 unit of positive charge (on the scale where the electron has 1 unit of negative charge), and a *mass* of almost 1 billion *electron volts* (actually 939 MeV, equivalent to 1.6726×10^{-24} gm), 1,836 times the mass of an electron.

proton–proton reaction (p–p chain)   The series of *nuclear fusion* reactions that generate *energy* inside a *star* like the Sun.

The modern understanding of this process depends upon a combination of measurements (using particle accelerators here on Earth) of the rates at which various fusion reactions occur (their so-called cross-sections) and *models* of conditions inside the Sun, especially the temperature and pressure at the heart of the Sun, based on observations of the Sun's *luminosity*, size and *mass*. The proton–proton chain was first proposed as the source of solar energy by Hans *Bethe* and his colleague Charles Critchfield in 1938, but it was not established as the best model until the 1950s, partly because in the 1930s and 1940s it was not fully appreciated that more than 95 per cent of the Sun is simply *hydrogen* and *helium*.

The p–p chain begins when two *protons* (hydrogen *nuclei*) get close enough to fuse as a result of the *tunnel effect*. They form a deuteron (a nucleus of *deuterium*), with one of the protons spitting out a *neutrino* and a *positron* in the process, to become a *neutron*. Another proton can then tunnel into the deuteron, making a nucleus of helium-3, containing two protons and one neutron. Finally, when two nuclei of helium-3 collide they form one nucleus of helium-4 (two protons plus two neutrons), ejecting the two extra protons. About 95 per cent of the helium-3 nuclei suffer this fate; the rest are involved in more complicated interactions that do not contribute much to solar energy generation, but which are relevant to the *solar neutrino problem*.

Everything hangs together if the Sun is made of 70 per cent hydrogen, 28 per cent helium and 2 per cent of *heavy elements*, and if it has its observed physical properties and a central temperature of about 15 million K. Each time four protons are combined to make one helium-4 nucleus, just 0.7 per cent of their mass is released as energy. And although only one collision between protons in every 10 billion trillion (1 in 10^{22}) inside the Sun initiates this process, there are so many protons (and so many collisions) there that overall about 5 million tonnes of mass are converted into pure energy every second. So far, the Sun has 'lost' about 4 per cent of its original stock of hydrogen nuclei in this way, through p–p reactions that have been going on inside it for some 4.5 billion years.

The proton–proton reaction is the main source of energy in stars on the *main sequence* that have roughly the same mass as the Sun, or less

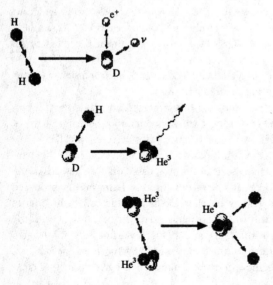

p–p chain. The chain of nuclear interactions that generates heat inside stars like the Sun.

mass. More massive main sequence stars get their energy primarily from the *carbon cycle*.

protoplanet   Disc of material around a star that is in the process of forming a planet by *accretion* of material within the disc. Infrared observations made by *IRAS* have shown evidence of protoplanetary systems around nearly 50 stars, and dusty 'cocoons' are common features of *T Tauri* systems.

protostar   Part of a *molecular cloud* of gas that has formed a clump and is collapsing to form a *star*, but in which *nuclear fusion* reactions have not yet begun. Many protostars can be identified in our Galaxy, especially by their *infrared* emission.

Proxima Centauri   The closest known star to the Sun, at present at a distance of 1.295 *parsecs*. Proxima Centauri is a faint *dwarf star*, with a *mass* only one-tenth that of the Sun. It is almost certainly physically associated with *Alpha Centauri*, orbiting that *binary star* system at a great distance.

Ptolemaic system   See *Ptolemy*.

Ptolemy (about AD100–170)   Greek astronomer who wrote the *Almagest*, a thirteen-volume description of the Universe as understood at the time. Ptolemy used a genuinely scientific approach, and developed the idea of *epicycles* to account for the observed motion of heavenly bodies within the framework of an Earth-centred universe. This Ptolemaic system held sway for more than 1,000 years.

pulsar   A rapidly spinning *neutron star* that emits beams of *radio waves* that flick round like the beams from a fast radio 'lighthouse', producing regularly timed pulses of radio noise in *radio telescopes* on Earth. The name is a contraction of 'pulsating radio source', chosen to echo the name *quasar*.

The first pulsars were discovered in 1967 by Jocelyn *Bell Burnell*, a radio astronomer working in Cambridge under the supervision of Antony *Hewish*. She was using a radio telescope specially constructed to look for rapid variations in the radio 'brightness' of quasars (the radio equivalent of the twinkling of light from stars), and found that there was a previously completely unknown (and largely unsuspected) kind of rapidly varying *radio source*.

The rapid variation of the first pulsars discovered, flicking on and off every second or so, showed that they must be coming from a very small source. Radiation from the surface of a star can move 'in step' only if some signal travelling at (or below) the speed of light spreads across the star to trigger the burst of radiation. So each burst must come from a source smaller than the distance that light can travel in the length of one burst, otherwise the radiation would get smeared out (this is rather like the *horizon problem* in *cosmology*). This immediately told the Cambridge team that the objects they had discovered were much smaller than *main sequence* stars, and were probably no bigger than a *planet* like the Earth. Together with the extreme precision of the timing of the pulses, this led the radio astronomers to consider seriously the possibility that they had detected signals from an extraterrestrial civilization; but the discovery of more pulsars soon showed that this could not be the case and that they must be a natural phenomenon.

If the pulsars were a previously unknown kind of star, their small size meant that they could only be either *white dwarfs* (each with about as much mass as the Sun packed into the same volume as the Earth) or neutron stars (with about as much mass as the Sun packed into a sphere less than 10 km across). In 1967 nobody had ever found direct evidence for the existence of neutron stars, so they were regarded as a hypothetical speculation. But white dwarfs had been detected using *optical astronomy*, so the first guess was that pulsars might be pulsating white dwarf stars, like miniature *Cepheid variables*.

Working independently of each other, two Cambridge theorists, John Gribbin and John Skilling, soon showed that white dwarf stars could be made to vibrate just rapidly enough to explain the periods of the first known pulsars. But they could not vibrate much faster without breaking apart. When much more rapid pulsars were indeed found, this showed that they could not be white dwarfs after all, but must be neutron stars. It was the discovery of pulsars and the proof that they could not be white dwarf stars that convinced astronomers that neutron stars really existed, and encouraged them to take seriously the idea of even more extreme versions of collapsed matter, the *black holes*.

Once it was clear that pulsars must be neutron stars, the 'lighthouse' model quickly gained wide currency among astronomers. With hindsight, we can see that such objects were very nearly predicted. Although few astronomers before 1968 took the idea of neutron stars seriously, one of the few who did, Fritz *Zwicky*, had pointed out in the 1930s (in a paper written with Walter *Baade*) that a neutron star might be left behind by the explosion of a *supernova*. Baade later pointed out that, if Zwicky's idea was right, the best place to look for a neutron star would be in the *Crab nebula*, the expanding cloud of debris left behind by a supernova observed on Earth in 1054. He even identified a particular star in the Crab nebula that he thought might be the neutron star left behind by the explosion.

For years, hardly anyone took the idea seriously, but in 1967, shortly before the announcement of the discovery of pulsars, Franco Pacini published a paper in which he pointed out that, if an ordinary star did collapse to form a neutron star, the collapse would make the star spin faster (like

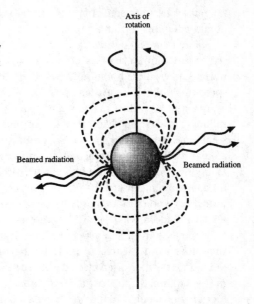

The standard model of a pulsar.
Strong magnetic fields (indicated by dashed lines) beam electromagnetic radiation out from the poles as the neutron star spins, like a cosmic lighthouse.

an ice skater drawing in her arms), and would strengthen the star's magnetic field as it was squeezed, along with the matter, into a smaller volume. Such a rotating magnetic dipole, said Pacini, would pour out electromagnetic radiation, and this source of energy could explain details of the way the central part of the Crab nebula is seen to be still being pushed away from the site of the explosion, almost 1,000 years after the supernova itself was observed.

If pulsars had not been discovered fortuitously just at the time Pacini's paper was being published, this might have led to a direct investigation of the Crab nebula that would have revealed the pulsar (which is indeed the star identified by Baade). But the specific speculation that the pulses of radio noise from pulsars might be produced by a lighthouse effect as the rapidly rotating neutron star flicks its beams past us was first published by Tommy *Gold* in 1968. This is now known as the Pacini–Gold model. The identification of the *Crab pulsar* early in 1969 established the model beyond any reasonable doubt; it is now generally accepted that the observed radiation from pulsars is produced by *synchrotron emission* from electrons moving in the very strong magnetic field of the rotating neutron star (about 10^8 tesla, roughly 1,000 million times the strength of the magnetic field at the surface of the Earth). This makes young, energetic pulsars, such as the one in the Crab nebula, visible as sources of X-rays and gamma rays (and light), as well as radio waves. A separate class of pulsars known as X-ray pulsars occur in some binary systems, and are powered by the energy of matter falling in from a companion via an *accretion disc*.

More than 650 pulsars are now known. Most have periods of about 1 second; the slowest one has a period close to 4 seconds, while the fastest pulsar known flicks on and off every 1.6 milliseconds. By and large, the speed of a pulsar is a measure of its age – like most of us, they slow down as they get older. The Crab pulsar, for example, is slowing by one part in a million per day; but like other young pulsars this steady slowdown is sometimes interrupted by *glitches* as the surface of the star cracks and adjusts itself to a new configuration. In some cases, though, a pulsar in a *binary system* can be spun up to greater speeds as it accretes matter from its companion. Pulsars have been detected at distances as great as 50 *kiloparsecs*, and they are mostly concentrated in the disc of the *Milky Way*, which is indeed where most supernovae occur. Only a few pulsars are known to lie within supernova remnants, however, which may be because they often do not 'switch on' until after the supernova debris has dispersed, or because they are often ejected from the supernova explosion and have moved far from the sites of their formation. In either case, we are extremely lucky to find a pulsar in the Crab nebula. But the number of pulsars found, together with estimates of their lifetimes based on theoretical *models*, suggests that one is formed every 20 years or so in our Galaxy, closely matching estimates of the frequency of supernovae, and suggesting that just about every supernova does leave a neutron star behind. Many pulsars have high velocities (the average is 450 km per second), carrying them far out of the plane of the Galaxy (in some cases, allowing them to escape from the Galaxy altogether). This could explain why they are not often seen near supernova remnants – presumably, some asymmetry in the supernova explosion sent them hurtling off into space.

Because of the way radiation is beamed out from pulsars, we should only be able to see about a fifth of all the pulsars active in the observable part of the Galaxy at any one time. But pulsars are such weak sources that astronomers estimate that only 1 per cent of all pulsars active in the Galaxy today are within range of our radio telescopes. There are probably a few hundred thousand active at any time in the Galaxy.

Some pulsars are found in *globular clusters*. This makes them relatively easy to detect, because a whole cluster containing thousands of stars can be in the beam of a radio telescope at one time. Many of these are very rapidly spinning, and are known as *millisecond pulsars*. It seems likely that these pulsars are old neutron stars which have been 'spun up' by the *accretion* of matter in a *binary system*. But only half of the more than 30 millisecond pulsars seen in globular clusters are in binary systems today, perhaps because the others have eaten up their companions (see *black widow pulsar*). Millisecond pulsars have now also been found in the plane of the Galaxy.

Out of the first 650 pulsars found, just 50 are known to be in binaries. Most have a normal white dwarf companion, five have another neutron star as their companion, three have very low-mass companions and may be active black widow systems, and just one has an ordinary, non-degenerate star as its companion. One pulsar, PSR 1257 + 12, has three planet-sized objects in orbit around it.

Although the Pacini–Gold model is well established as a broad picture of how a pulsar works, nobody has yet come up with a satisfactory detailed explanation of exactly how the radiation we detect is produced, and how exactly it is beamed out into space. Even without that complete understanding, however, pulsars have proved an invaluable 'test bed' for many theories. They are extraordinarily accurate timekeepers (precise, in many cases, to 1 part in 10^{10}), and the variations in the *binary pulsar* have provided the most accurate confirmation yet of the predictions of Albert Einstein's *general theory of relativity*. The first *millisecond pulsar*, discovered in 1982, has a period of just 1.6 milliseconds, which means that a star roughly as massive as our Sun is squeezed into a ball of material with the density of an atomic *nucleus*, spinning more than 600 times every second.

pulsating universe   See *oscillating universe*.

pulsating variables (pulsating stars)   *Stars* which change in *luminosity* as they pulsate in and out in a more or less regular fashion, as if they were breathing. In spite of their name, *pulsars* (short for pulsating radio sources) are not pulsating stars, although they were at one time thought to be.

Most stars pass through a phase of pulsation at some time during their lives, and there are different kinds of pulsating variable corresponding to stars with different *masses* at different stages in their evolution. These include *RR Lyrae* and *Cepheid variables* which have burned all of the nuclear fuel in their cores and have evolved away from the *main sequence* in the *Hertzprung–Russell diagram*.

The brightness of a pulsating star changes as the outer layers of the star expand and contract, while the *spectrum* of the star (its colour) changes in the same rhythmic fashion, as the temperature of the surface of the star varies. The pulsations can also be measured in terms of the changing *Doppler effect* in the star's spectrum, which shows the speed with which the surface moves in and out. Roughly speaking, the period of all these variations is inversely proportional to the square root of the density of the star, so more dense stars pulsate more slowly.

The energy that drives the pulsations comes ultimately from *nuclear fusion* reactions going on inside the star, which produce electromagnetic radiation that can be trapped by *ions* of helium near the surface. When the region of ionized helium is squeezed, it traps more of this energy and gets hotter, building up a pressure which pushes the material above it outward; as the outer layers expand, the density of the ionized helium region decreases, it becomes less effective at trapping energy, and the pressure drops, allowing the outer layers to fall back. The whole process then repeats.

Purple Mountain Observatory Possibly the observatory with the most romantic name in the world. It is run by the Chinese Academy of Sciences, on the Purple Mountain, 267 m above sea level, to the north-east of Nanjing. The main *telescopes* are a 60 cm reflector, a similarly sized *Schmidt camera*, and a *radio telescope* for *millimetre astronomy*. In spite of the romantic name of the location, the observatory is badly affected by light pollution from the nearby city.

Pythagoras (about 580–500 BC) Influential Greek philosopher who, among many other things, taught that the Earth is a sphere and that the planets move in circles – not for any truly scientific reason, but because of a mystic belief that circles were the 'perfect' form. Little is known about Pythagoras' life (even his dates of birth and death), but he is credited with discovering the famous property of right-angled triangles which bears his name, and which turns out to be an immensely useful tool in the investigation of relationships in *spacetime*, in the context of both the *special theory of relativity* and the *general theory of relativity*.

Pythagoras' theorem gives the length of the hypotenuse of a right-angled triangle in terms of the lengths of its other two sides. The length of the hypotenuse is the shortest distance between the two points at each end of the hypotenuse. So Pythagoras' theorem gives us a way to calculate the shortest distance between any two points, in terms of right-angled triangles drawn to make those points lie at the ends of the hypotenuse.

On an ordinary two-dimensional graph, the lengths of the two sides at right angles to each other are measured parallel to the x-axis and parallel to the y-axis. It doesn't matter where you put the origin of your graph, these two measurements will always be the same, and therefore they will always tell you, in turn, the length of the hypotenuse.

In three dimensions, the equivalent calculation involves three measurements at right angles, conventionally called the x-, y- and z-axes. Again, the measurements give you, from Pythagoras' theorem, the shortest distance between two points, defining a *geodesic*. The important point is that, no matter how you choose to set up your axes (your *frame of reference*), measuring the x's, y's and z's and using Pythagoras' theorem will give you the *same* measurement of the shortest distance between two points, a unique geodesic. And the theorem can even be extended to four (or more!) dimensions, to define geodesics in *spacetime* and the *metric* of the Universe at large or of the region of spacetime near, say, a **black hole**. Since the **event horizon** of a non-rotating black hole is the nearest thing that can exist to a perfect sphere, perhaps Pythagoras would have approved!

Pythagoras' theorem   See *Pythagoras*.

QSO = *quasar*.

quagma   See *quark–gluon plasma*.

quantum   The smallest possible component of a system, or the smallest change a system can make. Thus a **photon** is the smallest possible unit of **light**, and the essential features of a quantum leap are that it is usually the smallest possible change in a system, and that it is made at random (although obeying the strict rules of probability).

quantum chromodynamics (QCD)   See *grand unified theories*.

quantum electrodynamics (QED)   The *quantum theory* description of the behaviour of **electromagnetic radiation** (including **light**) and the way it interacts with matter that has electric **charge**, through the exchange of **photons**. The theory has been tested to extreme precision, comparing its predictions with measurements of the behaviour of electrons in **atoms**, and passed every test; it ranks with the **general theory of relativity** as one of the two most securely founded theories in all of science (see **binary pulsar**).

quantum field theory   See *field equations*.

quantum fluctuation   A temporary change in the state of empty space, allowed by the **uncertainty principle**. Quantum uncertainty allows small amounts of energy to appear out of nothing at all, provided they disappear in a very short time (the less energy there is in the fluctuation, the longer it can persist). This energy may take the form of short-lived pairs of **particles** and **antiparticles**, for example an **electron–positron** pair. See **free lunch universe, Hawking radiation**.

quantum gravity   General name for theories that attempt to combine the **general theory of relativity** and **quantum theory** (see **grand unified theories**). The key feature of any quantum theory of gravity is that the gravitational force must be carried by a **particle** called the **graviton** (a **boson**), equivalent to the role played by the **photon** in **quantum electrodynamics**.

quantum leap   See *quantum*.

quantum mechanics   See *quantum theory*.

quantum physics   See *quantum theory*.

quantum theory   Also known as quantum mechanics or quantum physics, quantum theory is the set of physical laws that apply primarily on the very small scale, for entities the size of atoms or smaller. At the heart of quantum theory lie the linked concepts of uncertainty and wave–particle duality.

In the quantum world, every entity has a mixture of properties that we are used to thinking of as distinctly different – waves and particles. For example, *light*, which is often regarded as an electromagnetic wave, behaves under some circumstances as if it were composed of a stream of particles, called *photons*. The discovery by Max *Planck*, at the end of the 19th century, that the nature of *black body radiation* can be explained only if light is emitted or absorbed by atoms in discrete quanta (photons), made physicists aware of the distinction between quantum physics and *classical mechanics*. The essential feature of Planck's discovery is that there is a limit to how small a change in *energy* an atom can experience. This corresponds, in modern terminology, to the emission or absorption of a single photon. The important feature of such a 'quantum leap' is that it is the smallest possible change; advertisers and politicians may therefore be being unwittingly honest when they describe progress as having occurred in a quantum leap.

Planck himself did not think in terms of photons, but interpreted black body radiation as a consequence of the inability of atoms to emit energy except in discrete lumps; he did not think of light itself as made up of particles. It was Albert *Einstein* who showed that light could be regarded in terms of particles, initially in a paper published in 1905 (which was the work he received his Nobel Prize for). This was developed into the description of light in terms of *bosons* in the 1920s. Also in the 1920s, it was established from experiments that the quintessential fundamental 'particles', *electrons*, could also behave as waves. The implications of wave–particle duality are best seen in terms of modern experiments which show electrons behaving as both waves and particles.

These experiments are based on the 'double-slit' experiment often used (for example, in school science classes) to prove that light travels as a wave. In such an experiment, light passes through a small hole in a screen, and on to a second screen in which there are two holes. Light from each of the two holes in the second screen continues on its way to fall on to a third screen, where it makes a pattern of light and dark stripes. The classical explanation of the stripy pattern is that waves from each of the two holes are arriving all over the final screen. Where the waves march in step, they add together to form a bright patch; where they are marching out of step, they cancel each other out and leave a dark patch. Exactly the same sort of thing happens to the ripples produced on a pond when you drop two pebbles in at the same time – in some places the ripples reinforce, while in other places they cancel out. So the experiment with two holes shows that light travels as a wave.

In the modern version of the experiment, carried out by Japanese researchers at the end of the 1980s, the light source is replaced by an electron 'gun' which fires electrons off one at a time. The role of the two holes is played by magnetic fields, and the final screen is a detector similar

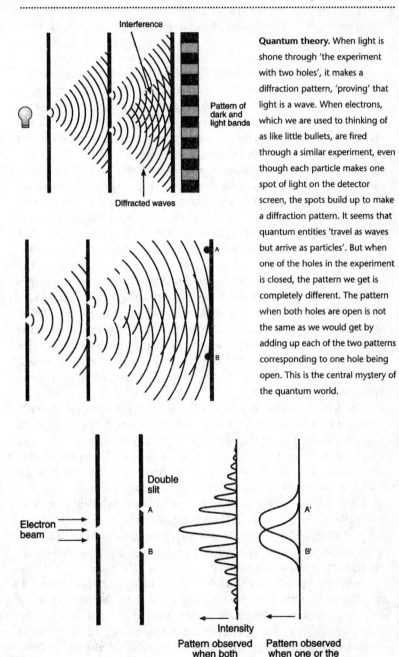

Interference

Pattern of dark and light bands

Diffracted waves

A

B

Double slit

A

B

Electron beam

Intensity

Pattern observed when both slits open

A'

B'

Pattern observed when one or the other slit is open

Quantum theory. When light is shone through 'the experiment with two holes', it makes a diffraction pattern, 'proving' that light is a wave. When electrons, which we are used to thinking of as like little bullets, are fired through a similar experiment, even though each particle makes one spot of light on the detector screen, the spots build up to make a diffraction pattern. It seems that quantum entities 'travel as waves but arrive as particles'. But when one of the holes in the experiment is closed, the pattern we get is completely different. The pattern when both holes are open is not the same as we would get by adding up each of the two patterns corresponding to one hole being open. This is the central mystery of the quantum world.

to the screen in a TV set. Each individual electron that passes through the experiment must go one way or the other (through one or other of the 'holes') to the detector screen. And, sure enough, when individual electrons are fired through the experiment, each one makes a definite spot of light on the screen, corresponding to the arrival of a single particle. But as more electrons are fired, one at a time, through the experiment, the spots of light made on the screen add up to make a distinctive pattern of light and shade. It is exactly the interference pattern associated with waves that pass through both holes at once en route to the screen.

The great physicist Richard Feynman (1918–88) said that the experiment with two holes encapsulates the 'central mystery' of quantum mechanics, and that nobody understands what is going on in it. It isn't just that quantum entities seem to travel as waves but arrive and depart as particles, but that they seem to know about the past and future as well. It is as if each electron leaves the gun as a particle, turns into a wave that travels both ways through the apparatus, then turns back into a particle to arrive at a definite spot on the screen. Not only that, but it chooses exactly the right spot to make its overall contribution to an interference pattern that is built up over a long period of time. How does it 'know' about all the other electrons and where they are going to end up in the pattern? And the classic double-slit experiment itself has even been carried out with light sources so weak that individual photons are passing through it, one at a time. Once again, they build up an interference pattern on the other side.

The standard explanation of all this, called the Copenhagen Interpretation (because it was largely developed by researchers in Copenhagen), is that quantum entities travel as waves which spread out and obey strict rules of probability, so that it is possible to calculate where the wave is strongest (that is, where the electron, or whatever, is most likely to be found) and where it is weakest. When an observation or measurement is made (for example, when the electron wave impinges on a detector screen), the 'wave function' collapses into a point-like particle. At that moment, the probability of finding the electron anywhere else vanishes, but as soon as the quantum entity is no longer being observed, the probabilities spread out from the point where it was last seen.

In spite of many unsatisfactory features, the Copenhagen Interpretation can be used to make predictions about the outcomes of experiments involving quantum entities such as electrons and photons, and it is the basis of the physics used, among other things, to develop lasers, computer chips and an understanding of complex biological molecules such as DNA. But the pre-eminent position of the Copenhagen Interpretation is as much a historical accident as anything else. Although the Copenhagen

Interpretation is the standard version of quantum theory used by physicists, because it was the first interpretation that could be made to work, it is only one of several different interpretations which all have unsatisfactory features but which all give exactly the same 'answers' when used to carry out these kinds of calculation. This suggests to many people that none of these interpretations is providing the correct insight into what goes on in the quantum world, and that a completely new understanding of physics may be needed before quantum theory can be put on a solid footing.

It is a mark of just what kind of intellectual leap might be required to do this that some interpretations of quantum mechanics require signals that travel backwards in time, while all of them require instantaneous communication between quantum particles even when they are widely separated.

Nevertheless, quantum theory can be used, like the recipes in a cookery book, to calculate the behaviour of atoms and other quantum systems. Just as you don't need to understand the physics going on inside an oven in order to bake a cake by following a recipe book, so you don't need to understand what is going on in the quantum world in order to calculate, say, the spectrum of hydrogen using the quantum rules. So the study of the Universe using *spectroscopy* directly depends on the understanding of atoms and molecules provided by quantum theory. The behaviour of atomic nuclei also depends on quantum processes, and our understanding of *nucleosynthesis* and the reactions that generate heat in the interiors of stars therefore depends on quantum theory. It is quantum uncertainty, for example, that explains both how an *alpha particle* escapes from a nucleus during *alpha decay* (through the *tunnel effect*), and how nuclei are able to overcome the repulsion caused by their positive electric charges and fuse together under the conditions that exist inside stars. Because the nuclei have uncertain positions, they are more spread out than the equivalent classical particles would be, and are able to 'overlap' and fuse even when classical mechanics says they are too far apart to do so. The success of *models* of how this happens inside the Sun in predicting many of the observed features of the Sun, including its central temperature, is one of the best large-scale indications that quantum physics is a good description (at least in the cookery book sense) of what is going on at this level.

The most important overlap between quantum physics and *cosmology* involves the *uncertainty principle* developed by Werner *Heisenberg* in the 1920s. This is related to the wave–particle duality, and can be most easily understood in terms of uncertainty in the position of an object and in its momentum – its knowledge of where it is going. Position is, clearly, a particle property. You can say exactly where a classical particle is. Equally, you cannot say where a classical wave is, only which region of space it is passing through, because by its very nature a wave is a spread-out thing. In the world of classical mechanics, waves do not have position in the same sense as particles, but they do have a direction – they carry momentum, and know where they are going.

Heisenberg showed that there is an intrinsic uncertainty associated with knowledge of position and momentum in the quantum world. You can never know *both* the position *and* the momentum of an entity such as an electron at the same time. If you try to measure the momentum precisely, this has the effect of enhancing the 'waviness' of the entity and making it spread out so that its position is uncertain. If you try to measure its position precisely, this creates uncertainty in its waviness, so that it is not sure where exactly it is going. The amount of uncertainty in position, multiplied by the amount of uncertainty in momentum, must always be more than a certain number, given by dividing *Planck's constant* by $2 \times \pi$ (the number is written ℏ, and is called 'h-cross').

This is *not* simply a consequence of the difficulty of making such measurements in experimental terms. Obviously, it is hard to measure the position and momentum of a single electron, and just by measuring it (perhaps by bouncing photons off it), you change the properties you are trying to measure, as the electron recoils from the impact of the photons. But quantum uncertainty is intrinsic to the very nature of the entities of the quantum world. An entity such as an electron literally does not have both a precise momentum and a precise position; it does not 'know' itself both exactly where it is and exactly where it is going at the same time.

The size of the effect is tiny in everyday terms – the critical number, h-cross, is about 10^{-34}, in standard units where mass is measured in grams; that is a measure of the uncertainty in position (in centimetres) of an object weighing about 1 gram. The more massive the object, the less the uncertainty. But for an electron, with a mass of just 10^{-27} g, the effect can be significant.

The importance of this uncertainty to astronomy is that the same sort of relationship links the energy of an object, or even of a region of empty space, to the length of time for which it is observed. If you take a long and careful look at something, you can measure its energy as accurately as you like. But if you only take a quick look, there is always going to be uncertainty in the energy – not just in the energy you measure, but in the amount of energy that is actually there. Just as a quantum entity does not 'know' its own position precisely, so it (and the Universe at large) does not 'know' exactly how much energy it has, for a short time interval. It is this quantum uncertainty that allows electron–positron pairs (and other particle–antiparticle pairs) to appear out of nothing at all (out of the vacuum), provided that they annihilate one another in the brief flicker of time allowed by quantum uncertainty. This is the source of the **Hawking radiation** associated with **black holes**. It is even possible that the entire Universe was created in this way, by **inflation** out of a quantum fluctuation of the vacuum (see **free lunch Universe**).

The ultimate hope of many physicists is to unify quantum theory with the **general theory of relativity** in a 'theory of everything' (see **grand unified theories**). The test bed for such theories is how well they can explain the behaviour of the very early Universe, when conditions were more extreme than those in the highest-energy collisions achieved in particle accelerators here on Earth.

Further reading: John Gribbin, *Schrödinger's Kittens*; Nick Herbert, *Quantum Reality*.

quark One of the two families of particles (see also **leptons**) from which all known matter is made. Quarks come in six varieties, which are related to one another in three pairs; they have been given arbitrary and somewhat whimsical names which merely serve as labels, and have nothing to do with their properties.

The first pair are called 'up' and 'down'; the up quark has an electric **charge** of $+\frac{2}{3}$ and the down quark a charge of $-\frac{1}{3}$, in units where the charge on an **electron** is -1. These are the counterparts in the quark family of the electron and its **neutrino** in the lepton family. A **proton** is composed of two up quarks and one down, while a **neutron** is composed of two down quarks and one up. Together with the electron and its neutrino, these are all the particles you need to describe the structure of ordinary matter made of **atoms**.

For reasons which are still not fully understood, the whole pattern is duplicated twice in nature, with successively heavier 'generations' of particles. The next pair of quarks is dubbed 'strange' and 'charm' (with lepton counterparts the **muon** and its neutrino), and the heaviest pair is made up of 'top' and 'bottom' (with lepton counterparts the **tau** particle and its neutrino). Together, these combine in threes to make up other **hadrons**.

Quarks, unlike leptons, have an additional property called 'colour' (again, an arbitrary name) which is analogous to electric charge but comes in three varieties. All hadrons have zero net colour, produced by suitable combinations of quarks in threes – this is why there are three quarks inside each of the **baryons**, including the proton and neutron. The colour force between quarks (equivalent to the electromagnetic force between electrically charged particles) is carried by **gluons** (equivalent to **photons**). The **strong nuclear interaction** is a side effect of the colour (or 'glue') force operating between quarks.

It is also possible for quarks and antiquarks to exist in paired combinations known as mesons (see *elementary particles*), but isolated quarks cannot exist on their own except under conditions of extremely high temperature and energy, such as those which existed in the *Big Bang* (see *quark–hadron phase transition*).

quark era   See *hadron era*.

quark–gluon plasma   See *quark–hadron phase transition*.

quark–hadron phase transition   A change in the physical state of matter at the end of the *quark era*, early in the *Big Bang*, when individual *quarks* could no longer roam freely through the primordial fireball, but became held together inside *hadrons* (including the *baryons* from which everyday atomic matter is made). This is analogous to the transformation of electrically neutral *atoms* into a mixture of positive *ions* and negative *electrons* – a *plasma* – at temperatures of a few thousand *Kelvin*.

The quark–hadron phase transition occurred between one-millionth of a second and one-hundred-thousandth of a second after the outburst from an initial *singularity*. At this time, the Universe cooled to the point where *quarks* could combine to form *hadrons*. Just before the quark–hadron phase transition, the Universe consisted of a soup of quarks and *gluons* (the particles which carry the so-called 'colour' forces between quarks) known as a 'quark–gluon plasma' (sometimes shortened to 'quagma'). The temperature in the very early Universe was so high that individual quarks had too much energy for the gluons to be able to hold them together. But as the temperature fell, the energy of the individual quarks dropped to the point where they could be confined by the colour force. The quark–gluon 'era' ended about one-hundred-thousandth of a second after the moment at which the Universe burst out from the initial *singularity* (or, more probably, from a state with the *Planck density*). The temperature at that time was about 10,000 billion Kelvin (10^{13} K).

Physicists are now attempting to recreate the conditions that existed in the quark–gluon plasma by colliding massive *nuclei* together in particle accelerators on Earth. These 'little bangs' may eventually involve smashing gold nuclei (each with 197 times as much *mass* as a *proton*) head-on into one another at 0.999957 times the *speed of light*. The first steps towards this have been carried out using nuclei of sulphur, each weighing 32 *atomic mass units*.

Such short-lived mini bangs release a flood of radiation, and particles of many kinds, manufactured by the conversion of energy into mass. It is difficult to unravel the complexity of all the particles produced. But measurements of the energy of individual *photons* produced in these events, carried out in 1994, unambiguously confirm that the photons are coming from a quark–gluon plasma. This is one of the most important experimental confirmations of the theory of the Big Bang fireball.

quasar The highly energetic core of an *active galaxy*. The first quasars to be found were detected by their radio emission, and were termed quasistellar radio sources; when other quasars were found which do not emit radio noise, the name was changed to quasistellar object, or QSO. Only one quasar in 200 is actually a *radio source*. In either case, the name 'quasar' can be regarded as a contraction of 'quasistellar'.

In the early 1960s, radio astronomers knew of many intense sources of radio noise which could not be identified with optical stars or galaxies. The British astronomer Cyril Hazard pointed out that in 1963 the Moon would pass in front of one of these sources, number 273 in the Third Cambridge catalogue. This *occultation* enabled astronomers to pinpoint the position of the source 3C 273 by timing the moment when the edge of the Moon cut off the radio noise. This led to the identification of 3C 273 with a star-like object. But the 'star' has a large *redshift*, placing it far outside our Milky Way Galaxy. Later studies showed that quasars lie at the hearts of *galaxies* which are themselves too dim to be seen clearly at such great distances.

Flickering variations in the brightness of quasars show that their energy is coming from a region about 1 light day across, roughly the size of our Solar System. But in order to be seen so brightly at such large distances, they must radiate about 1,000 times as much *energy* as all the stars in the Milky Way put together. The best explanation for this is that each quasar is a supermassive *black hole*, equivalent to perhaps 100 million solar masses of material, which is swallowing matter from its surrounding galaxy at a rate of about 1 solar mass per year. The energy released when this occurs can be as much as half of the theoretical maximum implied by Albert *Einstein*'s equation $E = mc^2$, ample to explain the energetic output of a quasar.

Some quasars are the most distant objects known, with redshifts greater than 4. They are seen by light which left them when the Universe was less than 20 per cent of its present age – more than 10 billion years ago on the standard *model* of cosmology, only 2 or 3 billion years after the *Big Bang* (see *look back time*).

In 1995 astronomers were puzzled when observations made with the *Hubble Space Telescope* failed to show any sign of the expected surrounding galaxy for eight out of a sample of fourteen quasars studied. Instead, they found that some of these quasars are associated with companion galaxies, which lie within 10 *kiloparsecs* of the quasar (less than the diameter of the Milky Way) and are being distorted by the intense gravitational pull of the quasar. One possible explanation is that the giant black holes associated with quasars form before star formation has switched on in the surrounding cloud of baryonic material, so that at first only the quasar is visible, with the galaxy growing up around it later. Another possibility is that material from smaller galaxies that venture too close to a giant black hole may ignite quasar activity as the matter falls into the black hole – or reignite a black hole that has swallowed up all of the original material that surrounded it. But details of the relationship between quasars and galaxies may need rethinking in the light of these observations.

radar astronomy Technique of bouncing *radio waves* off astronomical objects, including *meteors* and *planets*, and studying the reflections to investigate their properties. Radar (from the *Arecibo radio telescope*) was used to map Venus before it was visited by *spaceprobes*, and has been used to measure the distances to the inner planets extremely precisely, providing a calibration of the *astronomical unit*.

radial velocity The speed with which an object such as a *star* is moving directly towards or away from us. Radial velocity can be measured directly by the *Doppler effect*, but this does not tell us how the object is moving across the line of sight (see *transverse velocity*).

radian A unit of angular measurement. There are 2π radians in a complete circle of 360°, so one radian is 57.3°.

radiation Term widely used to describe any process which carries *energy* through space. The most familiar form is *electromagnetic radiation*, but the word is also used to describe energetic particles such as *alpha particles* or *beta particles*, and even for *gravitational radiation*.

radiation era The epoch in the early Universe that was dominated by *electromagnetic radiation*, from about 1 second after the birth of the Universe to 300,000 years later, the time when *recombination* occurred. See *Big Bang*.

radiation pressure *Pressure* exerted on an object by *electromagnetic radiation*, best thought of in this context as a stream of *photons* pushing on the object. Radiation pressure on dust grains in space can dominate over the force of gravity, and explains why the tenuous tail of a *comet* always points away from the Sun – it is pushed outward by radiation pressure.

radioactive decay Process whereby an unstable *nucleus* or *particle* spits out one or more particles and transforms into a stable nucleus or particle. The extreme case where a massive nucleus splits into two roughly equal parts is generally referred to as *nuclear fission* rather than decay, but also involves the ejection of other particles as well as the two 'half nuclei'.

The two main kinds of radioactive decay associated with atomic nuclei are *alpha decay* and *beta decay*. They have the effect of transforming a radioactive original nucleus (called the parent) into a nucleus of another element (called the daughter), which may or may not be radioactive itself. Decay happens on a characteristic timescale known as the *half life*. This kind of decay may occur in a chain several steps long before ending up with a stable nucleus. Decay may also involve the release of energy in the form of electromagnetic radiation.

Unstable particles decay in similar ways – the classic example is the beta decay of a lone neutron to produce a proton, an electron and a neutrino. According to the accepted *models* of particle physics (including the *grand unified theories*, or GUTs), the only reasonably stable particles are the lightest family of *quarks* and the electron. Everything else – more massive particles created out of energy in particle accelerators on Earth or in violent events in the Universe – will decay, ultimately into either quarks or electrons. The quarks form protons and neutrons, and the neutrons decay into protons and electrons. Even the proton ought to be unstable, changing itself into a positron and a particle called a pion, which itself decays into two *gamma rays*; the positrons produced in this way would annihilate with electrons to make more gamma rays, mirroring the way in which matter is thought to have been created out of energy in the *Big Bang*. The *supersymmetry* versions of the GUTs suggest that the lifetime of the proton is, however, about 10^{45} years, so there is no immediate prospect of the Universe disappearing in a puff of gamma rays.

radio astronomy   Study of the Universe using *radio waves*, especially *electromagnetic radiation* with wavelengths from the *microwave* band (a few centimetres) upwards. *Millimetre astronomy*, bridging the gap between *infrared astronomy* and conventional radio astronomy, strictly speaking also involves radio waves, but is usually regarded as a distinct sub-discipline. Radio astronomy plays a key role in the investigation of the Universe, especially the study of sources such as the *background radiation*, *HI regions*, *HII regions*, *pulsars*, *radio galaxies*, *quasars* and *supernova remnants*.

Radio waves from outside the Solar System were first detected (as a source of noise in radio communications equipment) by Karl *Jansky* in the early 1930s, but this roused little interest among astronomers at the time, and only Grote *Reber* followed up the discovery in the 1930s by building a dedicated *radio telescope* to investigate the discovery. During the Second World War, the development of radar provided the technological base for what was to become radio astronomy: astronomical sources of radio noise also showed up on wartime radar equipment, and after the war there were both scientists experienced in radar work and surplus radar equipment available to give radio astronomy a start. Pioneering development in these early years of radio astronomy was particularly strong in Britain, where Bernard *Lovell* was the guiding force behind the construction of the radio telescope at *Jodrell Bank* that now bears his name, and Martin *Ryle* headed a rival team of radio astronomers in Cambridge.

Radio astronomy from the surface of the Earth is practicable only because the atmosphere of the Earth lets radio waves through. The atmosphere actually blocks most wavelengths of electromagnetic radiation, with only visible light, some infrared radiation and millimetre wave radiation, and plenty of radio waves in the range from about 1 cm wavelength to about 30 m wavelength able to get through 'windows' in the atmosphere to reach the ground.

Although many sources of radio noise were found to be associated with objects in our own Galaxy, such as the *Crab nebula*, the key discovery made by radio astronomy in the 1950s was the existence of large numbers of radio galaxies, objects far beyond the *Milky Way* that are in many respects similar to radio-quiet galaxies but which produce far more radio noise than a galaxy like our own or the *Andromeda galaxy*. At about the same time, Soviet physicists developed the theory of *synchrotron emission*, which helped to explain how this intense radio noise is generated. It was radio astronomy that first showed the Universe to be a violent place in which huge amounts of energy are being released by events going on in the nuclei of some galaxies.

Radio galaxies are typically only faintly visible using optical astronomy, so the identification of radio sources with optical galaxies became an important way of locating interesting extragalactic objects. In 1960, one of these faint radio galaxies was found to have a *redshift* of 0.46, which made it the most distant object known at the time. This encouraged optical astronomers to search for more record-breaking redshifts in the light from the optical counterparts to radio galaxies, and led to the discovery of *quasars* a few years later. These objects have since been investigated in detail using techniques such as *interferometry*, which can pick out details in the structure of radio sources on as fine a scale as 0.001 of an *arc second*, equivalent to being able to measure the angular size of a baseball at a distance of 15,000 km. By the 1990s, the record redshift had been pushed out to above 4 (corresponding to looking back in time across 90 per cent of the history of the Universe since the *Big Bang*), and there were more than 20,000 extragalactic sources logged by radio astronomers, compared with something over 50,000 optical galaxies in the *catalogues* (many more optical galaxies have been photographed on wide-angle views of the sky, but have never been studied in detail; just over half of the catalogued extragalactic radio sources have been identified with optical counterparts).

The development of radio astronomy in the 1960s also saw the accidental discovery of the background radiation by Arno *Penzias* and Robert *Wilson*, and the (also fortuitous) discovery of pulsars by Jocelyn *Bell Burnell*. The first discovery transformed *cosmology* by showing astronomers that there really had been a Big Bang, and making the study of the early Universe respectable science; the second showed that violent events are also taking place in our own Galaxy, and pointed the way to investigations that led to the identification of stellar-mass *black holes*. Much larger black holes, containing many millions of times as much mass as there is in a star like our Sun, were invoked to explain the sources of energy in quasars and radio galaxies.

But the story of radio astronomy is not solely one of violence in the Universe. Early in the development of radio astronomy, in the 1950s, clouds of hydrogen in space were identified by their characteristic emission at a wavelength of 21 cm. These investigations provide a way to study not only the distribution of hydrogen around the Galaxy, but also its motion – since the radiation is emitted at a very precisely known wavelength, the motion of the emitting clouds shows up clearly through the *Doppler effect*. It was 21 cm mapping that first provided direct and unambiguous evidence that the Milky Way is indeed a rotating *disc galaxy* with *spiral arms*. Nearby galaxies such as the Andromeda galaxy can be mapped in the same way, and are very similar at radio wavelengths to our own Galaxy.

Radio astronomy is also a key tool in the identification of *interstellar molecules*. Before 1963 the only compounds identified in space were methylidyne (CH) and cyanogen (CN); the first interstellar compound to be identified by radio astronomy (the hydroxyl radical, OH), was detected in 1963 from its spectral trace at a wavelength of 18 cm, and scores of molecules have now been identified in *interstellar clouds*, mostly using millimetre astronomy.

Radio techniques are also useful in studying the Sun, where features such as *sunspots* and *solar flares* can be investigated by their emission at radio wavelengths. Jupiter is also a weak source of radio noise, but the other planets and moons in the Solar System are radio-quiet, although they can be studied by bouncing radio waves off them (*radar astronomy*). But the most important feature of the development of radio astronomy since the 1940s has been not its ability to provide more information about objects that are already familiar to us at optical wavelengths, but the way in which it has opened up the investigation of the Universe by revealing new and unexpected objects that could not be investigated in any other way.

radio galaxy Any *galaxy* that is a strong emitter of *radio waves*. About one galaxy in every million is classified as a radio galaxy, and a typical radio 'brightness' of such a galaxy is about 1 million times that of the *Milky Way*. This activity may be associated with the presence of *black holes* at the centres of these galaxies, which resemble *quasars* in many ways.

radio source Any source of radio noise is, strictly speaking, a radio source, but cosmologists use the term specifically to refer to *radio galaxies* and *quasars*.

radio telescope The whole system used to collect and examine *radio waves* from space. A radio telescope is made up of an *antenna*, which collects the

Radio telescope. The large dish antenna of the radio telescope at Lessives in Belgium.

radio noise, an amplifier and a detector/recorder. Radio telescopes come in many sizes and shapes. Some use huge steerable dishes like the *Lovell telescope*, others use long rows of wire spread out across large fields.

radio waves *Electromagnetic radiation* with *wavelengths* in the range from a few millimetres (on the long side of *infrared radiation*) up to hundreds of kilometres – in principle, indefinitely long.

The existence of what we now call radio waves was predicted by James Clerk *Maxwell* in the 1860s, when he discovered the equations that describe the way electromagnetic disturbances move through space, and found that they move at the *speed of light*, revealing that *light* itself is a form of electromagnetic radiation. The long-wavelength radiation predicted by Maxwell was first produced artificially by the German physicist Heinrich Hertz (1857–94) in 1888; it was not until the 1930s that radio waves from space were discovered (see *radio astronomy*).

RATAN 600 *Radio telescope* located in the Caucasus Mountains, made up of 895 panels each 2 m across, in a circle nearly 600 m across (actually 576 m across, but 600 makes a nicer name). The dishes can be linked electronically to work as a single parabolic *antenna*, or each quadrant of the circle can be used as a separate antenna system.

Reber, Grote (1911–) American radio enthusiast who became the first radio astronomer after he heard of Karl *Jansky's* discovery of radio noise from space. Reber built the first dedicated radio telescope in his backyard in Illinois, in 1937. Although years ahead of anyone else in studying the Universe at radio wavelengths, Reber never became part of the mainstream astronomical establishment and eventually settled in Tasmania (in 1954), where he could observe the southern skies, including the centre of the *Milky Way*, far from any sources of man-made radio noise. He was still actively involved in these observations in the mid-1990s.

recession of galaxies Motion of *galaxies* (strictly speaking, *clusters of galaxies*) away from one another in the expanding Universe, caused by the stretching of the space between the galaxies, not by motion of galaxies through space. See *redshift*.

recombination Process that occurred when the Universe was about 300,000 years old and it had cooled to the point where *electrons* could combine with *nuclei* to make electrically neutral *atoms*. This happened at about the temperature at the surface of the Sun today, some 6,000 *Kelvin*. Since neutral atoms had never existed before, this was strictly speaking 'combination', not 'recombination', but the name has been borrowed from studies of *plasmas* which are first heated until they form *ions* and are then allowed to cool and recombine. See *Big Bang*.

recombination era The time when *recombination* took place.

recurrent nova A *nova* which has been seen to flare up more than once. All novae are probably recurrent, but the others happen to have been seen to flare up only once since astronomers began looking at them.

reddening Effect of dust grains in space on the light from distant stars. In exactly the same way that dust in the air makes sunsets red, shorter (blue) wavelengths of light are scattered by the dust more effectively than longer (red) wavelengths, so the red light penetrates dust more easily. This is why *infrared* studies are often the best way to investigate dusty regions of the *Milky Way*. Reddening has nothing to do with the *redshift*.

red dwarf Name of a cult BBC TV science fiction sitcom of the 1990s. It was borrowed from the term used by astronomers to describe M and K stars at the cool end of the *main sequence*, with *masses* between about 20 per cent and 80 per cent of the *mass* of the Sun, and surface temperatures in the range from 2,500–5,000 K. Even smaller red dwarfs are allowed to exist according to standard stellar *models*, but none has been found during searches by the *Hubble Space Telescope*.

red giant Name used by astronomers to describe M and K stars (and some others) which have evolved off the *main sequence* and expanded to 10 or 100 times the diameter of the Sun. They have surface temperatures similar to those of *red dwarfs*, but radiate more energy because they have larger surface areas. Red giants can have a wide range of *masses*, up to tens of solar masses. See *stellar evolution*.

redshift *Light* or other electromagnetic radiation from an astronomical object may be stretched, making its wavelength longer, by any of three effects. Because red light has a longer wavelength than blue light, the effect of this stretching on features in the optical *spectrum* is to move them towards the red end of the spectrum. So all three processes go by the name 'redshift'.

The first kind of redshift was described by Christian **Doppler**, a professor of mathematics in Prague, in 1842. It is caused by motion. An object such as a *star* which is moving away from the observer will show a redshift in its spectrum, compared with the spectrum from a stationary star, because the moving star stretches out the light emitted behind it. In a similar way, light waves from a star moving towards us are squashed by the motion of the star. This means that the light has shorter wavelengths, and is said to be blueshifted.

Exactly equivalent changes occur in the wavelength (pitch) of sound waves emitted by a moving object. Sound waves from an object moving towards you are squashed, and have a higher pitch; sound waves from an object moving away are stretched, and are lower in pitch. Both effects are familiar to anyone who has heard an ambulance or other emergency vehicle race past with its siren wailing. Both for sound waves and for electromagnetic radiation, the process is known as the **Doppler effect**.

Measurements of redshifts and **blueshifts** caused by the Doppler effect enable astronomers to work out how fast stars are moving through space, and to measure, for example, the way that galaxies rotate. Astronomical redshifts are measured in terms of the fractional change in redshift, known as z. If $z = 0.1$, the wavelengths have been lengthened by 10 per cent, and so on. Provided the speeds involved are much less than the speed of light, then z is also equal to the speed of the moving object divided by the speed of light. So a redshift of 0.1 would mean that the star was receding from us at one-tenth of the speed of light.

In 1914, Vesto **Slipher**, working at Lowell Observatory, discovered that eleven out of fifteen of the objects then known as spiral nebulae (now called **galaxies**) showed redshifts in their light. In the 1920s, Edwin **Hubble** and Milton **Humason**, working at Mount Wilson Observatory, extended these observations. First, Hubble established that the nebulae are other galaxies, like our **Milky Way**. Then they found that the vast majority of galaxies have redshifts in their light. By 1929, Hubble had established, mainly by comparing the redshifts of galaxies with their apparent brightness, that the redshift is proportional to the distance of a galaxy from us (this is now known as Hubble's Law). The rule breaks down only for a handful of the nearest neighbours in space to the Milky Way, such as the **Andromeda galaxy**, which show blueshifts in their spectra.

At first, the redshifts in the light from distant galaxies were interpreted as a Doppler effect caused by their motion through space, as if they were all flying apart from an explosion centred on the Milky Way. But it was quickly realized that this kind of expansion was implicitly predicted by the equations of the *general theory of relativity*, which had been published more than ten years before the discovery of Hubble's Law. When Albert *Einstein* himself had first applied those equations to provide a description of the Universe (a cosmological *model*) in 1917, he had found that they required the Universe to be in motion – either expanding or contracting. The equations did not allow for the possibility of a stationary model. Because nobody knew at the time that the Universe was expanding, Einstein fiddled his equations by introducing an extra factor to hold the models still; he later described this as the 'biggest blunder' of his career.

With the fiddle factor removed, Einstein's equations exactly described the behaviour observed by Hubble. The equations said that the Universe should be expanding, not because galaxies were moving through space, but because the empty space between the galaxies (strictly speaking, *spacetime*) is expanding. The cosmological redshift results because the light from distant galaxies is stretched by the amount that space expands while the light is en route to us.

Because the redshift is proportional to distance, this provides cosmologists with a yardstick with which to measure the Universe. The measuring rod has to be calibrated by measurements on relatively nearby galaxies, and there is some uncertainty about the calibration (see *cosmological distance scale*), but this is still the single most important discovery in cosmology. Without a way to measure distances, cosmologists could not begin to understand the nature of the Universe, while the exact nature of Hubble's Law shows that the general theory of relativity is a good description of the way the Universe works.

For historical reasons, the redshifts of galaxies are still described in terms of velocities, even though astronomers know they are not caused by motion through space. The distance to a galaxy is equal to its redshift 'velocity' divided by a constant, known as Hubble's constant. It has a value of about 60 km per second per *Megaparsec*, which means that a galaxy shows a redshift velocity of 60 km/sec for every million *parsecs* that it is distant from us. This cosmological redshift is tiny for our nearest neighbours, and the blueshifts seen in the light from galaxies such as Andromeda are simple Doppler effect blueshifts caused by their motion through space. Galaxies in distant clusters (like swarms of bees) show a spread of redshifts around some central value; the central value is the cosmological redshift of the cluster, and the variations are caused by the Doppler effects of the motion of the galaxies within the cluster.

Hubble's Law is the only kind of redshift/distance law (except for a stationary universe) which 'looks the same' from any galaxy in the Universe. *Every* galaxy (except very close neighbours) is receding from every other galaxy in accordance with the law, and there is no 'centre' to the expansion. The usual analogy is with paint spots dotted over the surface of a balloon. As the balloon is inflated, the paint spots get further apart from one another, because the skin of the balloon expands, *not* because the paint spots move over the surface of the balloon. Measurements made from any spot will show all the other spots receding uniformly, in accordance with Hubble's Law.

For large redshifts, corresponding to velocities greater than about one-third of the speed of light, redshift calculations have to take account of the requirements of the *special theory of relativity*. So a redshift of 2 does not mean that an object has a cosmological 'velocity' of twice the speed of light. In fact, $z = 2$ corresponds to a cosmological velocity of 80 per cent of the speed of light. The most distant *quasars* known have redshifts a little greater than 4, corresponding to 'velocities' just over 90 per cent that of light; the record redshift for a galaxy is held by an object known as 8C 1435 + 63, with a redshift of 4.25. The cosmic microwave *background radiation* has a redshift of 1,000.

The third kind of redshift is produced by gravity, and is also explained by Einstein's general theory. Light moving outwards from a star is travelling 'uphill' in the star's gravitational field, and loses energy as a result. When a material object such as a rocket moves upward in a gravitational field, it loses energy and slows down as a result (which is why the rocket motors have to fire to boost it into orbit). But light cannot slow down; it always travels at the same speed, a fraction under 300,000 km/sec, denoted by c. Instead of slowing down, when light loses energy its wavelength increases. In other words, it is redshifted.

This gravitational redshift applies, in principle, to light escaping from the Sun, or even to light shone upwards from a torch on the surface of the Earth. But it is only big enough to be measurable for intense gravitational fields, like those at the surface of a *white dwarf* star. A *black hole* can be thought of as an object with such a large gravitational attraction that it produces an infinite redshift in light trying to escape from it.

All three kinds of redshift can be at work at the same time. If we had telescopes sensitive enough to see light from a white dwarf star in a distant galaxy, the overall redshift in that light would be due to a combination of Doppler, cosmological and gravitational redshifts.

Further reading: John Gribbin, *In Search of the Big Bang*; William Kaufmann, *Universe*.

redshift survey Map of *galaxies* across a large volume of space, which combines information about their distances (from *redshift* measurements) with their positions on the sky to give three-dimensional information.

Rees, Sir Martin John (1942–) British astrophysicist, born in York, who became the 15th Astronomer Royal (succeeding Arnold *Wolfendale*) at the beginning of 1995. Educated at Trinity College, in Cambridge, Rees has spent most of his career at the Institute of Astronomy in the same city, apart from a short time at the University of Sussex in 1972–3. He spent two spells as director of the Institute (1977–82 and 1987–91), and was Plumian Professor of Astronomy and Experimental Philosophy from 1973 to 1991, resigning from the post (and from the directorship) in order to free himself from administrative duties and devote more time to research and to communicating about science.

His research interests span almost the entire range of *astrophysics*, from the nature of *quasars* and *active galaxies* to the birth of the Universe, *black holes*, the mystery of *dark matter* and *anthropic cosmology*. In all of these areas – and more – Rees has made major contributions. His most significant contribution to astrophysics, however, may have been as a teacher and inspiration for younger scientists, many of whom have benefited from his advice and encouragement. He was knighted in 1992.

Further reading: John Gribbin and Martin Rees, *The Stuff of the Universe.*

reflecting telescope (reflector) A *telescope* that gathers light and magnifies an image primarily through the use of a curved mirror. The first reflector was developed by Leonard *Digges* in the 16th century, but the idea was never put to practical use until it was reinvented by Isaac *Newton* in the second half of the 17th century. The first reflectors used a polished metal mirror. Metal mirror telescopes were used to great effect by William *Herschel* and Lord *Rosse* in the 18th and 19th centuries, but at the end of the 19th century techniques for producing large silvered glass mirrors became available, and these formed the basis of the great *optical telescopes* of the 20th century. In a new development in the 1990s, Canadian researchers have developed a reflecting telescope system which uses a curved rotating dish covered with a thin layer of mercury as the mirror.

reflection nebula A *nebula* which shines by reflected *light*.

refracting telescope (refractor) A *telescope* that gathers light and magnifies an image primarily through the use of lenses. The first refractor was probably made by Leonard *Digges* in the second half of the 16th century. The invention was made independently a little later in Holland, and the Italian *Galileo* was the first person to use such a telescope for astronomical observations, in the first decade of the 17th century. When the principle of the *reflecting telescope* was rediscovered and put to practical use by Isaac *Newton*, refractors fell out of favour for astronomical work because (unlike reflectors) they suffer from a problem called chromatic aberration, which produces brightly coloured distortions in the images.

The problem of chromatic aberration was overcome by the invention of the achromatic lens in the early 1800s, and during most of the 19th century refractors became the mainstay of *optical astronomy*. But there is a limit to how big a lens can be without sagging under its own weight (a mirror, of course, can be supported from the back), and this, combined with the development of silver-coated glass mirrors, meant that refractors had been superseded by the beginning of the 20th century. The largest refractor is still the 102 cm instrument at **Yerkes Observatory**, built in the 1880s (and still in use).

Reissner–Nordstrøm solution Solution to Albert **Einstein**'s equations of the **general theory of relativity** that describes a **black hole** which is not rotating but has an electric **charge**. This is not likely to occur in the real Universe, because such an object would attract opposite charge from its surroundings and soon become an electrically neutral **Schwarzschild black hole**. The solution was developed by Heinrich Reissner in Germany in 1916, and by the Finn Gunnar Nordstrøm in 1918, but they never worked together.

relative sunspot number See **Wolf number**.

relativistic astrophysics The use of theoretical **models** to describe the nature of astronomical objects under conditions where either the **special theory of relativity** or the **general theory of relativity** (or both!) has to be taken into account. This tends to be in places where there are high energies, densities or velocities involved, and under these extreme conditions it is often the case that **quantum theory** has to be taken into account as well. Classic examples of the application of relativistic astrophysics involve the study of **white dwarfs**, **neutron stars** and **black holes**, as well as **quasars**, **active galaxies** and the very early Universe (see **inflation**).

relativistic mechanics The modification to **classical mechanics** (which is itself based on **Newton's laws of motion**) that is required when dealing with objects moving at a sizeable fraction of the **speed of light**. Relativistic mechanics was developed by Albert **Einstein** with his **special theory of relativity**. Among other things, the equations of relativistic mechanics correctly describe how to add up velocities, so that no matter how much you try to increase the speed of an object, it will never go faster than light.

In fact, the special theory also describes perfectly the behaviour of objects that are moving at slow speeds – if the speeds involved are much less than the speed of light, the equations of relativistic mechanics give exactly the same 'answers' (including addition of velocities) as the equations of classical mechanics. Newton's laws are included within the special theory as a special case applying for slow speeds, so classical mechanics is actually more special (in the sense of being restricted in scope) than the special theory of relativity.

There is one curious feature of the Universe that is not always fully appreciated in this context. The special theory of relativity says that *all* inertial frames are equivalent. But at any point in the Universe there is a preferred frame of reference, the one in which the Universe is seen to be expanding smoothly in all directions, and in which the **background radiation** looks the same in all directions. The Universe itself is not, in the jargon of relativity theory, 'Lorentz invariant', and at any place in the Universe, the expansion defines a preferred frame of reference, the 'Hubble frame'.

This was first pointed out by William **McCrea** in the 1950s. It does not mean that there is anything wrong with relativistic mechanics, but it may mean that relativistic mechanics is not the whole story – for example, when we are dealing with the question of **inertia** and **Mach's Principle**. Intriguingly, the formulation of **quantum electrodynamics** is much simpler if the calculations are carried through solely for the Hubble frame, when unpleasant infinities disappear from the equations.

relativity theory   See *special theory of relativity* and *general theory of relativity*.

resolution   A measure of the ability of a *telescope* or other instrument to record fine detail in an image. If two *stars*, for example, are close together on the sky, you need a high resolution in order to show them as two separate points of light. A telescope with low resolution will show them as one larger blob.

resolving power = *resolution*.

rest mass   The *mass* measured for an object that is stationary, in the same *frame of reference* as the person (or apparatus) making the measurement. So the rest mass is the mass the object has in its own frame of reference.

The **special theory of relativity** tells us (and many experiments have confirmed) that the mass of an object increases when it travels at high speed. But this increase in mass is not detectable to anyone travelling with the fast-moving object, only to observers in the frame of reference where they see the object to be moving at high speed. For them, the measured mass of the moving object increases in accordance with the equations of *relativistic mechanics* until, if the object could be made to move at the *speed of light*, it would have infinite mass. This is one way of understanding why no 'subluminal' object can ever be accelerated to the speed of light.

The only way in which objects which do move at the speed of light – such as the photons of light itself – can avoid having infinite mass is if they have zero rest mass. In a sense, the concept of a rest mass for a photon is meaningless, since a photon is never at rest in any frame of reference; the cornerstone of the special theory of relativity is that in any frame of reference you will measure the speed of light to be the same. But conventionally all entities that travel at the speed of light are described as having zero rest mass, which conveniently means that they still have zero mass when travelling at light speed.

Bernhard Riemann (1826–66).

Rhea The second largest *moon* of Saturn. Rhea, discovered by Giovanni *Cassini* in 1672, is 1,530 km in diameter and has a density 30 per cent greater than that of water. It is very heavily cratered, and orbits Saturn at an average distance of 527,040 km.

Riemann, (Georg Friedrich) Bernhard (1826–66) German mathematician who was the first person to develop a comprehensive mathematical description of curved space (*non-Euclidean geometry*), providing the mathematical framework later used by Albert *Einstein* in his *general theory of relativity*.

Riemann was born at Breselenz, in Hanover, on 17 September 1826, one of six children of a Lutheran minister. He showed a precocious aptitude for mathematics, but was encouraged by his father to study theology. He entered Göttingen University at the age of twenty with a theological career in mind, but soon persuaded his father to allow him to switch to the study of mathematics. Riemann was taught at Göttingen by Karl Gauss (1777–1855), one of the greatest of all mathematicians, who had himself investigated one form of non-Euclidean geometry as far back as 1799, but never published his discoveries (this was par for the course for Gauss, who kept notes of his work in a private mathematical shorthand; where this has been deciphered, it shows that he made many discoveries that were later rediscovered and pub-lished by others, while undeciphered entries in his notebooks may refer to mathematical discoveries that have still not been rediscovered).

After spending two years studying in Berlin, from 1847 to 1849, Riemann returned to Göttingen and worked for his PhD, which was awarded in 1851. The accepted way for a young academic to get established in a German university in those days was to work at first as a kind of lecturer known as a *privatdozent*; the key feature of this post was that it carried no salary, but the lecturer received an income from the fees paid by students who chose to take the course. After two years on this first rung of the academic ladder, Riemann hoped to obtain a more secure position in the university. To demonstrate his suitability for such an appointment, the applicant had to give a lecture to the faculty of the university, and the rules required him to offer three possible topics for this lecture, from which they would choose. It was also a tradition that, while three choices had to be offered, the professors always chose one of the first two on the list. The story is that when Riemann handed in his list it gave two topics which he had thoroughly prepared, and, just to make up the numbers, an after-thought entitled 'On the Hypotheses Which Lie at the Foundations of Geometry'.

Riemann was certainly interested in geometry, but had not prepared anything to go with the title, never expecting it to be chosen. Unfor-tunately for him, but fortunately for succeeding generations of math-ematicians and cosmologists, Gauss, now in his seventies but still a dominating figure at Göttingen, found the title, which echoed one of his own youthful interests, irresistible, and the 27-year-old prospective academic found that geometry was what he would have to impress the professors with if he wanted a proper job.

Probably partly because of the strain of having to prepare a talk on which his whole career depended at short notice, Riemann fell ill, missed the date set for the lecture, and did not recover until after Easter in 1854. Happily this gave him a breathing space. He now had seven weeks to prepare the talk before the revised date set for it – only for Gauss to fall ill and call another postponement. The talk was eventually given on 10 June 1854. The lecture was not published until 1867, a year after Riemann's early death (from tuberculosis). But it covers a breathtaking array of topics, including a workable definition of the curvature of *space* and how it could be measured, the first description of spherical geometry (including the speculation, presaging the modern version of the idea of *black holes*, that the space in which we live might be gently curved, so that the entire Universe is closed up, in three dimensions, in the same way that the surface of a sphere is closed in two dimensions), and, most important of all, the extension of geometry into more than three dimensions with the aid of algebra.

Needless to say, the professors were sufficiently impressed to give him a more secure appointment. In 1855, less than a year after Riemann gave his lecture on geometry, Gauss died. Then, in 1859, on the death of Gauss's successor, Riemann became the professor himself, less than five years after the famous lecture. He made many other contributions to mathematics during his short life (but none of any direct relevance to *cosmology*), and died only seven years later, at the age of 39, on 20 July 1866.

right ascension One of two coordinates used in astronomy to define the position of an object on the sky (see *declination*). Right ascension (RA) is the angular distance of the object eastward from a standard point, known as the vernal *equinox* – equivalent to celestial longitude. It is measured in hours, minutes and seconds; 1 h = 15 arc degrees.

ripples in time See *background radiation*.

Robertson–Walker metric A *metric* that describes the properties of *space-time* in a *universe* that is *homogeneous* and *isotropic* – that is, one which obeys the *cosmological principle*. The metric was developed independently by the American Howard Robertson and the British mathematician A. G. Walker in 1935. It is the basis of mathematical *models* of *cosmology* incorporating the *Big Bang*. The Robertson–Walker metric allows spacetime to be separated mathematically into two components – curved *space* and cosmic *time*; although this seems obvious in terms of common sense (because we live in a Universe where this metric applies), under more general conditions the distinction between space and time is blurred. That is why the region around a *black hole*, for example, has to be described using the *Kerr metric* or the *Schwarzschild metric*.

Roche, Edouard Albert (1820–83) French mathematician who studied at the University of Montpelier and in Paris, and who was professor of mathematics in Montpelier from 1852 to 1881. He calculated, in 1850, the *Roche limit* for a *moon* in *orbit* around a *planet*, and later carried out a mathematical analysis of the *nebular hypothesis*.

Roche limit If a moon in orbit around a planet with the same density approaches the planet to within 2.456 times the radius of the planet, it will be broken up by gravitational forces. This is the Roche limit. Alternatively, a ring of particles orbiting within the Roche limit can never form a moon by *accretion*. The rings of Saturn are within the Roche limit for the planet. Small *satellites* can orbit intact within the Roche limit because they are held together by the forces between *atoms* and *molecules*.

Roche lobe See *equipotential surfaces*.

Rømer, Ole (1644–1710) Danish astronomer who measured the speed of light in 1675, using observations of *eclipses* of the *moons* of Jupiter to reveal how long it takes *light* to cross the *orbit* of the Earth. He was working in Paris at the time, with Giovanni *Cassini*. The figure he came up with was equivalent, in modern units, to 225,000 km per second. Using the same calculation with the modern value for the size of the Earth's orbit gives 298,000 km/sec; the modern value for the speed of light is 299,792 km/sec.

Roque de los Muchachos Observatory   Observatory at an altitude of 2,400 m on the island of La Palma in the Canaries. This is one of the best observing sites in the world. *Telescopes* are operated there by many European countries, and include the *William Herschel Telescope* and the *Isaac Newton Telescope*.

ROSAT   Short for Röntgenstrahlen Satellit, a name honouring the discoverer of *X-rays*, Wilhelm Röntgen (1845–1923). It is a German/British/American *satellite* for *X-ray astronomy*, launched in June 1990. It carried out the first complete survey of the sky in the part of the *spectrum* bridging the far *ultraviolet* and X-rays (XUV).

Rosse, William Parsons (Third Earl of) (1800–67)   Irish astronomer, educated at Trinity College in Dublin and at Oxford University, where he graduated in 1822. He served as a Member of Parliament from 1822 to 1834, but resigned to have more time for astronomy. He built a series of *reflecting telescopes*, at the family seat of Birr Castle, in central Ireland, culminating in the 183 cm *aperture* 'Leviathan of Corkstown', contained in a tube nearly 18 m long supported between two stone pillars 15 m high and 7 m apart. The telescope had a polished metal mirror. Rosse began observations with the Leviathan in 1845. In spite of the bad weather for astronomy in Ireland, and the fact that the *telescope* could not be tracked across the sky, Rosse made some important observations, including the fact that some *nebulae* have a spiral structure. He also studied and named the *Crab nebula*, and noted the ring-like structure of *planetary nebulae*.

After Rosse's death, his son, the fourth Earl, continued the astronomical tradition at Birr Castle, but the telescope was dismantled when he died in 1908. (See photo on p. 418.)

rotation curve   A graph which indicates the way in which the speed of rotation of *stars* in a *disc galaxy* (that is, the speed with which they are moving in their orbits around the galaxy) depends on their distance from the centre of the galaxy. All rotation curves for disc galaxies show the same characteristic shape. The orbital speed of the stars (revealed by the *Doppler effect*) increases rapidly over the first few *kiloparsecs* out from the centre, then levels off and stays largely flat to the edge of the visible disc. This pattern of behaviour can be explained only if the entire visible disc of stars is being held in the gravitational grip of a much larger halo of unseen

Rosse's telescope. The great 72-inch (1.8 metre) reflecting telescope set up at Birr Castle in Ireland by the third Earl of Rosse in 1845.

material. Such observations provide one of the most direct and straight-forward indications of the presence of *dark matter* in the Universe.

Royal Greenwich Observatory (RGO)   Observatory founded by Charles II at Greenwich in 1675. It remains the main organization responsible for *optical astronomy* in Britain, although it has moved twice (first to Herst-monceux, in Sussex, and then to its present home in Cambridge), and the most important *telescopes* that are in its care are now overseas.

The original purpose of the RGO was to tackle the problem of deter-mining longitude at sea, a major necessity for a maritime nation in the 17th century. It was as a result of this work that the meridian through Greenwich became established as the zero point of latitude by the Wash-ington Conference of 1884. The RGO has also played an important his-torical role in timekeeping. During the 20th century, however, it also became a centre for *astrophysics*.

From the foundation of the RGO until 1972, the head of the observ-atory was given the title Astronomer Royal. After Sir Richard *Woolley* retired from these twin duties in 1971, Margaret *Burbidge* was appointed director of the RGO, but in a shameful split with tradition she was not made Astronomer Royal. Instead, the title became a purely honorary one and went to Sir Martin *Ryle*. No woman has ever been appointed Astronomer

Royal; the fifteen holders of the post to date have been: John *Flamsteed*, Edmund *Halley*, James *Bradley*, Nathaniel *Bliss*, Nevil *Maskelyne*, John *Pond*, George *Airey*, William *Christie*, Frank *Dyson*, Harold Spencer *Jones*, Richard Woolley, Martin Ryle, Francis Graham *Smith*, Arnold *Wolfendale*, and Martin *Rees*. Up to the end of 1994, the average incumbency was just under 23 years, and the longest two (Maskelyne and Airey) were 46 years each.

The RGO operates the *Isaac Newton Telescope* and the *William Herschel Telescope*, among others.

Royal Observatory, Edinburgh (ROE) Observatory founded privately in 1818 which received the 'Royal' title in 1822, from George IV, and became nominally the Scottish equivalent of the *Royal Greenwich Observatory*, although it was never funded on the same scale. It became part of the University of Edinburgh in 1834, and is now responsible for running the *UK Schmidt Telescope* in Australia and the *UK Infrared Telescope* in Hawaii.

r-process A process by which heavy *elements* are manufactured through *nucleosynthesis* when there is a large supply of *neutrons* available – for example, in a *supernova*. The name is short for 'rapid process', and involves a *nucleus* capturing two or more neutrons in quick succession. This is one of the ways in which elements heavier than those of the *iron* group are made (see also *s-process*; many *isotopes* are produced by both processes). Heavy nuclei built up by the r-process are often unstable, and form stable nuclei through chains of *beta decay*.

The important feature of the r-process is that a nucleus captures several neutrons before it has time to decay by spitting out an *electron* or in some other way. This requires a density of about 300 billion billion (3×10^{20}) neutrons in every cubic centimetre of stuff at the heart of a *star*, which is why the conditions suitable for the r-process occur only in supernovae, and then only briefly. A handful of stable, slightly neutron-rich nuclei are produced only by the r-process and subsequent beta decays. The r-process ends for very heavy nuclei which split apart through *nuclear fission* as soon as they are formed.

RR Lyrae stars *Variable stars* which resemble *Cepheids* but have shorter periods, typically in the range 9–17 hours. They are hot, old *giant stars*, members of *Population II* with spectral types A or F, and about half of the known 2,000 or so occur in *globular clusters*.

Russell, Henry Norris (1877–1957) American astronomer, born in Oyster Bay, New York, who developed the technique of plotting the *absolute magnitudes* of *stars* against their *colour*, independently of the work of Ejnar *Hertzsprung*. Russell's version of what is now known as the *Hertzsprung–Russell diagram* was published in 1913.

Ryle, Sir Martin (1918–84) British radio astronomer who was the twelfth Astronomer Royal (and first *not* to be director of the *Royal Greenwich*

Observatory), succeeding Sir Richard **Woolley** in 1972 and giving up the title in 1982. He was born on 27 September 1918, at Brighton, in Sussex.

Ryle was one of the pioneers who developed *radio astronomy* in the decades following the Second World War, and he played a major part in the development of *interferometry* and *aperture synthesis* techniques. Educated at Oxford University, he worked at the Telecommunications Research Establishment at Malvern from 1939 to 1945, then joined the Cavendish Laboratory in Cambridge. He spent all of his astronomical career at the University of Cambridge, where he became the first professor of radio astronomy in 1959. He was knighted in 1966, and received the Nobel Prize in 1974 for his contribution to the development of radio astronomy, specifically the aperture synthesis technique. Ryle and Antony **Hewish**, who shared the award in 1974, were the first astronomers of any kind to receive the Nobel Prize.

Ryle was an ardent supporter of the *Big Bang* theory, which led him into a sometimes bitter wrangle with Fred **Hoyle**, and he was delighted to find evidence from radio observations of distant objects that the Universe is not unchanging (see *look back time*) and that therefore the simple *Steady State hypothesis* is incorrect. He died in Cambridge on 14 October 1984, having spent the last ten years of his life vociferously opposing the spread of nuclear power stations, arguing that the dangers implied by their links with the manufacture of nuclear weapons outweighed their benefits.

Ryle telescope   A *radio telescope* near Cambridge, England, which uses the *aperture synthesis* technique. It was named after Martin **Ryle**. The telescope uses four fixed dish *antennae*, each with an *aperture* of 13 m, together with four similar antennae that can be moved on rails (along the line of a track abandoned by British Rail) along a baseline 4.6 km long. (See picture on page 32.)

SAGE   Acronym for Soviet–American Gallium Experiment, a *neutrino* detector in an underground laboratory in the Caucasus Mountains. The detector uses 60 tonnes of gallium. See *solar neutrino problem*, *GALLEX*.

Sagittarius A   Brightest component of a strong, complex *radio source* at the centre of the *Milky Way*. It is possibly the site of a *black hole* with a *mass* about 1 million times that of our Sun.

Sagittarius B   A huge *molecular cloud* located at the centre of the *Milky Way*. It may be part of an expanding shell of material pushed out from the centre by a large explosion.

Sakharov, Andrei Dimitrievich (1921–89)   Russian physicist who became well known as a dissident under the Soviet regime in the 1970s and 1980s (he was awarded the Nobel Peace Prize in 1975 for his efforts to obtain a nuclear test-ban treaty), but who had earlier been the leading scientist involved in the development of the Soviet hydrogen bomb. Among his many other contributions to physics, in the 1960s he proposed a mechanism for the formation of matter (in preference to *antimatter*) in the *Big Bang*. This suggestion went largely unnoticed at the time, but became a cornerstone of the standard *model* of the Big Bang in the 1980s.

Sakharov was born in Moscow on 21 May 1921, the son of a physics teacher. He started studying physics at Moscow State University in 1938, but the department was evacuated to Ashkebad, where he graduated in 1942. After graduation, he worked for three years as an engineer at an arms factory in Ulyanovsk, on the Volga.

In 1945, Sakharov joined the Lebedev Institute, in Leningrad, receiving a PhD for work on *cosmic rays* in 1947, and going on to work on the problem of achieving *nuclear fusion* for both civil and military use, developing the hydrogen bomb and proposing the use of a 'magnetic bottle' to trap *plasma* in a fusion reactor. This has been developed into one of the most promising designs for a fusion reactor, called Tokamak. Although in 1953 he had become the youngest person ever elected to the Soviet Academy of Sciences, by the 1960s he was campaigning both for the test-ban treaty and for civil rights, and became increasingly distanced from the establishment. He produced no scientific papers at all between 1958 and 1965.

Partly as a deliberate rejection of his earlier work, he became increasingly interested in the apolitical science of *cosmology*; he also worked on the theory of *quarks*, and on attempts to develop a quantum theory of gravity. Sakharov puzzled over the question of what came 'before' the Big Bang (espousing the idea of an infinite repetition of *universes* which finds an echo in some versions of *inflation*), and was one of the first people to suggest (in 1969) that there might be a large amount of *dark matter* in the Universe, in the form of what are now called *WIMPs*.

But Sakharov's most important contribution to cosmology was undoubtedly his investigation of baryon asymmetry – the fact that the matter in our Universe is in the form of *baryons*, not an equal mixture of matter and antimatter. The puzzle is that when matter is formed out of energy on Earth (in experiments using particle accelerators), each particle is accompanied by its antiparticle counterpart. The only way to 'make' an electron, for example, is by making a positron as well. But if matter and antimatter had been made in equal quantities in the Big Bang, all the particles and antiparticles would have met up and annihilated one another, leaving nothing but electromagnetic radiation behind. We are made of baryons, and we exist only because the Big Bang produced an excess of baryons over antibaryons, so that a little matter was left over to form stars, galaxies, planets and people after all the annihilation had finished.

In a paper published in 1967, Sakharov showed that there is a tiny asymmetry in the laws of physics, which actually says that under the conditions operating in the Big Bang roughly a billion and one baryons would have been produced for every billion antibaryons. We are made of some of the one-in-a-billion particles that didn't get annihilated; the rest (along with all the antiparticles) got turned into the *background radiation*.

But this could have happened only in a hot Big Bang which is in the process of cooling (and therefore has an inbuilt *arrow of time*). It is connected with the way the four *fundamental interactions* split apart from one another as the energy density of the Universe decreases (see *grand unified theories*). Sakharov was way ahead of his time, and his work received the attention it deserved only after a similar model was developed independently by a Japanese physicist, Motohiko Yoshimura, in 1978. Sakharov's insight, explaining the requirements that had to be met in order for matter to exist in the Universe today, made before the grand unified theories were developed and providing not only a powerful theoretical argument for the existence of a hot Big Bang, but also an actual (correct!) prediction of the strength of the background radiation (roughly 1 billion photons for every baryon) was itself worthy of a Nobel Prize, and is one of the most perceptive insights in all of cosmology. But, although he repeatedly returned to the study of the baryon asymmetry throughout the rest of his life, Sakharov was distracted from promoting the idea as he increasingly turned his attention to the more pressing political problems of the time in the 1970s.

In 1980, as part of a clampdown on dissidents, Sakharov was sent into internal exile in the city of Gorky, where he went on repeated hunger strikes to try to obtain permission for his wife, Yelena Bonner, to travel outside the Soviet Union for medical treatment. In spite of this, and his isolation from other scientists, he continued to work and publish important papers in cosmology, investigating (among other things) the properties of evaporating *black holes* (see *Hawking radiation*) and the nature of *shadow matter*. He was released by Mikhail Gorbachev in December 1986, and renewed his campaign for civil rights in the USSR, while still working on cosmology and in particular on the origin of the baryon asymmetry. He was elected to the Congress of People's Deputies in 1989, shortly before he died, on 14 December that year.

Further reading: John Gribbin, *In Search of the Big Bang*.

Salpeter, Edwin (1924–) American astrophysicist who suggested, in 1952, that *carbon* might be manufactured out of *helium* inside *stars*, through what is now known as the *triple alpha process*.

salpeter process = *triple alpha process*.

Sandage, Allan Rex (1926–) American astronomer and cosmologist who has made important contributions to many areas of *astrophysics*, but who is best known for his contribution to the debate about the value of the *Hubble constant*, which measures how fast the Universe is expanding.

Sandage was born in Iowa City on 18 June 1926. His father had been raised on a farm, but became the first member of his family to receive an academic education, gaining a PhD and working in economics at the University of Iowa. So Allan Sandage was brought up in an academic atmosphere. He became interested in astronomy in childhood, while the family was living in Philadelphia and a school friend let Sandage look through his *telescope*. He immediately decided that he wanted to become an astronomer, but his education was interrupted by wartime service in the navy in 1944 and 1945, before he studied at the University of Illinois and at CalTech – a member of the first group of graduate students to be admitted to study astronomy at CalTech. One of the key elements in the development of his interest in cosmology was a course taught by Fred *Hoyle* on a visit to CalTech while Sandage was still a graduate student.

Sandage obtained his PhD in 1953, a year after he had joined the staff of the *Mount Wilson and Palomar Observatories*. He had begun working there as an assistant to Edwin *Hubble* while still a student, and he remained there throughout his career. After Hubble died, Sandage inherited the task of measuring distances to remote objects in the Universe, and determining the rate at which the Universe is expanding. In addition to his duties there, he was a professor of physics at Johns Hopkins University from 1987 to 1989.

He was involved in the first optical identification of a *quasar* in 1960 (using the *Hale telescope* at Palomar to pin down the location of the *radio source* 3C 48) and in the discovery of radio-quiet quasars (now known to be much more common than radio-loud quasars) in the middle of the 1960s. Sandage is interested in the ultimate fate of the Universe, and at one time argued the case for an *oscillating universe model*. Together with Gustav *Tammann*, Sandage has carried out a long, in-depth analysis of measurements of the Hubble constant, concluding that it has a value close to (perhaps even below) 50 km per second per *Megaparsec*; unfortunately, another school of thought analysing exactly the same data, led by Gerard *de Vaucouleurs*, concludes that the value of the Hubble constant is at least 80, in the same units. Determination of the true value of the Hubble constant remains one of the prime objectives of cosmology today.

Late in his career, Sandage became concerned about the *horizon problem*, describing it at the end of the 1980s as 'the most important problem in the field [of cosmology]', and regarding the discovery of the ripples in the *background radiation* by the *COBE* satellite as crucial in giving a boost to the idea of *inflation* and resolving the puzzle of how *galaxies* formed from primordial fluctuations in the expanding Universe. He sees determination of the value of the *density parameter* and the nature of the *dark matter* as the remaining outstanding problems in cosmology.

Further reading: Alan Lightman and Roberta Brawer, *Origins*.

Sanduleak −69° 202   See *Supernova 1987A*.

satellites (artificial)   Any artificial object which has been placed in *orbit* around the Earth or some other astronomical body. The first of these, Sputnik 1, was launched from the Soviet Union on 4 October 1957; it weighed 84 kg and burned out in the atmosphere after 92 days in orbit. Many satellites are used for commercial/environmental purposes (communications satellites, weather satellites, and so on) or have military uses (such as providing photographs of military activity in otherwise inaccessible regions). The whole technology of satellite launching developed during the Cold War from the rocketry needed to deliver nuclear weapons to anywhere on Earth.

Scientific satellites and spaceprobes have been used to obtain information about the Sun and planets of the Solar System, and also to observe the Universe at wavelengths of electromagnetic radiation which cannot penetrate through the atmosphere to the surface of the Earth. These include infrared and microwave observations (at wavelengths longer than those of light) and observations using ultraviolet, X-rays and *gamma rays*, at successively shorter wavelengths than those of light. These astronomical satellites have transformed our understanding of the Universe, in particular showing that it is a much more violent place than used to be thought.

satellites (natural)   Any body in *orbit* around another body is a satellite. The term is often used as a synonym for *moons*, which orbit *planets*, but planets (together with *minor planets* and *comets*) are themselves satellites of *stars*, and even *galaxies* can be satellites of other galaxies (for example, the *Magellanic Clouds* are satellites of our own *Galaxy*).

Saturn   The sixth *planet* out from the Sun, one of the four *giant planets* and second in size only to Jupiter. Saturn's diameter (at the equator) is 9.4 times the diameter of the Earth and its *mass* is 95 times that of the Earth. This means that its overall density is only 70 per cent of that of water. It has a prominent ring system and at least twenty *moons*, with more still being discovered. Saturn *orbits* the Sun once every 29.46 years at a distance varying between 9 and 10 *astronomical units*.

scalar field   A field which does not have an inbuilt sense of direction. A commonly used example is a field that represents the temperature of the air at every point in a room. Obviously, the field 'fills' the room – there is a number corresponding to the temperature at every point. But although a thermometer placed in the field will record a low temperature near the door, where there is a draught coming in, and a high temperature by the radiator on the other side of the room, there is no force which always

Saturn photographed by Voyager 1 in 1980. As well as the famous ring system, three of the moons of Saturn are also just visible.

pushes thermometers towards (or away from) hot spots. Differences in temperature do set up convection currents, which flow because of unevenness in the field, but there is no equivalent of the way a tiny charged particle is forced to move in a certain direction along electric field lines (see *vector field*, *field theory*).

One of the peculiarities of a scalar field is that it may be very difficult to detect. A completely uniform scalar field will have no influence on its surroundings (for example, if every point in the room is at the same temperature, there will be no convection currents, no matter how high that temperature is). An imperfect (but insightful) analogy can be made with two physics laboratories, one in the basement of a tall building and the other on the top floor. The strength of *gravity* is not *quite* the same in the two laboratories, but it would be very hard to tell this from any experiments that involve collisions between balls moving about on a smooth pool table. In either laboratory, you would find *Newton's laws of motion* describing the behaviour of the pool balls in the same way. But if the pool table in the top lab were pushed out of the window, it would plummet to the ground and be smashed, releasing gravitational energy in the form of heat.

Something similar happened when the scalar fields that filled the entire Universe before and during the era of *inflation* 'fell down' into a state of lower energy (cosmologists prefer to think of the fields as 'rolling down', like a marble rolling down into the bottom of a large bowl, but the image is much the same). The one way in which a uniform scalar field of this particular kind would reveal itself would be as a form of antigravity, driving the rapid expansion of the Universe during inflation. If there is a tiny residual scalar field left over from inflation (most cosmologists regard this as a very big 'if' indeed), it could still be acting as a *cosmological constant*, weakening the ability of gravity to slow the expansion of the Universe, so that it is still expanding slightly faster than would otherwise be the case. This would affect the interpretation of the *age of the Universe* based on measurements of *Hubble's constant*, making the Universe older than those measurements suggest.

Schiaparelli, Giovanni Virginio (1835–1910) Italian astronomer, born at Savigliano, in Piedmont, on 14 March 1835, who is best known for his studies of Mars, and in 1877 described the markings of the surface of that planet as 'canali'. Mistranslated as 'canals' (the term actually just means 'channels'), the description led to wild speculation about the possible existence of intelligent life on Mars, and encouraged Percival *Lowell* to take up astronomy. Pictures from *spaceprobes* (notably the *Viking* missions) have shown that the channels on the surface of Mars are extensive natural drainage systems, probably formed by flowing water long ago when the *planet* had a thicker atmosphere and was warmer. They are essentially dried-up (*long* dried-up!) river beds.

Schiaparelli had been educated in Italy, Germany and Russia, returning to work at the Milan Observatory in 1860 and becoming the director of the observatory in 1862. He discovered a *minor planet* (Hesperia, in 1861), showed that the objects which form *meteors* travel around the Sun in the same orbits as *comets*, and made important studies of other planets, as well as Mars. After he retired (in 1900), he made an extensive study of the astronomy of Babylonian, Greek, biblical and medieval times. He died in Milan on 4 July 1910.

Schmidt, Bernhard Voldemar (1879–1935) Estonian optician who invented the *Schmidt camera* in 1930.

Schmidt, Maarten (1929–) Dutch-born American astronomer who was the first person to interpret the unusual *spectrum* of a *quasar* (3C 273) as being caused by a large *redshift*.

Schmidt camera A type of *optical telescope* mainly used for taking photographs of the sky – in effect, a wide-angle camera. This kind of telescope was developed by the Estonian Bernhard **Schmidt** in the 1930s. It uses a spherical mirror, which on its own would produce considerable distortions in the images; Schmidt corrected this by using a thin transparent 'plate' (a lens with a complicated shape) at the front of the telescope, so that the incoming **light** is modified before it reaches the main mirror in just the right way to cancel out the distortions caused by the mirror. The image is then focused on to a curved surface, across which the photographic film can be laid.

The overall result provides very sharp images over fields of view tens of square degrees in size – a popular size of Schmidt camera, with a correcting-plate aperture of 1.2 m (48 inches) covers 40 square degrees. A conventional **reflecting telescope** may have an effective field of view covering half a degree on the sky, about the size of the Moon seen from the Earth; a Schmidt can photograph patches of sky 16 degrees across, more than 30 times the apparent diameter of the Moon. This has made Schmidt cameras invaluable for surveys of the entire sky, providing photographic **catalogues** with millions of images on each plate, which can then be investigated using machines such as **COSMOS**. Telescopes using traditional parabolic mirrors provide even better images of objects exactly in the centre of the field of view, but cannot focus on all the objects in a wide field of view at the same time, so they come into their own once an interesting object (such as a **quasar**) has been identified from Schmidt plates, or in some other way.

Schwarzschild, Karl (1873–1916) German astronomer, born in Frankfurt on 9 October 1873, who discovered the solution to Albert **Einstein**'s equations of the **general theory of relativity** that describes what are now known as **black holes**.

Schwarzschild studied at Strasbourg University and Munich University, then worked at the Kuffner Observatory in Vienna before moving to the University of Göttingen in 1901. A year later, he became the director of the university's observatory, and was appointed a full professor, at the age of 28. He moved on to become director of the Astrophysical Observatory in Potsdam in 1909.

Schwarzschild was a skilled observer, who pioneered the development of photographic techniques in astronomy; a popular lecturer, who conveyed the excitement of his work to non-scientists; and a very able mathematician, who was one of several people to discuss the possibility that space might be **non-Euclidean** in the years before Einstein developed his general theory of relativity.

Although secure in his post at Potsdam, and already in his forties, Schwarzschild volunteered for military duties at the start of the First World War, and worked as a technical expert in Belgium and France before being transferred to the Eastern Front. There, he contracted a skin disease, pemphigus, that was incurable in those days. He was invalided out of the army, and died in Potsdam on 11 May 1916, later being awarded a posthumous Iron Cross. Even while serving in Russia, however, Schwarzschild kept in touch with scientific developments in Germany, and wrote several important papers, including those that describe the Schwarzschild solution to Einstein's equations. It was literally while lying on his death bed that he completed this work, which he sent to Einstein, who read the two papers to the meetings of the Academy of Sciences in Berlin on 16 January and 24 February 1916, a few weeks before Schwarzschild died.

Schwarzschild, Martin (1912–) German-born American astronomer, the son of Karl *Schwarzschild*. Martin Schwarzschild was born in Potsdam on 31 May 1912 and obtained his PhD in astronomy from the University of Göttingen in 1935, shortly before emigrating to the USA. His main work has been in the area of stellar structure and *stellar evolution*, including the nature of *pulsating variables*. His book *Structure and Evolution of the Stars* (Princeton University Press, 1958) is a standard text used by generations of students.

Schwarzschild black hole A spherical, non-rotating *black hole* with no overall electric *charge*, described mathematically by the *Schwarzschild metric*, which is the Schwarzschild solution to Albert *Einstein*'s equations of the *general theory of relativity*. This is the simplest possible kind of black hole.

Schwarzschild limit The maximum *density* possible for an object with a certain mass without it collapsing to form a *black hole*. The bigger the object is, the lower the Schwarzschild limit. For the Sun, the limit is 10^{16} times the density of water; for a mass of 100 million Suns, it is the same as the density of water.

Schwarzschild metric The *metric* that describes *spacetime* in the vicinity of the simplest kind of *black hole*, a *Schwarzschild black hole*. The metric, discovered by Karl *Schwarzschild* in 1915, actually describes spacetime in the vicinity of any spherical concentration of *mass*. See also *Kerr metric*, *Robertson–Walker metric*.

Schwarzschild radius The radius of the *event horizon* around a black hole, from within which not even *light* can escape. This is essentially the distance from the centre of the black hole where the *escape velocity* is equal to the *speed of light*. It is given by the expression $R = 2GM/c^2$, where G is the constant of *gravity*, M is the *mass* of the black hole, and c is the speed of light.

Schwarzschild solution See *Schwarzschild black hole*.

scientific notation To avoid writing out long strings of zeros for large or small numbers, scientists use a mathematical convention that gives the number of zeros as a 'power of ten'. So 100 can be written as 10^2 (meaning 'a 1 followed by 2 zeros') and 0.001 can be written as 10^{-3} ('a decimal point followed by 2 zeros and a 1'). The convention comes into its own when we are dealing with numbers like 10^{30} or 10^{-26}. A number such as 326 can be written as 3.26×10^2 ($= 3.26 \times 100$), and so on.

scintillation   Technical term for the 'twinkling' of starlight. It occurs because the atmosphere of the Earth acts like a variable lens, refracting the *light* from a *star* irregularly as it passes through the atmosphere. To the naked eye, scintillation shows up chiefly as rapid changes in brightness of a star, but through a telescope the image of the star seems to wander about, so that it becomes blurred on long-exposure photographs.

A similar effect distorts *radio waves* as they pass through the ionized layers of the Earth's atmosphere, and when they pass through clouds of ionized material in interplanetary and interstellar space. It was while using a radio telescope designed to study such rapidly flickering *radio sources* that Jocelyn *Bell* discovered *pulsars*.

Scorpius X-1   The archetypal *X-ray source* in our *Galaxy*, the first one to be discovered (using a rocket-borne detector in 1962) and the brightest in the sky. The discovery came as a surprise – the detectors were designed to search for X-rays produced by particles from the Sun striking the Moon – and was one of the first indications of the kind of violent activity now known to be commonplace in the Universe.

In 1967, Sco X-1 was identified with a faint variable star known as V818 Sco, and two Cambridge theorists, Paul Feldman and John Gribbin, were able to show from an analysis of the variability of the light from the star that it is a *neutron star* surrounded by an *accretion disk*. Together with the discovery of pulsars, this was one of the first direct pieces of evidence that neutron stars exist. More recent observations suggest that the accreting material comes from a companion which orbits the X-ray star once every 0.78 days. The distance to Sco X-1 is still not known accurately, but is probably somewhere between 300 and 600 *parsecs*.

seasons   Regular changes in the weather on Earth, caused by the tilt of the *planet* relative to the plane of its orbit around the Sun. An imaginary line through the Earth from the North Pole to the South Pole is tilted at 23.45° away from making a perpendicular to the orbital plane (the *ecliptic*). As a result, on one side of its orbit the Earth's Northern Hemisphere leans towards the Sun, while on the other side of the orbit (6 months earlier or later) the Southern Hemisphere leans towards the Sun. This makes the Sun rise high in the sky, as seen from the appropriate hemisphere, and brings summer; at the same time, the opposite hemisphere experiences winter. In between seasons are called spring and autumn (fall). Seasonal differences in weather are more pronounced at higher latitudes; equatorial regions experience little in the way of seasonal changes.

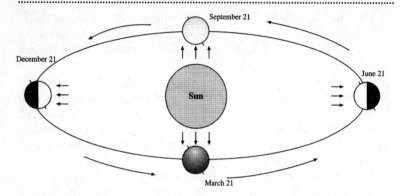

Seasons. Because of the tilt of the Earth, the amount of solar heat reaching the northern and southern hemispheres each day changes as the Earth orbits the Sun.

Seasons also occur on other planets which are tilted out of the vertical, including Mars, which has a similar tilt to Earth. The most extreme seasonal changes in the Solar System occur on Uranus, which lies almost 'on its side', so that the north and south poles alternately point almost straight at the Sun as Uranus moves around its orbit.

second Unit of time defined as the duration of 9,192,631,770 oscillations of the *electromagnetic radiation* corresponding to a particular *quantum* change in *energy* level of a caesium-133 *atom*.

second law of thermodynamics The law of nature which says that things wear out. One expression of the second law of thermodynamics is that heat cannot flow from a cold object to a hotter object of its own volition. Place an ice cube in a cup of warm water, and the ice melts as heat flows into it from the water, ending up with a cup of water slightly cooler than you had before. You never see ice cubes spontaneously forming in cups of water, as heat drains out of the cold ice into the hotter liquid. Ice cubes can only be made (for example, in a domestic freezer) by using *energy* to pump heat out. Another facet of the second law is the way in which a house left unattended for a long time will crumble away under the influence of wind and weather, whereas a pile of bricks left unattended will never spontaneously form itself into a house (see *arrow of time*).

In his book *The Nature of the Physical World*, Arthur *Eddington* said that:

> The second law of thermodynamics holds, I think, the supreme position among the laws of Nature. If someone points out to you that your pet theory of the universe is in disagreement with Maxwell's equations – then so much the worse for Maxwell's equations. If it is found to be

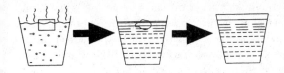

Second law of thermodynamics. Heat always flows from a hotter object (the boiling water) into a cooler object (the ice cube). This is another manifestation of the arrow of time.

contradicted by observation – well, these experimentalists do bungle things sometimes. But if your theory is found to be against the second law of thermodynamics I can give you no hope; there is nothing for it but to collapse in deepest humiliation.

There are actually three laws of thermodynamics, which can be para-phrased as summing up life and the Universe: 1. You can't win; 2. You can't even break even; 3. You can't get out of the game.

second quantization   See *field equations*.

self-sustaining star formation   Process which maintains the pattern of *spiral arms* in a *disc galaxy* as new waves of *star* formation are triggered by the blast waves from *supernova* explosions.

semi-detached system (semi-detached binary)   See *equipotential surfaces*.

SETI   Acronym for Search for Extraterrestrial Intelligence. It was changed from the earlier CETI (Communication with Extraterrestrial Intelligence) because some people (notably the politicians holding the scientific purse-strings in the USA) were frightened by the idea of actually speaking to our neighbours (if any) in the Universe; much better just to spy on them without letting them know we are here.

This 'spying' consists of so far unsuccessful attempts to eavesdrop on any radio communications from other civilizations which may come our way – either signals deliberately sent out in the hope of making contact with new members of the galactic club, or the routine equivalent of our domestic TV and radio communications. There have also been some limited attempts at genuine CETI, with signals beamed out towards distant *stars* from the *Arecibo radio telescope* (any politician who reads this and may worry about the implications can take comfort from the fact that it will take several thousand years for the signals to reach the target stars). An ongoing SETI programme run by *NASA* uses a sophisticated computer system to listen into the Universe with large *radio telescopes*, in the gaps between their regular astronomical observations.

Seyfert   A kind of *active galaxy*, named after the American astronomer Carl Seyfert (1911–60), who first drew attention to them in 1943. His original investigation of twelve *disc galaxies* with barely detectable *spiral arms* showed that the outer regions of these *galaxies* were being outshone by the bright, bluish central regions, and analysis of the *Doppler effect* in lines in the *spectra* of these objects showed the presence of hot clouds of gas moving at speeds of hundreds or even thousands of kilometres per second within the central regions of the galaxies. These were astonishing and unprecedented discoveries at the time, and the place of Seyferts in the cosmic hierarchy became clear only much later, following the discovery of *quasars* and *BL Lac objects*. Seyfert, who died on 13 June 1960, did not live to see those discoveries.

More than 150 Seyfert galaxies are now known. Their distinguishing feature is a very bright central nucleus in an otherwise normal-looking disc galaxy. As well as this visible light, they produce a great deal of infrared radiation, togther with X-rays and ultraviolet radiation, but they are not usually very bright at radio wavelengths. The brightness of a Seyfert can vary (across all of this wide range of *wavelengths*) on timescales of a few months, indicating that the energy comes from a region no more than a few *light months* across. They are very much like more modest versions of quasars (about 1 per cent as energetic as quasars), and are almost certainly powered by the infall of material on to a massive central *black hole* via an *accretion disc*.

About 10 per cent of all large disc galaxies are seen to be active as Seyferts today, and this is interpreted as implying that all large disc galaxies (including our *Milky Way*) spend about 10 per cent of their lives (perhaps in repeated short bursts of activity) as Seyferts.

shadow matter   A hypothetical form of matter that may have been formed in the very early Universe at the moment (just after the *Planck time*) when *gravity* split off from the other *fundamental interactions*. According to some versions of *supersymmetry* theory, when that happened some of the energy in the Universe ended up in the form of the particles that we know today (including the *baryonic matter* that our bodies and the stars are made of), while the rest ended up as a completely separate set of particles, which share nothing in common with our particles except that they also feel the force of gravity. You could walk right through a shadow person and neither of you would ever notice.

There is no reason to think that the two kinds of matter would have formed in equal quantities, and shadow matter could account for all of the *dark matter* required by some *cosmological models*. Alternatively, there could be a mirror-image relationship between the two worlds, with shadow-electrons, shadow-protons and shadow-neutrons making up a universe of shadow stars and shadow galaxies which even includes its own shadow dark matter (perhaps in the form of shadow-*axions*). Unfortunately, the theory of shadow matter is so vague that just about any science-fiction-like possibility you can dream up is allowed (or at least, not forbidden), and there is no difficulty tailoring shadow stuff to fit any pet *model* of the Universe. The idea is not taken seriously by cosmologists.

shadow universe   The speculative idea that an entire *universe* made of *shadow matter* may co-exist with our Universe in the same spacetime, undetectable except through gravity. In the extremely unlikely event that such a shadow universe exists, it could account for some of the *missing mass*.

Shapley, Harlow (1885–1972)   American astronomer, born in Nashville, Missouri, who used observations of *variable stars* in *globular clusters* to map our *Galaxy* and determine its size accurately, working at the *Mount Wilson Observatory* in the second decade of the 20th century. He became director of *Harvard College Observatory* in 1921, a post he held for 31 years, and made Harvard a centre of excellence in astronomy.

shepherd moons   Name for small *moons* which *orbit* alongside rings such as those of Saturn, and whose gravitational influence holds the rings in stable configurations.

Shklovskii, Iosif Samuilovich (1916–85)   Soviet astrophysicist who played a prominent part in the debate about the possible existence of extra-terrestrial life, and who was the first to suggest, in 1953, that *radio waves* and *X-rays* from the *Crab nebula* are produced by *synchrotron radiation*. He also predicted the existence of astronomical *masers*.

shooting star   See *meteor*.

sidereal time   Time measured in terms of the rotation of the Earth compared with the *fixed stars*. Because of the movement of the Earth in its orbit around the Sun, if one sidereal day (the time for one rotation of the Earth) is defined as 24 hours exactly, averaging over a whole year the time from noon to noon the next day is 24 hours 3 minutes and 56.55 seconds.

Siding Spring Observatory   The main observatory for *optical astronomy* in Australia. The main *telescopes* include the 3.9 m *Anglo-Australian telescope*, a 2.3 m *reflecting telescope*, a 50 cm *Schmidt camera* run by the Swedish Uppsala Observatory, and the *UK Schmidt*. See *Mount Stromlo and Siding Spring Observatories*.

singularity   A place where the laws of physics as we know them break down. Singularities are usually thought of as points, but in principle they could exist in the form of one-dimensional lines, or even as two-dimensional sheets. According to the equations of the *general theory of relativity*, whenever a non-rotating *Schwarzschild black hole* forms, the material within the *event horizon* of the hole must inevitably fall under the influence of *gravity* into a point of infinite density, the singularity (see *Penrose, Roger*). The uniform expansion of the Universe away from the *Big Bang* is a mirror image of the collapse of such a *black hole*, implying that the Universe was born in a singularity.

In both cases, though, the equations take no account of *quantum theory*. It seems likely that the known laws of physics, including the general theory of relativity, actually break down when we are dealing with objects smaller than the *Planck length* or with intervals of time shorter than the *Planck time*. This means that it is legitimate to suggest that matter collapsing towards a singularity will be affected by quantum processes on that scale, perhaps being 'bounced' into an expansion outward into another set of dimensions. It has been argued that the Big Bang 'singularity' is actually a bounce of this kind.

A Caltech professor of theoretical physics, Kip Thorne, has described the quantum singularity as a place where gravity 'unglues' space and time from each other, and then destroys time as a concept and destroys the definiteness of space, leaving a 'quantum foam' from which anything might emerge (*Black Holes and Time Warps*, Norton, New York, 1994, page 476–77). Singularities – especially those associated with rotating black holes and (if they exist) *naked singularities* – even allow for the possibility of *time travel*. See also *baby universes*.

Sirius   The brightest *star* in the sky, also known as the Dog Star and as Alpha Canis Majoris (as the names suggest, it is in the constellation of the dog, Canis Major). It is a *binary system* in which an A star (Sirius A) is orbited by a *white dwarf* (the first to be identified; see *Sirius B*) once every 50 years. Sirius is 8.7 *light years* away, the seventh nearest star to the Sun.

Sirius B   A faint *white dwarf* that forms a *binary system* with *Sirius*, the brightest *star* in the sky. Studies made by the German astronomer Friedrich Bessel (1784–1846) in the 1830s and early 1840s showed that regular 'side to side' movement of Sirius could be explained by the gravitational influence of a companion star orbiting Sirius every 50.09 years; the companion was found by an American telescope maker, Alvan Clark, in 1862. While testing a new instrument, he turned it on Sirius and found the companion, which is so faint that, if it were the same distance from us that the Sun is, it would be only one-four-hundredth as bright as we see the Sun.

In spite of its faintness, the orbital details of the binary system show that the mass of Sirius B must be about 90 per cent of that of the Sun (Sirius itself weighs in at just over 2 *solar masses*). Throughout the rest of the 19th century, the astronomical consensus was that it must be a star roughly the same size as the Sun, but fainter and cooler. But in 1915, when Sirius B was at its furthest separation from Sirius (as seen from Earth), the American astronomer Walter Adams (1876–1956) obtained a *spectrum* which showed that Sirius B is actually as hot as Sirius itself. In order to be both hot *and* faint, it had to be small – only a little bigger than the Earth. It was the first white dwarf to be discovered.

de Sitter   See under *de.*

size of the Universe   See *cosmic distance scale.*

Slipher, Vesto Melvin (1875–1969)   American astronomer, born in Mulberry, Indiana, who was the first person to identify the *redshifts* in the *light* from what are now known to be *galaxies.* This work, carried out in the years up to 1925, paved the way for Edwin *Hubble's* discovery that the Universe is expanding. Slipher worked at the *Lowell Observatory*, where he was director from 1926 to 1952, and instigated the successful search for Pluto at the end of the 1920s.

Small Magellanic Cloud   See *Magellanic Clouds.*

SMC = *Small Magellanic Cloud.*

Smith, Sir Francis Graham (1923–)   English astronomer, born at Roehampton, in Surrey, on 25 April 1923. He studied at Downing College in Cambridge, where his degree course was interrupted by war work at the Telecommunications Research Establishment in Malvern, so that he did not graduate until 1946. His PhD, awarded in 1952, was for work in *radio astronomy* at the Cavendish Laboratory in Cambridge.

After a year at the Carnegie Institute in Washington, DC, Smith returned to Cambridge where he worked in radio astronomy until 1964, before moving to *Jodrell Bank* for the next ten years. In 1974 he moved to the *Royal Greenwich Observatory* (then still at Herstmonceux in Sussex), and the next year he also became a visiting professor in astronomy at the nearby University of Sussex. In 1976, Smith became the first radio astronomer to become director of the Royal Greenwich Observatory, and he played a major part in the choice of the site for what became the *Roque de los Muchachos Observatory* and in the early stages of its development. He moved back to Jodrell Bank as its director in 1981, and in 1982 he was appointed Astronomer Royal in succession to Sir Martin Ryle. He was therefore the first (and so far only) person since 1971 to be both director of the RGO and Astronomer Royal, although he did not hold the two positions at the same time. He retired as Astronomer Royal in 1990.

As well as being one of the key figures in the development of radio astronomy techniques, such as *interferometry*, and an able administrator, Smith has made important contributions to the study of *pulsars*, and to the investigation of the magnetic fields in interstellar space.

Smithsonian Astrophysical Observatory   Observatory in Cambridge, Massachusetts, founded by the Smithsonian Institution in 1890, initially in Washington, DC. It moved to Harvard in 1955, and combined with the Harvard College Observatory (itself established in 1839) in 1973 to become the Harvard-Smithsonian Center for Astrophysics (CfA).

SN 1987A   See *Supernova 1987A*.

solar constant   The amount of energy from the Sun crossing each square metre of a sphere, centred on the Sun, with the radius of the Earth's orbit (1 *astronomical unit*). The accepted value is 1.367 kilowatts per square metre; but the most interesting thing about the solar constant is that it may not really be constant.

1.37 kilowatts sounds modest compared with even the average electric fire (typically at least 1 kilowatt). But that energy is crossing every square metre, every second, at the distance of the Earth from the Sun – like having the Sun completely surrounded by electric fires spaced just over 1 metre apart. It is enough to melt a complete shell of ice 1 inch (2.5 cm) thick around the Sun, at a distance of 1 AU, in 2 hours and 12 minutes. Imagine shrinking that shell of ice inwards towards the Sun, so that its area gets less and less but its thickness increases, so that it always contains the same amount of ice. By the time the inner surface was touching the Sun, the imaginary ice layer would be more than 1 mile (over 1.6 km) thick – but it would still be thawed in the same brief span of time. And this prodigious output of energy goes on almost unchanged for billions of years.

Almost, but not quite, unchanged. Over its long lifetime, the energy output of a star like the Sun increases slightly, increasing the solar constant to match, perhaps by 25 per cent in the 4 billion years or so since the Sun formed (see *stellar evolution*). But this is far too slow a change to bother us today. Even a change in the solar constant of a fraction of 1 per cent would affect the climate on Earth, however; roughly speaking, a change in solar output of 1 per cent corresponds to a change in average temperatures on Earth of 1–2 degrees Celsius. There have been suggestions that more modest changes, perhaps associated with changes in the size of the Sun of about 0.01 per cent per century, could account for extreme climate events such as the Little Ice Age of the 17th century (see *Maunder minimum*).

solar cycle   A roughly periodic variation in the level of activity of the Sun. The most obvious feature of the cycle is the varying numbers of *sunspots* visible on the surface of the Sun. These build to a peak, die away and then build to another peak over a cycle roughly 11 years long (the sunspot cycle); however, the Sun's *magnetic field* reverses after each sunspot cycle, so it takes about 22 years to get back to where it started from. This magnetic cycle, also known as the double sunspot cycle and occasionally as the Hale cycle, is regarded as the true measure of the Sun's cycle of activity. Individual sunspot cycles vary considerably in length (they can be as short as 8 years or as long as 13 years) and in strength (measured in terms of the *Wolf sunspot number*). Individual sunspots can be very large indeed – the record was set by a sunspot seen in April 1947, which covered an area of the Sun's surface of more than 18 billion square km, but even an ordinary sunspot may be 10,000 km across.

The first person to measure the periodicity of the solar cycle was the German chemist and astronomer Samuel Schwabe (1789–1875), who announced his discovery of what he thought was a 10-year cycle of sunspot activity in 1843. It attracted little attention until 1851, when the German explorer Friedrich von Humboldt (1769–1859) publicized the discovery of the sunspot cycle in his book *Kosmos*; soon afterwards, Rudolf **Wolf** carried out his first studies of the cycle, finding a period of 11.1 years.

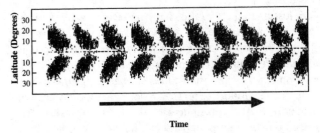

Time

Solar cycle. Schematic representation of the *butterfly diagram* obtained by plotting out the positions where sunspots form at different stages of the solar cycle. In each cycle, spots first appear at high latitudes, then closer and closer to the equator. New spots from the next cycle begin to appear while the last spots of the old cycle are still present. The whole pattern repeats roughly every 11 years.

During an individual sunspot cycle, the rise towards maximum activity is steeper than the subsequent decline in activity. Roughly speaking, the build-up to solar maximum takes about 4.5 years, and the subsequent decline to solar minimum takes about 6.5 years. Sunspots tend to occur in groups; large groups may persist for several months, although small individual spots may last for less than 1 day, and a typical group lasts for less than a single rotation of the Sun (just about 1 month). At first, sunspots appear at relatively high solar latitudes, between 30° and 45°, in both hemispheres of the Sun; but as the activity on the Sun's surface builds up, new spots form at other latitudes as well, closer and closer to the equator. At solar maximum, there are spots at many latitudes. Then, as the activity declines, sunspots stop appearing at high latitudes, until at the low point of the cycle they appear only near the equator, around latitude 7° (they never appear exactly on the equator). By that time, the first spots of the next cycle may already be appearing at high latitudes. The maximum is a sharp feature usually confined to a single year; the minimum, or quiet Sun, phase is spread out over 2 or 3 years.

This pattern of activity is thought to be associated with a 'winding up' of the Sun's magnetic field caused by the differential rotation of the outer layers of the Sun, with the equator rotating more rapidly than the material at higher latitudes. The equator rotates once every 26 days, but at a latitude of 30° the Sun's surface takes 31 days to complete one rotation. The electrically charged material (*plasma*) in the outer layers of the Sun drags the magnetic field round with it, winding it tighter towards the equator until something has to give, and the magnetic field lines break and reconnect to reverse the overall magnetism. This suggested pattern of behaviour makes sense in general terms, but there is no detailed understanding of how the process works.

Sunspots are just one aspect of the changing level of activity of the solar cycle. There are more *solar flares* when the Sun is more active, for example, and the *solar wind* is stronger when there are more sunspots. Because of the influence of the solar wind on *cosmic rays* (acting to shield the Earth from cosmic rays when the wind is stronger), this means that the solar cycle has a direct influence on the amount of radioactive carbon (carbon-14) produced in the atmosphere of the Earth each year. This radiocarbon gets into living things, in particular the new wood laid down in tree rings each year, and the amount of carbon-14 in tree rings from old wood can be used to reconstruct changing levels of solar activity in the past. Samples of other radioactive *isotopes* produced by cosmic rays and laid down in the polar icecaps also allow astronomers to infer the level of solar activity long before records began.

There are suggestions that the changing level of solar activity also affects the weather on Earth. By and large, it seems that the Earth is marginally warmer when the Sun is more active (see *Maunder minimum*). These claims are controversial, but traces of what seem to be the solar cycle's influence on weather have been found in very old geological deposits, formed from sediments laid down year by year at the bottom of still lakes. These ideas were lent weight by observations of the Sun made by Nimbus-7 and other satellites over a complete solar cycle, from 1980 to 1991. They showed that the amount of heat from the Sun arriving at the top of the Earth's atmosphere was 0.3 per cent less at the time of solar minimum than at the time of solar maximum.

The length of each sunspot cycle (measured in terms of the Wolf number) varies in what may be a regular way, with some evidence of 'supercycles' about 76 years and about 180 years long modulating the overall level of activity. A detailed analysis of all of the available measurements of the solar radius, carried out by Ronald Gilliland at the High Altitude Observatory in Boulder, Colorado, in the early 1980s, suggests that all of these changes, and the solar cycle influence on climate, may be linked to small, regular changes in the size of the Sun, as if it were breathing in and out. The records only go back to 1715, and the evidence is not accepted by all astronomers, but according to Gilliland there is a regular variation in which the Sun breathes in and out by about 140 km (just 0.02 per cent of its radius) with a period of 76 years, and an even smaller oscillation over the sunspot cycle.

In both cases, when the Sun is bigger there are fewer sunspots. The high levels of solar activity at the end of the 1970s and at the beginning of the 1990s match this pattern, since the Sun was then at its smallest on the 76-year cycle; if Gilliland is correct, the next solar maximum, due early in the 21st century, will be less pronounced.

The first spots of a new cycle were seen in August 1995.

solar day See *Earth*.

solar eclipse See *eclipse*.

solar flare See *flares*.

solar luminosity The *luminosity* of the Sun, 3.826×10^{26} joules per second. This is used as a unit of brightness, so the luminosity of a *star* or *galaxy* may be expressed as a multiple of the solar luminosity.

solar mass The *mass* of the Sun, 1.9891×10^{30} kg. This is used as a unit of mass, so the mass of a star or galaxy may be expressed as a multiple of the solar mass.

solar neutrino problem The fact that *neutrino* detection experiments on Earth do not find as many neutrinos coming from the Sun as standard *models* of the way energy is generated inside the Sun by *nuclear fusion* reactions (standard *astrophysics*) suggest ought to be present.

The fact that we know anything at all about solar neutrinos from observations is due to the enthusiasm of two men, the theoretical astrophysicist John **Bahcall** and the experimenter Ray **Davis**. Neutrinos were only definitely observed in experiments in 1956, and at the end of the 1950s there seemed little prospect of building a detector that could observe neutrinos from the Sun, while at the same time there seemed little point in such an experiment because 'everybody knew' from the standard solar models how many neutrinos it would detect. But Bahcall and Davis thought it might be worth the effort.

It is worth the effort because the neutrinos that reach us from the Sun come from the very heart of the Sun, not from its surface. All of the other radiation we receive from the Sun comes from its outer layer, and until the development of *helioseismology* the neutrinos offered the only hope of 'seeing' deep inside the Sun. Even with helioseismology, astronomers use observations of the surface layer to infer what is going on inside the Sun; *neutrino astronomy* potentially provides a window directly into the centre of the Sun, detecting the very particles produced in the nuclear fusion reactions going on there. Neutrinos produced in this way pass right through the Sun, across space, and right through the Earth; the snag is that all but a few of them also pass right through any detector designed to trap them.

Because many other particles, such as *cosmic rays*, could trigger any detector sensitive enough to capture neutrinos, and because neutrinos themselves pass essentially unaltered right through the Earth, the place to build a solar neutrino detector is not on top of a mountain (like a conventional telescope) but down a mine shaft, shielded from cosmic rays by layers of solid rock (see *neutrino astronomy*). The Davis detector was installed, beginning in 1964, 1,500 m below the ground in the Homestake Gold Mine at Lead, South Dakota. Some 7,000 tonnes of rock had to be removed to make room for the detector, a tank the size of an Olympic swimming pool, containing 400,000 litres of perchlorethylene (C_2Cl_4), commonly used in so-called dry-cleaning processes.

Bahcall and Davis reasoned that the chlorine in the cleaning fluid could be used to trap neutrinos from the Sun. About a quarter of all the chlorine *atoms* on Earth are in the form of the *isotope* chlorine-37. With four chlorine atoms in every molecule of perchlorethylene, that gives one chlorine-37 atom per molecule, a total of about 2×10^{30} targets for the neutrinos to hit. On the very rare occasions that a neutrino does interact with the nucleus of an atom of chlorine-37, one of the *neutrons* in the nucleus is converted into a *proton* and an electron is emitted, in a forced *beta decay* process. The nucleus has been converted into one of argon-37, which escapes into solution in the tank. If the team could count the number of argon-37 atoms in the tank, they would know how many neutrinos had interacted with chlorine nuclei in the tank.

The way this is done is a beautiful piece of work that deserves a Nobel Prize. First, the argon-37 is removed from the tank by bubbling helium and inert argon-36 or argon-38 gas through it. Then, all the argon, now including a few argon-37 atoms, is separated from the helium, and watched using special detectors. Because argon-37 happens to be radioactive, with a *half life* of about 34 days, the experimenters can count the decays of each radioactive atom to find out how many they had in the first place. After all that effort, each run of the experiment records about twelve counts, corresponding to the same number of neutrino interactions occurring in the tank over a period of two or three weeks.

The first results from the experiment emerged in 1968, and disagreed with the predictions of the standard solar models. At first, nobody worried much about this, suspecting teething problems with the experiment. But as the observations continued to build up during the 1970s and 1980s, it became increasingly clear that something was wrong. The experiment seemed to be working beautifully, but it found only about one-third of the expected number of solar neutrinos. This meant that astronomers did not fully understand how stars worked, or that calculations of the **proton–proton reaction** that keeps the Sun hot were wrong, or that there was something about neutrinos that was not understood.

The neutrinos actually captured in the Davis detector come from a side branch of the proton–proton reaction, mainly from an interaction in which a nucleus of beryllium-7 captures a proton, becomes a nucleus of boron-8, and very quickly emits a positron and a neutrino to become a nucleus of beryllium-8, then splits into two nuclei of helium-4. The neutrinos spat out by the boron-8 are particularly energetic, and can trigger the conversion of chlorine-37 into argon-37; the main flood of neutrinos produced in the proton–proton reaction each have less energy, and cannot do this.

The predictions of how many neutrinos from this reaction ought to be reaching the Earth are based on observations made at the Kellogg Radiation Laboratory at the California Institute of Technology, by researchers such as Willy **Fowler**. They found that, under the conditions at the heart of the Sun described by the standard solar model, the main p–p reaction should be producing a flood of 60 billion neutrinos crossing every square centimetre of the Earth each second. Unfortunately, none of these has enough energy to be recorded in the Davis experiment. The figures for high-energy neutrinos are very sensitive to the central temperature inside the Sun, which is 15 million **Kelvin** in the standard model. For that temperature, there should be just 3 million boron-8 neutrinos crossing each square centimetre of the Earth each second, and that should produce just 20 'events' in the tank each month. Together with a few events caused by another side chain of the p–p reaction (in which beryllium-7 captures an electron and gives off a neutrino, becoming lithium-7, which *then* captures a proton and splits into two helium-4 nuclei), the theory predicts 25 events per month in the Davis detector. In fact, the Davis experiment actually records 9 events per month, over more than twenty years.

Bahcall has invented a number he calls the Solar Neutrino Unit, or SNU, as a measure of neutrino activity from the Sun. In these units, the standard model predicts that the Davis detector should be recording between 6 and 8 SNU, allowing for anticipated uncertainties in the calculations. In fact, it observes 2 or 3 SNU. Whichever way you look at it, this is only one-third of what was expected. Why?

Since the early 1970s, astronomers and physicists have tried many desperate remedies to solve the solar neutrino problem. If the temperature at the heart of the Sun could be reduced by just 10 per cent, that would do the trick – at least, as far as these particular reactions are concerned. Stories of the solar neutrino problem do not always emphasize that this is, in many ways, a triumph. Neutrinos from the heart of the Sun *have* been detected, and the discrepancy between theory and experiment is, in this sense, only 10 per cent. The problem is that astrophysicists think they understand how stars work so well that they cannot see how to lower the temperature that much.

One idea which looked very attractive at the end of the 1980s was that there might be particles of *dark matter* (WIMPs) inside the Sun. Such particles, each with a few times as much mass as a proton, could provide the dark stuff needed to explain the rotation of our Galaxy, and if enough of them fell into the Sun's gravitational field, they would settle at the heart of the Sun and smooth out the temperature there, carrying a little heat out from the very centre to warm the region a little further out. There would be just as much heat generated inside the Sun overall, but at a lower temperature spread over a greater volume.

The situation had changed in the early 1990s, however, both because more solar neutrino experiments were now available and because theorists had devised better solar models. The WIMP solution could still be made to work, but it is not regarded as the front runner. The calculated solar models now only predict just under twice as many neutrinos as the Davis detector sees (5.3 SNU against 3.0 SNU), and there are even models which match the latest observations. Although Bahcall in particular argues vehemently that these models are not realistic descriptions of the Sun, there is a school of thought which says that the experimental observations should be taken at face value, and that the only 'good' solar model is one which matches the observations.

The new detectors, including *GALLEX* and *SAGE*, are important because they are sensitive to lower-energy neutrinos than the Davis detector and provide a different 'window' on the *spectrum* of solar neutrinos, even providing information about the p–p neutrinos. The *Kamiokande* experiment is also sensitive to high-energy neutrinos, but measures them using a completely different technique from the chlorine-based detector and comes up with essentially the same numbers. All of the experiments continue to show disagreement with the standard solar models.

There is a further complication, from the Davis data which have now accumulated for more than two decades. Over two 11-year *solar cycles* of activity, there is a hint of a pattern in which even fewer neutrinos are seen when the Sun is more active, and slightly more neutrinos are seen when the Sun is less active. This is utterly baffling, since the solar cycle is associated with activity on its surface, while the neutrinos come from the middle of the Sun. The apparent correlation is largely ignored by people working on the solar neutrino problem, while they keep half an eye on the observations to see if the pattern repeats over the present solar cycle.

The resolution of the basic solar neutrino problem that is most widely favoured today is that something happens to the neutrinos after they are produced in the nuclear reactions inside the Sun and before they reach the Earth. One widely publicized version of this process is known as the MSW effect, from the initials of the names of three physicists who played key roles in its development – S. P. Mikheyev and A. Yu. Smirnov, in Russia, working on the basis of a suggestion made by a US physicist, Lincoln Wolfenstein. The neutrinos produced by nuclear reactions inside the Sun are all electron neutrinos, and these are the only kind that the detectors can detect. But there are also two other varieties of neutrino, one associated with the *muon* and the other with the *tau* particle. If some of the electron neutrinos are converted into their counterparts en route to Earth, that could explain the shortfall of electron neutrinos seen by the detectors.

Theory says that this is possible, but only if the neutrinos involved have non-zero mass. The mass of the electron neutrino has to be very close to zero, and the mass of the muon neutrino no more than about 0.01 *electron Volts*. There are also other processes which (in theory) might allow oscillations of neutrinos from one kind to another, but all require the neutrinos to have some mass.

At present, the observations from several different experiments have narrowed down the range of options considerably. Only a limited range of masses is still allowed, for example. If you are an optimist, this is good news, telling us that the actual mass of the electron neutrino has been pinned down. If you are a pessimist, you may think that the next experiment or two will close even that loophole, leaving no room for any neutrinos with mass and ruling out this 'solution' to the problem. At present, the solar neutrino problem is still unresolved. At one extreme, a few theorists say there is no problem and what they regard as the best solar models now agree, within the limits of experimental error, with the observations. Most astronomers and particle physicists think there is a problem, but it is still not certain whether it will be resolved by a better understanding of astrophysics or a better understanding of particle physics. As more experimental results come in over the next few years, this is likely to be one of the continuing hot topics in astrophysics.

solar oscillations See *helioseismology*.

solar radius The *radius* of the Sun, 1.392×10^6 km. This is used as a unit of size, so the size of a *star* may be expressed as a multiple of the solar radius.

Solar System Overall name for the Sun, its attendant *planets*, *comets*, *minor planets*, gas and general debris – everything held in *orbit* around the Sun by its *gravitational field*. 99.86 per cent of the *mass* of the Solar System is concentrated in the Sun, and two-thirds of the remaining mass is locked up in Jupiter.

solar time   Time measured in terms of the rotation of the Earth, from noon on one day to noon the next day. See *sidereal time*.

solar units   System of units based on the dimensions of the Sun. See *solar luminosity, solar mass, solar radius*.

solar wind   A flow of electrically charged particles, mainly *electrons* and *protons*, outward from the Sun. This tenuous *plasma* moves at hundreds of kilometres per second, but contains only about eight particles in every cubic centimetre at the distance of the Earth from the Sun, and thins out as it moves through the Solar System.

space   Traditionally, the void between the *stars* and *planets*. In the context of the *special theory of relativity* and the *general theory of relativity*, however, even 'empty space' has to be thought of as having a well-defined structure and properties – it is the stage on which material events take place. The properties of space, and in particular the way it is curved, determine the way objects move and even the *fate of the Universe*. In relativity theory, three-dimensional space is united with time to make a four-dimensional continuum.

This picture breaks down, according to *quantum theory*, on the smallest scales, over distances comparable to the *Planck length*. There, both space and time lose their identity in a 'quantum foam'. The American physicist John *Wheeler* has suggested that the presence of what we regard as a real particle in space is no more significant in the context of the activity of the quantum foam than the presence of a cloud is to the dynamics of the atmosphere. We see the cloud, or the particle, but it is only a minor disturbance in a sea of activity. In this book, 'space' is usually used in the relativistic way.

space observatories   Unmanned (so far) artificial satellites of the Earth, specifically designed for astronomical work. Space observatories are often used to study the Universe using wavelengths of electromagnetic radiation which cannot penetrate through the atmosphere to the surface of the Earth. Even lifting an optical telescope (such as the *Hubble Space Telescope*) above the atmosphere does, however, dramatically improve the visibility. Data from space observatories and *spaceprobes* are returned to ground stations by radio links.

spaceprobes   Unmanned (so far) probes which are sent out of Earth *orbit* to explore other *planets* in the Solar System, the Sun itself and (one day) other *stars*. They are designed to probe things within space, rather than the structure of space itself. Some spaceprobes (such as the *Voyager* series) fly past the planets of interest and send back data from a single close encounter with each object of interest; others, such as *Galileo*, are designed to go into orbit around their targets and send back data from one planet for a longer period of time. And in some cases, such as *Viking*, the probes descend through the atmospheres of the planets to send back data from their surfaces.

spacetime The union of *space* and *time* into one four-dimensional whole, proposed by Hermann **Minkowski** in the context of the *special theory of relativity* and soon extended by Albert **Einstein** into the description of the *general theory of relativity*. In the context of Einstein's theory – which is the basis for our understanding of the Universe at large and the behaviour of everything in it under the influence of *gravity* – spacetime is a real and tangible structure, which can be likened to the surface of a stretched rubber sheet. It can be bent, stretched and squeezed, and can even have holes poked in it. There are even ripples in spacetime (see *gravitational radiation*) akin to sound waves moving through the air, and the overall shape of spacetime determines the ultimate *fate of the Universe*.

The relationship between two events in space and time may look different for observers moving at different speeds (in different *frames of reference*). For example, the two events might be the departure of a spaceship from the Solar System and its arrival at *Alpha Centauri*. The length of time the spaceship takes to reach Alpha Centauri is different in different frames of reference, and the distance between the Sun and Alpha Centauri is different in different frames of reference. But the correct measurement of the 'distance' between the two events in spacetime, given by Einstein's equations and known as the spacetime interval, is the same for all observers in all inertial frames.

spacetime diagram = *Minkowski diagram*.

spacetime interval See *spacetime*.

space travel Usually used to refer to manned spaceflight. So far, people have travelled only as far as the Moon. Travel to Mars would be feasible with present-day technology, and there have been discussions about the possibility of a joint Russian–American manned expedition to Mars for early in the 21st century. Many astronomers are ambivalent about the desirability of manned space travel. They like the idea of exploring the Solar System (and, perhaps, beyond), but given the limitations of present-day technology and budgets, they believe that it is much more cost effective (and safer) to send unmanned *spaceprobes* to explore the nearby Universe.

A few enthusiasts, however, have carried out calculations which show that, with technology only slightly more advanced than we possess, it would be possible to colonize the *Galaxy*. One approach could be to use large spaceships (perhaps hollowed-out *asteroids*) in which generations of astronauts could live while voyaging to another star. There are other suggestions, and the details are less important than the fact that star travel is definitely possible for civilizations only slightly more advanced than we are. It does not violate any laws of physics, or require any forms of matter or energy that we do not already know about. This leads to the *Fermi Paradox*. See also *time travel*.

Special Astrophysical Observatory Main observatory for both *radio astronomy* and *optical astronomy* in the former Soviet Union. It is located in the Caucasus Mountains, and its instruments include a 6 m optical *telescope* and the *RATAN 600 radio telescope*.

special theory of relativity Description of the relationships and inter-actions between moving objects, developed by Albert *Einstein* early in the 20th century. The theory was first published in 1905, in a mathematical form based on equations; its implications can be more clearly visualized, however, using a geometrical description of events taking place in a four-dimensional *spacetime*, first applied in this context by Herman *Minkowski* in 1908. The 'special' theory gets its name because it applies only to the special case of objects moving at constant speeds in straight lines – that is, at constant velocities. It does not deal with accelerated motions, including the acceleration caused by gravity. The later extension of Einstein's theory to deal with gravity and other accelerations (the *general theory of relativity*) developed the geometrical *model* of spacetime further.

The key features of the special theory are that, from the point of view of an observer who is regarded as stationary (in his or her own *inertial frame*), time recorded on a moving clock will run slow, a moving object will shrink in the direction of motion, and the moving object will gain mass. The speed of light is the same for any observer in any inertial frame, no matter how he or she is moving relative to the source of the light, and it is impossible to accelerate an object from below the speed of light up to the speed of light. All of these predictions have been tested and verified many times in experiments. It is the special theory that says that mass and *energy* can be interchanged in line with Einstein's equation $E = mc^2$; this too has been confirmed by experiments, including the explosion of atomic bombs. See also *relativistic mechanics*.

spectral classification of stars A refinement of the classification of *stars* by their *colour* (see *colour index*), using detailed studies of the *spectrum* of their starlight. The basic classification scheme developed by Henry *Draper* at Harvard was refined early in the 20th century into a classification which labelled stars according to their spectra and colours as O, B, A, F, G, K and M, in decreasing order of temperature. O stars, at one extreme of the classi-fication, are blue-white and show features due to ionized *helium* in their spectra; G stars, which are much cooler and orange-yellow in colour, show strong lines associated with ionized calcium, and lines of metals such as iron.

As spectroscopic techniques improved, it became possible to subdivide the seven main classes of the Harvard sequence, so that the Sun, for example, is not just a G star but is classified as G2V. And when it turned out that some cool stars have strong absorption features in their spectra that are not seen in other stars of the same colour, three new classes, R, N and S, were added to the cool end of the Harvard classification. Additional information about the star in question is given by other code letters in the modern development of the Harvard classification, called the MK system. This means that to an expert the code giving the classification of a star contains a wealth of information; for the armchair cosmologist, however, the original Harvard classification is all you need to worry about.

spectroscopic binary   A *binary system* in which the two stars are too close together to be seen separately, even with a telescope, but where the orbital motion can be inferred from the regularly changing *Doppler effect* in the *spectrum* of light from the system.

spectroscopy   The study of the nature of stars and other objects by analysing the light or other radiation they produce – their *spectrum*. Traditionally, spectroscopy dealt with visible light, but it has been extended to cover other wavelengths of *electromagnetic radiation* and even to measurements of the distribution of energy among particles, such as *cosmic rays*. Spectroscopy is the single most important tool used in astronomy and (especially) *astrophysics*, and without spectroscopy we would essentially know nothing about the Universe except for the positions of stars and galaxies on the sky.

The first thing spectroscopy does is to tell us what stars, galaxies and so on are made of. It does this because the *atoms* of each kind of *element* produce their own characteristic features in the spectrum, called lines. When atoms emit or absorb energy in the form of light, they do so only at very well-defined *wavelengths*, which correspond to changes in the arrangement of the *electrons* that surround the *nucleus* of the atom. A convenient way to picture this is to think of an electron in a particular energy state, sitting on one step on a staircase. If the electron jumps down to the next step, to a state of lower energy, it emits a *quantum* of electromagnetic radiation with a wavelength determined by the height of the step. An electron on a lower step can also jump up to a higher step, but only by absorbing precisely the right quantum of energy to make the jump. Emission produces a bright line in the spectrum, while absorption produces a dark line in the spectrum where electrons have 'stolen' energy from a background source of light. Some sources also produce a continuous spectrum of energy with a characteristic shape – the two most important examples are *black body radiation* and *synchrotron radiation*. Black body radiation is actually made up from a combination of many wavelengths of radiation added together in accordance with the quantum rules, while synchrotron radiation is produced by electrons moving freely in a magnetic field, not attached to atoms.

All of this behaviour is extremely well understood and is described beautifully by *quantum theory*. But you do not need quantum theory in order to make use of spectroscopy. All you need is the knowledge, gained from observations in the laboratory, that each atom absorbs and emits light of a particular colour (a particular wavelength), or colours. Sodium, for example, radiates strongly at two precise wavelengths in the orange part of the spectrum when the atoms are heated or stimulated by an electric discharge; this is what gives many street lights their characteristic orange-yellow colour. Equally, if white light is shone through a substance that contains sodium (perhaps dissolved in a liquid), there will be dark lines in the appropriate part of the spectrum of the light, where the sodium has absorbed energy.

The key discovery that led to the development of spectroscopy was made by the German physicist Josef von Fraunhofer (1787–1826) in 1814. He was the first person to study the rainbow pattern produced by passing light through a prism in detail under intense magnification. He was actually interested in the properties of the glass in the prisms, and how it affected the light, but to his surprise he discovered that there are many dark lines in the spectrum of white light, including light from the Sun. A few of these dark lines in the solar spectrum, now known as Fraunhofer lines, had been noticed earlier by the English physician and physicist William Wollaston (1766–1828) in 1802, but their significance had not been appreciated then and Fraunhofer knew nothing of Wollaston's discovery. Fraunhofer soon counted 574 lines in the solar spectrum, and found many of the same lines in light from Venus and from many stars.

An explanation of the Fraunhofer lines as due to absorption of light by different elements present in the Sun's atmosphere was published by the German scientist Gustav Kirchoff (1824–87) in 1859. He went on to formulate the basic principles of spectroscopy, working together with Robert Bunsen (1811–99) at the end of the 1850s. It is no coincidence that this is the same Bunsen whose name is linked with that of the eponymous burner (although the 'Bunsen Burner' is actually a modification, made by one of Bunsen's assistants, of a device invented by Michael Faraday). The burner provides a clean, hot flame in which different substances can be heated until they glow (or burn), radiating light at their own characteristic spectral wavelengths, which can be studied and analysed. The technique soon led to the discovery of previously unknown elements, and the value of spectroscopy to astronomy was spectacularly demonstrated a few years later, when Norman *Lockyer* found a 'new' element, helium, by analysing the spectrum of light from the Sun. For a particular element, the bright lines produced by a hot sample are at exactly the same wavelengths as the dark lines produced when light passes through a cold sample.

Spectra of stars and galaxies are obtained by using prisms attached to telescopes to split the incoming light into its rainbow pattern – a technique which goes right back to Isaac **Newton**'s discovery that light can be split into its component colours in this way. The spectrum can then be photographed and studied in detail. The positions and strengths of the lines in the spectrum can also be determined electronically, using suitable detectors attached directly to telescopes or designed to pull out information from the photographic plates. Such spectra may show many bright lines, corresponding to emission of light by atoms in a hot region at the surface of the star, and also many dark lines, corresponding to absorption by atoms in cooler regions, further out from the surface of the star or in clouds of gas and dust in space. These can be compared with spectra obtained in the laboratory to find out exactly which elements are doing the absorbing and emitting. This directly reveals which elements are present in the objects being studied. The pattern of lines produced by each element is as distinctive as a fingerprint, and gives an unambiguous identification.

By measuring the strengths of the different lines in the spectrum, astrophysicists can work out how hot (or cold) the material producing the lines is (as well as what it is made of), while by measuring the displacement of the pattern of spectral lines towards the blue end of the spectrum or towards the red end of the spectrum they can use the **Doppler effect** to work out how fast a star is moving towards or away from us. Applying this technique to galaxies seen edge on, they can work out how fast the galaxies are rotating, and the cosmological **redshift** tells us how fast the Universe is expanding. A single spectrum can tell us what an astronomical object is made of, how hot it is, and how it is moving.

Molecules also produce characteristic spectral signatures, often in the millimetre part of the spectrum, while very energetic sources produce characteristic spectral signatures at **X-ray** and **gamma ray** wavelengths.

spectrum Any representation of how the strength of the **electromagnetic radiation** from a source depends on its **wavelength**. The most familiar example is the rainbow spectrum of visible **light**, which can be displayed using a prism, or seen in a rainbow itself. White light is made up of a mixture of wavelengths. The spectrum of colours seen by the human eye covers the range from red (with longest wavelength) through orange, yellow, green, blue, and indigo to violet (with shortest wavelength). The brightness of each colour in a spectrum shows how strongly that component of white light contributes to the overall brightness of the source. The properties of the spectrum can be determined accurately either photographically or using electronic detectors.

The spectrum extends beyond the visible range at both ends, into the ultraviolet and beyond, and into the infrared and beyond. The strength of radiation at different wavelengths outside the range of human eyes can be recorded by suitable instruments (such as *radio telescopes*) and displayed either as a set of numbers or as a graph. In both cases, there are places where there is a strong peak in the energy (corresponding to emission lines in the optical spectrum) and places where the energy dips (corresponding to absorption lines in the optical spectrum). The entire electromagnetic spectrum is broken up for convenience into separate bands, just as the optical spectrum is divided naturally into different colours. The full spectrum goes from radio waves (with longest wavelength) through microwaves, infrared, visible light, ultraviolet and X-rays to gamma rays (with shortest wavelength).

The term 'spectrum' has been borrowed for particle physics, where it is used to describe the pattern of the distribution of the number of particles with different amounts of energy. Sound waves can also be analysed in terms of a spectrum of different wavelengths contributing to the overall sound of, say, an orchestra, and if and when astronomers detect *gravitational radiation*, it too will be analysed in terms of the spectrum of waves with different wavelengths. But if the word 'spectrum' appears without qualification, it refers to the archetypal electromagnetic spectrum, and probably simply to the spectrum of visible light.

See also *spectroscopy*.

spiral arms   Pattern often made by bright stars in a *disc galaxy* as a result of *self-sustaining star formation*. The pattern was first noticed by Lord *Rosse* in 1845.

spiral galaxy   Alternative name for *disc galaxy*. Disc galaxy is a better term, because not all disc galaxies have prominent *spiral arms*, although all galaxies with spiral arms are disc galaxies.

Spörer, Gustav Friedrich Wilhelm (1822–95)   German astronomer, born in Berlin on 23 October 1822, who discovered the way in which *sunspots* appear at different solar latitudes at different stages of the *solar cycle*. His historical studies independently revealed the dearth of sunspots in the 17th century now known as the *Maunder minimum*; a similar dearth of sunspots in the 15th century is now known as the Spörer minimum.

s-process   A process by which heavy *elements* are manufactured through *nucleosynthesis* when there is a steady but relatively modest supply of *neutrons* available. The name is short for 'slow process', and involves a *nucleus* capturing one neutron at a time, and perhaps undergoing *beta decay* before it captures another neutron. This is one of the ways in which elements heavier than those of the *iron* group are made (see also *r-process*; many *isotopes* are produced by both processes). Heavy nuclei built up by the s-process may have to wait years or decades between successive neutron captures, but the process goes on slowly all the time inside stars where there is a constant supply of neutrons as by-products from *nuclear fusion* reactions.

The important feature of the s-process is that it goes on steadily inside most bright stars. Just 28 isotopes are produced only by the s-process. The s-process stops at the isotope bismuth-209, because if this nucleus captures an energetic neutron, the resulting nucleus rapidly undergoes *alpha decay.*

SS433 An intriguing star-sized object in our own Galaxy that produces energetic jets of material and behaves like a miniature *quasar.* It lies about 5 *kiloparsecs* from us, in the direction of the constellation Aquila, within a 40,000-year-old *supernova remnant.* The source radiates X-rays, radio waves and gamma rays, as well as visible light.

SS433 gets its name because it is object number 433 in a *catalogue* of stars with strong emission lines (see *spectroscopy*) compiled by two astronomers at Case Western Reserve University, Bruce Stephenson and Nicholas Sanduleak, in 1977. A year later, in June 1978, studies of the spectrum of light from SS433 showed an unprecedented pattern of behaviour. Strong emission lines in the spectrum move backwards and forwards over a wide range of wavelengths (corresponding to a wide range of *Doppler effect*) with a period of 164 days, while other features corresponding to the presence of hydrogen and helium in the source show a smaller Doppler effect varying with a period of 13.1 days.

The accepted explanation of this phenomenon is that SS433 itself is a hot O or B star with between ten and twenty times as much mass as our Sun, orbiting a *neutron star* (or possibly a *black hole*) in a *binary system* with a period of 13.1 days. The neutron star is stripping material from the large star by gravity, forming an *accretion disk* which feeds matter on to the neutron star. The powerful release of energy from the infalling material is driving gas out from the neutron star in two jets, which sweep around as the neutron star rotates, once every 164 days, producing strong *redshifts* and strong *blueshifts* from the same source. The jets travel at 80,000 km per second (about a quarter of the speed of light), sometimes pointing roughly towards us and sometimes more or less away from us, explaining the huge variations in the observed Doppler shift. Radio studies show that the jets extend out to 0.05 parsecs (0.16 *light years*) from the central 'engine', and X-ray studies show emission from 30 parsecs on either side of SS433, where the jets interact with the surrounding supernova remnant.

This is reminiscent of the way radio beams are thought to be emitted by *pulsars*, although no accretion disc is involved in that case. It is even more reminiscent of the way jets are emitted on a much larger scale from the huge black hole/accretion disc systems that are thought to power *active galaxies* and quasars. The activity seen in SS433 is probably typical of a temporary phase in the evolution of most X-ray binaries, and less extreme versions of this pattern of behaviour have now been seen in two other sources embedded in supernova remnants.

standard model of the Universe   See *Big Bang*.

starburst galaxies   Irregular *galaxies* (sometimes showing traces of spiral structure) containing a messy distribution of dust, in which a huge outburst of star formation is going on. Most starburst galaxies are interacting with (or even colliding with) other galaxies. They radiate very strongly in the *infrared*, where they may be 50 times brighter than in visible light. See *galaxy formation and evolution*.

star clusters   See *clusters of stars*.

star formation   *Stars* form when cool, relatively dense clouds of gas and dust in *space* shrink in upon themselves as a result of *gravitational collapse*. This mainly happens in giant *molecular clouds*, where the density is about 1 billion or 10 billion *atoms* per cubic metre.

It is actually very difficult to make a cloud like this collapse. It is held up by *pressure* resulting from the heating of the gas by radiation from nearby stars, by magnetic fields and by the *centrifugal effect* of any rotation. In a *disc galaxy* like our own *Milky Way*, star formation is triggered when clouds of gas are squeezed in the spiral density wave (see *origin and evolution of galaxies*; very little star formation goes on in *elliptical galaxies*). Clouds may also collapse when they feel the blast from a *supernova* explosion.

Once the cloud starts to collapse, it breaks up into fragments in accordance with the *Jeans criterion*. Their continuing collapse makes the fragments warm up, as gravitational energy is converted into heat. At first the infrared radiation produced can escape fairly easily, but as the fragments become more dense, they become opaque, holding in the radiation and causing the temperature inside the fragment to rise more dramatically. Each fragment of the original cloud is now a protostar, which continues to collapse until (after about 100,000 years for a star with the same mass as the Sun) a hot core, still gaining its energy from gravitational collapse, forms. Gradual collapse then continues in accordance with the *Kelvin–Helmholtz timescale*. At this stage, the protostar may be surrounded by a disc of material (especially if it is an isolated star, not in a *binary system*) from which planets can form. When the temperature at the heart of the protostar rises above about 10 million *Kelvin*, *nuclear fusion* reactions begin in its interior, and it settles down as a stable *main sequence* star. It takes a star like the Sun about 50 million years to reach the main sequence; more massive stars get there more quickly, lighter stars more slowly.

Starlink   Computer network which links British astronomers. Set up in 1979, it now includes connections between Britain and the UK *telescopes* at *Siding Spring Observatory* in Australia, *Roque de los Muchachos Observatory* in the Canary Islands, and *Mauna Kea Observatory* in Hawaii.

starquake   See *glitch*.

stars   The stars you can see in the sky are almost all hot balls of gas (strictly speaking, *plasma*), held together by *gravity*, which shine because heat is generated in *nuclear fusion* reactions in their interiors. The Sun is a typical star.

Some young stars, which are still contracting and have not yet joined the *main sequence*, are hot (and therefore shine) because gravitational energy is being released as they contract (see *Kelvin–Helmholtz timescale*); they have not yet started to 'burn' hydrogen in *nuclear interactions*. At the opposite extreme, some old stars are no longer kept hot by nuclear burning, and have shrunk down to become *white dwarfs* or *neutron stars*; such old stars may still glow for a time (even for millions of years) like a dying ember, and they are still called stars, even though they are no longer generating heat. A white dwarf will eventually cool into a burnt-out cinder, a cold *black dwarf*.

There may be a kind of faint star, intermediate between a star like the Sun and a *planet*, which generates heat and light for a time through slow *gravitational collapse*, but never becomes warm enough in its centre to ignite nuclear burning. These stars are known as *brown dwarfs*, and have masses intermediate between the mass of the Sun and the mass of Jupiter.

Main sequence stars, and *giant stars*, which do generate heat and light through nuclear interactions, all started out as larger, cooler clouds of gas which got hot in the middle as they shrank and released gravitational energy. Nuclear burning begins only when the central temperature of the star reaches a critical value – about 15 million *Kelvin* in the case of a star like the Sun, which is kept shining by the *proton–proton reaction*.

Astronomers sometimes highlight the role of nuclear burning in stars by asking newcomers to *astrophysics* to explain the role of nuclear inter-actions in determining the temperature at the heart of a star. The obvious answer is that nuclear interactions keep the star hot; but the obvious answer is wrong. When the protostar was collapsing towards the main sequence, it got hotter and hotter inside as gravitational energy was released. When nuclear burning began, it generated enough heat to create an outward pressure which stopped the collapse and stabilized the star at the appropriate point on the main sequence. Without nuclear burning, the protostar would have carried on collapsing, releasing *more* gravitational energy and getting hotter and hotter at the centre. The key role of nuclear burning inside a star is that it keeps the centre of the star *cool* (or at least, cooler than it would otherwise get).

If, for example, you could magically turn off the proton–proton inter-action inside the Sun, the central pressure would drop and the star would start to shrink and get hotter in the middle. But as the central temperature reaches about 20 million K, the series of interactions known as the **carbon cycle** dominate the generation of energy, increasing the pressure and halting the collapse. The fact that nuclear interactions stop the temperature inside a star rising higher is of key importance to an understanding of **stellar evolution**.

This also explains why stars are so stable. If anything happened to make a star expand slightly (for example, if the nuclear interactions began to go faster and generated more heat, increasing the pressure), it would get cooler in the middle as energy was used to drive the expansion. As it got cooler, the nuclear interactions would shut down, so less energy would be released, the pressure would drop, and the star would shrink again, back to its 'right' size. But if anything happened to make the star shrink slightly (for example, if the nuclear interactions began to go slower and generated less heat, reducing the pressure), gravitational energy would be converted into heat as the star shrank, the nuclear interactions would be driven more strongly as a result, and pressure would build up again, making the star expand back towards its equilibrium size. This is a so-called negative feedback process, which always acts to maintain the status quo whenever there is a disturbance.

There are several hundred billion stars in our **Galaxy**, but only a few thousand are ever visible to the naked eye from the surface of the Earth (no more than 3,000 under ideal conditions at any one time from one place on Earth). More than 1 million stars have been listed in **catalogues**, but for the great majority of these all that is known is their **apparent magnitude** and their position on the sky. Hundreds of millions of stars have been photographed in surveys carried out using **Schmidt cameras**, but most of these haven't even been catalogued.

Distances to stars are estimated by **parallax** and other techniques (see **cosmic distance scale**). The nearest star to the Sun is **Proxima Centauri**, at a distance of 1.3 **parsecs** (about 40 trillion km). It is because stars are so distant from us that they all (except the Sun) appear only as points of light on the sky (even through a telescope), even though main sequence stars have diameters similar to that of the Sun (just over 100 times the diameter of the Earth) and some giant stars have diameters 100 times greater than that of the Sun. The sizes of stars are worked out from their **colours** and **luminosities**. The colour tells astronomers how hot the star's surface is, and the luminosity tells them how much energy is getting out of the star in each second. From these observations, it is straightforward to calculate how large the star must be in order to put out that much energy at that particular **temperature**.

Information about stars comes chiefly from **spectroscopy**, which reveals their composition, their surface temperature and the way they are moving.

Stars are made almost entirely of hydrogen and helium; the composition of the Sun (which is typical) is, in terms of numbers of atoms (strictly speaking, nuclei), 90.8 per cent hydrogen, 9.1 per cent helium and 0.1 per cent *heavy elements*. These precise figures come partly from observations of the proportion of different elements revealed by spectroscopy in the surface layers, and partly from calculations using stellar *models* with different interior compositions to see which ones best match the observed properties of stars, including size and luminosity.

Most stars are in *binary systems* (or *multiple star* systems); the Sun's existence as an isolated star with a family of planets and no stellar companion puts it in the minority of main sequence stars, but this is a large minority, perhaps 15 per cent of the total. It is fortunate that most stars are in binary systems, because astronomers can use studies of the way the stars are moving in binary systems to infer their masses. A binary star system is held together by *gravity*, and the *orbits* of the stars around one another (strictly speaking, around their mutual *centre of mass*) obey *Kepler's laws*. By measuring the orbital period of a binary (relatively easy) and the separation of the two stars (difficult, but not impossible, at least for some binaries), it is possible to use Kepler's third law to work out the total mass of the two stars in the system put together.

Painstaking observations of the way the two stars are moving against the background of more distant *fixed stars* can then, in a few cases, reveal how far each star is from the centre of mass. Just like two children, one large and one small, balancing on a see-saw, for the system to be in balance the more massive star must be closer to the centre of mass, and the less massive star must be further out from the centre of mass. The ratio of the distances of each star from the centre of mass (which can be measured) is the inverse of the ratio of the masses of the stars, so this reveals the ratio of their masses.

With these two pieces of information (combined mass and mass ratio), it is straightforward to work out the actual mass of each star in the binary. After many years of careful observations, astronomers gleaned enough information in this way to identify the *mass–luminosity* relation and to be able to estimate the masses of stars which cannot be measured directly using this technique. Masses can also be estimated using models of stars, in which the known laws of astrophysics relate the size, mass and luminosity of the star. The masses indicated by the models for different kinds of star (such as *red giants*, main sequence stars or white dwarfs) broadly match the appropriate masses where they can be measured accurately for real stars, giving astronomers confidence in the accuracy of the models.

The lifetime of a star on the main sequence depends only on its mass, ranging from 3 million years for a star 25 times as massive as the Sun, through

10,000 million years for the Sun itself, to 200,000 million years for a star half as massive as the Sun. Partly because of this, there are many many more low-mass stars 'alive' at any one time than there are high-mass stars.

statistical parallax See *cosmic distance scale*.

Steady State hypothesis The idea that the Universe presents the same overall appearance to observers everywhere in it at all times, in accordance with the *perfect cosmological principle*. Since the Universe is seen to be expanding, with *clusters of galaxies* moving apart from one another, this would mean that new *galaxies* have to be created continuously to fill the gaps as old galaxies move apart.

The idea was originally put forward by Herman *Bondi*, Tommy *Gold* and Fred *Hoyle* in the 1940s. Although they came up with the idea together (after watching a film called *The Dead of Night*, a set of four linked ghost stories in which the end of the last story looped back to repeat the beginning of the first story), Bondi and Gold worked together in presenting a philosophically based discussion of the implications, while Hoyle alone tried to put the idea on a more scientific footing, invoking the *C-field* to describe the continuous creation of matter. The initial idea was that matter would appear quietly in the form of new atoms of hydrogen throughout intergalactic space (you would only need one new atom in every 10 billion cubic metres every year); later developments of the idea (on which Hoyle worked with the Indian astronomer Jayant Narlikar) envisage matter creation as a localized, energetic event going on in regions of intense gravitational field, such as in the nuclei of *active galaxies* and *quasars*.

In its simple original form, the Steady State hypothesis has been proved wrong by observations of objects at high *redshift* which show that the Universe as a whole has changed as it has aged (see *look back time*). The huge success of the *Big Bang* theory, and especially the discovery of the *background radiation* and the explanation of *nucleosynthesis* of the lightest *elements* in the Big Bang (ironically, largely achieved by Hoyle), meant that the Steady State idea was largely discarded by the end of the 1960s. But the recent development of theories of *inflation*, which see the entire visible Universe as just one bubble in an eternally expanding, self-replicating foam, strongly echoes the original philosophical basis of the Steady State hypothesis. In particular, it removes the need for a unique original event in the Big Bang *singularity*, which was the unacceptable face of the Big Bang theory that encouraged supporters of the Steady State idea in the first place.

Further reading: Fred Hoyle, *Home is Where the Wind Blows*.

stellar diameters The sizes of stars cover a wide range, from *supergiants* with diameters of hundreds of millions of kilometres (bigger than the diameter of the *orbit* of Mars around the Sun) down to about 10 km for a *neutron star*. The diameter of the Sun is 1,392,000 km, and a *white dwarf* has a diameter of a few thousand kilometres, roughly the same size as the Earth. Sizes of stars are usually expressed in terms of the size of the Sun (the *solar radius*); *main sequence* stars range in size from about ten times the size of the Sun to about one-tenth of the size of the Sun. See also *stellar evolution*.

stellar enegy sources   See *proton–proton reaction* and *carbon cycle*.

stellar evolution   When astronomers talk about the evolution of objects such as stars, they do not use the term in quite the same way as biologists do. In biology, evolution involves changes from one generation to the next; but in astronomy the evolution of a star refers to what biologists would call the life cycle of an individual, from birth to death.

Stars are born out of clouds of gas and dust in space, collapsing under the pull of *gravity* and getting hot inside as a result, until the temperature is so high that *nuclear fusion* reactions begin to take place in their centres (see *star formation*). This releases heat and produces an outward pressure which holds the star up against further collapse – at least, for as long as the supply of nuclear fuel lasts. All stars being born now start out with roughly the same mix of raw materials, just under 75 per cent hydrogen (by mass) and just over 25 per cent helium, with a smattering of *heavy elements* (earlier generations of stars contained slightly more hydrogen and less helium and heavy elements; see *nucleosynthesis*). Once a star has started burning nuclear fuel (converting hydrogen into helium) and has joined the *main sequence*, the rest of its life (its 'evolution') is determined solely by its mass. Although stars live for much longer than human civilization has been around, so that no star has been watched throughout its evolution, the life cycles of stars have been worked out by studying many different stars with different ages (rather like the way you could work out the life cycle of a tree by studying trees of all ages in a large forest), and by comparing these observations with computer *models* based on the known laws of physics.

It turns out that the Sun is a typical main sequence star. It has been radiating energy at almost the same rate for the past 4.5 billion years, and will continue to do so for about the same length of time again as hydrogen is converted into helium in its core, through the *proton–proton reaction*, at a temperature of about 15 million *Kelvin*. During the course of its lifetime on the main sequence, a star like the Sun gradually gets hotter as it adjusts to the continuing change in its internal composition caused by the steady conversion of hydrogen into helium. This change is not dramatic in astronomical terms (amounting to a warming of perhaps 20 per cent over the past 4 billion years), but it is interesting to geologists and climatologists, who have to explain why the Earth was warm enough for liquid water to flow and for life to evolve even when the Sun was younger and slightly fainter; the standard explanation is that a stronger *greenhouse effect* operating in the thicker atmosphere of the young planet did the trick.

The Sun is about as massive as a star can be and still generate heat mainly by the proton–proton reaction. For a more massive star, even the p–p process cannot provide enough energy to stop it collapsing a little more and getting a little hotter inside as a result. If the central temperature is more than about 16 million K, an alternative process for converting hydrogen into helium, the *carbon cycle*, begins to become important. The carbon cycle does contribute some energy production even inside the Sun: main sequence stars with more than a few times the mass of the Sun have central temperatures of around 20 million K and gain their energy largely through the carbon cycle. But whether the energy generated comes chiefly from the p–p process or mainly from the carbon cycle, in both cases the production of energy by nuclear fusion stabilizes the star and stops it collapsing further during its main sequence life.

Things change when the star has used up all of the hydrogen at its centre. The more massive a star is, the more rapidly it has to burn its fuel to provide enough pressure to hold itself up. Although the Sun has a main sequence lifetime of about 10 billion years, a star with three times the Sun's mass will stay on the main sequence for 500 million years, and one with twenty times the Sun's mass will stay on the main sequence for only 1 million years. Lighter stars than the Sun can stay on the main sequence for correspondingly longer – a star with half the Sun's mass stays about twenty times as long on the main sequence.

Sooner or later, though, the star runs out of hydrogen fuel in its heart. It now has a core consisting chiefly of helium, and containing most of the mass of the star, surrounded by an envelope (most of the volume of the star) which is still chiefly hydrogen. When the conversion of hydrogen into helium in the core (by either process) stops, the star begins to shrink and get hotter inside as gravitational energy is released. At first, the region just around the helium core will get hot enough to begin converting hydrogen into helium, generating energy which makes the outer layers of the star expand (moving it off the main sequence, to the right in the *Hertzsprung–Russell diagram*), while the core continues to get more massive and shrink inwards as more hydrogen is converted into helium at the top of the core. Even for a star with the same mass as the Sun, however, at this point in its life the central temperature is so high that this hydrogen shell burning involves the carbon cycle rather than the p–p process.

If the star is massive enough, eventually the core becomes so hot inside (about 100 million K) that helium *nuclei* begin to fuse to make carbon nuclei through the *triple alpha process*. If a star has more than about twice the mass of the Sun, core helium burning switches on gradually as the temperature in the core increases. In lower-mass stars, helium burning switches on quite suddenly, in a process known as the helium flash; this is because the temperature rises high enough to trigger helium burning only when the core has been squeezed into the form of *degenerate matter*, so that when the critical temperature is reached, quantum processes trigger a wave of energy production which lasts for a few seconds before the hot core expands slightly and stabilizes. Either way, once again, the star does stabilize, but now it has a helium-burning core surrounded by a hydrogen-burning shell and by a hugely extended outer envelope. The star has become a *red giant*. When the Sun becomes a red giant, its atmosphere will expand until it has a diameter greater than that of the orbit of Mercury (some accounts of the Sun's 'future history' mistakenly suggest that in its red giant phase the Sun will even engulf the Earth; those accounts do not allow for the fact that, by the time it becomes a red giant the Sun will have lost about a quarter of its mass by ejecting material into space).

The lifetime of a star as a relatively stable red giant is much less than the time it spent on the main sequence – typically about 5–20 per cent of its life as a simple hydrogen-burning star. The Sun itself will be a red giant for only about 1 billion years. If the star is massive enough, further stages of nuclear burning are possible, building up successively heavier elements by nucleosynthesis. A star may even have several different shells in which nuclear burning is taking place, arranged like onion skins around the core. But each stage of nuclear burning is over quicker than the one before. All this activity changes the outward appearance of the star, and may make it pass through phases of *variable star* activity, including *Cepheid* activity and *RR Lyrae* activity (the exact pattern of variability followed by a particular star depends on its mass).

Towards the end of its active life, a star will shed its outer layers to form a *planetary nebula*, leaving behind a dense core of material in which all fusion reactions have ceased, a *white dwarf*, slowly radiating its heat away into space over millions of years, and eventually cooling to become a *black dwarf*. At least, this is what happens for stars with up to about 8 solar masses of material. More massive stars run through the various stages of nuclear burning increasingly quickly, and then end their lives in dramatic *supernova* explosions.

In low-mass stars, including the Sun, the last stage of evolution prob-ably involves a dense core of carbon and oxygen, surrounded by a shell in which helium is being burnt and another shell in which hydrogen is being burnt. The fate of such a star is to become a white dwarf with about half the present mass of the Sun, the rest having been lost during its lifetime. It will be rich in carbon (literally a cinder!) with an outer shell of degenerate helium and possibly a tenuous atmosphere of hydrogen.

All of this discussion deals with isolated stars, but most stars actually occur in **binary systems**. Even in binaries, many stars are far enough away from their companions for the broad picture painted here still to hold. In close binaries, however, the evolution of the stars (particularly in the later stages) may be profoundly affected by the presence of a companion. The heavier star evolves more rapidly to the red giant stage, and gas from its swollen outer layers may be pulled on to the companion by tidal forces (see **symbiotic stars**). At a later stage in evolution, what was originally the more massive star may have become a white dwarf (or even a **neutron star** or **black hole**), and is now pulling matter on to itself from its companion. Such processes lead to the formation of **cataclysmic variables**, **X-ray binaries** and **recurrent novae**, and even to a second kind of supernova.

Further reading: William Kauffmann, *Universe* (2nd edn, W.H. Freeman, New York, 1988).

stellar masses The mass of a star is usually given in terms of the mass of the Sun, which is 1.9891×10^{30} kg (roughly 2 billion billion billion tonnes). According to stellar **models**, stars with as little as 8 per cent of the mass of the Sun could exist, but observations with the **Hubble Space Telescope** in the 1990s have failed to find a significant number of stars with less than about 20 per cent of the mass of the Sun (so-called **red dwarfs**). At the other extreme, the most massive known stars have about 100 times as much mass as the Sun.

Low-mass stars are defined as those with less than about 2.5 times the **solar mass**, intermediate-mass stars are those with about 2.5 to 8 times the solar mass, and high-mass stars are those with more than about 8 solar masses of material. Except for the as yet unexplained cut-off at 20 per cent of the solar mass, lower-mass stars are more common than higher-mass stars. **Main sequence** stars range in mass from a few tenths of the solar mass to more than 20 solar masses. The most massive stars seem to be unstable and highly variable, and do not fit on to the main sequence. See also **stellar evolution**.

stellar nucleosynthesis See **nucleosynthesis**.

stellar temperatures The temperature at the surface of a star is usually calculated in terms of the temperature that a **black body** would have if it had the same total **luminosity** as the star. This is known as the effective temperature, and is a good guide to the actual temperature of the star, since stars do radiate very much like black bodies. A slightly different (and less accurate) measure of a star's surface temperature is found by comparing its **spectrum** over a particular range of **wavelengths** with that of a black body. This is known as the colour temperature, and has the advantage that it is easily calculated from the standard **colour index**. The Sun has an effective temperature of 5,780 **Kelvin**, and a colour temperature of 5,700 K.

The total range of effective temperatures for visible *main sequence* stars is from about 40,000 K down to about 2,500 K – of course, even cooler stars may exist but be too faint to be observed from Earth. Some unusual hot stars have effective temperatures of around 100,000 K. An old, dead *white dwarf* will eventually cool until it is in equilibrium with the *background radiation*, but this will take a very long time indeed.

The temperature at the heart of a star is much greater than the temperature at the surface (unless it is an old, dead star). Except for stars made of *degenerate matter*, the temperature at the heart of a star is determined by the kind of *nuclear fusion* reactions going on to generate heat and prevent the star from shrinking as a result of *gravitational collapse* (see *stellar evolution*). So the range of central temperatures, especially for main sequence stars, is quite small. The temperature at the centre of the Sun is about 15 million K, typical of stars sustained by the *proton–proton reaction*, while more massive stars, kept hot by the *carbon cycle*, have central temperatures not much in excess of 20 million K – a much smaller proportional difference than the spread of surface temperatures.

stellar winds   Streams of material ejected into space by many stars. The *solar wind* is an example of a stellar wind.

string theory   See *grand unified theories*.

strong nuclear interaction (strong force)   See *fundamental forces*.

submillimetre astronomy   Astronomy carried out using electromagnetic radiation in the wavelength range 0.3–1 mm (or even slightly longer wavelengths). This can be regarded either as extremely long-wave infrared radiation, or as extremely short-wave radio waves. Either way, the observations at these wavelengths are especially important because many *interstellar molecules* radiate in the submillimetre band. Submillimetre observations can also probe the nature of regions of *star formation*, and provide information about the *background radiation*.

Telescopes for submillimetre astronomy typically use a dish antenna rather like that of the most familiar kind of radio telescope, cooled by liquid helium so that heat from the antenna does not swamp the observations. The observations are best made at high altitudes, to minimize the absorption of incoming radiation by water vapour in the atmosphere. The *James Clerk Maxwell Telescope* at the *Mauna Kea Observatory* in Hawaii is an archetypal example of an instrument designed to operate at submillimetre wavelengths.

Sun   The Sun is an ordinary star, roughly halfway through its lifetime on the *main sequence* of the *Hertzprung–Russell diagram*. It is a ball of hot gas (strictly speaking, *plasma*) with a *mass* of 1.9891×10^{30} kg (roughly 330,000 times the mass of the Earth) and a diameter of 1.392×10^6 km (roughly 109 times the diameter of the Earth). Since the Sun therefore has a volume roughly 1 million times that of the Earth (volume is proportional to radius, or diameter, cubed, and 100 cubed is 1 million), but a mass equivalent to only a third of a million Earths, its average density is only one-third of that of the Earth, about 1.4 times the density of water. But that average disguises a very wide range of densities, from a superdense core to the tenuous outer layers.

In terms of its overall appearance as a star, the Sun has a *luminosity* of 3.83×10^{26} watts and an *absolute magnitude* of 4.8. It is a yellow G2 *dwarf star*, with an *effective temperature* of 5,800 *Kelvin*, at an average distance of 149,597,870 km (499.005 *light seconds* or 1 *astronomical unit*) from the Earth, which is in *orbit* around the Sun. It is made up of 71 per cent hydrogen and 26 per cent helium by mass, with a smattering of *heavy elements*. The size of the Sun's disc on the sky covers 32 *arc minutes*, almost exactly the same as the angular size of the Moon as seen from Earth, a curious coincidence which makes for spectacular *solar eclipses*. (*Health warning*: Never look directly at the Sun, even during an eclipse; it can permanently damage your eyesight.) Because it is so close to us compared with other stars (see *cosmic distance scale*), the Sun is the brightest object in the sky as seen from Earth, with an *apparent magnitude* of −26.7. It rotates once every 25.4 days (on average; the equator rotates faster than the higher latitudes) and travels once around the *Milky Way* every 200 million years. The movement of the Earth in its orbit around the Sun means that the Sun has to rotate an extra bit to 'catch up' with the Earth, so that from Earth one rotation of the Sun seems to take 27.27 days. The rotation of the Sun flattens it slightly into an oblate shape, differing from a perfect sphere by 0.001 per cent, a difference of just 6 km in the radius measured around the equator or pole to pole. Although small, it was important to measure this flattening because any larger amount of oblateness (even 0.005 per cent) could affect the way the Sun's gravity influences the orbit of Mercury, and would have cast into doubt tests of the *general theory of relativity* based on the *advance of the perihelion* of Mercury.

As the closest star, the Sun has been studied intensively by astronomers, who have used a combination of observations and models based on the known laws of physics to work out what goes on inside the Sun. The accuracy of these models of the Sun's interior has been confirmed by *helioseismology*, and by studies of *neutrinos* from the heart of the Sun. Although there is a slight disagreement between the predictions of the models and the number of solar neutrinos observed (see *solar neutrino problem*), it is often not appreciated that the neutrino observations do, in fact, confirm the accuracy of the models (in, for example, determining the central temperature of the Sun) to within 10 per cent. It is a sign of how good the present understanding of stellar structure is that a disagreement

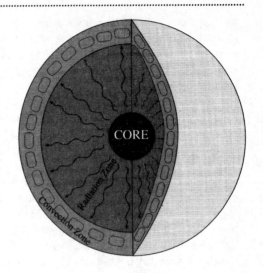

The Sun. An idealised representation of the internal structure of the Sun, showing its three main regions – the core, the radiation zone and the convection zone.

of a few per cent is a cause for concern among the experts; but it is of no significance when sketching a broad outline of how the Sun works, from the inside outwards.

The structure of the Sun can best be described in terms of a series of layers, or shells. The heart of the Sun, the core in which energy is produced by *nuclear fusion* processes (chiefly the *proton–proton reaction*), extends a quarter of the way from the centre of the Sun to the surface. This means that it represents only 1.5 per cent of the volume of the Sun. But within that core, electrons are completely stripped from the nuclei of their atoms, and the nuclei are packed together so closely that the core has a density twelve times that of solid lead (160 times the density of water). Because nuclei are so much smaller than atoms, though, even at this density they roam freely through the core, bouncing off each other in repeated collisions and behaving in exactly the same way that atoms do when they form a gas. The core of the Sun is a perfect gas with a density twelve times that of lead.

Because of the very high density in the core, this 1.5 per cent of the Sun's volume contains half of its mass. The temperature at the centre of the core (at the heart of the Sun itself, where the nuclear reactions are going on most vigorously) is about 15 million K; the temperature at the outer edge of the core is about 13 million K. The pressure in the core is 300 billion times the pressure of the atmosphere at the surface of the Earth, and under these conditions the radiation produced by the nuclear reactions in the form of high-energy photons (*gamma rays*) only travels a short distance before it collides with a charged particle – either a negatively charged electron or a positively charged nucleus. The gamma rays interact with the charged particles and are degraded into slightly less energetic *X-rays*, making their way slowly out of the core and up through the outer layers of the Sun.

The passage of the X-ray photons outward from the core to the surface really is slow, even though each photon travels at the speed of light, because they are constantly being bounced around by collisions with charged particles, like balls in some frenetic cosmic pinball machine. In each interaction, the photon may bounce off in any direction, including back the way it has come. The result is that it moves in an erratic, zig-zag path known as a 'random walk', with each step in the walk taking just about 1 cm. In fact, the walk is not quite random. Over a range of 1 cm, there is very little difference in temperature in this part of the Sun, above the core, which is called the radiation zone. But there is a tiny difference, ultimately caused by the fact that photons are being 'lost' from the surface of the Sun, and this tiny difference ensures that just a few more photons work their way outwards at each level inside the radiation zone than work their way inwards.

If a photon could fly in a straight line from the centre of the Sun to its surface, the journey would take just 2.5 seconds; but on average it actually takes a photon 10 million years to get from the centre to the surface. During all that time, it has been travelling at the speed of light, so its zig-zag path is literally 10 million *light-years* long – if it could be straightened out, it would stretch five times further than the distance from Earth to the *Andromeda galaxy*. Looking at this another way, it means that what we see going on at the surface of the Sun today is a result of what *was* going on in the core 10 million years ago. We cannot, just by looking at the surface of the Sun today, be absolutely sure that the nuclear interactions didn't switch off (or slow down) some time in the past few million years (this was, indeed, one suggested resolution of the solar neutrino problem, but it has now been ruled out by continuing neutrino observations and by helioseismology).

The radiation zone extends out to a distance of about 1 million km, 85 per cent of the way from the centre of the Sun to the surface. All the way, the plasma is getting cooler and thinner. Halfway from the centre of the Sun to the surface, the density is the same as that of water; two-thirds of the way out, it has dropped to the density of the air that we breathe. At the outer edge of the radiation zone, the density is only 1 per cent of that of water, and the temperature has dropped to a mere 500,000 K. Under these conditions, nuclei can begin to cling on to electrons to form more or less stable atoms and *ions*, although the electrons may get knocked off by encounters with energetic photons. But at the same time, the photons from the core have been degraded still further on their long, zig-zag journey outwards, shifting to longer wavelengths. This means that each photon carries correspondingly less energy, and interacts less violently with the particles it encounters. At the top of the radiation zone, the atoms and ions are just stable enough, and the photons are just weak enough, for many of the photons that collide with the atoms and ions to be absorbed without being re-radiated. The result is that the atoms and ions absorb energy and get hot. The material in this region of the Sun positively seethes with energy, dumped at the bottom of a layer known as the convection zone by radiation that has almost literally run into a brick wall.

The convection zone lives up to its name. Like water in a pan that is being heated from below on a stove, the gas in the convection zone, heated from below by radiation, convects. The hot material rises upwards through the zone and cools, being replaced by cooler material from above that is moving downwards. The cooler material has lost its heat (becoming denser, and therefore sinking) as radiation from the surface of the Sun. This convection produces a seething activity over the outer 15 per cent or so of the Sun, from a depth of about 150,000 km below the surface up to the visible surface. The details of the convection zone are not precisely known, but it is thought to consist of three main layers of convection, one on top of the other, together covering a distance a bit less than half of the distance from the Earth to the Moon.

The top of the convection zone corresponds to the visible bright surface of the Sun, called the photosphere, where the effective temperature is 5,800 K (the temperature actually falls from about 6,000 K at the base of the photosphere to 4,000 K at the top) and the density is less than 1 millionth of the density of water. But the force of gravity at the top of the photosphere is 27 times as strong as at the surface of the Earth, and so the pressure in the photosphere is still one-sixth of the atmospheric pressure at the surface of the Earth. At last, atoms and ionized material can no longer block the outward flow of radiation, and the hot material emits photons which stream freely out into space, crossing the distance to Earth in 8.3 minutes. The light we see all comes from this layer, only 500 km deep, representing no more than 0.1 per cent of the Sun's radius.

The surface of the Sun is marked by features such as **sunspots** and **solar flares**, which are associated with changes in its **magnetic field** over the **solar cycle** of activity. Above the visible surface, there is a tenuous, transparent solar atmosphere. In the lowest layer of the atmosphere, known as the chromosphere, the temperature rises once again from about 4,000 K to

50,000 K at the top of the layer, a few thousand kilometres above the visible surface of the Sun. Clearly, the chromosphere is absorbing energy flowing out from the Sun; this heating continues above the chromosphere, but no completely satisfactory explanation of just how it happens has yet been found. A transition region a few hundred kilometres thick separates the chromosphere from the corona, an even more tenuous layer where the temperature reaches a peak of 2 million K at a height of about 75,000 km above the solar surface. The corona extends for millions of kilometres into *space*, and blends in with the *solar wind* that carries a stream of particles outward from the Sun.

See also *stellar evolution*.

sunspot A dark patch on the surface of the Sun. Sunspots are usually between 1,000 and 40,000 km in diameter. They look dark because they are about 1,500 K cooler than the surrounding bright surface of the Sun. Spots are caused by strong magnetic fields which suppress convection. (See *solar cycle*.)

sunspot number See *Wolf number*.

Sunyaev–Zel'dovich effect A distortion in the *background radiation* caused by the way the radiation is affected by the material in rich *clusters of galaxies* along the line of sight. The material that actually interacts with the background radiation is the hot *plasma* in which the *galaxies* in the cluster are embedded. This plasma is tenuous by terrestrial standards, but thick enough to produce a noticeable change in the background radiation. The *photons* of the background radiation are scattered to slightly higher energies by electrons in the hot plasma, so there are fewer photons left in

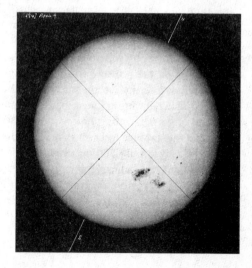

Sunspots. A large group of sunspots, photographed on 9 April 1947 from the Royal Observatory at the Cape of Good Hope. The sunspot group covers an area of a million square miles. (The crosswires are in the telescope.)

the part of the *spectrum* corresponding to radio waves to go on to reach our radio telescopes. Although this means that there are more photons at higher energies (with shorter wavelengths), as far as radio observations are concerned the result of the Sunyaev–Zel'dovich effect is to make the background radiation from that part of the sky look slightly cooler.

The effect, which was predicted by Rashid Sunyaev and Yakov *Zel'dovich* (working in Moscow) in 1972, changes the temperature of the background radiation by only about one part in 10,000 (0.01 per cent), but has actually been measured in a few clusters. This is important for two reasons. First, the fact that the background radiation is affected by the clusters shows that it really does come from far away across the Universe, behind those clusters of galaxies – that it really is 'background' radiation. Second, by comparing the measured size of the effect with the brightness of the cluster at X-ray wavelengths it is possible to work out an estimate of the value of the *Hubble constant*. The strength of the Sunyaev–Zel'dovich effect tells you how much hot plasma there is in the cluster to make X-rays (typically, this works out to be about the same as the amount of mass present in all the bright stars of the cluster galaxies), and that gives you the absolute *luminosity* of the cluster in X-rays. Measuring the apparent X-ray luminosity of the cluster therefore reveals how far away it is. And comparing this distance with the *redshift* gives the Hubble constant. Unfortunately, because of the small size of the effect, these measurements are very difficult, and the resulting estimates of the Hubble constant are still uncertain. But they tend to lie in the low end of the range indicated by other techniques, around 40–50 km per second per *Megaparsec*.

supercluster See *clusters of galaxies*.

supergiants The largest and brightest *stars*. They have masses up to 500 times that of the Sun, and *absolute magnitudes* of –5 to –10, bright enough to be seen in some other galaxies. Because there is an upper limit on the brightness of the brightest supergiants, their *apparent magnitudes* can be used as a measure of distances.

supergravity A class of unified theories that attempts to describe *gravity* and the other *fundamental forces* in one set of equations. The combination of supergravity and string theory led to the development of *superstring theory*. See also *grand unified theories*.

superluminal source A *radio source* or other object in which two components seem to be moving apart from one another faster than the speed of light. This is always a kind of optical illusion, a geometrical projection effect caused when jets of material travelling at a sizeable fraction of the speed of light are moving across the sky at a suitable angle to our line of sight (heading almost straight towards us). The effect does not even depend on *relativistic mechanics*, and can be entirely explained using *Newton's* *laws of motion*; it was predicted by Martin **Rees** shortly before the first superluminal sources were discovered in the 1970s. The jet in the *quasar* 3C 273 is among those which show this effect.

Supernova 1987A   The nearest *supernova* to have been seen since the invention of the astronomical telescope, and the first visible from Earth to the naked eye since the one observed by **Kepler** in 1604. SN 1987A was first seen from Earth in the southern sky on the night of 23–4 February 1987. It was the explosion of a star about 160,000 *light years* away from us, in the *Large Magellanic Cloud*; the precursor star was later identified on old photographic plates as a *supergiant* known as Sanduleak –69° 202.

SN 1987A, which is classified as a Type II supernova, was studied by just about every astronomical instrument in the Southern Hemisphere, and neutrinos from the blast were detected in the Northern Hemisphere (see *neutrino astronomy*). The wealth of information provided by these observations showed that the supernova behaved very much in the way predicted by theoretical *models* of such events, and the presence of features in the post-supernova *spectrum* of the object, corresponding to the *radioactive decay* of *isotopes* of cobalt and nickel to make iron, has provided important confirmation that explosive *nucleosynthesis* creates such *heavy elements* in supernovae. Photographed before, during and after the actual stellar explosion, Supernova 1987A has become the archetype of such an event. Before it exploded, Sanduleak –69° 202 was faintly visible on those photographs as a 12th *magnitude* object; at its peak brightness, seen in May 1987, it reached an *apparent magnitude* of 2.8, roughly 10,000 times brighter.

supernovae   A supernova is the explosive death of a star in an event so violent that for a brief period that single star shines as brightly as a whole galaxy of more than 100 billion ordinary stars like the Sun. This is a relatively rare event. Most stars end their lives in much quieter fashion, and only a few supernovae occur in a galaxy like the Milky Way every century. But such events are of key importance in the evolution of a galaxy and for the existence of life forms like ourselves, because supernovae both manufacture all the elements heavier than iron and scatter these and other *heavy elements* through space when they explode. A great deal of the material in your body consists of atoms that have been processed inside stars which have then exploded as supernovae, spreading the elements into the *interstellar matter* from which new generations of stars, planets and people can form. We are literally made of stardust.

All supernovae generate the enormous amounts of energy involved in these explosions in essentially the same way, when the core of a star suddenly collapses all the way down to the size of a *neutron star* (or possibly, in some cases, into a *black hole*); there are, though, two different ways in which this collapse can be triggered, and these produce supernovae with two somewhat different types of appearance (there are also more subtle differences between individual supernovae, since no two stars are identical, but these are not as important as the main distinction). The two kinds of supernova, Type I and Type II, were originally distinguished on the basis of *spectroscopy* – the *spectra* of Type II supernovae show features, caused by the presence of hydrogen, which are absent from the spectra of Type I supernovae. Continuing studies of supernova spectra and comparison with computer *models* can now explain this in terms of the way in which the two types of supernova are formed.

Type I supernovae occur in both *elliptical galaxies* and *disc galaxies*, and show no preference for being located in *spiral arms*. They are formed from the remnants of old, relatively low-mass *Population II* stars, and occur in *binary systems* where one star has evolved to the stage where it has become a *white dwarf* (see *stellar evolution*), and is gaining material from its companion by *accretion*. As the mass of the white dwarf increases, it eventually rises above the *Chandrasekhar limit* for a stable white dwarf (about 1.4 solar masses), and the star collapses under its own weight, releasing gravitational energy in the form of heat and triggering a wave of *nuclear reactions* that produce a flood of neutrinos.

Type I supernovae are divided into other sub-catagories, the main distinction being between Type Ia events, which show strong features due to silicon in their spectra, and Type Ib, which do not. It is thought that a Type Ia supernova produces the complete disruption of the collapsing white dwarf, which is blown apart by the energy released, spewing out a cloud of material containing about the same mass as the Sun to form an expanding shell (a *supernova remnant*), moving outward at tens of thousands of kilometres per second. All Type Ia supernovae seem to have much the same *luminosity* (corresponding to a peak *absolute magnitude* of –19), which makes them useful 'standard candles' that can be used to estimate distances to nearby galaxies.

Type Ib supernovae, which are more common than Type Ia, are triggered in much the same way, but are thought to involve white dwarfs left behind by relatively massive stars that have lost their outer layers in a strong *stellar wind*. The key difference with Type Ia is that a Type Ib supernova does leave behind a remnant in the form of a neutron star or a black hole. In either case, though, the binary system is likely to be disrupted by the explosion, leaving the companion to the original white dwarf hurtling through space as a so-called 'runaway star'. In one interesting example, three runaway stars known as 53 Arietis, AE Aurigae and Mu Columbae seem to have been shot out from a single point in the constellation Orion, and are almost certainly left over from a supernova

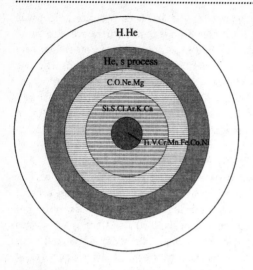

H.He

He, s process

C.O.Ne.Mg

Si.S.Cl.Ar.K.Ca

Ti.V.Cr.Mn.Fe.Co.Ni

Supernovae. The internal structure of a massive star just before it explodes to become a supernova.

explosion that occurred in what was then a quadruple star system about 3 million years ago.

Type II supernovae may also occur in binary systems (after all, most stars are in binaries), or in isolated stars. They are produced by explosions of young, massive *Population I* stars, rich in heavy elements, and occur mainly in the spiral arms of disc galaxies. They involve stars which still contain at least eight times as much mass as the Sun when they have exhausted all of their nuclear fuel. They are so big that even the ejection of material in a stellar wind cannot reduce their remaining mass below the Chandrasekhar limit, and even without the benefit of accretion their cores must collapse. Type II supernovae show more individual variety than Type Ia (Type Ib are more like Type II), and are slightly less bright – they reach absolute magnitudes of around –17. But their behaviour is reasonably well understood, and most of the details of the following description have been confirmed by studies of *Supernova 1987A* (although, as it happens, that supernova was not entirely typical because the precursor star seems to have lost some of its atmosphere before the final collapse occurred).

The key theoretical insight dates back to 1934, less than two years after the discovery of the neutron, when Walter *Baade* and Fritz *Zwicky* suggested that 'a supernova represents the transition of an ordinary star into a neutron star'. But this idea began to be fully accepted only in the 1960s, when *pulsars* were identified as *neutron stars* and the *Crab pulsar* was found at the sight of a supernova explosion that had been observed from Earth in AD 1054. Since then, different researchers have developed slightly different models of how a supernova works, but the essential features are the same. The outline given here is based on calculations carried out by Stan Woosley and his colleagues at the University of California, Santa Cruz, and describes the death throes of a star like the one that became Supernova 1987A.

The star was born about 11 million years ago, and initially contained about eighteen times as much mass as our Sun, so it had to burn its nuclear fuel furiously fast in order to hold itself up against the tug of gravity. As a result, it shone 40,000 times brighter than the Sun, and in only 10 million years it had converted all of the hydrogen in its core into *helium*. As the inner part of the star shrank and got hotter, so that helium burning began, the outer parts of the star swelled, making it into a *supergiant*. But helium burning could only sustain the star for about another million years.

Once its core supply of helium fuel was exhausted, the star ran through other possibilities at a faster and faster rate. For 12,000 years, it held itself up by converting carbon into a mixture of neon, magnesium and oxygen; for 12 years, neon burning did the trick; oxygen burning held the star up for just 4 years; and in a last desperate measure, *fusion* reactions involving silicon stabilized the star for about a week. And then, things began to get interesting.

Silicon burning is the end of the line even for a massive star, because the mixture of nuclei it produces (such as cobalt, iron and nickel) are among the most stable it is possible to form. To make heavier elements requires an input of energy (see *nucleosynthesis*). Just before the supernova exploded, all of the standard nuclear reactions leading up to the production of these iron-group elements were going on in shells around the core (with the *s-process* also at work). But as all the silicon in the core was converted into iron-group elements, the core collapsed, in a few tenths of a second, from about the size of the Sun into a lump only tens of kilometres across. During this initial collapse, gravitational energy was converted into heat, producing a flood of energetic photons which ripped the heavy nuclei in the core apart, undoing the work of 11 million years of nuclear fusion. This 'photo-disintegration' of the iron nuclei was first suggested by Willy *Fowler* and Fred *Hoyle* in the 1960s. As the nuclei broke apart into smaller nuclei and even individual *protons* and *neutrons*, *electrons* were squeezed into nuclei and into individual protons, reversing *beta decay*. Gravity provided the energy for all this. All that was left was a ball of neutron material, essentially a single 'atomic nucleus', perhaps a couple of hundred kilometres across and containing about one and a half times as much mass as the Sun.

The squeeze caused by this collapse was so intense that at this point the centre of the neutron ball was compressed to densities even greater than those in a nucleus, and it rebounded, sending a shock wave out into the ball of neutron stuff and into the star beyond. Material from the outer layers of the star (still at least fifteen times as much mass as there is in the Sun!), which had had the floor pulled from under it when the core collapsed, was by now falling inward at roughly a quarter of the speed of light. But when the shock wave met this infalling material, it turned the infall inside out, creating an outward-moving shock front that blew the star apart – but not before a flood of neutrons emitted during all this activity had caused a considerable production of very heavy elements through the *r-process*.

The shock wave was followed, but soon overtaken, by a blast of *neutrinos* from the core, produced as it shrank, in a second and final stage of collapse, all the way down to become a neutron star just 20 km across. This leisurely process took several tens of seconds (not tenths of a second) to complete. By that time, the outgoing shock wave was trying to shove 15 solar masses of material out of the way, and had begun to stall. But as the shock stalled, the density of material in the shock front became so great that even some of the neutrinos (a few per cent of the total), overtaking the shock at the speed of light, were absorbed in it, dumping enough energy into the shock that it was able to start moving outward again and complete its job of blowing the outer layers of the star away. The rest of the neutrinos, carrying a couple of hundred times the energy that the supernova eventually radiated as visible light, went right through the outer layers of the star and on across the Universe; in the case of SN 1987A, just a handful of them were eventually detected on Earth.

This was a key discovery, because astrophysicists calculated that without the extra push from neutrinos the shock wave would give up the ghost and a supernova would never explode out into space. The presence of the neutrino flood was a crucial prediction of the models, and many theorists breathed a sigh of relief when the neutrinos from SN 1987A were indeed found. Even with the neutrino boost, the shock, now moving at about 2 per cent of the speed of light, took a couple of hours to push the outer layers of the star into space and to light up the star as a visible supernova – which is why the neutrinos were recorded by detectors on Earth shortly before the star brightened visibly.

While all this was going on, even though the original iron core of the star had been converted into a ball of neutrons, according to theory a massive burst of nuclear reactions in the hot, high-pressure shock wave would have produced many heavy elements, up to and including the iron group. One of the main products of this activity would have been nickel-56, which is unstable and changes through successive *radioactive decays* first into cobalt-56 (with a *half life* of just over 6 days) and then from cobalt-56 into iron-56 (with a half life of 77 days); the iron-56 is stable.

So, at least, the theory said. Observations of the decline in brightness of SN 1987A after its initial outburst showed that, during the first 100 days, 93 per cent of the energy was indeed being provided by the decay of cobalt-56, and the pattern continued as the supernova continued to fade, in a triumphant confirmation of the theoretical models. Roger Tayler, of the University of Sussex, described these observations as 'the most important and exciting ones concerned with the origin of the elements, confirming that the theoretical model [of nucleosynthesis] is broadly correct'. Spectroscopic studies showed that as much nickel-56 as the equivalent of 8 per cent of the mass of the Sun was produced in SN 1987A.

The detailed behaviour of SN 1987A thus confirmed the accuracy of models built up on the basis of observations of hundreds of supernovae since the mid-1930s, and a handful of such spectacular events recorded by astronomers in past centuries, including Tycho **Brahe** in 1572 and Johannes **Kepler** in 1604. About ten supernovae are detected in other galaxies each year, but none has been identified in our **Galaxy** since the invention of the astronomical **telescope**. In a supernova, although the visible light is the most obvious feature to our senses, ten times as much energy is carried by the material blown off from the star in the explosion, and 100 or 200 times as much energy in the form of neutrinos, produced when the core of the supernova reached a temperature of about 48 billion **Kelvin** (equivalent to 4.2 million **electron Volts**). All of this comes from the gravitational energy released when the core of the star collapses. Even though the visible light from a supernova is a relatively small proportion of the energy released, for a week or so the star still outshines all of the other stars in its parent galaxy put together, having increased in brightness to this peak by about 15–20 magnitudes in less than a day. It then fades away slowly, as energy continues to be released through radioactive decays of the unstable nuclei, notably cobalt-56, produced in the explosion. It gets back down to its former faintness only after several years.

supernova remnant (SNR) An expanding shell of material formed from the outer layers of a star blown apart during a *supernova* explosion. Some supernova remnants can be seen through the glow of visible light that they radiate; some are seen only at X-ray and radio wavelengths. As the shock wave from the supernova races out through the *interstellar matter*, it heats the gas between the stars and produces a reflected shock which bounces back and heats the material in the supernova remnant as well. It is this heating that raises the material to temperatures where X-rays can be emitted. Within the shock waves, electrons are accelerated and emit radio waves in the form of *synchrotron radiation*.

Although the shell is actually essentially spherical, we see the radiation most clearly where we are looking through the thickness of the 'skin' of

Supernova remnant. The Veil nebula, also known as NGC 6979, is made up of debris from a star which exploded 30 to 40 thousand years ago.

the shell, in a ring around the site of the supernova explosion; but the material in the remnant often breaks up into clumps, so the visible radiation does not always show up as a perfect ring.

Most supernova remnants are slowing down as they plough into the interstellar material. But in a few cases (most notably the *Crab nebula*), the shell is still being driven outward by electrons moving at a sizeable fraction of the speed of light, emitted from a central *pulsar* left by the supernova. These electrons are detected by the synchrotron radiation they produce from the volume of space within the shell of the supernova remnant. The compression of interstellar material by the expanding supernova remnant triggers new bursts of *star formation*. Supernova remnants can become very large, extending across diameters of hundreds of light years before dissipating into the interstellar material. There is strong evidence that the Sun formed, along with many other stars, in a loose association in part of the expanding shell of a supernova remnant about 5 billion years ago.

superstring theory   A development of string theory that takes on board ideas from *supergravity* and the notion of *supersymmetry*. It is regarded by some mathematicians as the most likely route towards a *theory of everything*. See *grand unified theories*.

supersymmetry (SUSY) A development of *grand unified theories* which suggests there is a symmetry between *bosons* and *fermions*, so that in the particle world there should be a bosonic counterpart for every fermion, and a fermionic counterpart for every boson. All of these 'supersymmetric partners' are as yet undiscovered; the known bosons are not the supersymmetric counterparts of the known fermions. The hypothetical photino is the counterpart to the *photon*, and the hypothetical counterparts to quarks are dubbed squarks. Out of all these 'new' particles, only the 'lightest supersymmetric partner' (probably the photino) would be stable; but if such particles do exist they could provide some of the *dark matter* in the Universe.

SUSY See *supersymmetry*.

symbiotic stars Term used to describe stars in which the *spectrum* includes features attributable to a cool star (such as a *red giant*) and emission lines typical of a hot object. This indicates the presence of material at temperatures below about 4,000 *Kelvin* and material with temperatures above 20,000 K in the visible region of the same object. The most likely explanation is that the 'star' is actually a close *binary system* (perhaps even a semi-detached pair) in which matter from the cool star is falling on to a *white dwarf* companion (or possibly on to a *neutron star* or even a *main sequence* star) and forming a hot spot. Radiation from the hot spot could produce *ionization* in the infalling gas, which then radiates the characteristic emission line spectrum of hot gas.

The term has become something of a grab-bag for a variety of objects, and includes many *variable stars* but also some which do not vary noticeably, and some *recurrent novae*. One particularly intriguing object that is classed as a symbiotic star is the system R Aquarii, which has a narrow jet of gas about 1,500 *astronomical units* long, moving out from the central star at a speed of 2,000 km per second; it seems to be a scaled-down version of the jets seen in *SS433*. Symbiotic stars are sometimes known as Z Andromedae stars, after one of the first of their kind to be studied.

synchrotron radiation Characteristic *electromagnetic radiation* produced by charged particles (such as *electrons*) moving at high speed (a sizeable fraction of the speed of light) in a magnetic field. The faster the electrons move, the shorter the wavelength of the radiation. Synchrotron emission is seen in *supernova remnants*, extragalactic *radio sources* and *pulsars*.

synthesis telescope A *radio telescope* based on the principle of *aperture synthesis*.

syzygy Overall term to describe an alignment of the Sun, the Earth and another astronomical object in a straight line. It doesn't matter in which order the objects are placed along the line, so syzygy includes the alignments responsible for both *solar eclipses* and *lunar eclipses* (and approximately the alignments that occur every Full Moon and New Moon). The largest tides on Earth (the spring tides, so named because of the way they 'spring up', and nothing to do with the *seasons*) are produced twice a month at the times of syzygy. The term is also used to refer to the *conjunctions* and *oppositions* of other *planets* with the Earth and Sun, and sometimes even (strictly speaking incorrectly) to alignments of two or more planets with the Sun even when the Earth is not in the same straight line.

tachyons   Hypothetical particles that always travel faster than the speed of light. Although the *special theory of relativity* tells us that no particle can ever be accelerated from a speed below that of light to a speed above that of light, there are, in fact, solutions to the equations of the special theory corresponding to particles which *always* travel faster than light.

These 'superluminal' particles, if they exist, behave in a way which mirrors the behaviour of ordinary 'subluminal' particles, so that they can never, for example, slow down below the speed of light. Whereas an ordinary particle moving at high speed will tend to lose energy and slow down, a tachyon will tend to lose energy and *speed up* – it is as if both subluminal and superluminal particles are repelled from the speed of light. So if a tachyon were created in some violent event in space, it would radiate energy away furiously (perhaps in the form of *gravitational radiation*, or as *electromagnetic radiation*) and go faster and faster, until it had zero energy (and was therefore undetectable) and was travelling at infinite speed.

Most physicists believe that tachyons do not exist, and that the solutions of the equations which describe them are as meaningless as the 'negative roots' that come out of simple quadratic equations. For example, it is possible (if rather silly) to describe someone's height as being 'the square root of 4 metres'. The square root of 4 may be either $+2$ or -2, but it is obvious from the context that the person being described is not -2 metres tall.

If tachyons did exist, however, they would have one other curious property – they would travel backwards in time. This has encouraged some physicists studying *cosmic rays* to look for 'precursor' events, in which a particle from space arrives just before a cosmic ray burst. The argument is that, if the cosmic ray burst originates when a high-energy particle from space strikes the top of the Earth's atmosphere, then any tachyons created at the same time will reach the ground just before the cosmic ray burst is created, having travelled backwards in time during their journey down to the ground. Unfortunately, no conclusive evidence for tachyonic precursors of this kind has been found. See also *time travel*.

Tammann, Gustav (1932–)   Swiss-based astronomer who, together with Allan *Sandage*, has been one of the leading proponents of the arguments in favour of a relatively low value for the *Hubble constant*, implying a relatively large age for the Universe. Tamman was born in Göttingen on 24 July 1932 and studied at the universities of Basle, Göttingen and Freiburg before completing his PhD at Basle in 1961. He has also worked at the *Mount Wilson and Palomar Observatories*, the *Hale Observatories*, Caltech and the *European Southern Observatory*; since 1976 he has been professor of astronomy at the University of Basle.

tau particle   A *lepton* that is a heavy counterpart to the *electron*. See *elementary particles*.

Taurus A   Source of *radio waves* associated with the *Crab nebula*. This is mostly *synchrotron radiation* from the *nebula* itself, with a small proportion of the emission coming from the *Crab pulsar*.

Taurus X-1   Source of *X-rays* associated with the *Crab nebula*. The second *X-ray source* (apart from the Sun) to be discovered, in 1963.

Taylor, Joseph Hooton, Jr (1941–)   American astrophysicist and radio astronomer who, together with Russell *Hulse*, discovered the *binary pulsar* in 1974. In 1993, Taylor shared the Nobel Prize with Hulse for their investigation of this object.

Taylor was born in Philadelphia, on 29 March 1941. He studied at Haverford College and at Harvard University, and worked at the University of Massachusetts before becoming professor of physics at Princeton University in 1980. It was while he was at the University of Massachusetts that he and his student Hulse discovered the binary pulsar, PSR 1913+16.

TD-1   European *satellite*, launched in 1972, that made the first survey of the sky in the *ultraviolet*.

telescope   Any instrument used to collect radiation from beyond the Earth's atmosphere is called a telescope by astronomers. This includes *radio telescopes*, *infrared*, *X-ray* and *gamma ray* detectors, *cosmic ray* telescopes, and even instruments designed to detect *neutrinos* or *gravitational radiation*. But the archetypal telescopes, on which all of the other kinds of detector are ultimately based, are the ones used in *optical astronomy*, essentially the kind of everyday telescope conjured up in the minds of most people by the use of the word.

The important feature of such a telescope is that it gathers and concentrates *light* from a faint object, making it bright enough to see, or to be photographed or recorded by electronic detectors, such as *charge coupled devices*. The ability of a telescope to magnify an image is of lesser importance in astronomy (although useful for studying objects within the Solar System), because even through a telescope a *star* will only appear as a point of light. Studies of the structure of galaxies do, though, benefit to some extent from the magnifying ability of telescopes. But it is much more important that a telescope should be capable of good *resolution* – able to distinguish between two objects that are close together on the sky, or to pick out fine details in the structure of, say, a distant galaxy.

Refracting telescopes use a large lens as the principle light-gathering implement, while *reflecting telescopes* use a large mirror. In either case, further arrangements of smaller lenses and/or mirrors are used to focus the light from the object of interest and concentrate it at the place where it is studied. The size of an optical telescope is usually given in terms of the diameter of its main mirror, or main lens. The world's largest refractor, at the *Yerkes Observatory*, has an *aperture* of 101 cm; reflectors can be made much bigger, because the main mirror can be supported from behind without the support interfering with the mirror's light-gathering ability, and several reflectors with apertures greater than 3.5 m are now in operation. Refracting telescopes are just like the ones used for observing distant objects on Earth, but bigger; the three main types of reflecting telescope are the *Newtonian telescope*, the *Cassegrain telescope* and the *Schmidt camera*. The Schmidt configuration is also used as an 'ordinary' telescope, not just for astronomical photography, by many amateurs.

The most important optical telescopes in use today are all reflectors. Their primary mirrors are sometimes made from glass, but more usually from a ceramic material that does not change its shape as its temperature changes. The mirror surface is prepared to follow a precise shape, either part of a sphere or part of a paraboloid, smooth to an accuracy measured in fractions of the *wavelength* of light; a thin layer of aluminium is then deposited on the mirror (like a thin coat of paint) to provide the actual reflecting surface. In the latest optical telescopes, mirrors thinner and lighter than in older telescopes are used, and their shape is constantly controlled by so-called active optics systems, in which computers constantly adjust the supports of the mirror to maintain the desired shape; this makes it possible to build bigger telescopes. In a further development of this technique, large telescopes are now being built which will automatically compensate, by changing the shape of the mirror, for changing distortions in the light from a star (or other astronomical source) produced by the Earth's atmosphere. A version of this adaptive optics technique is being used, for example, in the 6.5 m telescope derived from the *Multiple Mirror Telescope* at the Steward Observatory, in Arizona.

The classic type of radio telescope is conceptually very similar to an optical reflector, and uses a large bowl-shaped *antenna* to carry out exactly the same role as the primary mirror in an optical reflector. But because the wavelengths of *radio waves* are longer than those of light, although the antenna must still be accurately shaped to a fraction of the appropriate wavelength, it need not be as smooth in absolute terms as the mirror of an optical telescope.

In a classic trade-off, however, in order to obtain good resolution you need an aperture much larger than the wavelengths of the radiation being studied. A good optical telescope gives a resolution of about 0.5 *arc seconds*, an accuracy requiring a mirror 50,000 wavelengths in diameter, smooth to a precision of a tenth of a wavelength. To achieve the same accuracy with a radio telescope operating at a wavelength of 21 cm only requires the dish to be smooth to an accuracy of 2 cm – provided that the dish has an aperture of 100 km! This is why radio astronomers have to resort to techniques such as *aperture synthesis*.

Many radio telescopes do not even have the classic dish-shaped antenna, but rely on other kinds of antenna to pick up radio waves from *space* (for example, see *pulsars*). At other wavelengths, some instruments, such as infrared telescopes, closely resemble optical telescopes, while others, such as the instruments used to detect cosmic rays, or the tanks of liquid that form the bulk of some neutrino detectors, bear no resemblance to optical telescopes at all. But they are all telescopes to astronomers.

temperature   Temperature is a measure of the amount of heat there is in a body – that is, of how fast the *atoms* and *molecules* that form the object are moving about. *Electromagnetic radiation* also has a temperature, which is a measure of how energetic the radiation is, and is related to the number of *photons* with different *wavelengths* present in the radiation (see *black body*). Although various temperature scales have been used historically, the one that is the scientific basis for temperature measurements today is the *Kelvin temperature scale*, which is based on measurements of the rate at which heat flows between objects with different temperatures. If there is no heat flow, there is no temperature difference; the bigger the measured heat flow, the bigger the temperature difference, by definition.

A fundamental feature of the Kelvin scale is that there is an *absolute zero* of temperature at which no body can give up any more heat. The object still contains heat at 0 K (roughly –273 °C), but it cannot give it up, for reasons which are explained by *quantum theory*. It is a fundamental law of nature that heat cannot flow from a colder object to a hotter object (see *second law of thermodynamics*), and always flows from a hotter object to a colder object.

terrestrial planets   The four rocky planets closest to the Sun – Mercury, Venus, Earth and Mars. These resemble the Earth in varying degrees, but they are all 'terrestrial' by comparison with the gaseous *giant planets* (which are sometimes referred to as the jovian planets). Pluto is a special case and does not belong in either of the two main groups of planets.

Thales of Miletus (about 624–548 BC)   The first scientist. Thales was a Greek philosopher who can be considered as inventing both western philosophy and the study of physical *cosmology*, through his teaching that the order inherent in the world should be explained not by invoking divine or mystical processes, but in terms of natural causes that could be understood by people. Thales' work is known to us almost entirely from the writings of *Aristotle*, who tells us that Thales proposed that the world is made of water and that the Earth is a disc which floats on the water. He explained earthquakes in terms of this idea of a floating Earth. What matters is not the details of his *model*, which have been superseded, but the fact that he was trying to explain natural phenomena as consequences of physical events and laws. He was also a pioneer of geometry, and is credited with defining the *constellation* Ursa Minor and writing a treatise explaining the value of the constellation to navigation.

Thales was not afraid to put his money where his mouth was. He seems to have been the original absent-minded professor, and a tale was told of him falling down a well while wandering along looking at the stars. But Aristotle says that, when Thales was taken to task for being an impractical dreamer, he correctly predicted that weather conditions for the year ahead would bring a good olive harvest, bought up all the olive presses and made a fortune from his monopoly of the market.

theory of everything   See *grand unified theories*.

Thomson, William (1824–1907)   See *Kelvin, Lord*.

Thorne, Kip S. (1940–)   American theoretical physicist who has made a major contribution to the study of *gravity* and its effects in the Universe, including the nature of *black holes* and the search for *gravitational radiation*, within the context of the *general theory of relativity*. He was also a leading member of the team at Caltech which established, in the late 1980s, that there is nothing in the laws of physics as at present understood to forbid *time travel*.

Thorne was born in Logan, Utah, on 1 June 1940. He studied at Caltech and at Princeton University, where he was one of John *Wheeler*'s students and was awarded a PhD in 1965. In 1962, when he was about to graduate from Caltech, Thorne asked a famous astronomer there about his career possibilities. He was told that under no circumstances should he study the general theory of relativity, because it 'had so little connection with the rest of physics and astronomy' (quoted by Clifford Will, *Was Einstein Right?*). Fortunately, Thorne ignored the advice. He was soon back at Caltech, where he has worked ever since and now (1995) holds the post of Feynman Professor of Theoretical Physics. He was one of the leading lights in the revival of interest in the general theory in the 1960s and 1970s, and together with John Wheeler and Charles Misner (1932–), he wrote the definitive book *Gravitation*, published in 1973, which has since been the standard textbook for serious (postgraduate) students of the subject.

Thorne's interest in black holes led him to make a famous bet with Stephen **Hawking**, when the tentative identification of **Cygnus X-1** as a black hole encouraged Hawking to wager that it was *not* a black hole – in the hope that he would lose the bet. If the black hole identification proved positive, Hawking promised to send Thorne a year's subscription to *Penthouse*; if it were proved that Cygnus X-1 is not a black hole, Thorne would send Hawking a four-year subscription to *Private Eye*. Hawking paid up (rather belatedly, in the opinion of most astrophysicists) in June 1990.

Three Degrees K radiation   See *background radiation*.

tidal forces   Tides occur when a large object is moving in an **orbit** (or any trajectory) in a **gravitational field**. The object behaves, as far as the gravitational field is concerned, as if all of its mass were concentrated at a point, the **centre of mass**. So the centre of mass moves in exactly the right orbit. But because we are dealing with an extended object, every part of the object that is not at the centre of mass is not quite moving in the correct orbit. If the object is a planet in orbit around the Sun, for example, bits of the planet that are further out from the Sun than the centre of mass is, are being dragged around faster than the correct orbital speed for their distance from the Sun, and bits of the planet that are closer to the Sun than the centre of mass is, are being held back from their correct orbital speeds. The overall effect of this is to produce tidal forces, which stretch the planet in both directions outwards from the centre of mass along a line joining the centre of mass to the Sun, and squeeze it sideways along a line through the centre of mass at right angles to the line joining the centre of mass to the Sun.

These forces produce tides even in the solid surface of a planet like the Earth (as a result, you move up and down by about half a metre every day, when the Moon's influence is included). But they have a much more pronounced effect on the atmosphere and on the oceans, which can flow around the planet in response to the forces. The result is that there are two bulges of water in the oceans, one on the side of the Earth facing the Sun and one on the opposite side, with two low points on either side.

In fact, the tides raised on Earth by the Sun are only half as big as the tides raised by the Moon, because although the Moon is much smaller than the Sun, it is much closer to us. We don't usually think of the Earth as being in orbit around the Moon, but the correct description of the double system is of both the Earth and the Moon being (like any other **binary system**) in orbit about the mutual centre of mass. The centre around which the Earth 'orbits' is actually below the surface of the Earth, which complicates the picture slightly, but still leads to the occurrence of two high tides, one on the side of the Earth facing the Moon, and the other on the opposite side.

The lunar and solar tides combine with each other to produce the overall pattern of tides on Earth, with spring tides occurring at New Moon and Full Moon (when the two effects add together), and neap tides at Half Moon (when the two effects are working against each other). The tides are modified by the flow of water around the land masses of the Earth, and as our planet rotates, each point on Earth sweeps through the whole pattern of two high and two low tides every day.

time Everybody knows what time is, but nobody can *explain* what it is. In physics, the important thing about time is that it provides a reference system (a set of coordinates) in which events can be ordered. One event comes before or after another in this system. But it is important that, although this defines an ***arrow of time***, there is no suggestion anywhere in the laws of physics that time actually flows from the past through the present and into the future. All times have equal status.

This shows up most forcefully in the ***special theory of relativity***, where time is regarded as a fourth dimension, on an equal footing with the familiar three dimensions of ***space***. You can imagine all of space and time represented as a four-dimensional ***spacetime*** map, on which all of history, the present and the future of the universe can be represented. This raises interesting questions about the nature of destiny and free will – is the future 'already there' in some sense, just waiting for our consciousness to move over it? But the uncertainty inherent in ***quantum theory*** suggests that a better theory of space and time, merging relativity and quantum theory, may restore the vagueness of the future to the description of spacetime.

Scientists are on more secure footing when using time simply as a measure of the interval that has elapsed (in ***seconds*** or some other suitable units) between two events, or the time taken for some process to happen. Happily, this is the straightforward usage of the term in most of this book, and the deeper philosophical issues can be ignored except where they are explicitly referred to.

time dilation The slowing down of clocks which are moving at high speed relative to the observer, or which are sitting in a strong ***gravitational field***. See ***special theory of relativity***, ***general theory of relativity***, ***redshift***.

time travel To the surprise of most physicists, and the delight of most science fiction writers, research carried out in the late 1980s showed that genuine time travel is not forbidden by the laws of physics as at present understood. This does not mean that it will be easy to build a working time machine; but it does mean that it may not be impossible. More importantly, it means that there may be naturally occurring objects in the Universe that act as time machines.

There are, in fact, at least two ways to build a time machine. The first possibility was pointed out by Frank Tipler, then at the University of Maryland, in 1973, and published in the highly respectable journal *Physical Review D* in 1974 (volume 9, pages 2203–6). The trick involves making a **naked singularity** that rotates extremely rapidly (in effect, the rotation flings away the **event horizon** and exposes the **singularity**). In the region close to the singularity, which would be hidden behind the event horizon if the singularity were not rotating rapidly, **spacetime** is extremely distorted by the singularity's strong **gravitational field**. The effect of the rotation is to twist spacetime in this region, tipping it over so that one of the **space** dimensions is replaced by the **time** dimension. The result is that a carefully piloted spaceship could approach close to the singularity and follow an orbit through what seemed to all on board the spaceship to be ordinary space, but was in fact a journey through time. When the spaceship emerged from the region of intense gravitational field, the occupants would find themselves in a different time from when they had started their journey.

According to Tipler's calculations, you could do the trick if you had a cylinder about 100 km long and about 10 km across, made of material compressed to just over the density of a **neutron star**, and rotating twice every millisecond. It would be like ten neutron stars joined pole to pole and given a strong twist. The main problem with building such a time machine would be preventing gravity from crushing the cylinder down along its length. Curiously, though, there are objects in the Universe which nearly fulfil the other requirements – so-called **millisecond pulsars** are known which contain almost the right density of matter and spin once every 1.5 milliseconds, at one-third of the speed needed to make a time machine. Such objects are so close to being time machines already that they hold out the tantalizing possibility that an advanced civilization might be able to tweak them up in the right way to allow time travel.

The second kind of time machine involves **wormholes** – those tunnels through spacetime which may, according to the equations of the **general theory of relativity**, connect a **black hole** in one part of spacetime with a black hole in another part of spacetime. This is such an obvious idea that it has been freely used by science fiction writers for decades, but until the mid-1980s it was generally accepted by physicists that such objects could not 'really' exist, and that a better understanding of Einstein's equations would prove this. They had to change their tune as a result of a careful investigation of wormholes carried out by Kip **Thorne** and his colleagues at Caltech – an investigation triggered by a science fiction story.

It happened like this. Carl Sagan, a well-known astronomer, had written a novel in which he used the device of travel through a black hole to allow his characters to travel from a point near the Earth to a point near the star known as *Vega*. Although he was aware that he was bending the accepted rules of physics, this was, after all, a novel. Nevertheless, as a scientist himself, Sagan wanted the science in his story to be as accurate as possible, so he asked Thorne, an established expert in gravitational theory, to check it out and advise on how it might be tweaked up. At the end of 1985, musing on the problem of how to hold a wormhole open, Thorne realized that this could be done by threading the wormhole with so-called exotic matter. The critical factor is that this exotic matter has an enormous tension, sufficient to hold the wormhole open in spite of the inward tug of gravity. Nobody has ever found anything that could do the job – but, so far from being forbidden by the laws of physics, such exotic matter almost exactly matches the description of *cosmic string*. If anything like cosmic string exists, and could be captured, it would be capable of holding wormholes open (it would also, incidentally, be capable of stabilizing the kind of time machine Tipler described).

Sagan gratefully accepted Thorne's modification to his fictional 'star gate', and the wormhole duly featured in the novel, *Contact*, published in 1985. But this was still only presented as a short-cut through *space*. Neither Sagan nor Thorne realized at first that what they had described would also work as a shortcut through *time*. Thorne seems to have been unaware of Tipler's work, and not to have given any thought to the time travel possibilities opened up by wormholes until, in December 1986, he went with his student, Mike Morris, to a symposium in Chicago, where one of the other participants casually pointed out to Morris that a wormhole could also be used to travel backwards in time.

The point is that space and time are treated on an essentially equal footing by Einstein's equations – that, indeed, is why Tipler's time machine works. So a wormhole that takes a short-cut through spacetime can just as well link two different times as two different places. Indeed, any naturally occurring wormhole would most probably link two different times.

Although fascinated by the implications as a purely scientific puzzle, Thorne became (to judge from the story as he eventually told it in his book *Black Holes and Time Warps*) almost paranoid about the possibility that, if anyone found out he was studying the physics of time travel, they would think he had gone crazy, and he would become a laughing stock. Even though publishing papers on time travel had done no harm to Frank Tipler's career, Thorne (perhaps unaware, at the time, of Tipler's work, which is only mentioned in a footnote in Thorne's book) worried that, if he let his students publish papers about time travel, their careers would be over before they had started. Working slowly and secretively with Morris and another student, Ulvi Yurtsever, Thorne persuaded himself that Einstein's equations really did allow for the existence of wormholes that link different times, and could be used as time machines. It wasn't until 1988 that Morris, Thorne and Yurtsever published their conclusions in the journal *Physical Review Letters* (volume 61, page 1446). Encouraged by his students, Thorne allowed the paper to appear under the title 'Wormholes, Time Machines, and the Weak Energy Condition'; but, still nervous about adverse publicity, he told the staff of the Caltech Public Relations Office not just to keep quiet about the paper, but to actively suppress any mention of the research!

This was, of course, impossible. As word spread, the effect was exactly the opposite of what Thorne had feared. His own career received a considerable boost from the time travel work, and the careers of his students were kick-started in spectacular fashion. Other physicists who were interested in the exotic implications of pushing Einstein's equations to extremes (including Igor **Novikov**) were encouraged to go public with their own ideas once Thorne was seen to endorse the investigation of time travel, and the Caltech work led to the growth of a cottage industry of time travel investigations at the end of the 1980s and into the 1990s. The bottom line of all this work is that while, as with Tipler's time machine, it is hard to see how any civilization could build a wormhole time machine from scratch, it is much easier to envisage how a naturally occurring wormhole might be adapted to suit the time-travelling needs of a sufficiently advanced civilization. This raises all kinds of interesting paradoxes and possibilities, which are discussed in the books cited below.

Further reading: John Gribbin, *In Search of the Edge of Time*; Kip Thorne, *Black Holes and Time Warps*.

Titan The largest *moon* of Saturn, discovered by Christiaan **Huygens** in 1655. Titan has a diameter of 5,150 km, making it the second largest natural *satellite* (after Ganymede) in the Solar System. It orbits Saturn once every 15.95 days, at an average distance of 1,222,000 km. Titan is almost twice as dense as water, suggesting that there is a rocky core beneath the thick, cold nitrogen atmosphere. It resembles a miniature frozen Earth, and some people have speculated that it might be thawed into life when the Sun becomes a *red giant* (or even sooner; see Arthur C. Clarke, *2010: Odyssey Two*).

Titania The largest *moon* of Uranus, discovered by William **Herschel** in 1787. It is an icy, cratered body with a diameter of 1,524 km, and orbits Uranus at a distance of 435,840 km every 8.71 days.

Titius, Johann Daniel (1729–96) German mathematician who first put forward (in 1772) the mathematical relationship describing the spacing of the *orbits* of the *planets*, now known as *Bode's Law*.

Titius–Bode Law   See *Bode's Law*.

TOE = *theory of everything*.

top down theory   See *galaxy formation and evolution*.

Townes, Charles Hard (1915–)   American physicist, born on 28 July 1915 in Greenville, South Carolina, who developed the *maser* in the 1950s and received the Nobel Prize for this work in 1964. He helped to develop both *microwave astronomy* and *infrared astronomy*, and led the team at the University of California, Berkeley, that discovered the first polyatomic *molecules* in interstellar space, water and ammonia, in 1968. He retired in 1986.

transverse velocity   The speed with which an object, such as a *star*, is moving across (at right angles to) the line of sight. This can be measured for many stars by patient observations over a long time, but it tells us nothing about the *radial velocity* of the object. If both radial and transverse velocities are known, they can be combined to reveal the true motion of the object through space. See *proper motion*.

trigonometrical parallax   See *parallax*.

triple alpha process   The process by which three *nuclei* of *helium* (also known as *alpha particles*) combine during *nucleosynthesis* to make one nucleus of *carbon*. This is an essential step in the manufacture of *heavy elements* inside stars, because the nucleus produced when two alpha particles combine, beryllium-8, is so unstable that it blasts itself apart within 10^{-19} seconds. The only way to make carbon-12 (and therefore anything heavier than carbon-12) is if a third alpha particle arrives on the scene and combines with the beryllium-8 nucleus during the tiny split-second of its existence. This triple alpha process was proposed by Edwin *Salpeter* at the beginning of the 1950s, but at first it seemed that the collision of the third alpha particle with the short-lived beryllium-8 nucleus would help to blow it apart, rather than encouraging the formation of a stable carbon-12 nucleus.

The puzzle was resolved when Fred *Hoyle* predicted the existence of a so-called 'resonance' in carbon-12 which would allow the energy of the incoming third alpha particle to be absorbed; the resonance was found, exactly where Hoyle had predicted, in experiments at the Kellogg Radiation Laboratory in California. The triple alpha process takes place inside *stars* that have left the *main sequence*, and in which the temperature has risen to about 100 million *Kelvin*; it is the dominant source of energy in *red giants*.

Triton   The largest *moon* of Neptune, discovered in 1846. Triton orbits in the opposite direction to the way Neptune rotates (that is, in a retrograde orbit), and is spiralling inward so that in about 100 million years it will cross inside Neptune's *Roche limit* and break up. It has a cratered and cracked surface, and seems to be made of a jumble of pieces of ice and rock. Triton's diameter is 2,700 km and it orbits Neptune at a distance of 354,800 km once every 5.88 days.

Trojan groups Two families of *asteroids* that follow the same orbit around the Sun as Jupiter, grouped around the *Lagrangian points* 60° ahead and 60° behind the *planet*. Hundreds of them are known, but only a small proportion have been named, most of them in the Lagrangian point ahead of Jupiter. They are named after warriors in Greek tales of the Trojan wars.

T Tauri stars Very young stars that are still contracting and moving towards the *main sequence*. Often found in groups, always in clouds of gas and dust, they have strong *stellar winds* and surface temperatures of 3,500–7,000 Kelvin.

Tully–Fisher method An approximate method for determining distances to distant *disc galaxies*. The *absolute magnitude* of a disc galaxy turns out to be related to the speed with which it is rotating (presumably because more massive galaxies which contain more stars can rotate faster without breaking up), revealed by studies of the characteristic emission of hydrogen at a wavelength of 21 cm. So measuring the width of the hydrogen line in the radio *spectrum* of a galaxy indicates the absolute magnitude of the galaxy, and then measuring its *apparent magnitude* indicates its distance. The relationship was discovered by Brent Tully and Richard Fisher in 1977, and is calibrated from the known distances to nearby disc galaxies, based mainly on observations of *Cepheid variables*. The Tully–Fisher method gives relatively high values of the *Hubble constant*, around 80 km per second per *Megaparsec*.

Tunguska Event An explosion in central Siberia that occurred at 7.17 a.m. on 30 June 1908, and is now attributed to the violent detonation of a fragment of icy material from a *comet* in the atmosphere of the Earth. An expedition which visited the region in 1927 found that the Siberian forest had been devastated over a region 40 km in radius. In the inner region, 30 km across at the centre of the circle of destruction, the trees still stood, but had been stripped of their branches; further out, the trees lay on the ground, pointing outwards from the site of the explosion. The energy released by the explosion was about 50 Megatons, equivalent to a large hydrogen bomb, at an altitude of about 8.5 km; it was seen as a bright light in the sky by eye witnesses 500 km away. See also *Doomsday asteroid*.

tunnel effect A result of the *uncertainty principle* of *quantum theory*, which allows particles to penetrate barriers, such as the barriers around nuclei. When two protons, for example, approach one another inside a star, they are repelled by the positive electric charge they each carry, which stops them touching. But they can come so close that their quantum *wave functions* overlap, enabling them to interact (see *proton–proton reaction*). It

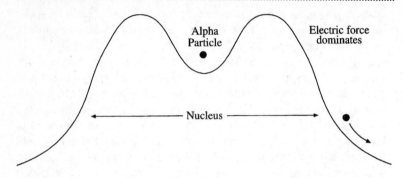

Tunnel effect. An alpha particle held in place within an atomic nucleus by the strong force is like a ball trapped in a valley between high mountain peaks. If the positively charged 'ball' were outside the valley, the electric force would push it away from the nucleus. Quantum uncertainty allows some alpha particles to escape, as if they had tunnelled through the 'mountain', even though they never have enough energy to climb out of the valley.

is as if one proton 'tunnelled' through the electric barrier between them. Similarly, the tunnel effect explains how *alpha particles* can escape from nuclei during *alpha decay*, even though they are restrained by the *strong nuclear interaction*. This explanation of alpha decay was first proposed by George *Gamow* at the end of the 1920s.

turn-off point See *globular cluster.*

Tycho See *Brahe, Tycho.*

Type I See *supernovae.*

Type II See *supernovae.*

UBV system See *colour.*

Uhuru The first *satellite* for *X-ray astronomy*, launched by *NASA* in 1970 from a platform just off the coast of Kenya. The name is the Swahili word for 'freedom'. Uhuru made the first detailed map of the *X-ray* sky, discovering X-ray binaries such as *Centaurus X-3* and *Hercules X-1*, and the extended X-ray emission associated with *clusters of galaxies*.

UK Infrared Telescope (UKIRT) A 3.8 m *telescope* for *infrared astronomy*, located on Mauna Kea in Hawaii (at an altitude of 4,200 m) and run by the *Royal Observatory, Edinburgh*. It began observing in 1979.

UK Schmidt Telescope A 1.2 m (48-inch) *Schmidt camera* located at the *Anglo-Australian Observatory*. It began observing in 1973 and was originally run by the *Royal Observatory, Edinburgh*, but now by the Anglo-Australian Telescope Board.

ultraviolet astronomy The study of the Universe using *electromagnetic radiation* with wavelengths in the range 90–320 nanometres, on the short-wavelength side of the *spectrum* of visible light. Ultraviolet radiation (sometimes referred to as ultraviolet light) is largely absorbed in the Earth's atmosphere (especially in the ozone layer, or stratosphere), but the small amount that does reach the ground causes sunburn and is responsible for some skin cancers and eye cataracts. Some ultraviolet observations have been carried out using balloons, but most of the information about the Universe we have from this part of the spectrum comes from instruments flown on rockets and satellites.

Ultraviolet observations are important because there are many spectral lines, produced by both atoms and molecules, in this wavelength band (see *spectroscopy*). The lower limit on the range of wavelengths available to study in this way is determined by the presence of large amounts of hydrogen in space, which absorbs almost all the ultraviolet radiation at wavelengths below 91.2 nm. Almost every astronomical object can be studied at ultraviolet wavelengths to yield additional information about its properties, but from a cosmological point of view the most interesting feature of ultraviolet astronomy is the investigation of redshifted lines of the *Lyman series* in *quasars* and clouds of intergalactic material (see *Lyman forest*).

ultraviolet radiation (UV) *Electromagnetic radiation* with wavelengths shorter than those of violet light, but longer than those of X-rays, from 10–320 nanometres. UV is absorbed in the atmosphere, and UV astronomy requires the use of balloons, rockets and *satellites*.

ultraviolet stars Hot *stars*, with surface temperatures above 10,000 *Kelvin*, which radiate energy strongly in the *ultraviolet* part of the *spectrum*.

Ulysses A *spaceprobe* launched by *NASA* (from the Space Shuttle) for the *European Space Agency* in 1990 to investigate the Sun, flying out of the plane of the *ecliptic* to observe the Sun's polar regions, which cannot be seen from Earth. In 1994 Ulysses made the surprising discovery that there is no well-defined magnetic pole in the southern hemisphere of the Sun.

Umbriel One of the *moons* of Uranus, discovered in 1851. Umbriel has a diameter of 1,174 km and a relatively dark surface. It orbits Uranus every 4.14 days at an average distance of 265,970 km.

uncertainty principle A fundamental feature of *quantum theory*, which says that for any object certain pairs of properties, such as position and momentum (which is a combination of the mass and velocity of an object), are linked in such a way that they cannot both be precisely determined at the same time. The principle was first expressed by Werner *Heisenberg* in the 1920s, and is often referred to as Heisenberg's uncertainty principle.

The key feature of quantum uncertainty is that it has nothing to do with the ability (or inability) of our instruments to make accurate measurements; it is an intrinsic property of the quantum world. This is why an entity such as an *alpha particle* is able to 'tunnel through' the barrier of electrical repulsion when it combines with the *nucleus* during *nuclear fusion* reactions. The electrical barrier keeps the alpha particle out of range of the attractive *strong nuclear force*, according to the classical image of particles and forces. But the uncertainty relation means that the alpha particle is in effect spread out over a small volume of space, not concentrated in a point. The fuzzy, spread-out edges of the alpha particle 'cloud' may extend beyond the electrical barrier into the heart of the nucleus, allowing the strong force to take a grip on it and tug it in, as if the alpha particle had tunnelled right through the barrier.

The uncertainty principle also applies to *energy* and *time*, and is the reason why pairs of particles (such as an *electron* and a *positron*) can appear briefly out of nothing at all. The Universe itself is not certain how much energy there is around, on short enough timescales, and this provides the leeway to make short-lived particles out of energy, in line with Albert *Einstein*'s famous equation $E = mc^2$. See also *free lunch universe*.

The uncertainty principle does not affect large objects noticeably because the amount of uncertainty is determined by *Planck's constant*, which is 6.6×10^{-34} joule seconds; it comes into its own for particles such as electrons, which have a mass of 9×10^{-28} grams. For any object, the amount of uncertainty in position is inversely proportional to the mass, so for anything much bigger than an atom the uncertainty in position is very small indeed, although in principle it does still exist.

unified theories See *grand unified theories*.

Universal Time (UT) Essentially the same, for everyday purposes, as *Greenwich Mean Time*. UT is actually calculated from *sidereal time*, and is the basis for civil timekeeping. Coordinated Universal Time (UTC) is the time used for broadcast time signals, and is kept in step with *International Atomic Time* by introducing occasional 'leap seconds' into the broadcast time signals.

universe Used with a lower-case 'u', refers to a particular *cosmological model*, not necessarily to the real Universe.

Universe With the capital 'U', the term used for everything that we can ever have knowledge of, the entire span of *space* and *time* accessible to our instruments, now and in the future. This may seem like a fairly comprehensive definition, and in the past it has traditionally been regarded as synonymous with the entirety of everything that exists. But the development of ideas such as *inflation* suggests that there may be something else beyond the boundaries of the observable Universe – regions of space and time that are unobservable *in principle*, not just because light from them has not yet had time to reach us, or because our telescopes are not sensitive enough to detect their light.

This has led to some ambiguity in the use of the term 'Universe'. Some people restrict it to the observable Universe, while others argue that it should be used to refer to all of space and time. In this book, we use 'Universe' as the name for our own expanding bubble of *spacetime*, everything that is in principle visible to our telescopes, if we wait long enough for the light to arrive. We suggest that the term 'Cosmos' can be used to refer to the entirety of space and time, within which (if the inflationary scenario is correct) there may be an indefinitely large number of other expanding bubbles of spacetime, other *universes* with which we can never communicate.

Uranus The seventh *planet* out from the Sun, and the first to be discovered using a *telescope* (by William **Herschel** in 1781). Uranus has an equatorial diameter of 51,200 km and orbits the Sun once every 84 years at a distance varying between 18.3 and 20.1 *astronomical units*. It rotates once every 17.24 hours, and has a mass 14.5 times that of the Earth (just 5 per cent of the mass of Jupiter), giving it a density 1.3 times that of water.° Unlike every other planet in the Solar System, Uranus lies almost on its side in its orbit, pointing first one pole and then the other towards the Sun as it moves around its orbit. This may be the result of the impact of a large *comet* with Uranus long ago. It has at least eleven moons.

Ussher, Archbishop James (1581–1656) The Archbishop of Armagh who added up the dates implied by biblical genealogies, back to Adam, and came up with the conclusion that the Earth was created at precisely 10 a.m. on 26 October in the year 4004 BC.

valley of stability See *nuclear fission*.

variable star Any *star* whose brightness, *colour* or some other property varies with time. Almost always, such stars vary in brightness, even if other properties change as well. The term covers *cataclysmic variables* (up to and including *supernovae*) as well as regularly varying *pulsating variables*, such as the *Cepheids*, and even *eclipsing binaries*, in which the variation is caused by light from one star being obscured by another, not by any intrinsic variation in *luminosity*. Some variable stars change in brightness because they are rotating and the brightness of their surface is not uniform. In 1786 just twelve variable stars were known; today, tens of thousands of variable stars of all kinds have been listed in various *catalogues*. Very few of these have been studied in detail, and amateur astronomers make an important contribution by monitoring known variables and passing on information about the variations to their professional colleagues. See also *recurrent novae*.

Vaucouleurs, Gerard de See *de Vaucouleurs, Gerard*.

vector field A field (see *field equations*) which has an inbuilt sense of direction. The electric field is an example of a vector field – a tiny charged particle (such as an electron) placed in an electric field will move in a certain direction (in the case of an electron, towards positive charge, away from negative charge), accelerating at a rate which depends on the strength of the field. A tiny charged particle will move along a line of force even in a perfectly uniform vector field. See also *scalar field*.

Vega   The brightest star in the constellation Lyra, also known as Alpha Lyrae. The fifth brightest star in the sky, Vega is an A star with *magnitude* 0.03, and was the pole star 14,000 years ago. It is 7.5 *parsecs* from the Sun.

Vega spaceprobes   Two *spaceprobes* launched from the Soviet Union in 1984 to fly past Venus (dropping probes into the planet's atmosphere) and on to encounter *Halley's Comet*. The name comes from Venera + Galley (the Russian version of Halley).

Vela pulsar   A *pulsar*, also known by its coordinates as PSR 0833 – 45, associated with a *supernova remnant* 10,000 years old in the *constellation* Vela. It has a period of 89 milliseconds, and was one of the first pulsars discovered, in 1968. The pulsar undergoes *glitches*, and was the second to be identified as an *optical pulsar*, in 1977. It is about 500 *parsecs* away.

velocity   A measure of *both* the speed with which an object is moving *and* its direction. Strictly speaking, a velocity cannot be specified only in terms of speed (such as the number of kilometres per hour), but must also tell you the direction (50 km per hour in a direction north-north-east). Like most of us, astronomers are often careless about this usage and sometimes use the two terms 'speed' and 'velocity' interchangeably. But the distinction can be important – the speed of light through space, for example, is an absolute constant, but the velocity of light through space is not, because the light can move in different directions at that speed.

Venus   Second *planet* out from the Sun. In size, Venus is almost a twin of the Earth, with 82 per cent of the Earth's mass. It orbits the Sun once every 225 days at an average distance of just over 0.72 *astronomical units*, in a nearly circular orbit. It rotates backwards (retrograde) once every 243 days, relative to the *fixed stars*, so in a sense a 'day' on Venus is longer than a 'year'. Because of the motion of Venus in its orbit, however, if you could see the Sun from the surface of the planet, the time from noon to noon would be 116.8 of our days, so there would be just under two Venusian days in each Venusian year.

The equatorial diameter of Venus is 12,104 km. Unlike the Earth, Venus has a very thick atmosphere rich in carbon dioxide. This produces a strong *greenhouse effect*, which, combined with the fact that it is closer to the Sun than we are, raises the surface temperature to 730 *Kelvin*. The pressure of this atmosphere at the surface of Venus is 90 times the atmospheric pressure at sea level on Earth.

Venus is completely cloud covered and has a high *albedo*, reflecting

Venus. The cloud-covered atmosphere of Venus, photographed by Mariner 10 from a distance of 720,000 km on 6 February 1974. This image was taken using ultraviolet light.

79 per cent of the incoming sunlight. As a result, after the Sun and Moon, Venus is the brightest object in the sky, seen close to the Sun as the evening star or as the morning star, depending on the position of Venus and Earth in their orbits. It has no *moons*.

very early universe See *inflation.*

Very Large Array (VLA) Large *radio telescope* based on the *aperture synthesis* principle, located near Socorro, in New Mexico, and run by the *National Radio Astronomy Observatory.* It uses 27 dish *antennae*, each with an *aperture* of 25 m, spaced along a Y-shaped track with two arms 21 km long and the other 18.9 km long. It has a best *resolution* of 0.05 *arc seconds* at a *wavelength* of 1.3 cm, and 1 arc second at 6 cm wavelength. (See picture on page 494.)

Very Large Telescope (VLT) A set of four linked 8 m *optical telescopes* being built by the *European Southern Observatory* at an altitude of 2,600 m on Cerro Paranal in Chile, to mimic the power of a single telescope with an *aperture* of 16 m.

Very Long Baseline Array (VLBA) A chain of ten identical *radio telescopes* (each with an *aperture* of 25 m) from St Croix in north-eastern Canada to Hawaii in the Pacific, which can be combined to act as an interferometer with a baseline 8,000 km long and a *resolution* of 0.0002 *arc seconds.* The

The Very Large Array, near Socorro, in New Mexico. This view, from the centre of the array, shows one arm of the system.

system is controlled from the home of the *Very Large Array* in Socorro, New Mexico.

very long baseline interferometry (VLBI) Technique of linking *radio telescopes* thousands of kilometres apart to form an *interferometer.* See *Very Long Baseline Array.*

V404 Cygni The most accurately 'weighed' *black hole.* Interest in V404 Cygni was stirred in 1989 when the Japanese X-ray astronomy satellite *Ginga* found a burst of X-rays from an object labelled at the time GS2023+338. Astronomers soon realized that the X-rays were, in fact, coming from alongside a star which had been seen as a *nova* in 1938. When optical records were checked, they showed that the old nova had brightened considerably at the same time that the burst of X-rays was observed, confirming that they came from the same source. The burst of X-rays was too strong to be explained in terms of the presence of a *white dwarf* in a *recurrent nova*, so they had to be coming from a *neutron star* or a black hole.

Because V404 Cygni is relatively close to us (estimated to lie between 1.5 and 4 *kiloparsecs* away), it is much easier to study than other black hole candidates. Over the next few years, V404 Cygni was investigated in great detail, and by late 1994 astronomers using the *William Herschel*

Telescope and the *UK Infrared Telescope* had enough information to pin down the properties of the two stars in this *binary system*. The visible star orbits an unseen companion once every 6.5 days with a speed of 210 km per second. The details of the orbit show that the unseen companion is seventeen times more massive than the visible star. The final piece of the puzzle slotted into place when the observers were able to measure the angle at which the orbit of the visible star is tilted on the sky, which reveals that the actual masses of the two objects are 0.7 solar masses and 12 solar masses. The lighter object is the visible star, which means that the unseen companion has twelve times as much mass as our Sun, which is about four times as much as the *Oppenheimer–Volkoff limit* for the maximum possible mass for a neutron star. It has to be a black hole, and is the closest one to our Solar System that we know of.

Viking spaceprobes   Two unmanned Mars missions, launched by *NASA* in 1975. Each consisted of an orbiter and a lander, and returned a wealth of information about Mars and its *moons*.

Virgo A   The most powerful *radio source* in the *constellation* Virgo. The radio noise comes from the *galaxy* M87, a giant *elliptical* about 17 *Megaparsecs* away at the heart of the *Virgo cluster*, and is associated with a *jet* 4,000 *light years* long. This activity is probably linked with the presence of a supermassive *black hole* in M87. The source is also known as 3C 274.

Virgo cluster   The nearest rich *cluster of galaxies*, about 17 *Megaparsecs* away, in the direction of (but far beyond!) the *constellation* Virgo. Attempts to determine the exact distance to the centre of this cluster are a key step in determining the *cosmic distance scale*. About 2,500 *galaxies* have been identified in the Virgo cluster, three-quarters of them *disc galaxies*.

visual binary   A *binary system* in which the two *stars* can be distinguished by eye, unaided or using a *telescope*.

visual magnitude (V)   The apparent brightness of an object to the human eye, which is most sensitive at a wavelength of 560 nanometres. These V magnitudes are now measured electronically (or photographically) at that wavelength.

voids   Regions of the *Universe* that do not contain many visible bright *galaxies*. The *large scale structure* of the Universe seems to consist of sheets and filaments of *clusters of galaxies* spread around large voids, like a heap of bubbles clinging together. The first of these voids to be discovered lies in the direction of the constellation Bootes, and was noticed in 1981. It is about 100 *Megaparsecs* across. Voids take up about 98 per cent of the volume of the Universe, with bright stuff concentrated in the skin of the 'bubbles' making up the other 2 per cent. But this does not necessarily mean that the voids are empty. *Models* of the Universe based on the presence of *dark matter* suggest that most of the mass of the Universe may

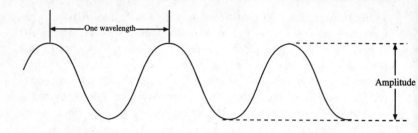

Wavelength. A wave.

be in the form of dark stuff in the voids, while the visible bright galaxies form only in regions where the density of *baryons* is at its highest. This overall picture was confirmed, but slightly modified, in 1994, when astronomers found 50 very faint galaxies in the Bootes void itself. This is only about a third of the number of galaxies found in a typical volume of the Universe the size of the void (that is, in a typical bright part of the Universe), and their faintness suggests that they contain less baryonic matter than ordinary galaxies in other parts of the Universe.

Voyager spaceprobes   Two *spaceprobes* launched by *NASA* in 1977 to send back pictures and other data from the outer planets of the Solar System. Each probe weighed just 825 kg, and used the gravity of the planets it encountered to speed it on its way to its next encounter.

Voyager 1 visited Jupiter and its moons in 1979, and Saturn in 1980, and then moved out of the Solar System. Voyager 2 visited the Jupiter system in 1979, Saturn and its rings in 1981, Uranus in 1986 and Neptune in 1989, before also going on out of the Solar System. The information from the two probes revolutionized astronomers' understanding of the outer planets and their moons.

wave function   The description of a quantum entity, such as an *electron*, in terms of a wave. Everything we think of as being a *particle* also has a wave function associated with it. See *quantum theory*, *uncertainty relation*.

wavelength   The distance from peak to peak (or from trough to trough) in a wave.

wave mechanics   An old name for quantum mechanics; see *quantum theory*.

weak force = *weak nuclear interaction*.

weak nuclear interaction   See *fundamental forces*.

weight   See *mass*.

weightlessness   See *free fall*.

Westerbork Radio Observatory   Dutch observatory near Groningen in the Netherlands. It was opened in 1970 and uses an *aperture synthesis* telescope made up of fourteen *antennae*, each with an *aperture* of 25 m, on a line 2.8 km long.

Wheeler, John Archibald (1911–) American theoretical physicist who worked with Richard Feynman (1918–88) as his PhD supervisor, at one time espoused the idea of parallel *universes* (the 'many worlds' version of *quantum theory*), has contributed to the study of *black holes*, and has suggested that the Universe exists only because we are watching it.

Wheeler was born in Jacksonville, Florida, on 9 July 1911. He studied at Johns Hopkins University, receiving his PhD in 1933, and then worked for two years in Copenhagen with many of the pioneers of the development of quantum theory. Back in the United States, he worked at the University of North Carolina until 1938, then joined the faculty at Princeton University. Wheeler worked on the theory of *nuclear fission*, contributing indirectly to the Manhattan Project to develop the atomic bomb, and in 1949–50 he was directly associated with the hydrogen bomb project, involving *nuclear fusion*. From 1947 he was professor of physics at Princeton University, moving in 1976 to the equivalent post at the University of Texas, from which he retired in 1986.

From the 1950s to the 1980s, Wheeler was probably the leading expert on the *general theory of relativity* in the United States. He helped to develop the mathematical description (the 'equation of state') of cold *stars* that have exhausted all of their nuclear fuel, which helped to confirm that a dead star with more than a certain amount of *mass* must collapse to form a black hole (see *Oppenheimer–Volkoff limit*). In fact, for some time Wheeler argued against Robert *Oppenheimer's* suggestion that black holes must exist, but once he had persuaded himself that they did exist, he became a leading supporter of the idea; in December 1967 he was the first person to use the term 'black hole' in its modern astrophysical connection, at a meeting of the American Association for the Advancement of Science.

Together with Charles Misner (1932–) and Kip *Thorne*, Wheeler wrote the definitive gravitational textbook, *Gravitation*, which has been the standard text used by students of the general theory of relativity since the book was first published (by W. H. Freeman in 1973). He is the originator and chief proponent of the idea that at the *Planck length* the fabric of *spacetime* dissolves into a quantum foam, so that *singularities* do not actually exist; the idea of a foam of quantum activity going on on the scale of the Planck length underlies many versions of the theory of *inflation*, which describes the behaviour of the very early Universe.

Wheeler has also thought long and hard about quantum theory, and made major contributions to its interpretation. One of his strangest speculations takes the standard Copenhagen Interpretation at face value, and applies it to the entire Universe. The Copenhagen Interpretation says that quantum phenomena exist only as waves of probability until they are observed, when they 'collapse' into a definite state. Taking this to its logical extent, Wheeler suggests that the 'wave function' of the entire Universe (right back to the **Big Bang** itself) only 'collapsed' into reality when somebody looked at the Universe and noticed its existence (it was partly in an effort to get round this kind of non-commonsensical conclusion that Stephen **Hawking** and others developed the idea of a timeless Universe which had no beginning in the everyday sense of the term; see **no boundary condition**). If Wheeler is correct, cosmology is the most important of all human activities, because by observing the Universe we have brought both it and ourselves into existence. Without wishing to spoil the fun, we should mention that there are other interpretations of quantum theory!

Further reading: John Gribbin, *In Search of The Edge of Time*; John Gribbin, *Schrödinger's Kittens*.

Whipple, Fred Lawrence (1906–) American astronomer, born in Red Oak, Iowa, on 5 November 1906. Whipple's research concentrated on understanding the evolution of the Solar System, and he is best known for his proposal (in 1950) of the 'dirty snowball' **model** of **comets**.

Whipple Observatory See **Fred L. Whipple Observatory**.

whirlpool galaxy A particularly striking and photogenic example of a **disc galaxy** with well-defined **spiral arms**. Also known by its **catalogue** numbers as M51 or NGC 5194, this was the first galaxy (then known as a **nebula**) in which spiral structure was seen, by Lord **Rosse** in 1845. The whirlpool galaxy appears almost face on to us on the sky, near the end of the tail of the **constellation** of Ursa Major (the Great Bear, or Plough). It is about 6 **Megaparsecs** away, and weighs in at rather less than half the mass of our own **Galaxy**. A small, less well-developed disc galaxy known as NGC 5195 seems to be attached to the end of one of the spiral arms of the whirlpool galaxy; it is actually in orbit around the whirlpool galaxy (in much the same way that the **Magellanic Clouds** orbit our own Galaxy). Computer simulations show that the kind of spiral structure seen in the whirlpool galaxy can be produced from a uniform disc of material by gravitational interaction with a companion like NGC 5195 during a close encounter.

white dwarf **Star** with about the same mass as our Sun, but occupying a volume about the same as that of the Earth. One cubic centimetre of white dwarf material would have a mass of about 1 tonne – 1 million times the density of water. White dwarfs form from the collapse of stars like the Sun at the end of their lives, when they are no longer supported by **nuclear fusion** reactions going on in their cores. Such a star is like a hot ember left by the original star, cooling and radiating the last of its energy away into space; it will be made of helium, carbon and other elements produced by **nucleosynthesis**, and will eventually cool to become a **black dwarf**.

The Whirlpool galaxy, also known as M51, which lies in the direction of (but far beyond) the constellation Canes Venatici.

The first white dwarf to be discovered was found in 1862. It orbits around the bright star *Sirius*, and is known as *Sirius B*. Many white dwarf stars have since been discovered. The gravity at the surface of a white dwarf is so strong – tens of thousands of times greater than the gravity at the surface of the Earth – that it is often possible to discern a measurable *redshift* in the light from such a star.

In 1931 the Indian astronomer Subrahmanyan *Chandrasekhar* calculated that any white dwarf with more than about 1.4 times as much mass as our Sun could not hold itself up against the pull of gravity, but must collapse further. Later studies showed that, although dead stars with masses between 1.4 and 3 solar masses could form even more compact *neutron stars*, stars which end their lives with more than about three times the mass of our Sun must collapse indefinitely, to form *black holes.*

white hole Hypothetical counterpart to a *black hole*; a white hole is a region where matter and energy are emerging from a *singularity* into the Universe at large. No white hole has ever been conclusively identified, though some people argue that the outburst of the Universe from the *Big*

Bang singularity is an example of this kind of behaviour.

Shortly after *quasars* were discovered, it was suggested that the violent outpouring of energy associated with these objects (and also with *active galaxies* and *radio sources*) might be an example of white hole activity, but this idea has never been taken seriously by the majority of astronomers. The great drawback to all white hole models is that the region around a white hole singularity, like the region around a black hole singularity, would have a strong *gravitational field* which would attract matter towards it. Calculations suggest that anything falling in towards a white hole (matter or radiation) would gain so much energy, becoming highly *blue-shifted* as it fell into the strong gravitational field, that it would smother the singularity in what is known as a *blue sheet*, quickly turning it into a black hole.

One speculative way around this problem is the possibility that processes unknown under low-gravity conditions (such as on Earth) operate at the superdensities occurring near singularities, creating a 'bounce'. One well-developed variation on this theme has been put forward by Fred *Hoyle* in the context of the *Steady State hypothesis* (this also draws on work by William *McCrea*). Hoyle regards any theory which contains equations that predict the existence of singularities as flawed, and suggests that the collapse of matter towards a singularity is halted and then reversed at very high densities by the action of a *C-field*, which effectively acts like antigravity. It is certainly a curious fact, worthy of more attention than it is usually given, that although a black hole is by definition an object that things fall *into*, all of the violent activity seen in the Universe on the scale of quasars and active galaxies is associated with matter and energy bursting *outward* from a compact region.

A rather more vague speculation suggests that white holes may be the 'other ends' of black holes, linked to them by *wormholes*. See *time travel*.

Wien's Law   A relationship which gives the *temperature* of a *black body* in terms of the *wavelength* at which it radiates the maximum amount of *energy* in its *spectrum*. It was named after the German physicist Wilhelm Wien (1864–1928), who was awarded the Nobel Prize, in 1911, for his work on the laws governing the radiation of heat.

A graph of the amount of energy radiated by a black body at different wavelengths rises smoothly from lower energies at shorter wavelengths to a peak at some intermediate wavelength, then slides down smoothly again towards lower energies at longer wavelengths. The position of the peak moves towards shorter wavelengths at higher temperatures, and the temperature of the black body (in *Kelvin*) is given simply by dividing the wavelength of the peak emission (in micrometres) into the number 2,900; this is Wien's Law. So if the peak in the black body curve is at, say, 4 micrometers (0.004 mm), the temperature of the object is 725 K.

This is a useful way of measuring the temperature of an object (provided that it is radiating roughly like a black body) from a few measurements of the intensity of its emission at different wavelengths around the peak in the spectrum. It is because the peak in the spectrum of the *background radiation* occurs at 1,060 micrometers (that is, just above 1 mm) that we know the temperature of the background radiation is 2.73 K.

Wilkinson, David T, (1935–) American astrophysicist whose career has been largely devoted to investigations of the *background radiation*. Wilkinson was born in Hillsdale, Michigan, on 13 May 1935. He studied at the University of Michigan, receiving a PhD in physics in 1962, and stayed there for a further year before moving to Princeton University, where he stayed for the rest of his career, becoming professor of physics in 1972. Early in 1965, he was a member of Robert *Dicke*'s team actually carrying out an experiment to detect the background radiation when they heard of the serendipitous discovery of this radiation by Arno *Penzias* and Robert *Wilson*.

Once the background radiation had been discovered, Wilkinson became deeply interested in developing experiments to measure any irregularities (anisotropies) in the radiation, and was largely responsible for developing improved detectors and observing techniques. Nobody made a greater contribution to the study of the background radiation over the next 30 years, and a lot of his expertise went into the *COBE* mission; Wilkinson was a founding member of the COBE team. While continuing to be involved with ground-based and balloon-borne experiments, in the mid-1990s Wikinson was actively promoting a new *satellite* mission to measure the anisotropy of the background radiation more accurately.

William Herschel Telescope A 4.2 m *reflecting telescope* operated by the *Royal Greenwich Observatory*, but based at the *Roque de los Muchachos Observatory*. It began operating in 1987, and observing time is shared with Spain and the Netherlands. (See picture on page 502.)

Wilson, Robert Woodrow (1936–) American radio astronomer who (together with Arno *Penzias*) was the accidental discoverer of the *background radiation*.

Wilson was born in Houston, Texas, on 10 January 1936, the son of a chemical engineer. He gained straight As in all his science courses at Rice University in Houston, where he graduated in 1957, and was offered places at graduate schools in both Caltech and MIT, the two premier scientific research institutions in the United States. He chose Caltech, where he was strongly influenced in favour of the *Steady State hypothesis* by attending a course of lectures given by Fred *Hoyle*. He stayed on at Caltech for a year after completing his PhD, in 1962, and in 1963 moved to the Bell

The dome of the **William Herschel Telescope**, at the Roque de los Muchachos Observatory, on La Palma, in the Canary Islands.

Laboratories' research centre in Holmdel, New Jersey, where he teamed up with Penzias. Apart from the discovery of the background radiation, Wilson's main research interest has been the search for, and identification of, *interstellar molecules* using *radio astronomy* techniques, and determinations of the relative abundances of different *isotopes* in *interstellar matter*. He remained with Bell and became head of the Radiophysics Research Department in 1976, two years before he was awarded the Nobel Prize.

At Crawford Hill, Wilson initially shared the single radio astronomy post offered by the Bell Labs with Arno Penzias, each of them spending half his time on other projects. They were allowed to use an *antenna* designed for work with early communications satellites, and found a source of cosmic radio noise which they could not explain, but which Robert *Dicke* and colleagues at nearby Princeton University interpreted as the echo of the *Big Bang*.

Further reading: Jeremy Bernstein, *Three Degrees Above Zero*.

WIMPs   See *cold dark matter*.

Wolf, (Johann) Rudolf (1816–93)   Swiss astronomer, born in Fällenden on 7 July 1816, who made a study in the 1850s of solar observations going back to the 17th century and showed that there had been a *solar cycle* of activity with an average period of 11.1 years since that time. From 1855 onwards (as well as being a professor of astronomy at *both* the University of Zurich *and* the city's Institute of Technology), he was director of the then-new Federal Observatory in Zurich, where he established a consistent method of recording solar activity in terms of the 'Zurich relative sunspot number', better known today as the *Wolf number*. Nothing to do with Max *Wolf*.

Wolf, Maximilian Franz Joseph (1863–1932) Always known as Max Wolf, he was a German astronomer (born in Heidelberg on 21 June 1863) who specialized in photographic and spectroscopic observations, made early observations of spiral *nebulae*, and discovered 232 *asteroids*. Nothing to do with Rudolf *Wolf*.

Wolfendale, Arnold Whittaker (1927–) The 14th Astronomer Royal, Wolfendale succeeded Francis *Graham-Smith* in 1991 but, appointed only shortly before the now-accepted retiring age for the post, served only until the end of 1994, the shortest 'reign' of any Astronomer Royal except Nathaniel *Bliss*, who died in office. He was knighted in 1995.

Wolfendale was born on 25 June 1927 and studied at the University of Manchester, where he graduated in 1948 and received his PhD in 1953. By then, he had already been appointed assistant lecturer in physics at the university. In 1956 he moved to Durham University, where he remained for the rest of his career except for a spell as visiting professor in Hong Kong in 1977–8. Wolfendale's research involved studies of *cosmic rays*.

Wolf number (Wolf sunspot number) A parameter used as a measure of the Sun's activity, introduced by Rudolf *Wolf* at Zurich Observatory in 1848. The definition of the Wolf number is based on counting the number of *sunspots* and the number of groups of sunspots visible on the surface of the Sun. The number, R, is equal to $k(f + 10g)$, where g is the number of sunspot groups, f is the total number of sunspots and k is a number close to 1 that depends on the instruments being used to make the observations.

The Wolf number is an imperfect measure of solar activity, not least because different observers will assign spots to groups in different ways. There are better techniques for measuring solar activity today, but the Wolf number has the advantage that sunspots and groups have been monitored by observers since the time of *Galileo*, and so historical records can be used to estimate past sunspot numbers and reconstruct the changing level of solar activity since the early 17th century. This is the technique which shows, for example, that the Sun was very quiet during the cold period of the late 17th century on Earth (see *Maunder minimum*).

In 1994 a new analysis of the old records, based only on the number of groups of sunspots, calibrated the historical observations against measurements made at the *Royal Greenwich Observatory* after 1874. Another series of observations, made in Greece between the 1840s and 1883, overlaps with the Greenwich record, and in the overlapping years the Greek record consistently counted fewer sunspot groups than the Greenwich observers, so a correction factor can be applied. Older records overlap with the beginning of the Greek observations, and can be corrected in line with the improved version of the Greek record, and so on back to the early 1600s.

This analysis suggests that the Sun was rather less active in the 18th century than indicated by earlier analyses of the old records (but still considerably more active than during the Maunder minimum). This discovery should lead to a better understanding of short-term changes in solar activity. The improved measure of solar activity is called the group sunspot number; the Wolf number itself is also known as the relative sunspot number, as the Zurich relative sunspot number, and since 1981 as the International Sunspot Number. It is now monitored by the Sunspot Index Data Centre, in Brussels.

Woolley, Sir Richard van der Riet (1906–86)   British astronomer, born in Weymouth, Dorset, on 24 April 1906, who worked at several observatories before becoming the 11th Astronomer Royal in 1955, succeeding Harold Spencer *Jones*. He played a large part in developing the astrophysical side of the work of the *Royal Greenwich Observatory*, which had just been moved to Herstmonceux in Sussex. He was knighted in 1963 and retired in 1971, but then spent five years as the first director of the South African National Observatory.

world line   The path of an object through *spacetime*. In a *Minkowski diagram*, the entire life history of an object can be represented by a wiggly line, the world line. This line always moves from the past into the future ('up the page', on a standard Minkowski diagram), and movement through *space* is represented by the way in which the line wiggles to left and to right. But the line must always stay within its own future *light cone*, corresponding to movement through space at less than the speed of light. In *string theory*, the fundamental strings and loops sweep out two-dimensional equivalents of world lines, traces in spacetime known as world sheets; the way the world sheets interact is then thought to be responsible for the way *elementary particles* behave.

world model   Obsolete term for *cosmological model*.

wormhole   A hypothetical tunnel through the fabric of *spacetime*. A wormhole can be thought of as a short-cut through space-time, a cosmic subway connecting two *black holes* or (more speculatively) one black hole and a *white hole*. The 'other end' of a wormhole could be anywhere in space, and also anywhere in time, allowing an object that passed through the wormhole to appear instantaneously in some other part of the Universe –

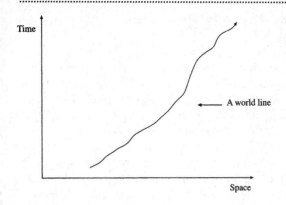

The *world line* shows the life history of a particle in a Minkowski diagram.

not just in a different place, but also in a different time.

Solutions to the equations of the *general theory of relativity* that describe wormholes were actually found in 1916, shortly after the theory was developed, although they were not interpreted in this way at the time. Albert *Einstein* himself, working with Nathan Rosen at Princeton in the 1930s, discovered that the *Schwarzschild solution* actually represents a black hole as what they called a bridge (now known as an Einstein–Rosen bridge) between two regions of flat spacetime. Although these equations were studied as mathematical curiosities (notably by John *Wheeler* and his colleagues), before 1985 they were not regarded as real features of the Universe, because every example investigated mathematically opened up only very briefly, snapping shut again (according to the equations) before anything, even light, could traverse the tunnel.

Although the idea was loved by science fiction writers, it was generally accepted by scientists that there must be some law of nature preventing the existence of wormholes. But when relativists working at Caltech tried to prove this in the 1980s, they found that they could not. There is nothing in the *general theory of relativity* (the best theory of *gravity* and spacetime that we have, which has passed every test that it has been subjected to) which forbids the existence of wormholes. What's more, Kip *Thorne* and his colleagues found that there are, after all, solutions to Einstein's equations which allow for the existence of long-lived wormholes.

The 'mouth' of such a wormhole would look like the *event horizon* of a spherical black hole, but with one important difference. The event horizon is a one-way surface, and nothing can come out of it. But the surface of a wormhole mouth allows two-way traffic. If we looked into a wormhole whose other end was located near the star *Vega*, we would see light from Vega coming out of the tunnel – and an observer near Vega, looking into the same wormhole from the other end, would see light from the Sun.

It would still be extremely difficult, and for all practical purposes it may be impossible, to build a large wormhole through which people could travel (see *time travel*). But physicists are intrigued by the possibility that naturally occurring wormholes may exist on the scale of the *Planck length*, providing the basic foam-like structure of spacetime, weaving the fabric of spacetime itself (to mix the metaphor) out of wormhole strands.

If so, there are many curious possibilities. For example, such tiny (ultra-sub-microscopic) wormholes may link far distant regions of the Universe, allowing information to leak through and ensuring that the laws of physics are the same here on Earth as they are in a distant *quasar*. Or a small wormhole may pinch off from our Universe, and begin to grow into a separate universe, through *inflation* (see *baby universes*).

Further reading: John Gribbin, *In Search of the Edge of Time*; Kip Thorne, *Black Holes and Time Warps*.

Wright, Thomas (1711–86) Thomas Wright of Durham (born at Byer's Green, near that city, on 22 September 1711) was an English instrument maker and philosopher who proposed (in his book *An Original Theory or New Hypothesis of the Universe*, published in 1750) that the *Milky Way* is a slab of *stars*, with the Sun inside it. He realized that the Sun is not at the centre of the system (which he likened to the grinding wheel of a mill) and suggested that *nebulae* lie outside the Milky Way. Similar ideas did not become fully accepted until the 1920s, with the work of Vesto *Slipher*, Milton *Humason* and Edwin *Hubble*. Wright later described the Universe in terms of systems of spheres and rings of stars, which influenced Immanuel *Kant*'s thinking about the nature of the Universe, and suggested that the rings of Saturn are not solid objects but are made up of many small bodies.

W Virginis stars Another name for *Cepheid variables* found among *Population II stars*. See *period–luminosity relation*.

WZ Sagittae A *recurrent nova* (also similar to the *dwarf novae*) that has been observed to flare up three times, in 1913, 1946 and 1978. The system is an *eclipsing binary* with a period of 81.5 minutes, and the masses of the two stars in the binary are estimated to be 0.59 and (very roughly) 0.03 times the mass of the Sun. Unusually (but not uniquely), both stars in the *binary system* are *white dwarfs*. The primary (more massive) star is orbited by a ring of gas moving with a speed of 720 km per second. Studies using *spectroscopy* suggest that the secondary (less massive) star is overflowing its *Roche lobe*, with matter flowing on to the primary (via the disc) at a rate of 20 billion kg per second. Even with mass of only 3 per cent of that of the Sun, the loss of mass from the secondary is only a tiny fraction of its total mass (about 6×10^{28} kg), and it can keep doing this for millions of years without being significantly affected by the loss.

X-ray astronomy   The study of the Universe using *electromagnetic radiation* with wavelengths in the X-ray part of the spectrum. Because the atmosphere of the Earth absorbs X-rays, X-ray astronomy proper began only in the 1960s, when astronomers were able to lift X-ray detectors above the obscuring layers of the atmosphere on rockets, balloons and (eventually) satellites. X-ray astronomy provides a picture of the Universe as a violent place in which high-energy events are common occurrences, and in particular it picks out the places where *black holes* are swallowing up matter, either in *binary systems* with masses a few times that of our Sun or on a huge scale at the hearts of *galaxies* and *quasars*.

None of this was anticipated by astronomers, who discovered the existence of X-ray sources beyond the Solar System serendipitously in 1962. X-rays from the Sun had been discovered in the late 1940s, using detectors carried on V2 rockets, and were studied by instruments on board Ariel I, a UK satellite launched by NASA in 1962. The Sun was such a feeble source of X-rays, however, that it would be undetectable (using the technology available in the early 1960s) in this part of the spectrum at a distance of a few *light years*, and astronomers assumed that there was no point in looking for X-ray sources beyond the Solar System. But on 18 June 1962, a team of researchers led by Riccardo *Giacconi* launched a rocket carrying an X-ray detector, which they hoped would be able to pick up X-rays from the surface of the Moon, produced by the action of particles from the Sun striking the surface of the Moon. No evidence was found for any such emissions on that rocket flight; indeed the Moon was only 'discovered' as an X-ray source in 1990, by instruments on board the satellite ROSAT. But as the detector on board the spinning rocket scanned across the sky, it recorded intense X-ray emission (at wavelengths in the range 0.2–0.8 nanometres), far stronger than the X-ray emission from the Sun, coming from a point source in the constellation Scorpius. In addition, the detector picked up a faint background of X-ray 'noise' from all of the directions scanned in its spin.

Theorists were caught on the hop by the discovery of this source (dubbed Sco X-1), an X-ray *star* emitting 10^{16} times as much energy as the Sun in the form of X-rays – a staggering 100,000 times more energy in the form of X-rays as the *total* radiation from the Sun at all wavelengths put together. The source of this energy was later explained (by Paul Feldman and John Gribbin, working in Cambridge) as a result of *accretion* on to a *neutron star* in a binary system. By the end of the 1960s, after dozens of rocket flights which had together provided a few hours of observation of the Universe at X-ray wavelengths, almost 50 individual *X-ray sources* had been discovered, including the *Crab nebula* and its *pulsar*, several other *supernova remnants*, a couple of peculiar galaxies, and the quasar 3C 273. Some X-ray detectors were also flown on satellites which had other primary roles. Then, in December 1970, X-ray astronomy was transformed by the launch of *Uhuru*, the first X-ray astronomy satellite, an orbiting observatory dedicated to the study of the Universe at X-ray wavelengths.

Uhuru (another product of Giacconi's team) carried out a complete survey of the sky in the X-ray band, with a sensitivity sufficient to detect any source with at least one-thousandth of the X-ray brightness of the Crab nebula, and found more than 300 separate sources. Many of these objects were identified with optically visible counterparts, and studies of the variability of these systems confirmed that the X-rays were often being produced by accretion on to a neutron star or a *white dwarf* in a binary system. A few of the sources turned out to be X-ray pulsars. In one of the systems studied by Uhuru in the early 1970s, known as *Cygnus X-1*, a combination of X-ray and optical data provided the first firm evidence for the existence of a black hole, rather than a neutron star, as the powerhouse for the X-ray emission.

Since the success of Uhuru, there have been several other orbiting X-ray observatories. The next major step forward came with the *Einstein satellite*, launched by NASA in 1978, which had much better *resolution* and sensitivity than earlier X-ray observatories. It used proper X-ray telescopes, which focus the radiation by reflecting it at a grazing angle from appropriately shaped mirrors, and found several thousand 'new' X-ray sources as well as providing detailed maps of already known sources such as supernova remnants and *clusters of galaxies*. The Einstein satellite also showed that many ordinary stars radiate much more strongly in the X-ray band than was expected, and that all quasars are X-ray emitters.

The observations of gas between the galaxies in clusters (first hinted at by Uhuru) have proved particularly important for cosmology. It is now clear that the hot gas in galaxies, revealed by its X-ray emission, contains several times as much mass as all of the visible galaxies put together; the X-rays provide an important indication of the total cluster mass, and studies of this hot gas are also important in the *Sunyaev–Zel'dovich effect*, one of the most promising techniques for determining the expansion rate of the Universe (the *Hubble constant*). See also *baryon catastrophe*.

Since the end of the 1970s, X-ray astronomy has been on a par with *radio astronomy* and *optical astronomy*, able to study all kinds of objects in the Universe, from individual stars to clusters of galaxies, in detail. Recent satellites, such as EXOSAT, Ginga and ROSAT, have made X-ray observations almost routine. ROSAT has catalogued more than 60,000 separate X-ray sources, and has shown, for example, that the familiar constellation of the Pleiades contains not only the seven stars visible to the naked eye (the 'Seven Sisters'), but about 500 hot, young stars that are emitting large amounts of X-radiation. Future orbiting observatories are planned which will have even better resolution and sensitivity.

X-ray background   A diffuse background of *X-rays* from all over the sky. *ROSAT* observations show that at least 60 per cent of this background actually comes from *active galaxies* at high *redshifts*.

X-ray binary   A *binary system* in which the *accretion* of material on to a compact object (a *white dwarf, neutron star* or *black hole*) releases energy in the form of *X-rays*. See *Centaurus X-3, Cygnus X-1*.

X-ray burster   Sources within our *Galaxy* that occasionally produce bursts of *X-rays*. Fewer than 100 are known, most of them in the *Milky Way* but a few in *globular clusters*. They are thought to be *binary systems* that resemble *recurrent novae*, but with a *neutron star* rather than a *white dwarf* as the star on to which matter is falling.

X-rays   *Electromagnetic radiation* with wavelengths shorter than those of ultraviolet radiation but longer than those of gamma rays, in the range from 12 billionths of a metre (12 nanometres, or 12×10^{-9} m) to about 12 trillionths (12×10^{-12}) of a metre. At these very short wavelengths, corresponding to high-energy radiation, X-rays are more usually described in terms of the energy carried by individual photons, which ranges from about 100 *electron Volts* for longer-wavelength X-rays up to about 100,000 eV for shorter wavelength X-rays. The distinction between X-rays and gamma rays is arbitrary and not clear cut; what some people call very high-energy X-rays others regard as low-energy gamma rays. There is a similar overlap between low-energy X-rays and high-energy *ultraviolet radiation*. Lower energy X-rays are sometimes called soft X-rays, while higher-energy X-rays are correspondingly referred to as hard X-rays.

X-rays were discovered by the German physicist Wilhelm Röntgen in 1895 (he received the Nobel Prize for his discovery in 1901); they used to be known as Röntgen rays. On Earth, X-rays are produced when a stream of energetic electrons bombards the surface of a material object.

In astronomical sources, X-rays may be produced by very high-temperature gas (at temperatures of 1–100 million *Kelvin*), in the same way that visible light is produced by gas at a temperature of a few thousand K (see *black body radiation*). X-rays are also produced when electrons interact with magnetic fields (see *synchrotron emission*) and when electrons interact with low-energy photons and give them a boost. X-rays do not penetrate the Earth's atmosphere (they can penetrate a little bit of air, but not the whole thickness of the atmosphere), and X-rays from space can only be studied in detail using instruments carried to altitudes above 150 km on rockets or satellites, and to a lesser extent using high-altitude balloons.

X-ray sources Objects in space which emit *X-rays*. The first astronomical X-ray source to be discovered was *Scorpius X-1*, in 1962. Since then, tens of thousands more have been found, most of them since 1991 in an all-sky survey carried out by *ROSAT*. Some, like Sco X-1, are star-sized systems within our own Galaxy. Others are associated with activity on larger scales, including *quasars* and *active galaxies*. X-rays are even found coming from the hot gas (strictly speaking, hot *plasma*) that fills the space between the galaxies in many *clusters of galaxies*.

The brightest X-ray sources on the sky as seen from Earth are all in our Galaxy, and are typically associated with *binary systems* such as Sco X-1 itself and *Cygnus X-1*. Some of them, including Cygnus X-1 and *V404 Cygni* probably (in the case of V404 Cygni, certainly) harbour *black holes*. More distant sources such as *Virgo A* (better known as a *radio source*) are intrinsically brighter at X-ray wavelengths than these local sources, but look fainter because they are further away. Many of them are thought to be associated with supermassive black holes.

XUV astronomy Observations of the Universe in the part of the *spectrum* bridging the range from extreme *ultraviolet* to relatively low-energy ('soft') *X-rays*. This corresponds to 10–100 nanometres in terms of wavelength, or 0.01–0.1 kilo *electron Volts* (keV) in terms of energy. Many XUV sources are hot *white dwarfs*. See *ROSAT*.

Yagi array An *antenna* made up of a row of parallel rods (dipoles). Many domestic TV antennae are Yagi arrays; they are also used in some *radio telescopes*. Named after Hidetsugu Yagi.

year The time taken for the Earth to move once around the Sun in its orbit. Relative to the *fixed stars*, this takes 365.25636 days (1 sidereal year); relative to the Sun, it takes 365.24219 days (the solar, or tropical, year).

Yerkes Observatory Observatory near Chicago founded by George Ellery *Hale* in 1897, using funds provided by the millionaire Charles T. Yerkes. It is now run by the University of Chicago. The 40-inch (101 cm) *refracting telescope* is still the largest of its kind in the world; other instruments include a 104 cm *reflecting telescope* and a 61 cm reflector.

ylem George *Gamow's* name for the primordial material of the *Big Bang* fireball.

Z Andromeda stars See *symbiotic stars*.

Zel'dovich, Yaakov Borisovitch (1914–87) Soviet astrophysicist who originally worked in nuclear physics, but turned to *cosmology* and particle physics in the 1950s and became one of the pioneers of the theoretical investigation of the kind of particle interactions that went on in the *Big Bang*, and of calculations of the amount of *helium* they should have produced.

Zel'dovich was born in Minsk on 18 March 1914. He studied at the

Yerkes Observatory. The 40-inch refractor at Yerkes Observatory, still the largest telescope of its kind in the world.

University of Leningrad, graduating in 1931 and going to work in Moscow at the Institute of Chemical Physics of the Soviet Academy of Sciences. During the 1930s, Zel'dovich worked on projects investigating the behaviour of conventional explosives, and this work continued into the 1940s.

Alongside these studies, however, Zel'dovich worked on the theory of *nuclear fission*, and in 1939 and 1940 he published (together with Yulvi Khariton, the 'father of the Soviet atomic bomb') calculations of the chain reaction involved in the fission of uranium. He later worked on the Soviet hydrogen bomb project. His contributions to the Soviet war effort gained Zel'dovich the Stalin Prize in 1943, but his inside knowledge of the Soviet nuclear programme ensured that he was never allowed out of the Soviet sphere of influence during his subsequent career. After the war, Zel'dovich still worked under the auspices of the Soviet Academy of Sciences (he became a Corresponding Member in 1946 and a full Academician in 1958), but at the Institute of Cosmic Research, part of the Space Research Institute in Moscow. He became the leading Soviet astrophysicist and mentor for generations of students.

Among his many contributions to cosmology and *astrophysics*, Zel'dovich was one of the first people to suggest that *quasars* and *radio galaxies* are powered by supermassive *black holes*, predicted the existence of *X-ray*-emitting black holes in *binary systems*, and proved (with Alexei Starobinsky) that spinning black holes radiate energy as a result of quantum effects (but at first he did not accept the suggestion that even non-rotating black holes produce *Hawking radiation*).

In 1964, shortly before the discovery of the *background radiation*, Zel'dovich predicted the existence of this radiation and was the inspiration for work by several Soviet scientists who produced a flurry of papers on the subject. But in those days communications between Russia and western scientists were not always as swift as they are today, and the Soviet work was not known to Arno *Penzias* and Robert *Wilson*, even though one of the Soviet papers actually suggested using the *antenna* at Bell Labs to search for the radiation. In 1972, together with Rashid Sunyaev, Zel'dovich also predicted an influence of *clusters of galaxies* on the background radiation which could be used as a probe of the early Universe. This *Sunyaev–Zel'dovich effect* was found some 20 years later and is now beginning to be used to measure the value of the *Hubble constant*.

Zelenchukskaya Astrophysical Observatory See *Special Astrophysical Observatory*.

zero-age main sequence (ZAMS) Representation of the *main sequence* of the *Hertzprung–Russell diagram* for a group of *stars* with different *masses*, all beginning to burn *hydrogen* through *nuclear fusion* at the same time. This is what the main sequence of a *globular cluster*, for example, would have looked like just as the cluster formed.

zero gravity See *free fall*.

zodiac An imaginary band around the sky, extending out to about 9 degrees of arc on either side of the *ecliptic*. This band includes the visible paths of the Sun, Moon and all the *planets* except Pluto across the sky, and this drew the attention of observers to it in ancient times. The Greeks divided the zodiac into twelve equal parts (each being a segment of sky 30 arc degrees wide) and gave them names corresponding to prominent *constellations* that they saw in each of the subdivisions. These names are the 'signs of the zodiac'.

Since the subdivisions of the zodiac were named, the *precession of the equinoxes* has shifted the zodiacal band of the sky as seen from Earth, but the old names for the signs of the zodiac have never been modified, even though they no longer correspond accurately to the actual constellations seen in the zodiac itself – for example, the Sun now 'passes through' Ophiuchus, which is not one of the original zodiac constellations. Astrologers happily cling to the old names, as if the Sun still did pass through, say, Pisces between 19 February and 20 March (in fact, the Sun is just entering Pisces when, according to astrology, it is supposed to be entering Aries); anybody who is born when the Sun is 'in' Ophiuchus cannot possibly have an accurate horoscope cast, even by the rules of the astrologers themselves (although an entertaining book published under the pseudonym 'Richard Tilms' did purport to make allowance for the thirteenth sign in astrology, it turned out to be an elaborate hoax). Anyone who has got this far in this book surely cannot be in any doubt that astrology is nonsense – but many early astronomical observations were made for astrological purposes, and only later incorporated into the science of astronomy.

Zond Overall name for a series of Soviet *spaceprobes* launched between 1963 and 1970. Zond 1 flew by Venus and Zond 2 past Mars, but neither sent back any data. Zonds 3 and 5–8 investigated the Moon without landing. Zond 4 failed. The last three Zonds (5–8) went round the Moon and back to Earth, where they were recovered.

Zurich relative sunspot number See *sunspot number.*

Zwicky, Fritz (1898–1974) Swiss astronomer who made many contributions to *astrophysics*, studied *supernovae, compact galaxies* and *clusters of galaxies*, and was one of the few people to take the idea of *neutron stars* seriously before the discovery of *pulsars*.

Although Zwicky was born in Varna, in Bulgaria (on 14 February 1898), his parents were Swiss, and he retained his Swiss nationality even when he later went to live and work in the United States. He studied in Switzerland, graduating in 1920 and receiving his PhD in 1922 (both from the Federal Institute of Technology in Zurich), but he left Switzerland to move to Caltech in 1925. He was originally taken on to work on crystallography and the study of the structure of atoms, but became seduced by the exciting new developments in astronomy being made at Caltech and the nearby observatories at the time. He stayed there until he retired in 1968, rising to become professor of astronomy, and then lived in Switzerland in retirement until he died on 8 February 1974. Apart from his astronomical work, Zwicky was also involved in (among other things) research into jet propulsion, rocketry and studies of the behaviour of crystals.

Zwicky was an ebullient character, who had an unshakable faith in his own genius and managed to annoy many of his colleagues by tossing out ideas that he had arrived at largely intuitively, without always backing them up with thorough analysis. But one of the most annoying of these wild ideas eventually proved to have hit the bull's-eye.

Early in his astronomical career, Zwicky carried out a search for supernova explosions in nearby galaxies. When he began this work, it was not yet clear how far away the other galaxies are, and there was still considerable doubt among astronomers about whether there really was a separate class of 'superluminal' stellar outbursts, even exceeding the brightness of *novae*. In the 1930s, Zwicky's interest in superluminal outbursts led him to write a prescient paper with Water *Baade* in which they introduced the term 'supernova' and confidently stated that these are indeed a different category of phenomena from novae, and, chiefly at Zwicky's instigation (but with the first three cautionary words inserted by Baade), they commented:

> With all reserve we advance the view that a super-nova represents the transition of an ordinary star into a *neutron star*, consisting mainly of neutrons. Such a star may possess a very small radius and an extremely high density ... A neutron star would therefore represent the most stable configuration of matter as such.

This paper was published in 1934 (*Proceedings of the National Academy of Sciences*, volume 20, pages 259–63), just a couple of years after the discovery of the neutron, and was a breathtaking leap of the imagination for its time (see also Lev *Landau*). So breathtaking, indeed, that after some discussion of the implications in the 1930s and 1940s, most astrophysicists shied away from the idea, preferring to believe, even into the 1960s, that even an explosion as violent as that of a supernova would leave behind nothing more compact than a *white dwarf*. In their defence, remember that white dwarfs had been identified by the 1930s, but that neutron stars were not detected (as pulsars) until the mid-1960s. Zwicky's personality didn't help his cause, either. He even quarrelled with Baade in the 1940s, so violently that Baade genuinely feared for his life, and Robert *Oppenheimer*, although well aware of Zwicky's work, deliberately avoided making any reference to it in his own publications on neutron stars and what are now known as *black holes*, while pointedly including many references to Landau's work.

In retaliation, when Zwicky's own theory of collapsed stars was published in its fullest form in 1939, the paper made no reference to the work of Oppenheimer and his colleagues. And all this in spite of (or perhaps because of?) the fact that Zwicky and Oppenheimer both worked at Caltech!

In the 1930s, there seemed no hope of identifying neutron stars directly, and Zwicky (after establishing that a supernova may briefly be 100 million times brighter than the Sun, a figure now revised upwards by a factor of 100 because of the revision in the *cosmic distance scale*) turned the main focus of his attention to other topics, although continuing to investigate supernovae whenever possible.

His life's work was a major study of clusters of galaxies, using photographs obtained with the *Schmidt camera* at *Mount Palomar Observatory*, establishing that most galaxies do indeed occur in clusters, which may contain thousands of individual galaxies. He was one of the first people to appreciate the value of the Schmidt design for this kind of large-scale survey of the Universe.

Zwicky's analysis of the way galaxies move within clusters (revealed by studies of the *Doppler effect* on the *spectra* of individual galaxies after the average cosmological *redshift* for the whole cluster had been subtracted out) provided the first direct evidence, as long ago as the 1930s, for the existence of large amounts of *dark matter* in the Universe. At the time, and for decades afterward, this was widely regarded as another of Zwicky's crackpot ideas. It was largely ignored, in the hope that the problem would disappear when the models were improved; in fact, as the models and observations got better, the evidence for dark matter mounted until it became incontrovertible. Some of the 'extra' mass in clusters can now be explained by the presence of hot gas between the galaxies (see *baryon catastrophe*), but a great deal cannot be in the form of *baryons*.

Zwicky's work on clusters of galaxies culminated in a six-volume catalogue which lists 10,000 clusters (mostly in the Northern Hemisphere sky), published in the 1970s. In 1963, Zwicky had also been the first person to identify compact galaxies as a separate class of compact, bright objects in the Universe, but the term is seldom used today.

ZZ Ceti stars A type of *pulsating variable*, all of them *white dwarfs*, named after the archetype, ZZ Ceti. They have surface temperatures around 12,000 *Kelvin* and pulsate with periods from less than 1 minute up to 30 minutes long.

BIBLIOGRAPHY

....

Details of all books cited in the text as 'further reading' are given here, together with those of some additional books of interest.

John Barrow and Frank Tipler, *The Anthropic Cosmological Principle* (Oxford University Press, Oxford and New York, 1986)

Jeremy Bernstein, *Hans Bethe* (Dutton, New York, 1980)

Jeremy Bernstein, *Three Degrees Above Zero* (Scribner's, New York, 1984)

Subrahmanyan Chandrasekhar, *Eddington* (Cambridge University Press, Cambridge, 1983)

Marcus Chown, *The Afterglow of Creation* (University Science Books, Mill Valley, CA, 1993)

Ronald Clark, *Einstein: The life and times* (World Publishing, New York, 1971)

Arthur C. Clarke, *2010: Odyssey Two* (Dell Ray, New York, 1984)

Frank Close, *The Cosmic Onion* (Heinemann, London, 1983)

Francis Crick, *Life Itself* (Macdonald, London, 1982)

Paul Davies, *The Runaway Universe* (Dent, London, 1978)

Paul Davies, *The Forces of Nature* (Cambridge University Press, 2nd edn, 1986)

Philip K. Dick, *Counter-Clock World* (Berkley-Medallion, New York, 1967)

Timothy Ferris, *The Red Limit* (Quill, New York, revised edn, 1983)

Richard Feynman, *Six Easy Pieces* (Addison-Wesley, Boston, MA, 1995)

Robert Forward, *Dragon's Egg* (Dell Ray, New York, 1983)

George Gamow, *The Birth and Death of the Sun* (Viking, New York, 1940)

George Gamow, *The Creation of the Universe* (Viking, New York, 1952)

George Gamow, *Mr Tompkins in Paperback* (Cambridge University Press, Cambridge, 1967)

George Greenstein, *Frozen Star* (Freundlich, New York, 1984)

John Gribbin, *In Search of Schrödinger's Cat* (Black Swan, London, and Bantam, New York, 1984)

John Gribbin, *In Search of the Big Bang* (Black Swan, London, and Bantam, New York, 1986)

John Gribbin, *In Search of the Edge of Time* (Penguin, London, and Harmony, New York, 1992)

John Gribbin, *In the Beginning* (Penguin, London, and Little, Brown, New York, 1993)

516

John Gribbin, *Schrödinger's Kittens and the Search for Reality* (Weidenfeld and Nicolson, London, 1995)

John and Mary Gribbin, *Fire on Earth* (Simon and Schuster, London, 1996)

John Gribbin and Martin Rees, *The Stuff of the Universe* (Penguin, London, 1995)

Mary and John Gribbin, *Being Human* (Dent, London, 1993)

Edward Harrison, *Cosmology* (Cambridge University Press, Cambridge, 1981)

Edward Harrison, *Darkness at Night* (Harvard University Press, Cambridge, MA, 1987)

Stephen Hawking, *A Brief History of Time* (Bantam, New York, 1988)

Nigel Henbest and Heather Couper, *The Guide to the Galaxy* (Cambridge University Press, Cambridge, 1994)

Nick Herbert, *Quantum Reality* (Rider, London, 1985)

Fred Hoyle, *The Black Cloud* (Heinemann, London, 1957)

Fred Hoyle, *Home is Where the Wind Blows* (University Science Books, Mill Valley, CA, 1994)

John Imbrie and Katherine Palmer Imbrie, *Ice Ages* (Macmillan, London, 1979)

Robert Jastrow, *God and the Astronomers* (Norton, London, revised edition, 1994)

William Kaufmann, *Universe* (W. H. Freeman, New York, 2nd edn, 1988; also available on CD ROM, 1994)

Alan Lightman and Roberta Brawer, *Origins* (Harvard University Press, Cambridge, MA, 1990)

Jay Pasachoff, Hyron Spinrad, Patrick Osmer and Edward Cheng, *The Farthest Things in the Universe* (Cambridge University Press, Cambridge, 1994)

Ilya Prigogine and Isabelle Stengers, *Order Out of Chaos* (Heinemann, London, 1984)

Michael Rowan-Robinson, *The Cosmological Distance Ladder* (W. H. Freeman, San Francisco, CA, 1985)

J. F. Scott, *The Scientific Work of René Descartes* (Taylor and Francis, London, 1952)

Joseph Silk, *A Short History of the Universe* (Scientific American Library/W. H. Freeman, New York, 1994)

George Smoot and Keay Davidson, *Wrinkles in Time* (Little, Brown, New York and London, 1993)

Kip Thorne, *Black Holes and Time Warps* (Norton, New York, and Picador, London, 1994)

Robert Weber, *Pioneers of Science: Nobel Prize Winners in Physics* (Adam Hilger, Bristol and Philadelphia, 2nd edn, 1988)

Steven Weinberg, *The First Three Minutes* (Deutsch, London, 1977)

Richard Westfall, *Never at Rest* (Cambridge University Press, Cambridge, 1980)

Richard Westfall, *The Life of Isaac Newton* (Cambridge University Press, Cambridge, 1993)

Michael White and John Gribbin, *Stephen Hawking: A life in science* (Penguin, London and Plume, New York, 1992)

Michael White and John Gribbin, *Einstein: A life in science* (Simon and Schuster, London, and Plume, New York, 1993)

Clifford Will, *Was Einstein Right?* (Basic Books, New York, 1986)

PICTURE ACKNOWLEDGEMENTS

Additional credits where necessary are given in brackets after the page number.

Galaxy Picture Library: 32, 250, 327, 425.

Image Select/Ann Ronan Picture Library: 135, 215 (NASA), 255, 280.

Mansell Collection: 138, 282, 345, 414.

Mary Evans Picture Library: 155, 178, 238.

Royal Astronomical Society: 28 (Hale Observatories), 153, 162, 245 (Hale Observatories), 303, 367, 466, 511.

Science Photo Library: 22 (NRAO/AUI), 54 (Dr Rudolph Schild), 58 (M. F. Marten), 61 (Los Alamos National Laboratory), 86 (Harvard College Observatory), 95, 100, (NOAO), 131 (Brookhaven National Laboratory), 147 (NOAO), 169 (Roger Ressmeyer, Starlight), 228 (Royal Greenwich Observatory), 230 (Dr Rob Stepney), 247 (A. Barrington Brown), 253 (NASA), 276 (James Stevenson), 309 (Royal Observatory, Edinburgh), 312 (NASA), 318 (Dr Rudolph Schild), 331 (NASA), 334 (Roger Ressmeyer, Starlight), 378 (David Hardy), 406 (World View/Maarten Udema), 474 (NOAO), 502 (Royal Greenwich Observatory).

Starland Picture Library: 50 (NOAO), 103 (Yerkes Observatory/University of Chicago), 120 (NOAO), 150 (NASA), 176 (NOAO), 199 (NASA), 207 (NOAO), 266 (NOAO), 270 (NOAO), 275 (NOAO), 317 (NASA), 335 (NOAO), 364 (NOAO), 376 (NASA), 381 (ESA), 418 (Yerkes Observatory/University of Chicago), 493 (Dr Bruce C. Murray/NSSDC), 494 (NRAO/AUI), 499 (Yerkes Observatory/University of Chicago).

Weidenfeld Archives: 106, 195.

TIMELINE 1

....

Birth dates of scientists who made significant contributions to our understanding of the Cosmos.

TIMELINE 2

....

Key dates in the development of physical science and (especially) of our understanding of the Cosmos.

TIMELINE 3

....

Key dates in history, with emphasis on events important in a scientific context.

About **15000–10000 BC**: The world warms out of the latest Ice Age. Cave paintings.

About **12000 BC**: Dog domesticated.

10000 BC: World human population 3 million.

About **8000 BC**: Invention of trade tokens (early 'money'), agriculture. Pottery.

About **6000 BC**: Development of irrigation. Weaving.

About **4500 BC**: Copper smelting.

About **4000 BC**: Sumerian city of Ur founded.

About **3500 BC**: Invention of writing and the wheel.

About **3000 BC**: Egyptians use sailing ships. Bronze in widespread use. Pyramid of Giza built. Stonehenge first stage built. Population 100 million.

Birth dates of scientists	*Key dates in science*	*Key dates in history*
		About **2500 BC**: Beaker people spread across Europe.
		About **2000 BC**: Rise of Babylon. Stonehenge stone circles added. Spoked wheels developed in Asia Minor. Minoan civilization flowers in Crete.
		About **1800 BC**: The first alphabet. Babylonians use multiplication tables.
		1700 BC: Judaism founded by Abraham.
		About **1650 BC**: Volcanic island of Santorini (Thera) explodes; probable source of Atlantis legend. Bronze ploughs in use in Vietnam.
		About **1500 BC**: Chinese writing.
		About **1350 BC**: Exodus of Jews from Egypt.
		About **1200 BC**: Iron working developed. Trojan Wars.
		878 BC: Carthage founded.
		About **800 BC**: Olmec pyramids built in Mexico. Homer writes *The Iliad*.
		776 BC: First Olympiad.
		710 BC: Egypt conquered by Ethiopian invaders.

Birth dates of scientists	*Key dates in science*	*Key dates in history*

624 BC: Thales of Miletus. The first scientist; a Greek philosopher who proposed that the world is made of water, and that the Earth is a disc which floats on water.

About **600 BC**: Greek science begins. Aesop writes his fables. Chinese invent fumigation of houses to destroy pests.

611 BC: Anaximander of Miletus. The first philosopher to suggest that the surface of the Earth is curved.

605–562 BC: Nebuchadnezzar creates Hanging Gardens of Babylon.

580 BC: Pythagoras. Taught that the Earth is a sphere and that the planets move in circles – not for any truly scientific reason, but because of a mystic belief that circles were the 'perfect' form. He is also credited with discovering the famous property of right-angled triangles which bears his name, and which turns out to be an immensely useful tool in the investigation of relationships in space-time, in the context of both the special theory of relativity and the general theory of relativity.

585 BC: Thales of Miletus correctly predicts a solar eclipse.

565 BC: Taoism founded.

538 BC: Persians conquer Babylon.

528 BC: Beginning of Buddhism in India.

525 BC: Persians conquer Egypt.

About **520 BC**: Anaximander devises a cylindrical model of the Earth.

500 BC: Anaxagoras of Clazomenae. The first astronomer to give the correct explanation for eclipses of the Sun and Moon.

500 BC: Pythagoreans teach that the Earth is a sphere.

About **500 BC**: Pythagoreans argue that the Earth is a sphere. Steel manufactured in India.

490 BC: Greeks defeat Persians at Marathon.

About **490 BC**: Empedocles. One of the earliest proponents of the idea that everything is made up of four 'elements', fire, air, water and earth.

478 BC: Athenian empire established.

470 BC: Socrates born.

Birth dates of scientists	Key dates in science	Key dates in history

About 427 BC: Plato. Greek philosopher who based much of his thinking on the idea of perfection, arguing that the Earth must be a perfect sphere and that all other objects in the Universe moved in perfect circles around the Earth.

430 BC: Democritus of Abdera develops the idea that everything is made of atoms.

427 BC: Plato born.

400 BC: First settlement at site of London.

384 BC: Aristotle born.

388 BC: Heraklides of Pontus. Greek philosopher and astronomer who taught that the Earth turns on its axis once every 24 hours – an idea that did not become widely accepted for 1,800 years.

348 BC: Plato writes *The Republic*.

340 BC: Philip II of Macedonia rules in Greece.

336 BC: Alexander the Great succeeds Philip.

384 BC: Aristotle. Among his many interests, Aristotle wrote about cosmology; building on the ideas of his predecessors, he came up with the model of the Universe as a series of concentric spheres, centred on the Earth and rotating about it.

323 BC: Alexander dies after conquering most of the known world.

312 BC: First aqueduct built to bring water to Rome.

306 BC: Euclid born.

Early third century BC: Aristarchus of Samos. The first person to attempt to work out the relative distances to the Sun and Moon after it was realized that the Earth is round. Aristarchus realized that the apparent movement of the stars across the sky is caused by the Earth's rotation.

Third century BC: Euclid gathers together the knowledge of his time, and writes it down; this knowledge includes the geometric ideas now known as Euclidean geometry, which formed the basis of mathematical teaching in many places until well into the 20th century.

300 BC: Alexandria Museum built.

280 BC: Colossus of Rhodes completed.

265 BC: Archimedes discovers the law of specific gravity (an object displaces its own weight when floating in water).

264 BC: First Punic War (between Rome and Carthage) begins.

About **273 BC**: Eratosthenes of Cyrene. Carried out the first reasonably accurate calculation of the size of the spherical Earth.

About **260 BC**: Construction of Great Wall of China begins.

Birth dates of scientists	*Key dates in science*	*Key dates in history*
		240 BC: Chinese astronomers observe Halley's Comet – the earliest recorded visit of the comet.
		About **235 BC**: Eratosthenes correctly calculates the size of the Earth.
		218 BC: Hannibal crosses the Alps.
2nd century **BC**: Hipparchus of Nicaea. Measured the length of the year to an accuracy of 6.5 minutes, and made the first realistic estimates of the distances to the Sun and Moon.	**165 BC**: Chinese astronomers record sunspots – first accurately dated observations.	About **200 BC**: Growth of the Roman Empire.
		146 BC: Rome destroys Carthage after taking over Greece.
		About **140 BC**: Invention of paper in China. Venus de Milo sculpted; artist unknown.
		About **100 BC**: Great Wall of China completed. Chinese ships visit east coast of India, navigating with the aid of the magnetic compass (lodestone).
	46 BC: Julian calendar brought into use in the Roman Empire by Julius Caesar. In order to bring the calendar back in line with the seasons, the year 46 BC has 445 days in it!	**54 BC**: Julius Caesar invades Britain.
		46 BC: Julian calendar introduced in Rome.
	44 BC: Major eruptions of Mount Etna. Volcanic pollution blocks sunlight and cools the Earth, causing crop failures in China.	**44 BC**: Julius Caesar killed.
		40–4 BC: Reign of Herod the Great.
		5 BC: Christ born. World population 250 million.

Birth dates of scientists	**Key dates in science**	**Key dates in history**
		28 AD: Jesus crucified.
		66 AD: Mark's Gospel written.
	79 AD: Pliny the Younger provides the first detailed written description of a major volcanic eruption, the Vesuvian outburst that destroyed Pompeii.	**79 AD**: Pompeii and Herculaneum destroyed by eruption of Vesuvius.
About **100 AD**: Ptolemy. Greek astronomer who wrote the *Almagest*, a 13-volume description of the Universe as understood at that time. Ptolemy used a genuinely scientific approach, and developed the idea of epicycles to account for the observed motion of heavenly bodies within the framework of an Earth-centred universe. This Ptolemaic system held sway for more than 1,000 years.		About **100 AD**: Paper first used for writing.
		122 AD: Romans construct Hadrian's Wall.
	132 AD: In China, Zhang Heng invents the first seismograph. It indicates the direction of an earthquake by dropping balls shaken from the mouths of bronze dragons by the earthquake wave.	**286 AD**: Roman Empire divided into Eastern and Western halves, initially for administration.
		330 AD: Constantinople founded.
		389 AD: Library at Alexandria destroyed.
		395 AD: Roman Empire completely split, with two separate emperors.
	400 AD: The term 'chemistry' used for the first time by scholars in Alexandria.	**410 AD**: Visigoths sack Rome.
		432 AD: St Patrick sets out to convert the Irish.
		433 AD: Attila becomes leader of the Huns.
		455 AD: Vandals sack Rome.
		476 AD: Last remnant of Roman Empire in Italy destroyed by German invaders.
		About **500 AD**: The abacus.

Birth dates of scientists	Key dates in science	Key dates in history
		517 AD: Buddhism introduced into China.
		525 AD: Introduction of the Christian calendar.
		570 AD: Mohammed born.
		About **600 AD**: Invention of printing press – Chinese woodblocks. Windmills built in Persia.
		613 AD: Mohammed begins to teach openly.
		616 AD: Visigoths conquer Spain.
		620 AD: Flight of Mohammed from Mecca (the hegira); this event marks the start of the Islamic calendar.
		632 AD: Spread of Arab Empire begins following the death of Mohammed.
	635 AD: Unknown Chinese scholar writes down the rule that the tail of a comet always points away from the Sun.	**670 AD**: Venerable Bede born.
		697 AD: Carthage destroyed by Arabs.
		699 AD: *Beowulf* completed.
		711 AD: Arab invasion of Spain.
		731 AD: Mayan Empire at the start of its greatest flowering.
		732 AD: Defeat of Arabs at Poitiers marks the limit of the western expansion of Islam.

Birth dates of scientists	Key dates in science	Key dates in history
		748 AD: First printed newspaper, in Beijing.
		768 AD: Charlemagne becomes King of the Franks.
		About **790 AD**: Irish monks settle in Iceland.
	827 AD: Ptolemy's *Megale syntaxis* is translated into Arabic as the *Almagest*.	About **800 AD**: Spread of Vikings begins.
	840 AD: First Arab record of observations of sunspots.	**849 AD**: Alfred the Great born.
		863 AD: Invention of Cyrillic alphabet.
	880 AD: Arab chemists distil alcohol from wine.	**900 AD**: Greenland discovered by Norse.
		924 AD: Death of Alfred the Great of England.
		929 AD: Good King Wenceslas murdered.
		975 AD: Modern arithmetical notation introduced into Europe by the Arabs.
		982 AD: First Viking settlement in Greenland.
		About **1000**: Islamic science flourishes. Chinese use coal as fuel. In India, the mathematician Sridhara realizes the importance of the zero.
		1005: Science library founded in Cairo.

Birth dates of scientists	*Key dates in science*	*Key dates in history*
		1036: Modern musical notation introduced by Guido d'Arrezzo.
		1050: First use of moveable type to print books in China.
	1054: Chinese astronomers witness the supernova explosion that created the Crab nebula. This explosion is temporarily brighter than Venus, being visible in daylight for 23 days.	
	1066: Large comet (now known to be a visit of Halley's Comet) seen.	**1066**: William of Normandy defeats Harold at the Battle of Hastings.
		1086: Domesday Book.
		1099: Jerusalem falls to the Crusaders.
		About **1100**: Magnetic compass in use in China. Sinchi Roca becomes the first King of the Incas.
		1150: First rockets used in China. University of Paris founded.
		1155: Oldest known printed map produced in China.
		1157: Richard the Lionheart born.
		1167 or **1168**: Formal foundation of the University of Oxford.
		1174: Construction of the Tower of Pisa begins.
		1204: Crusaders sack Constantinople.

Birth dates of scientists	Key dates in science	Key dates in history
		1206–27: Genghis Khan conquers vast area of Eurasia.
		1213: Beginnings of the University of Cambridge.
		1215: Magna Carta.
		About **1250**: Invention of the quill pen.
		1260: Genghis Khan's grandson, Kublai Khan, becomes Emperor of China.
		1271: Marco Polo sets out on his great journey to the east.
About **1285**: William of Ockham. Developer of the idea of 'Ockham's razor'. This idea states that if there are two possible explanations for something, and one explanation is simpler than the other, then the simpler explanation should be preferred.		**1294**: Kublai Khan dies.
		1295: Marco Polo returns to Italy.
		About **1300**: Beginning of the Ottoman Empire, which will rule a large part of the Mediterranean and Middle East until 1923.
		1305: Giovanni Pisano completes his sculpture, the *Madonna and Child*.
		1307: Dante starts work on his *Divine Comedy*.
		About **1310**: First mechanical clocks in Europe.
		1311: Notre-Dame Cathedral completed in Paris.
		1314: Battle of Bannockburn.

Birth dates of scientists	Key dates in science	Key dates in history
		1331: Black Death emerges in China and spreads, eventually to Europe.
		1338: Beginning of the 'Hundred Years' War' between England and France.
	1350s: Jean Buridan develops the idea of 'impetus', a forerunner of the modern concept of inertia. He rejects the idea that planets and other 'heavenly bodies' are pushed along by angels, and says that impetus is all that is needed to do the job.	**1356**: First use of cannon in warfare, in China.
		1366: Petrarch writes *Canzoniere*.
		1386: Heidelberg University founded.
		1387: Chaucer starts work on *The Canterbury Tales*.
		1388: Chaucer writes *Troilus and Criseyde*.
		1398: Delhi destroyed by Tamburlaine.
		1409: Donatello completes his sculpture, *David*.
		1419: Publication of Baccaccio's *Decameron*.
		1426: Masaccio's painting, the *Virgin Enthroned*.
		1431: Joan of Arc burned at the stake.
		1440s: Printing press using movable type developed in Europe by Gutenberg.
		1452: Completion of the Medici Palace in Florence. Richard III of England born. Leonardo da Vinci born.

Birth dates of scientists	Key dates in science	Key dates in history
		1453: Constantinople falls to the Turks and becomes part of the Ottoman Empire.
		1454–5: Gutenberg Bible printed.
		1455–85: Wars of the Roses.
1473: Copernicus, Nicolaus (Mikolaj Kopernigk). Polish astronomer (and doctor) who set out the idea that the Sun, and not the Earth, is at the centre of the Solar System.		**1473**: Michelangelo paints the ceiling of the Sistine Chapel.
		1474: Caxton prints the first book in English.
		1478: Botticelli paints *Primavera*. Establishment of the Spanish Inquisition.
About **1480**: Magellan, Ferdinand. First European to describe the Magellanic Clouds in detail.		**1489**: First use of + and − signs in mathematics.
	1490: Leonardo da Vinci studies capillary action of liquids in narrow tubes.	**1491**: Henry VIII born.
		1492: Columbus discovers the islands off the east coast of Central America. First globe made by geographer Martin Behaim.
		1498: Vasco da Gama voyages to India round the Cape of Good Hope.
	16th century: The first reflecting telescope, a telescope that gathers light and magnifies an image primarily through the use of a curved mirror, is developed by Leonard Digges.	**1513**: Machiavelli publishes *The Prince*.
		1516: Thomas More publishes *Utopia*.
		1517: Martin Luther begins the Protestant movement.

Birth dates of scientists	Key dates in science	Key dates in history
1520(?): Digges, Leonard. Inventor of the telescope.	**1519**: Ferdinand Magellan describes the Magellanic Clouds in detail.	**1519**: Ferdinand Magellan commences the voyage that will end with one of his ships (but not Magellan, who is killed en route, in the Philippines) completing the first circumnavigation of the globe.
		1532: Pizarro conquers Peru.
		1533: Elizabeth I born.
1546(?): Digges, Thomas. Promoted the use of the telescope; also proposed that the Universe is infinite in extent.	**1540**: The German astronomer Peter Apian records the first European discovery of the fact that comet tails always point away from the Sun.	**1543**: Publication of Nicolaus Copernicus's book, *On the Revolutions of Heavenly Bodies*.
1546: Brahe, Tycho. Made accurate measurements of the positions of stars and the movements of planets, paving the way for Kepler to discover the laws of planetary motion.		
1548: Bruno, Giordano. Italian monk who was an early supporter of the proposal made by Copernicus that the Earth moves round the Sun. Burned at the stake for heresy in Rome.	Second half of the 16th century: Leonard Digges probably makes the first refracting telescope.	**1550**: Tobacco growing begins in Spain.
		1551: Leonard Digges invents the theodolite. Titian paints Prince Felipe of Spain.
		1558: Elizabeth I becomes Queen of England.
1564: Galileo (Galilei, Galileo). First person to make systematic observations of the		**1564**: Shakespeare born. Horse-drawn carriage introduced into England from the continent.

heavens using a telescope. Disproved commonly held theory that heavy objects fell faster than lighter objects. Supported the Sun-centred model of the Solar System put forward by Copernicus.

1571: Kepler, Johannes. Discovered the three laws of planetary motion that helped lead Newton to his universal law of gravity.

1581: Ussher, Archbishop James. The Archbishop of Armagh who added up the dates implied by biblical genealogies back to Adam, and came up with the conclusion that the Earth was created at precisely 10 a.m. on 26 October in the year 4004 BC.

1572: Supernova observed by Tycho Brahe.

1576: Thomas Digges discards the Ptolemic idea of the stars being attached to a single crystal sphere surrounding the Earth, and suggests that the stars are distributed into an endless infinity of space.

1581: Galileo investigates the behaviour of pendulums.

1582: Gregorian calendar, the calendar widely used on Earth today, is introduced by Pope Gregory XIII. It replaces the less accurate Julian calendar. The Gregorian calendar is introduced first to Catholic countries.

1582: Joseph Scaliger devises the Julian Date, what will become the standard system for coordinating events recorded on different calendars. This system arbitrarily sets 1 January 4713 BC as the starting date, from which the Julian Date is given by the

1568: Mercator introduces his eponymous map projection.

1582: Pope Gregory XIII reforms the calendar. As a result, this is the shortest year on record.

Birth dates of scientists	Key dates in science	Key dates in history
	number of days that have elapsed.	
		1586: Russian expansion east of the Urals begins. Walter Raleigh introduces tobacco smoking to England.
	1590s and 1600s: Mira becomes the first variable star to be discovered, through a series of observations by the Dutch astronomer David Fabricus.	
1596: Descartes, René. Invented the techniques of coordinate geometry (Cartesian geometry).		
		1588: Spanish Armada defeated.
	First decade of the 17th century: Galileo becomes the first person to use a refracting telescope for astronomical observations.	**1599**: First performance of Shakespeare's *Much Ado about Nothing*.
	1604: Supernova observed by Kepler.	**1600**: First performance of Shakespeare's *Hamlet*.
		1605: In *Advancement of Learning*. Francis Bacon encourages the scientific investigation of the world. Gunpowder Plot fails to blow up the English Parliament. Cervantes writes *Don Quixote*.
		1609: Galileo uses a telescope to study the Moon and planets. Kepler's laws published.
	1610: Johannes Kepler becomes the first person to realize that the darkness of the night sky directly conflicts with the idea of an infinite Universe filled with bright stars. He concludes that the Universe must therefore be finite – that, in effect, when we look through the gaps between the stars, we see the dark end of the Universe.	**1610**: French colony established in Quebec. **1611**: King James Bible published.

Timelines

....

| Birth dates of scientists | Key dates in science | Key dates in history |

Birth dates of scientists

1625: Cassini, Giovanni Domenico. Determined the rotation periods (the 'length of day') for the planets, and calculated tables of the motions of the satellites of Jupiter. In 1675 he discovered a gap in the rings of Saturn, still known as the Cassini division.

1629: Huygens, Christiaan. Invented the first successful pendulum clock; designed and built improved astronomical telescopes; developed a complete wave theory of light; discovered Titan, and recognized the nature of Saturn's rings.

1642: Newton, Isaac. Discovered the law of gravity and the laws of motion that bear his name; invented the mathematical technique of calculus; carried out important

Key dates in science

1613: Neptune may have been seen and sketched by Galileo, without him realizing it was a planet.

1631: Pierre Gassendi is the first person to observe a transit of Mercury across the face of the Sun.

Key dates in history

1615: Completion of St Peter's in Rome.

1618: The Thirty Years' War begins.

1620: Pilgrims land at Plymouth Rock.

1632: Christopher Wren born.

1636: Harvard College founded.

1637: Descartes' *Discourse on the Method of Rightly Conducting Reason and Seeking Truth in the Sciences* published.

1638: Birth of Louis XIV.

1641: First pendulum clock completed by Galileo's son.

1642: English Civil War begins.

Birth dates of scientists	*Key dates in science*	*Key dates in history*

studies in optics, including the design of a new kind of reflecting telescope.

1643: Torricelli makes the first barometer.

1644: Rømer, Ole. Measured the speed of light in 1675, using observations of eclipses of the moons of Jupiter to reveal how long it takes light to cross the orbit of the Earth.

1646: Flamsteed, John. Became the first Astronomer Royal in 1675. The Royal Greenwich Observatory was set up for Flamsteed to prepare accurate astronomical tables.

1647: First map of the Moon made by Johannes Hevelius.

1649: Charles I beheaded.

About **1650**: Cassegrain. The inventor of the system of mirrors used in a Cassegrain telescope.

1650: First binary star identified by the Jesuit astronomer Joannes Riccioli, who discovers, with the aid of a telescope, that the star Zeta Ursae Majoris is actually a double star.

1650: Bishop Ussher sets the date of the Creation at 4004 BC. Cyrano de Bergerac suggests seven ways of flying to the Moon.

1651: In *Leviathan*, Thomas Hobbes says that man's life is 'solitary, poor, nasty, brutish, and short'.

1654: Grand Duke Ferdinand II of Tuscany invents the thermometer.

1655: Titan, the largest moon of Saturn, is discovered by Christiaan Huygens.

1656: Halley, Edmund. Realized that the comet which now bears his name is in a regular orbit around the Sun under the influence of gravity, in accordance with the law of gravity established by Newton. Also compiled a star catalogue, detected

1656: Huygens identifies the true nature of the rings of Saturn.

1659: Huygens observes surface markings on Mars.

1656: Huygens develops an accurate pendulum clock.

Birth dates of scientists	Key dates in science	Key dates in history

Birth dates of scientists

the proper motion of some stars using historical records, and initiated a research programme that led to a good estimate of the distance from the Earth to the Sun.

Key dates in science

1663: John Gregory proposes the design of reflecting telescope that becomes known as the Gregorian telescope.

1664: Robert Hooke discovers the Great Red Spot of Jupiter.

1665: Plague closes Cambridge University and sends Isaac Newton back home to Woolsthorpe, where he makes many of his great discoveries.

1666: Giovanni Cassini observes the polar caps of Mars.

1667: Paris Observatory (Observatoire de Paris) founded.

1668: Isaac Newton reinvents the reflecting telescope first invented by Leonard Digges in the 16th century, and becomes the first person to put the invention to practical use. When the principle of the reflecting telescope is perfected, refractors fall out of favour for astronomical work, because (unlike reflectors) they suffer from a problem called chromatic

Key dates in history

1660: Restoration of the monarchy in England.

1661: Construction of Versailles Palace begins.

1663: Royal Society receives its Charter.

1664: Descartes' *Treatise on Man* says that animals and people are 'mechanical' objects with no 'vital force', or soul.

1665: Royal Society starts publication of its *Philosophical Transactions*. Rembrandt van Rijn paints *Juno*. Plague strikes England.

1666: Great Fire of London. French Royal Academy of Sciences founded.

Birth dates of scientists	*Key dates in science*	*Key dates in history*

aberration, which produces brightly coloured distortions in the images.

1670: Molière writes *Le Bourgeois Gentilhomme.*

1671: The distance to Mars is first measured reasonably accurately, by a team of French astronomers observing the position of the planet on the sky from Cayenne, in French Guiana, while a team in Paris note its position at the same time.

1672: Cassegrain publishes the design of the telescope which becomes known as a Cassegrain telescope. The design was not put into practice until the 18th century.

1672: Rhea, the second largest moon of Saturn, is discovered by Giovanni Cassini.

1675: Charles II founds the Royal Greenwich Observatory.
   Cassini discovers the gap in the rings of Saturn now known as the Cassini division.

1676: The finite speed of light is determined by Ole Rømer.

1676: British Museum founded.

1680: The first clocks with minute hands (previously, they only showed the hours).

1683: Turks besiege Vienna.

Birth dates of scientists	*Key dates in science*	*Key dates in history*
1685: Berkeley, (Bishop) George. Argued that all motion is relative, and must be measured against something.		**1685**: Birth of Johann Sebastian Bach.
	1687: Newton's great work, the *Principia*, which gives the three fundamental laws describing the dynamical behaviour of objects, is published.	**1688**: In the 'Glorious Revolution' in England, the Catholic James II is replaced by the Protestant William and Mary, of the Dutch House of Orange.
		1690: Locke publishes his *Essay Concerning Human Understanding*.
		1692: Salem witchcraft trials.
1693: Bradley, James. Discovered the aberration of starlight and used this to determine the speed of light, arriving at a figure equivalent to 308,300 km/sec, close to the modern value of 299,792 km/sec.		**1697**: Birth of Canaletto.
		1698: Steam-powered pump to remove water from mines patented. Place Vendôme completed in Paris.
	1700s: Immanuel Kant suggests that distant nebulae might be complete star systems beyond the Milky Way.	**1701**: Yale University founded.
1701: Celsius, Anders. Suggested a temperature scale based on two fixed points: 0 degrees for the boiling point of water, and 100 degrees for the melting point of ice. Soon after his death this scale was reversed, so that the boiling point of water was 100 degrees, and the melting point of ice 0 degrees.	**1700s**: Johannes Kepler discovers the three laws of planetary motion, using data gathered by Tycho Brahe. These laws become known as 'Kepler's laws'.	**1702**: The first daily newspaper, the *Daily Courant*, published in London. Jethro Tull invents the machine drill, for planting seeds.
	1704: Isaac Newton publishes his *Opticks*.	**1704**: The *Boston Newsletter* is the first American newspaper.
	1705: Edmund Halley publishes his prediction of the return of the comet that now bears his name.	**1706**: Benjamin Franklin born.
		1707: Union of England, Wales and Scotland to form Great Britain.

Birth dates of scientists

Key dates in science

Key dates in history

1710: Completion of St Paul's in London.

1711: Wright, Thomas. Proposed, in his book *An Original Theory or New Hypothesis of the Universe*, that the Milky Way is a slab of stars, with the Sun inside it. Wright later described the Universe in terms of systems of spheres and rings of stars, which influenced Immanuel Kant's thinking about the nature of the Universe, and suggested that the rings of Saturn are not solid objects but are made up of many small bodies.

1712: First volume of John Flamsteed's star catalogue published.

1714: Gabriel Fahrenheit devises a mercury thermometer and uses the temperature scale that will later be named after him.

1718: Halley becomes the first person to realize that stars move across the sky, and are not really 'fixed'.

1721: Edmund Halley reads two papers to the Royal Society on the puzzle of why the sky is dark at night.

1711: David Hume born. Alexander Pope writes his *Essay on Criticism*.

1712: Newcomen develops an improved steam engine.

1714: British government offers a prize of £20,000 for a technique to find longitude at sea.

1715: First Jacobite rebellion.

1719: Daniel Defoe writes *Robinson Crusoe*.

1720: Collapse of the 'South Sea Bubble', a speculative venture in England.

1723: Adam Smith born.

1724: Michell, John. First person to come up with the notion of black holes.

1724: Kant, Immanuel. German philosopher who was also interested in cosmology and who, in an essay published in 1755, made one of the first suggestions that the planets formed from a cloud of material around the Sun.

1725: Flamsteed's complete catalogue published posthumously.

1725: Birth of Robert Clive ('Clive of India'). Vivaldi writes *The Four Seasons*.

1726: Jonathan Swift writes *Gulliver's Travels*.

1729: Titius, Johann Daniel. First put forward, in 1772, the mathematical relationship describing the

1729: James Bradley discovers aberration, an apparent shift in the position of a star caused by

spacing of the orbits of the planets, now known as Bode's Law.

1730: Messier, Charles. Compiled an important catalogue of faint astronomical objects, many of which are now known to be galaxies.

1736: Lagrange, Joseph Louis. Italian-born French mathematician whose contributions to the development of the theory of dynamics are important in calculations of the orbits of planets, satellites and moons.

1738: Sir William Herschel
1750: Caroline Herschel
1782: John Herschel
In 1781, William and Caroline Herschel discovered Uranus. They also extended Charles Messier's catalogue of 100 nebulae, eventually to more than 2,000 objects. John Herschel extended and revised their work.

1743: Paley, William. Argued that living things are far too complicated to have arisen by chance, and inferred that therefore they

the finite speed of light and the motion of the Earth in its orbit around the Sun. Through this he is able to determine the speed of light, arriving at a figure equivalent to 308,300 km/sec, close to the modern value of 299,792 km/sec.

1733: Chester Hall invents the achromatic telescope, using a lens made from two different kinds of glass, but it is not immediately a practicable tool.

1735: John Harrison builds his first marine chronometer in an attempt to win the prize offered by the British Board of Longitude for a way of keeping time accurately at sea.

1742: Anders Celsius invents the temperature scale which now bears his name (formerly the Centigrade scale).

1732: George Washington born.

1737: Göttingen University founded. William Hogarth paints *The Good Samaritan*.

1739: Royal Society of Edinburgh founded. Hume writes his *Treatise on Human Nature*.

Birth dates of scientists	Key dates in science	Key dates in history

had been created by the Christian God.

1744: Jean-Phillipe Loÿs de Chésaux estimates that there would be a star visible in every direction we look into space, provided that the Universe was (in modern terms) 10^{15} light-years or more across. To explain why this is not so, he suggests that empty space simply absorbs the energy in the light from distant stars, so that the light gets fainter and fainter as it travels through the Universe.

Pierre de Maupertius states the principle of least action, a formal version of the idea that objects (including light rays) follow the quickest possible paths.

1745: Second Jacobite rebellion.

1746: Leonhard Euler uses Christiaan Huygens' wave theory of light to explain refraction.

1746: Princeton University founded.

1747: Samuel Johnson's dictionary published.

1748: First blast furnace constructed at Bliston in England.

1749: Laplace, Pierre Simon, Marquis de. Updated Newton's investigations of planetary movements, to account for perturbations in the strict elliptical orbits propounded by Kepler. One of the first people to consider the possibility of black holes.

1749: Benjamin Franklin invents the lightning rod.

1749: Bode, Johannes Elert. Best known for the law named after him, although he did not discover the law, but popularized the earlier work of Johann Titius.

Birth dates of scientists	Key dates in science	Key dates in history
	1750: Thomas Wright's *An Original Theory and New Hypothesis of the Universe* published.	About **1750**: Population of China reaches 225 million.
	1752: The Gregorian calendar is introduced to Britain and some other parts of the world. To catch up with the Gregorian calendar, 11 days were omitted, so that the day after 3 September became 14 September. Benjamin Franklin carries out his famous kite experiment.	**1751**: Calendar reform in Britain makes 1 January the official first day of the year.
	1753: George Richmann is killed by lightning while copying Franklin's kite experiment.	
	1755: Immanuel Kant suggests, in his book *Universal Natural History and Theory of the Heavens*, that the planets condensed out of a cloud of primordial matter.	**1755**: University of Moscow founded. **1756**: Birth of Wolfgang Amadeus Mozart.
1758: Olbers, Heinrich Wilhelm Matthäus. Publicized the puzzle of why the sky is dark at night, developed an improved method of calculating the orbits of comets, and discovered two of the minor planets.	**1758**: Halley's Comet reappears, as predicted. **1760**: Daniel Bernoulli discovers that electricity obeys an inverse square law similar to the law of gravity. **1762**: James Bradley completes a catalogue of 60,000 stars, shortly before his death.	**1758**: The 'Imperial' system of weights and measures formally established in Britain. **1761**: Rousseau publishes *La nouvelle Héloïse*. John Harrison's 'Number Four' chronometer taken to the West Indies under test. **1763**: Boston Massacre. **1764**: Richard Arkwright patents his spinning jenny.

Birth dates of scientists	Key dates in science	Key dates in history

1766: Wollaston, William. Physician and physicist who first noticed dark lines in the solar spectrum, although he did not appreciate their significance at the time.

1766: 'Bode's Law' of planetary orbits proposed by Johann Titius.

1765: James Watt develops an improved steam engine. John Harrison receives the first half of his prize.

1767: Pond, John. The sixth Astronomer Royal, he made many much needed improvements to both the equipment and the observing techniques used at the Royal Greenwich Observatory, and introduced (in 1833) the practice of lowering a 'time ball' on a pole on top of the Observatory at 1 p.m. each day – the first public time signal.

1767: John Michell suggests that stars which appear close together on the sky are really physically associated in space, and not the result of a chance juxtaposition at quite different distances along the line of sight – binary stars hypothesized.

1768: Publication of the *Encyclopaedia Britannica* starts, initially in weekly instalments.

1769: von Humbolt, Friedrich. German explorer who publicized Samuel Schwabe's discovery of the sunspot cycle in his book *Kosmos*, in 1851.

1769: Transit of Venus across the face of the Sun observed by (among others) Captain James Cook, in Tahiti.

1769: First meeting of the American Philosophical Society.

1770: Ludwig van Beethoven born.

1771: The first major catalogue of nebulae, compiled by Charles Messier, is published. Messier went on to update his original list of 45 objects twice, adding 23 more objects in 1780, and finishing his final list of 105 objects in 1781.

1771: Discovery of oxygen. Tobias Smollett writes *The Expedition of Henry Clinker*.

1772: Lagrangian points, points of stability in the orbital plane of a two-body system, first identified theoretically by Joseph Lagrange.

545

Birth dates of scientists	*Key dates in science*	*Key dates in history*

Bode's Law popularized by Johann Bode – a mathematical relationship which relates the distances of the planets from the Sun to a simple numerical sequence.

1773: From studies of the 'proper motion' of stars, William Herschel infers that the Sun and Solar System are moving in the direction of the constellation Hercules.

1773: John Harrison receives the second half of his prize at the age of 80, after the British Board of Longitude is told by King George III to stop delaying the award. Boston Tea Party.

1774: Nevil Maskelyne determines the mass of the Earth by measuring the amount by which a mountain deflects a plumb line from the vertical.

1775: James Watt's steam engine patented.

1776: Pierre Simon de Laplace claims that, if all the forces acting on all objects at any one time were known, then the future would be completely predictable.

1776: American Declaration of Independence. Adam Smith's *Wealth of Nations* published.

1777: Gauss, Karl Friedrich. German mathematician and astronomer who pioneered the development of non-Euclidean geometry (important in the general theory of relativity) and did important work in calculating orbits of planets, as well as extensive work in mathematics and physics.

1777: First performance of Mozart's Concerto no. 9.

1778: James Cook discovers Hawaii.

1779: World's first iron bridge built in Coalbrookdale, England.

1781: Uranus, the seventh planet out from the Sun, is discovered by William Herschel.
Charles Messier completes a catalogue of more than 100 star clusters and other objects that might be mistaken for comets.

1781: James Watt patents a system for developing rotary motion from a steam engine.

Birth dates of scientists	Key dates in science	Key dates in history
	1783: John Michell becomes the first person to suggest that there might exist 'dark stars' whose gravitation is so strong that light cannot escape from them, presenting his ideas to the Royal Society. Basing his calculations on Newton's theory of gravity, and on the corpuscular theory of light, he assumed that the particles of light would be affected by gravity in the same way as any other objects.	**1783**: Montgolfier brothers build and fly hot-air balloons. On 21 November, Jean de Rozier and François Laurent, in a Montgolfier balloon, become the first humans to fly.
1784: Bessel, Friedrich. Studied the regular 'side to side' movement of Sirius, and showed that it could be explained by the gravitational influence of a companion star orbiting Sirius every 50.09 years.	**1784**: John Goodricke identifies Delta Cephei as a variable star.	**1784**: Benjamin Franklin invents bifocal spectacles.
	1785: The term 'planetary nebula', defining a nebula associated with a star, and looking like a disc through a telescope or in astronomical photographs, is first used by William Herschel. Herschel also gives the first approximately accurate description of the shape of the Milky Way, but mistakenly locates the Sun near the centre.	**1785**: First balloon crossing of the English Channel. Seismograph invented.
	1786: Herschel publishes his catalogue of nebulae, which is later expanded into the New Galactic Catalogue.	**1786**: First experiments with gas lighting.
1787: von Fraunhofer, Joseph. The first person to study in detail, and under intense magnification, the rainbow pattern produced by passing light through a	**1787**: Oberon, one of the larger moons of Uranus, is discovered by William Herschel.	

Birth dates of scientists	Key dates in science	Key dates in history
prism. To his surprise, he discovered that there are many dark lines in the spectrum of white light from the Sun. Fraunhofer counted 574 lines in the solar spectrum, and found many of the same lines in light from Venus and from many stars.	**1787**: Titania, the largest moon of Uranus, is discovered by William Herschel.	
1789: Schwabe, Samuel. The first person to measure the periodicity of the solar cycle, who announced his discovery of what he thought was a 10-year cycle of sunspot activity in 1843.	**1789**: Mimas, one of the moons of Saturn, is discovered by William Herschel. Herschel completes his large reflecting telescope, with a mirror 49 inches (124.5 cm) across.	**1789**: The storming of the Bastille on 14 July triggers the French Revolution.
	1790: Herschel discovers planetary nebulae.	
		1791: Metric system introduced in France; it is officially adopted in 1795. Michael Faraday born.
	1794: Ernst Chladni proves that meteors come from beyond the Earth's atmosphere.	**1792–1815**: Napoleonic Wars.
	1796: Pierre Laplace suggests, independently of John Michell, that there might exist 'dark stars' whose gravitation is so strong that light cannot escape from them. He also proposes the 'nebular hypothesis' for the origin of the Solar System.	**1796**: Edward Jenner develops vaccination for smallpox.
		1797: First use of iron railways, for horse-drawn waggons.
	1798: Henry Cavendish determines the mass of the Earth, establishing that it has an average density 5.5 times that of water.	**1798**: Thomas Malthus publishes (initially anonymously) his *Essay on the Principle of Population*.
		1799: Discovery of the Rosetta Stone.

Birth dates of scientists

1803: Doppler, Christian Johann. Predicted what is now known as the Doppler effect, in 1842.

1809: Poe, Edgar Allan. Speculated that the reason why the sky at night is dark – although it was believed that the Universe was infinite in extent, and therefore, logically, that the sky should be a mass of solid stars – is that the distances involved are so immense that the light from many stars has yet to reach us.

1811: Le Verrier, Urbain Jean Joseph. Predicted, in 1846, the existence of a planet beyond Uranus, by studying perturbations in the orbit of Uranus.

Key dates in science

1800: Herschel discovers infra-red radiation.

1801: First known asteroid, Ceres, discovered.

1802: Thomas Young publishes the first of his papers on the wave theory of light.
   William Wollaston becomes the first person to notice dark lines in the solar spectrum, although he does not appreciate their significance at this time.

1807: Young introduces the concept, and the word, 'energy'.

Early **1800s**: With the problem of chromatic aberration in refracting telescopes overcome by the invention of the achromatic lens, improved lenses now became practicable, and during most of the 19th century refractors become the mainstay of optical astronomy.

Key dates in history

1800: Richard Trevithick builds a high-pressure steam engine. The following year, he builds a steam-powered vehicle. World population now some 870 million.

1803: Successful trials of Robert Fulton's steam-powered boat on the Seine.

1804: Napoleon becomes Emperor of France.

1805: Battle of Trafalgar.

1807: First use of gas to light London streets. Improved steamboat tested in the East River off New York by Robert Fulton.

1808: Humphrey Davy invents the electric arc light. Richard Trevithick builds a passenger railway in London.

1809: Charles Darwin born.

1810: Foundation of the University of Berlin.

Birth dates of scientists

Key dates in science

Key dates in history

1811: Bunsen, Robert. Worked with Gustav Kirchoff in formulating the basic principles of spectroscopy. One of Bunsen's assistants modified a device invented by Michael Faraday, coming up with what is now known as the Bunsen Burner. This provides a clean, hot flame in which different substances can be heated until they glow, radiating light at their own characteristic spectral wavelengths, which can be studied and analysed.

1812: Napoleon invades Russia.

1814: Ångström, A. J. Swedish pioneer of spectroscopy.

1814: Joseph von Fraunhofer becomes the first person to study, in detail and using intense magnification, the rainbow pattern produced by passing light through a prism. To his surprise he discovers (independently of Wollaston) that there are many dark lines in the spectrum of white light from the Sun. Fraunhofer soon counts 574 lines in the solar spectrum, and finds many of the same lines in light from Venus and from many stars.

1814: George Stephenson's first steam locomotive starts work.

1815: Battle of Waterloo. First steam-powered warship built in America. Humphrey Davy invents the coal miners' safety lamp.

1816: Wolf, (Johann) Rudolf. Made a study, in the 1850s, of solar observations going back to the 17th century, and showed that there had been a solar cycle of activity with an average period of 11.1 years since that time.

1816: Augustin Fresnel develops his version of the wave theory of light.

1816: Rossini writes *The Barber of Seville*.

Birth dates of scientists	Key dates in science	Key dates in history
1818: Mitchell, Maria. The first American woman astronomer of note; she discovered a comet in 1847.	**1818**: Edinburgh Observatory founded.	
1819: Adams, John Crouch. In 1845, predicted the existence of a planet beyond Uranus by studying perturbations in the orbit of Uranus.		**1819**: Paddle steamer *Savannah* crosses the Atlantic. Birth of Queen Victoria.
1820: Roche, Edouard Albert. Calculated, in 1850, the Roche limit for a moon in orbit around a planet, and later carried out a mathematical analysis of the nebular hypothesis.	**1820**: Royal Astronomical Society founded. **1821**: Catholic Church lifts its ban on teaching the Copernican theory.	**1821–32**: Greek wars of independence (from Ottoman Empire).
1822: Spörer, Gustav Friedrich Wilhelm. Discovered the way in which sunspots appear at different solar latitudes at different stages of the solar cycle. His historical studies independently revealed the dearth of sunsports in the 17th century now known as the Maunder minimum; a similar dearth of sunspots in the 15th century is now known as the Spörer minimum.	**1822**: Edinburgh Observatory receives the 'Royal' title from George IV.	**1823**: Charles Macintosh invents a waterproof coat.
1824: Kelvin, (Baron Kelvin of Largs, William Thomson). Made many contributions to thermodynamics and the theory of electromagnetic radiation, and was responsible for the success of the first transatlantic cable communications		

system. Kelvin's main contribution to astronomy dealt with the ages of the Earth and the Sun. Also proposed (in 1848) an 'absolute' scale of temperatures.

1824: Kirchoff, Gustav. Explained the Fraunhofer lines as due to absorption of light by different elements present in the Sun's atmosphere, and went on to formulate the basic principles of spectroscopy, working together with Robert Bunsen.

1825: First passenger steam train.

1826: Riemann, (Georg Friedrich) Bernhard. First person to develop a comprehensive mathematical description of curved space (non-Euclidean geometry), providing the mathematical framework later used by Einstein in his general theory of relativity.

1826: Heinrich Olbers publishes a landmark paper outlining the problem with having an infinitely large universe, and a dark night sky. He reaches a conclusion similar to that reached by de Chésaux – that the light from distant stars is absorbed in a thin gruel of material between the stars.

1826: Delacroix paints Greece in the *Ruins of Missolonhi*.

1829: Pogson, Norman. Proposed the magnitude scale, also known as the Pogson scale, used by astronomers as a measure of the brightness of astronomical objects.

1830: Joseph Henry discovers the principal of the dynamo, but does not publish his discovery. He publishes after the same discovery is announced by Michael Faraday.

1830: Spread of English-speaking Americans west begins.

1831: Maxwell, James Clerk. His many achievements included developing the three-colour theory of colour vision, and the equations which describe the

1831: Michael Faraday and Joseph Henry independently discover electromagnetic induction.

Birth dates of scientists	Key dates in science	Key dates in history
behaviour of electromagnetic radiation.		
		1832: Reform Bill extends the franchise in Britain.
		1833: Karl Gauss and Wilhelm Weber develop an electric telegraph which sends signals between two stations 2 km apart.
	1834: John Herschel begins his survey of the southern skies. Earliest version of the second law of thermodynamics developed by Benoit Clapeyron.	**1834**: Louis Braille perfects system of reading for the blind.
1835: Schiaparelli, Giovanni Virginio. Astronomer who is best known for his studies of Mars. In 1877 he described the markings on the surface of Mars as 'canali'. Mistranslated as 'canals', the description led to wild speculation about the possible existence of intelligent life on Mars.	**1835**: Halley's Comet makes its second return since Halley's death. Gutave Coriolis describes the force which now bears his name.	**1835**: Samuel Colt patents his revolver. John Constable paints *The Valley Farm*.
1837: Draper, Henry. One of the pioneers of photographic spectroscopy. After his death, Draper's widow established a fund to continue his work; this eventually led to the publication of the Henry Draper Catalog, which lists stars according to their spectroscopic properties.		**1837**: Samuel Morse patents his electric telegraph.
1838: Mach, Ernst. Philosopher and physicist	**1838**: The first stellar distance determined by	

whose ideas influenced Einstein's thinking at the time he was developing the general theory of relativity.

1838: Morley, Edward Williams. Carried out an experiment with Albert Michelson which famously failed to find any evidence for an 'ether' through which light was transmitted.

parallax, the distance to 61 Cygni, is published. Parallax is the apparent movement of an object across the sky when it is seen from two different points, and can be used to calculate the distance to the object by triangulation.

1839: Harvard College Observatory founded – the first official observatory in the United States.

1841: Friedrich Bessel shows, using observations going back to 1834, that the regular 'side to side' movement of Sirius can be explained by the gravitational influence of a companion star orbiting Sirius every 50.09 years.

1842: The Doppler effect first predicted by Christian Doppler. This effect shows how a change may be made in the frequency of light, or in the pitch of a sound, by the motion of the object emitting the light or making the sound.
  Julius Mayer is the first person to state the law of conservation of energy.

1843: Samuel Schwabe announces his discovery of what he thinks is a 10-year cycle of sunspot activity.

1839: Louis Daguerre describes a technique for making photographs. Charles Goodyear develops a technique for 'vulcanizing' rubber.

1843: First tunnel under the Thames opened.

1844: Friedrich Engels writes *The Condition of the Working Class in England*.

1844: Boltzmann, Ludwig. Suggested that the Universe might be a gigantic statistical freak. In a uniform Universe, according to Boltmann's interpretation of Poincaré's

Birth dates of scientists	Key dates in science	Key dates in history

work, it will very occasionally happen that all the particles in one part of the universe will be moving in just the right way to create stars, galaxies, or a Big Bang. A region of order would, by chance, form out of some of the more permanent state of disorder.

1845: Clifford, William Kingdom. One of the first people to suggest that non-Euclidean geometry discussed by Riemann might be the correct description of the Universe in terms of curved space.

1845: Röntgen, Wilhelm. Discoverer of X-rays.

1845: a 73-inch reflecting telescope is built at Birr Castle, in central Ireland, by the Earl of Rosse. This telescope is for many years the most powerful in the world.

The first galaxy (then known as nebula) in which spiral structure is seen is discovered by Lord Rosse. This is the Whirlpool galaxy, a particularly striking and photogenic example of a disc galaxy with well-defined spiral arms, and is also known by its catalogue numbers as M51 or NGC 5194.

Hippolyte Fizeau and Leon Foucault obtain the first good photographs of sunspots, using the daguerreotype technique.

Faraday suggests that light is a form of electromagnetic wave.

1845: First publication of *Scientific American*.

1846: Lockyer, (Joseph) Norman. Discovered helium from its spectroscopic fingerprint in sunlight. He was the first scientist to notice the link between solar activity and weather cycles, and

1846: Neptune discovered, on the basis of predictions made by John Adams and Urbain Le Verrier.

Triton, the largest moon of Neptune, discovered.

1846: Foundation of the Smithsonian Institution in Washington, DC.

1846–51: Potato famine in Ireland.

founded the journal
Nature.

1846: Pickering, Edward
Charles. Director of the
Harvard College
Observatory from 1876 to
1918, Pickering dominated
American astronomy in the
last quarter of the 19th
century, encouraging many
young scientists to take up
astronomy, including
many women. He devoted
an enormous effort to
cataloguing, and
pioneered the
development of the colour
index technique for
classifying stars.

1847: The Gould Belt, a
belt of stars and gas which
is tilted at about 16 degrees
to the Galactic plane and
contains a large number of
giant stars, is first noted by
Sir John Herschel.
　James Joule
independently discovers
the law of conservation of
energy.

1848: Hyperion, one of the
moons of Saturn, is
discovered by W. C. Bond.
　The Wolf number, a
parameter used as a
measure of the Sun's
activity, is introduced by
Rudolf Wolf. The definition
of the Wolf number is
based on counting the
number of sunspots and
the number of groups of
sunspots visible on the
surface of the Sun.
　Julius Mayer calculates
that the Sun would cool in
only 5,000 years if it had
no source of energy.
　Hippolyte Fizeau
suggests that the Doppler
effect should also apply to
light, and predicts a
redshift in light from an
object moving away from
the observer.
　Edgar Allan Poe
speculates that the sky

1848: Karl Marx and
Friedrich Engels produce
the *Communist Manifesto*.

Birth dates of scientists	*Key dates in science*	*Key dates in history*
	might be dark at night, even in a Universe that is infinite in extent, if the distances involved are so immense that the light from many stars has yet to reach us.	
	1849: Fizeau measures the speed of light to within 5 per cent of the accepted modern value. William Thomson (later Lord Kelvin) coins the term 'thermodynamics'.	**1849**: California gold rush.

1851: Fitzgerald, George Francis. Best known for his suggestion that moving objects are shortened in the direction of their motion (the Fitzgerald contraction).
 	
1851: Maunder, Edward Walter. Found, from historical records, that there was hardly any sunspot activity between 1645 and 1715, a time now known as the Maunder minimum.
	1851: Jean Foucault makes use of a long pendulum to demonstrate the rotation of the Earth – such a pendulum becomes known as Foucault's pendulum. As the Earth rotates, the pendulum remains swinging in the same plane. This means that the pendulum appears to drift slowly, relative to the ground.
	  Friedrich von Humboldt publicizes Samuel Schwabe's discovery of the solar cycle, in his book *Kosmos*.
	  William Thomson develops the concept of absolute zero of temperature and shows that it corresponds to −273 °C.
	  Umbriel, one of the moons of Uranus, is discovered.
	
	Soon after **1851**: Rudolf Wolf carries out his first studies of the sunspot cycle, finding a period of 11.1 years.
		1851: Crystal Palace opened by Queen Victoria.

Birth dates of scientists	Key dates in science	Key dates in history
1852: Michelson, Albert Abraham. Made many determinations of the speed of light and carried out an experiment with Edward Morley which famously failed to find any evidence of an 'ether' through which light was transmitted.		
1853: Lowell, Percival. Famous for his ideas about life on Mars, he also founded the Lowell Observatory and was instrumental in encouraging Vesto Slipher to carry out the work that led to the discovery of the redshifts of galaxies.		
1854: Poincaré, Henri. Showed that an 'ideal' gas, trapped in a box, must eventually pass through every possible arrangement of particles that is allowed by the laws of thermodynamics. However, even a small box of gas might contain 10^{22} atoms, and it would take that many atoms a time much longer than the age of the Universe to pass through every possible arrangement.	**1854**: Hermann von Helmholtz suggests that the Sun is kept hot by gravitational energy, released as it shrinks slowly under its own weight.	**1854**: First electric telegraph link between London and Paris. **1854–6**: Crimean War.
	1855: William Parsons observes the spiral structure of some 'nebulae'.	**1855**: Louvre opens in Paris.
1856: Thomson, J. J. Worked at the Cavendish Laboratory in Cambridge, and found that a form of radiation produced by atoms could be explained as a stream of tiny charged particles, now called electrons, which broke off	**1856**: Norman Pogson proposes a new form of scale for measuring the brightnesses of astronomical objects.	**1856**: Henry Bessemer invents a cheap process for manufacturing steel.

Birth dates of scientists	*Key dates in science*	*Key dates in history*

from the atoms themselves. Until Thomson's work, it had been believed that atoms were the fundamental building blocks of matter.

1857: Hertz, Heinrich. German physicist who first produced long-wavelength radiation artificially, in 1888.

1857: Leon Foucault begins to manufacture glass mirrors coated with a thin film of silver to use in astronomical telescopes.

1858: Planck, Max Ernst Karl Ludwig. Proposed, in 1900, that electromagnetic radiation can only be emitted or absorbed in definite units, which he called quanta. This explained the nature of radiation from a black body, and laid the foundations for quantum theory.

1859: Maxwell publishes a paper proving that the rings of Saturn must be made up of many small particles, each in their own orbit around the planet, and cannot be solid objects.

Gustav Kirchoff and Robert Bunsen develop the use of spectroscopy for chemical analysis.

1858: First Atlantic telegraph cable.

1859: Construction of the Suez Canal begins; it will be completed in 1869. Publication of the *Origin of Species*. First internal combustion engine (using gas as its fuel) developed by Jean Lenoir. First oil well drilled in Titusville, PA.

1859: Arrhenius, Svante August. Suggested the Earth might have been 'seeded' with life from space, in the form of micro-organisms riding on dust particles.

1860s: The development of spectroscopy shows that some nebulae are in fact clouds of gas.

1860: First vehicle driven by an internal combustion engine developed by Jean Lenoir.

1861: Unification of Italy. First telegraph linking San Francisco to New York. American Civil War begins.

1862: Charlier, Charles. Proposed the hierarchical model of the Universe – the idea that matter in the Universe is grouped in successively larger clumps, like the way Russian dolls nest inside each other. Stars group together to form galaxies, galaxies are grouped into clusters, clusters of galaxies form superclusters, and so on.

1862: The companion star to Sirius, predicted by Friedrich Bessel, is found by American telescope maker Alvan Clark. This is the first white dwarf to be discovered, and is known as Sirius B. A white dwarf is a star with about the same mass as the Sun, but occupying a volume about the same as that of the Earth.

1862: Richard Gatling invents the machine gun.

559

1863: Cannon, Annie Jump. American astronomer who carried out pioneering work in the classification of stars using spectroscopy.

1863: Wolf, Maximilian Franz Joseph. Specialized in photographic and spectroscopic observations, made early observations of spiral nebulae, and discovered 232 asteroids.

1864: Minkowski, Hermann. One of Einstein's teachers, Minkowski came up with the idea of four-dimensional space-time as the stage on which physical interactions take place.

1864: Wien, Wilhelm. Awarded the Nobel Prize, in 1911, for his work on the laws governing the radiation of heat.

1868: Leavitt, Henrietta Swan. Worked at the Harvard College Observatory with Edward Pickering, and discovered the period–luminosity relation of Cepheid variables.

1871: Rutherford, Ernest. Showed that the positive charge inside atoms –

1863: William Huggins uses spectroscopy to show that the same chemical elements exist in the atmospheres of stars as exist on Earth.

1864: James Clerk Maxwell publishes a set of equations describing how the electric and magnetic fields can be thought of as oscillating at right angles to one another as they move together through space. The changing electric field, he suggests, produces the changing magnetic field, and the changing magnetic field produces the changing electric field. This leads him to predict the existence of radio waves.

1863: National Academy of Sciences founded in USA.

1865: Lewis Carroll (Charles Lutwidge Dodgson) writes *Alice's Adventures in Wonderland*. American Civil War ends.

1867: Publication of *Das Kapital*.

1869: First publication of the science journal *Nature*. First railway line linking the west and east coasts of the USA completed.

1870: Vladimir Ilyich Lenin (originally Ulyanov) born.

Birth dates of scientists	*Key dates in science*	*Key dates in history*

logically necessary,
because of the known
negative charge of
electrons – is concentrated,
together with most of the
mass of an atom, in a tiny
central nucleus.

1872: de Sitter, Willem.
One of the first people to
apply Einstein's equations
of the general theory of
relativity to provide a
mathematical model of the
Universe.

1872: Sarah Bernhardt
begins to work with the
Comédie-Française in Paris.

1873: Schwarzschild, Karl.
Discovered the solution to
Albert Einstein's equations
of the general theory of
relativity that describes
what are now known as
black holes.

1873: Richard Proctor
suggests that the craters
on the Moon were formed
by meteorites.

1873: Hertzprung, Ejnar.
Published the first
presentation of what
became known as the
Hertzprung–Russell
diagram in 1905, and later
worked with Karl
Schwarzschild.

1874: Lyman, Theodore.
Studied the nature of the
spectral features in
hydrogen, now known as
the Lyman lines.

1874: Potsdam
Observatory founded.
Cavendish Laboratory
completed.

1874: First exhibition of
Impressionist paintings in
Paris. Levi Strauss invents
blue jeans with rivets.

1875: Slipher, Vesto
Melvin. The first person to
identify the redshifts in the
light from what are now
known to be galaxies. This
work, carried out in the
years up to 1925, paved
the way for Edwin Hubble's
discovery that the Universe

1875: Georges Bizet writes
Carmen.

Birth dates of scientists	*Key dates in science*	*Key dates in history*

is expanding. He also instigated the successful search for Pluto at the end of the 1920s.

1876: Adams, Walter. American astronomer who obtained a spectrum showing that Sirius B, although far fainter, is actually as hot as its companion star, Sirius. In order to be both hot and faint, the star must be small – little bigger than the Earth. It was the first white dwarf to be discovered.

1876: The Meudon Observatory founded in Paris.

1877: Jeans, James Hapwood. Although Jeans' own contributions to astrophysics were rather technical and specialized, he also wrote many popular works and made radio broadcasts about the Universe, and was largely responsible for publicizing the idea of the heat death of the Universe.

1877: The two moons of Mars found by Asaph Hall. Giovanni Schiaparelli reports that he has seen channels (mistranslated as 'canals') on Mars.

1877: Russell, Henry Norris. Developed the technique of plotting the absolute magnitude of stars against their colour, independently of the work of Ejnar Hertzsprung.

1878: Telephone invented.

1879: Schmidt, Bernhard Voldemar. Invented the Schmidt camera in 1930.

1879: Einstein, Albert. Developed both the special theory of relativity and the general theory of

1879: Forerunner of the modern electric light bulb invented independently by Joseph Swan in England and Thomas Edison in America. Lev Davidovich Bronstein (Leon Trotsky) born. First performance of

Birth dates of scientists	Key dates in science	Key dates in history

relativity, as well as making major contributions to the development of quantum theory.

1879: Milankovitch, Milutin. Developed the theory that the cycle of Ice Ages on Earth is caused by changes in the Earth's orbit which alter the balance of heat between the seasons.

1882: Eddington, Sir Arthur Stanley. The first astrophysicist – carried out the crucial test of Einstein's general theory of relativity, developed the application of physics to an understanding of the structure of stars, and was a great popularizer of science in the 1920s and 1930s.

1882: Goddard, Robert. Developed the first liquid-fuelled rocket in 1926.

1885: Shapley, Harlow. Used observations of variable stars in globular clusters to map our Galaxy and determine its size accurately, working at the Mount Wilson Observatory in the second decade of the 20th century.

Eugene Onegin by Tchaikovsky. Mary Cassatt paints *The Cup of Tea*.

1880s: Algol becomes the first eclipsing binary star to be identified, using spectroscopic techniques.

1880s: Bishop Berkeley's suggestion that all motion is relative, and must be measured against something, is taken up by Ernst Mach, who suggests that if we want to explain the equatorial bulge of the Earth as due to centrifugal forces, 'it does not matter if we think of the Earth as turning round on its axis, or at rest while the fixed stars revolve around it' – it is the relative motion that is responsible for the bulge. This principle is later to be dubbed 'Mach's Principle' by Einstein.

1884: An international committee establishes the meridian through the Royal Greenwich Observatory as the 'prime meridian' from which longitude is to be measured.

1880: Ned Kelly captured.

1881: First practical electric generation and distribution system.

1882: Electric light introduced in New York. Franklin Delano Roosevelt born.

1883: System of four time zones officially adopted in the USA. Gottlieb Daimler invents the first version of the modern internal combustion engine, and tests it in a boat.

1885: Automobile and motorbike invented.

Birth dates of scientists	*Key dates in science*	*Key dates in history*

1887: Michelson–Morley experiment carried out, in an attempt to detect the motion of Earth through the ether, by measuring differences in the speed of light determined along the line of the Earth's motion and at right angles to that line. These experiments show, through demonstrating that no effects attributable to the motion of the Earth can be seen in any measurements of the speed of light, that the ether does not exist.

Joseph Lockyer's spectroscopic study of the Sun leads to the discovery of helium.

1888: Friedmann, Aleksandr Aleksandrovich. Worked in practical applications such as hydromechanics and meteorology, but is best remembered for his solutions to Albert Einstein's equations of the general theory of relativity.

1888: The long-wavelength radiation predicted by Maxwell is first produced artificially by the German physicist Heinrich Hertz.

First measurements of the velocities of stars using the Doppler effect.

The Lick Observatory, the first observatory in the United States to be built on a mountain top, is completed.

Late 19th century: Henri Poincaré shows that an 'ideal' gas, trapped in a box, must eventually pass through every possible arrangement of particles that is allowed by the laws of thermodynamics. However, even a small box of gas might contain 10^{22} atoms, and it would take

1888: Wiliam Burroughs patents his adding machine. Oscar Wilde writes *The Happy Prince*.

Birth dates of scientists | *Key dates in science* | *Key dates in history*

1889: Hubble, Edwin Powell. Proved that many objects classified as nebulae are other galaxies; discovered the relationship between redshift and distance, and inferred that the Universe is expanding.

1889: Fowler, Ralph. Made pioneering theoretical investigations into the structure of stars, working with Edward Milne.

that many atoms a time much longer than the age of the Universe to pass through every possible arrangement of atoms.

Ludwig Boltzmann suggests that the Universe might be a gigantic statistical freak. It is possible, he suggests, that we are simply a chance collection of atoms – a state of temporary order within a more permanent disorder.

1890s: Edward Pickering establishes the first standard system of measuring the temperature of a star by taking measurements at two separate wavelengths of light from that star, the comparison of which enabled him to work out the star's temperature.

1890: Smithsonian Astrophysical Observatory founded by the Smithsonian Institution.

1889: Eiffel Tower completed.

1891: Humason, Milton Lasell. Worked with Edwin Hubble on the redshift survey which led to the discovery that the Universe is expanding.

1891: Chadwick, James. Discovered the neutron, a particle with no electric charge and almost the same mass as the proton.

1892: Compton, Arthur. Carried out pioneering experiments with X-rays.

1893: Baade, (William Heinrich) Walter.

1893: While studying old records of solar activity,

1893: Mormon Temple completed in Salt Lake

Birth dates of scientists

Responsible for a major revision in the cosmic distance scale in the 1940s and early 1950s.

1893: Öpik, Ernst. First suggested the existence of the Oort cloud of comets, an idea later to be developed by Jan Hendrik Oort.

1894: Lemaître, Georges Édouard. Developed the first version of what became the Big Bang description of the Universe.

1896: Milne, Edward Arthur. Made pioneering theoretical investigations into the structure of stars; in the early 1930s, showed, with William McCrea, that the expanding Universe predicted by the general theory of relativity could actually be described very well in the context of classical mechanics and Isaac Newton's theory of gravity.

1896: Maksutov, Dmitri Dmitrievich. Expert in astronomical optics who designed and built several large reflecting telescopes, and invented the Maksutov telescope.

Key dates in science

Edward Maunder discovers the 'Maunder minimum'.
Wilhelm Wien discovers the relationship between the wavelength of maximum intensity of 'black body' radiation and its temperature; this enables astronomers to measure the temperatures of stars.

1894: The Lowell Observatory is founded by Percival Lowell.

1895: X-rays discovered by Wilhelm Röntgen.

1896: First photographic atlas of the Moon published by Lick Observatory.
Radioactivity discovered.

1897: The Yerkes Observatory, near Chicago, is founded using funds provided by the millionaire Charles T. Yerkes. The 40-inch refracting telescope

Key dates in history

City. Zip fastener invented. Mao Tse-tung born.

1894: Manchester Ship Canal completed. Guglielmo Marconi develops a radio transmitter that will ring a bell at a distance of 10 metres.

1895: Lumière brothers present moving pictures to the public.

1896: Discovery of radioactivity.

1897: Camille Pisarro paints *Boulevard des Italiens*.

Birth dates of scientists	*Key dates in science*	*Key dates in history*

Key dates in science

remains the largest of its kind in the world.
J. J. Thomson discovers the electron.

1898: Zwicky, Fritz. Made many contributions to astrophysics, studied supernovae, compact galaxies and clusters of galaxies, and was one of the few people to take the idea of neutron stars seriously before the discovery of pulsars.

1898: Atkinson, Robert D'escourt. Atkinson's chief contribution to science was to show, with Fritz Houtermans, how energy could in principle be produced inside stars by fusion.

1898: Phobos, the innermost of the two moons of Mars, is discovered by Asaph Hall.
Radioactivity named by Marie Curie.

1899: Ernest Rutherford discovers that there are two forms of radioactivity, which he calls alpha and beta radiation; a third form, gamma radiation, is identified later.

End of the 19th century: Techniques for producing large silvered glass mirrors become available, and these form the basis of the great optical telescopes of the 20th century.

1900: Oort, Jan Hendrik. One of the pioneers of the use of radio astronomy to study the Galaxy, using the characteristic emission of hydrogen at a wavelength of 21 cm. In the early 1950s, he proposed the existence of what is now known as the Oort cloud of comets, developing this from an idea put forward in the 1930s by Ernst Öpik.

1900: Pauli, Wolfgang. Made many contributions to the development of quantum theory, including the Pauli exclusion principle, which is crucial

1900: Max Planck makes the discovery that the nature of black body radiation can be explained only if light is emitted or absorbed by atoms in discrete quanta (photons). The essential feature of Plank's discovery is that there is a limit to how small a change in energy an atom can experience – this corresponds, in modern terminology, to the emission or absorption of a single photon.

Early 20th century: Charles Charlier proposes the hierarchical model of the

1900: First offshore oil wells. Ferdinand von Zeppelin builds his first dirigible airship.

to an understanding of white dwarfs and neutron stars. He was also the first to propose the existence of the neutrino.

1900: Payne-Goposchkin, Cecilia Helena. Worked with Harlow Shapley at Harvard and became, in 1956, the first female professor of astronomy at Harvard University. Her observations in the 1920s helped to establish the dominance of hydrogen and helium in the composition of stars.

Universe; this model suggests that the Universe is composed of larger and larger clusters, each containing the other many times. So, for example, we see stars grouping together to form galaxies, galaxies grouping into clusters, clusters into superclusters. This theory, were it correct, would involve the Universe extending into infinity.

Theodore Lyman studies the nature of the spectral features in hydrogen now known as the Lyman lines, or Lyman series.

Mount Wilson Observatory is established by George Hale.

1901: Fermi, Enrico. Took up Wolfgang Pauli's idea of a 'new' particle, used to explain anomalous behaviour observed during beta decay, and coined the name 'neutrino' for this particle. He is best remembered as the leader of the group that built the first nuclear reactor, and is known in astronomy for his claim that the fact that the Earth has not been colonized by aliens 'proves' that we are the only civilization in the Galaxy.

1901: Annie Jump Cannon completes the Harvard Classification of stars.

1901: First electric typewriter. First transatlantic radio communication. Paul Gauguin paints *The Gold in their Bodies*.

1901: Heisenberg, Werner. Developed the uncertainty principle, during the 1920s. This shows that there is an intrinsic uncertainty associated with knowledge of position and momentum in the quantum world.

1902: New York's 'Flatiron' building completed.

1903: Wright brothers' first powered flight.

Birth dates of scientists

1904: Gamow, George
('Joe'). Made the first
calculations of conditions
in the Big Bang, predicted
the existence of the
background radiation, and
played a part in cracking
the genetic code of DNA,
the molecule of life.

1904: McCrea, Sir William
Hunter. Cosmologist,
mathematician and
astrophysicist who has
made fundamental
contributions to many
areas of 20th-century
science. Used
spectroscopy to show that
there is vastly more
hydrogen than any other
element in the Sun. Also
studied the evolution of
galaxies and the formation
of molecules in interstellar
clouds, and investigated
the Steady State
hypothesis.

1904: Oppenheimer,
(Julius) Robert. Worked on
quantum theory in the
1920s and 1930s, and is
best known for the part he
played in the production of
the atomic bomb – he was
the first director of the Los
Alamos Scientific
Laboratories, where the
bomb was developed.

1905: Kuiper, Gerard Peter.
Played a major part in
reviving the interest of
astronomers in the study of
the Solar System, and
contributed to the
development of NASA's

Key dates in science

1904: Hendrik Lorentz
develops the equations
now known as Lorentz
transformations, to
describe how
electromagnetic fields
would look to observers
moving at different
velocities.
    Lord Kelvin publishes his
thoughts on the problem of
the darkness of the night
sky.
    Mount Wilson
Observatory established.

1905: Einstein's special
theory of relativity unites
space and time in one
mathematical description,
dealing with the dynamical
relations between objects
moving at constant speeds

Key dates in history

1904: Construction of the
Panama Canal begins; it
will be completed in 1914.

1905: Special theory of
relativity.

Birth dates of scientists	*Key dates in science*	*Key dates in history*

programme of unmanned exploration of the planets.

1905: Jansky, Karl. An American radio engineer who was the first person to detect radio emission from the Milky Way, and therefore was arguably the first radio astronomer.

in straight lines. Einstein suggests that light (previously regarded purely as an electromagnetic wave) could be described in terms of particle-like quanta – what are now called photons. This is a key step in the development of quantum theory.

Ejnar Hertzprung publishes the first presentation of what becomes known as the Hertsprung–Russell diagram.

1906: Bethe, Hans Albrecht. Worked out the mechanisms by which stars derive their energy from nuclear fission reactions.

1906: William Morgan suggests that the Milky Way has a spiral structure.

1906: Great San Francisco Earthquake. Kellogg's cornflakes go on sale for the first time.

1906: Tombaugh, Clyde. Discovered Pluto.

1906: Bok, Bart. In 1947, Bok discovered small, dark, circular clouds of material in space, which show up against the background of stars or luminous clouds. These Bok globules are thought to be stars like our Sun in the process of forming.

1906: Gödel, Kurt. Austrian-born American mathematician who found a solution to Einstein's equations that permits time travel.

1906: Whipple, Fred Lawrence. Whipple's research concentrated on

Birth dates of scientists	*Key dates in science*	*Key dates in history*

understanding the evolution of the Solar System, and he is best known for his proposal (in 1950) of the 'dirty snowball' model of comets.

1906: Woolley, Sir Richard van der Riet. Worked at several observatories before becoming the 11th Astronomer Royal in 1955. He played a large part in developing the astrophysical side of the work of the Royal Greenwich Observatory, which had just been moved to Herstmonceux in Sussex. He was knighted in 1963.

1907: Einstein's equivalence principle is formulated, stating that acceleration and gravity are equivalent.
Fournier d'Albe, drawing on the work of Lord Kelvin, explains the fact that the night sky is dark by pointing out that, in a Universe of infinite extent, the distances involved may be so immense that the light from many stars has yet to reach us.

1907: Invention of colour photography. Pablo Picasso displays *Les Demoiselles d'Avignon*.

1908: Landau, Lev Davidovich. Awarded the Nobel Prize in 1962 for his work on the nature of liquid helium.

1908: Ambartsumian, Victor Amazaspovich. A pioneer, in the mid-1950s, of the idea that the energetic events observed in radio galaxies were not a result of collisions between galaxies, but were caused by violent explosions taking place at the centre of individual galaxies.

1908: Greenstein, Jesse Leonard. American astronomer, who was one of the leading pioneers of radio astronomy, and was

1908: The Tunguska Event, an explosion in central Siberia, occurs at 7.17 a.m. on 30 June 1908. It is now attributed to the violent detonation of a fragment of icy material from a comet in the atmosphere of the Earth.
Henrietta Leavitt, working under the supervision of Edward Pickering at the Harvard College Observatory, publishes observations that the brighter a Cepheid is, the more slowly it goes through its cycle of variations. Leavitt went on (in 1912) to work out the exact relationship between brightness and period by studying Cepheids in the Small Magellanic Cloud;

involved in the identification of the redshift of the first known quasar, 3C 48.

she was then able to tell the relative distances to Cepheids in our own galaxy. However, she could not tell the actual distances, because none of the distances to individual Cepheids had been measured.

1908: Hermann Minkowski shows that the uniting of space and time in one mathematical description is mathematically equivalent to a description in terms of four-dimensional Euclidean geometry, so that the combination of space and time in space-time could be regarded as the four-dimensional equivalent of a flat sheet of paper.

1909: Hey, James Stanley. Worked on radar during the Second World War and discovered that the Sun is a strong source of radio noise, which caused interference with radar equipment in the 4–8 m wavelength band.

1909: Karl Bohlin proposes that the Sun is not at the centre of the Milky Way.

1909: Robert Peary and Matthew Henson are the first modern explorers to reach the North Pole. First electric toaster on sale. Louis Blériot flies from Calais to Dover.

1910: Chandrasekhar, Subrahmanyan. Best known for his theoretical investigations of stars at the end point of their lives, especially white dwarf stars, and the mathematical theory of black holes.

1910: Protons are identified by the British physicist J. J. Thomson, who also discovered the electron.

1911: Fowler, William Alfred. Contributed to the development of an

1911: Ejnar Hertzprung measures the distances to several of the nearer

1911: Roald Amundsen's party are the first people to reach the South Pole. First

Birth dates of scientists	Key dates in science	Key dates in history

understanding of the way elements are manufactured inside stars (nucleosynthesis) and in the Big Bang.

1911: Wheeler, John Archibald. At one time espoused the idea of parallel universes; has contributed to the study of black holes; has suggested that the Universe exists only because we are watching it.

1911: Seyfert, Carl. First drew attention to the kind of active galaxies which are now known as Seyfert galaxies. Analysis of the Doppler effect in lines of the spectra of these Seyfert galaxies showed the presence of hot clouds of gas moving at hundreds or even thousands of kilometres per second within the central regions of the galaxies. The place of Seyferts in the cosmic hierarchy only became clear later, following the discovery of quasars and BL Lac objects.

1911: Reber, Grote. American radio enthusiast who became the first radio astronomer, after he heard of Karl Jansky's discovery of radio noise from space, and built the first dedicated radio telescope in his backyard in Illinois, in 1937.

1912: Schwarzschild, Martin. German-born American astronomer, whose main work has been

Cepheids in our own Galaxy, using a version of the parallax technique.

Ernest Rutherford suggests that the positive particles inside the atom – logically necessary, because of the known negative charge of electrons – are concentrated, together with most of the mass of an atom, in a tiny central nucleus.

escalators introduced, at Earl's Court station in London. Henri Matisse paints *The Red Studio*.

1912: Cosmic rays discovered by Victor Hess on balloon flights.

1912: *Titanic* sinks.

Birth dates of scientists	*Key dates in science*	*Key dates in history*

in the area of stellar structure and stellar evolution, including the nature of pulsating variables.

1913: Henry Norris Russell publishes his version of what is now known as the Hertzprung–Russell diagram, the technique of plotting the absolute magnitudes of stars against their colour.

Niels Bohr completes his 'orbital' model of the atom.

1913: Henry Ford uses an assembly line to speed up production of his cars. Geiger counter invented. First performance of Stravinsky's *Rite of Spring*.

1914: Davis, Ray, Jr. Devised and built the first experiment to measure the flux of neutrinos coming from the sun.

1914: Herman, Robert. Physicist who worked with George Gamow and Ralph Alpher in the 1940s on the model of the origin of the Universe in a hot Big Bang.

1914: Zel'dovich, Yaakov Borisovitch. One of the pioneers of the theoretical investigation of the kind of particle interactions that went on in the Big Bang and of calculations of the amount of helium they should have produced.

1914: Vesto Slipher, working at Lowell Observatory, discovers that 11 out of 15 of the objects then known as spiral nebulae (now called galaxies) showed redshifts in their light.

Arthur Eddington suggests that the spiral nebulae are galaxies.

1914–18: First World War.

1914: Brassière patented.

1915: Hoyle, Sir Fred. Explained how elements are manufactured inside stars by nucleosynthesis, and developed the Steady State model of the universe.

1915: Townes, Charles Hard. Developed the maser in the 1950s and received the Nobel Prize for this work in 1964. He

1915: The astronomer Walter Adams obtains a spectrum which shows that Sirius B is as hot as Sirius itself. It is, however, far fainter. In order to be both hot and faint it must be small – only a little bigger than the Earth. This is the first white dwarf to be discovered.

Albert Einstein's general

1915: First transcontinental telephone conversation in the USA. Marc Chagall paints *Birthday*.

Birth dates of scientists	*Key dates in science*	*Key dates in history*

helped to develop both microwave astronomy and infrared astronomy, and led the team at the University of California, Berkeley, that discovered the first polyatomic molecules in interstellar space, water and ammonia, in 1968.

1916: Dicke, Robert Henry. Best known for his investigations of the cosmic background radiation.

1916: Shklovskii, Iosif Samuilovich. Played a prominent part in the debate about the possible existence of extraterrestrial life, and was the first to suggest, in 1953, that radio waves and X-rays from the Crab nebula are produced by synchroton radiation. He also predicted the existence of astronomical masers.

theory of relativity presented to the Prussian Academy of Science. The theory describes what happens when the combination of space and time is distorted by the presence of matter.

1916: Solutions to the equations of the general theory of relativity that describe wormholes are found shortly after the theory is developed, although they are not interpreted in this way at the time.

Edward Barnard discovers Barnard's Star.

Heinrich Reissner becomes the first to develop the Reissner–Nordstrøm solution to Albert Einstein's equations of the general theory of relativity, which describes a black hole that is not rotating but has an electric charge. This is not likely to occur in the real Universe, because such an object would attract opposite charge from its surroundings and soon become an electrically neutral Schwarzschild black hole. The solution was developed by Reissner independently of Nordstrøm, who later arrived at the same conclusions.

Einstein presents Karl Schwarzschild's analysis of the implications of the

1916: Frank Lloyd Wright completes work on the Imperial Hotel in Tokyo.

general theory of relativity to the Prussian Academy of Sciences. Schwarzschild calculated the exact mathematical description of the geometry of space-time around a spherical mass, showing that for any mass there is a critical radius, now called the Schwarzschild radius, which corresponds to such an extreme distortion of space-time that, if the mass were to be squeezed inside the critical radius, space would close around the object and pinch it off from the rest of the Universe, becoming, in effect, a self-contained Universe in its own right, from which nothing (not even light) can escape. This is the simplest kind of black hole – a Schwarzschild black hole – although the metric actually describes space-time in the vicinity of any spherical concentration of mass.

1917: Dominion Radio Astrophysical Observatory established near Victoria, in British Columbia (Canada).

Applying his general theory of relativity to the world, Einstein discovers that the equations say that the Universe must be either expanding or contracting, but not stable. To stabilize the models, Einstein introduces his 'cosmological constant' as a parameter which makes the model of the Universe,

1917: Russian Revolution. Clarence Birdseye invents frozen food. John Singer Sargent paints John D. Rockefeller. John F. Kennedy born.

Birth dates of scientists	*Key dates in science*	*Key dates in history*

described in terms of the general theory of relativity, static. He later calls this 'cosmological constant' his greatest mistake.

Willem de Sitter also finds an 'expanding universe' solution to Einstein's equations.

100-inch (2.5 m) telescope becomes operational at Mount Wilson.

1918: de Vaucouleurs, Gerard Henri. Specialized in investigations of the distribution of galaxies and attempts to determine the value of the Hubble constant, and therefore the age of the Universe.

1918: Ryle, Sir Martin. British radio astronomer who was the twelfth Astronomer Royal. He was one of the pioneering developers of radio astronomy in the decades following the Second World War, and played a major part in the development of interferometry and aperture synthesis techniques.

1918: Feynman, Richard. Nobel Prize-winning quantum physicist and great teacher. He said that the two-hole experiment encapsulates the 'central mystery' of quantum mechanics.

1918: Gunnar Nordstrøm develops the Reissner–Nordstrøm solution to Albert Einstein's equations of the general theory of relativity, which describes a black hole that is not rotating but has an electric charge. This is not likely to occur in the real Universe, because such an object would attract opposite charge from its surroundings and soon become an electrically neutral Schwarzschild black hole. The solution was developed by Nordstrøm independently of Reissner.

Harlow Shapley provides the first accurate estimate of the size of the Milky Way, and correctly locates the Sun in the outer part of the Galaxy.

1918: First radio link between Britain and Australia.

Birth dates of scientists	*Key dates in science*	*Key dates in history*

1919: Bondi, Sir Hermann. One of the three original proponents of the Steady State model of the Universe, along with Tommy Gold and Fred Hoyle.

1919: The bending of starlight caused by the gravity of the Sun is measured, during a solar eclipse, and is found to match Einstein's prediction rather than Newton's.

1920: Gold, Tommy. Austrian-born American physicist who was one of the originators, in the 1940s, of the Steady State hypothesis, and was the first person to suggest in print, in 1968, that pulsars are rapidly rotating neutron stars.

1920: A. A. Michelson and colleagues pioneer a form of aperture synthesis. Michelson measures the diameter of Betelgeuse – the first time the diameter of any star other than the Sun has been measured.

William Draper proposes the existence of a neutral counterpart to the proton – the neutron.

1920: John Thompson patents the submachine gun.

1920s: Radio broadcasting takes off.

1920–33: Prohibition in the USA.

1920: Hayashi, Chushiro. Japanese astrophysicist who made contributions to the Big Bang theory in the 1950s and to the understanding of stellar evolution in the 1960s. The evolutionary track of a young star as it approaches the main sequence on the Hertzprung–Russell diagram is named after him.

1920s: Edwin Hubble identifies individual stars in the outer parts of the Andromeda galaxy, a distant nebula, proving Kant's hypothesis that such nebulae are in fact distant complete star-systems.

The work of Edwin Hubble, and other observers, shows not only that our Milky Way is just one galaxy among many in the Cosmos, but that galaxies are moving apart from one another as the Universe expands. This dispels the need for Einstein's cosmological constant.

1921: Sakharov, Andrei Dimitrievich. The leading scientist involved in the development of the Soviet hydrogen bomb. Sakharov also proposed a mechanism for the formation of matter in the Big Bang.

1921: Mao Tse-tung involved in forming the Chinese Communist Party.

1921: Alpher, Ralph Asher. Together with George Gamow and Robert Herman, he made the first calculations of how

1921: Theodor Kaluza points out that, by extending the equations of the general theory of relativity into five

Birth dates of scientists	*Key dates in science*	*Key dates in history*

elements could have been made in the Big Bang, and predicted the existence of the cosmic microwave background radiation.

dimensions, they automatically yield Maxwell's equations of electromagnetism.

1922: Burbidge, (Eleanor) Margaret. The first woman to be director of the Royal Greenwich Observatory. Often worked with her husband, Geoffrey Burbidge. The team are best known for their work, with Fred Hoyle and Willy Fowler, on nucleosynthesis.

1922: Aleksandr Friedmann finds a family of solutions to Einstein's equations, describing different model universes.

1922: Benito Mussolini comes to power in Italy. Tomb of Tutankhamun discovered.

1923: Smith, Sir Francis Graham. Has been both director of the RGO and Astronomer Royal, although he did not hold the two positions at the same time. As well as being one of the key figures in the development of radio astronomy techniques, such as interferometry, and an able administrator, Smith has made important contributions to the study of pulsars, and to the investigation of the magnetic fields in interstellar space.

1923: Louis de Boglie comes up with the idea of particle–wave duality.

1923: Great earthquake in Tokyo; more than 130,000 killed.

1924: Hewish, Antony. Involved in the discovery of pulsars in 1967.

1924: Salpeter, Edwin. Suggested, in 1952, that carbon might be manufactured out of helium inside stars through what is now known as the triple alpha process.

1924: Arthur Eddington discusses the implications of the mass–luminosity relation. Roughly speaking, the more massive a star is, the hotter it must be inside in order to generate enough pressure in its interior to hold it up against gravitational collapse.

1924: First use of insecticides.

Birth dates of scientists

1925: Burbidge, Geoffrey. Often worked with his wife, Margaret Burbidge. The team are best known for their work, with Fred Hoyle and Willy Fowler, on nucleosynthesis.

1925: Cameron, Alastair Graham Walter. Independently came up with ideas about stellar nucleosynthesis similar to those of the B²FH team, in the mid-1950s.

1926: Sandage, Allan Rex. Best known for his contribution to the debate about the value of the Hubble constant, which measures how fast the Universe is expanding.

Key dates in science

Eddington suggests that white dwarf stars are made of degenerate matter.

Statistical rules outlining the behaviour of bosons ('particles' similar in nature to the photons of light) developed by Satyendra Bose and Albert Einstein.

1925: Wolfgang Pauli formulates the Pauli exclusion principle, an expression of a law of nature which prevents any two electrons (or other members of the particle family now known as fermions) from existing in exactly the same quantum state.

1926: The rules by which fermions behave are developed by Enrico Fermi and Paul Dirac, demonstrating that the total number of fermions – particles such as electrons, protons and neutrons, which make up 'solid matter', together with more exotic particles that can be produced only through energetic interactions at particle accelerators or in energetic astronomical sources – can never be changed in isolation in a particle interaction; if a fermion is produced, it must always be accompanied by the appropriate antiparticle. So the total number of fermions in the Universe is always the same, and was determined by conditions

Key dates in history

1925: Publication of Franz Kafka's *The Trial*.

1926: Talking pictures ushered in with *The Jazz Singer*. John Logie Baird invents a television system. Robert Goddard launches a liquid-fuelled rocket to a height of 56 m.

Birth dates of scientists	Key dates in science	Key dates in history
	that operated in the Big Bang.	

Gilbert Lewis introduces the term 'photon' for the particle of light. Robert Goddard develops the first liquid-fuelled rocket.

The five-dimensional unification of gravity and electromagnetism is extended by Oskar Klein to include quantum effects, and becomes known as Kaluza–Klein theory, after the two researchers who pioneer the work. This unification is largely ignored for many years, because it requires not just one but several 'extra' dimensions to include the effects of the more complicated weak and strong interactions, which were discovered just after the initial triumph of the Kaluza–Klein theory.

1927: Arp, Halton Christian ('Chip'). Suggested that the usual interpretation of the cosmological redshift is not always applicable.

1927: Wolfendale, Arnold. The 14th Astronomer Royal, Wolfendale succeeded Francis Graham-Smith in 1991, but served only until the end of 1994.

1927: Werner Heisenberg develops the uncertainty principle, which shows that there is an intrinsic uncertainty associated with knowledge and momentum in the quantum world. You can never know both the position and the momentum of an entity such as an electron at the same time – if you try to measure the momentum, this has the effect of enhancing the 'waviness' of the entity and making it spread out so that its position is uncertain; if you try to measure its position

1927: Charles Lindbergh flies the Atlantic.

Birth dates of scientists	*Key dates in science*	*Key dates in history*
	precisely, this creates uncertainty in its waviness. Electrons observed behaving as waves.	
	1928: The tunnel effect, which enables particles to penetrate barriers, such as the barriers around nuclei, is proposed as an explanation of alpha decay by George Gamow. The theory argues that when two protons (for example) approach one another inside a star, although they are repelled by the positive electric charge they each carry, which stops them touching, they can nevertheless come so close that their quantum wave functions overlap, enabling them to interact. This may also be applied to alpha particles.	**1928:** First Mickey Mouse cartoon.
1929: Schmidt, Maarten. The first person to interpret the unusual spectrum of a quasar as being caused by a large redshift.	**1929:** Edwin Hubble measures the distance to the Andromeda galaxy and other galaxies using the Cepheid technique. Through the discovery of the redshift in light from galaxies beyond the Local Group, he infers the Universe is expanding, removing the need for Einstein's cosmological constant in its original form.	**1929:** FM radio broadcasts begin. Foam rubber put on the market.
1930: Drake, Frank Donald. Best known for his contributions to the search for extraterrestrial intelligence (SETI).	**1930:** Pluto discovered, by the American astronomer Clyde Tombaugh. Wolfgang Pauli proposes the existence of the particle later called the neutrino.	**1930:** Foundation of the Institute for Advanced Study, in Princeton. Frank Whittle patents the jet engine.

Birth dates of scientists	*Key dates in science*	*Key dates in history*

Key dates in science

1930s: Astronomers first realize that there is at least some dark matter in our own Milky Way Galaxy.

William McCrea and Edward Arthur Milne show that the expanding Universe, predicted by the equations of the general theory of relativity, can actually be described very well in the context of classical mechanics and Isaac Newton's theory of gravity.

Albert Einstein and Nathan Rosen discover that the Schwarzschild solution actually represents a black hole as what they called a bridge (now known as an Einstein–Rosen bridge) between two regions of flat space–time. Although this was considered a mathematical curiosity at the time, it was generally accepted by scientists that there had to be some law of nature preventing the existence of wormholes.

Einstein predicts the 'Einstein ring', a gravitational lens effect in which light or other electromagnetic radiation from a distant point source (such as a quasar) is spread into a ring on the sky by the gravity of an intervening object (such as a galaxy) along the line of sight.

The Schmidt camera, a type of optical telescope mainly used for taking photographs of the sky – in effect a wide-angle

Key dates in history

1930s: Worldwide population boom begins.

camera – is developed by
Bernhard Schmidt.

Observations using
spectroscopy indicate that
the Universe has a
temperature close to 3 K.
The importance of this is
not appreciated until the
1960s.

1931: Penrose, Roger.
Made major contributions
to the development of the
theory of black holes.

1931: Giacconi, Riccardo.
Italian-born American
physicist who pioneered
the development of X-ray
astronomy.

1931: Subrahmanyan
Chandrasekhar discovers
that there is no way for a
white dwarf star with more
than about 1.4 times the
mass of the Sun to hold
itself up against collapse
once its nuclear fuel is
exhausted. The implication
is that any star left with
more mass than this
Chandrasekhar limit at the
end of its lifetime would
collapse indefinitely,
forming what is now called
a black hole.

Karl Jansky detects radio
emission from the Milky
Way, while carrying out
research for the Bell
Telephone Laboratories
into the origin of static.

1932: James Chadwick
discovers the neutron, a
particle with no electric
charge and almost the
same mass as the proton.
The nucleus is then
explained as a collection of
protons and neutrons held
together by the strong
nuclear interaction, or
strong force.

Carl Anderson discovers
the positron.

John Cockroft and Ernest
Walton use the first particle

1932: Oil found in Arabia.
First production of Noel
Coward's *Design for Living*.

Birth dates of scientists	Key dates in science	Key dates in history
	accelerator to 'split the atom'. The asteroid Apollo-1862 is discovered when it comes within 0.07 astronomical units of the Earth.	
1933: Penzias, Arno Allan. Together with Robert Wilson, Penzias was the accidental discoverer of the background radiation.	**1933:** The international Astronomical Union rationalizes the system of constellations by dividing the sky into 88 areas.	**1933:** Adolf Hitler becomes Chancellor of Germany.
1934: Kerr, Roy Patrick. Has made important contributions to the understanding of black holes. He developed the Kerr solution to the equations of the general theory of relativity.	**1934:** Louis de Broglie introduces the term 'antiparticle'. Walter Baade and Fritz Zwicky suggest that 'a supernova represents the transition of an ordinary star into a neutron star'.	
1935: Peebles, (Phillip) James Edwin. Has been especially interested in the physics of the Big Bang, the large-scale structure of the Universe, and the background radiation.	**1935:** The Robertson–Walker metric is developed, a metric that describes the properties of space-time in a universe that is homogeneous and isotropic – that is, one which obeys the cosmological principle. The metric is the basis of mathematical models of cosmology incorporating the Big Bang, and was developed independently by Howard Robertson and A. G. Walker.	**1935:** Development of the beer can.
1936: Wilson, Robert Woodrow. Together with Arno Penzias, Wilson was the accidental discoverer of the background radiation.	Rupert Wildt detects methane and ammonia on the giant planets.	**1936:** First paperback books published by Penguin. Edward VIII abdicates. Buddy Holly born. **1936–9:** Spanish Civil War.
	1937: Grote Reber follows up Janski's discovery by building a dedicated radio telescope.	**1937:** Japanese invasion of China. Whittle develops a working jet engine.

Birth dates of scientists	Key dates in science	Key dates in history
	1938: The CNO cycle, the process of nuclear fusion reactions that provides the energy source inside hot, massive stars, worked out by Hans Bethe, and, independently, Carl von Weizsäcker. The proton–proton chain first proposed as the source of solar energy by Hans Bethe and Charles Critchfield. Otto Hahn splits the uranium nucleus.	**1938**: Lazlo Biro patents the ballpoint pen. Nylon commercially available in the USA. Jean-Paul Sartre writes *La Nausée*.
	1939: George Volkoff and Robert Oppenheimer publish a paper showing that there is an upper mass limit (the Oppenheimer–Volkoff limit) above which no stable neutron star can exist. Robert Oppenheimer and Hartland Snyder publish what is now regarded as the first clear description of the astrophysics of black holes.	**1939–45**: Second World War. **1939**: First flight of a jet airplane, the He 178, in Germany.
	1940s: The term 'Big Bang' is coined by Fred Hoyle, as a mark of derision for the theory. Some comologists attempt to extend the principle of terrestrial mediocrity, proposing a 'perfect cosmological principle' which said that the Universe should look much the same not only from any place within it, but at any time. This leads to the Steady State hypothesis, first put forward by Herman Bondi,	**1940**: First antibiotics. First use of freeze drying to preserve foods. John Lennon born.

Birth dates of scientists	*Key dates in science*	*Key dates in history*

Tommy Gold and Fred Hoyle, but this has since been refuted by evidence that the Universe is changing as time passes and almost certainly originated in a Big Bang a finite time ago.

The importance of geosynchronous orbit for communications first appreciated by Arthur C. Clarke, long before the technology exists to make this a practical reality.

1940s: George Gamow and Ralph Alpher become the first people to attempt to describe the conditions in the Big Bang quantitatively. Applying the developing understanding of quantum physics, they investigate the kind of nuclear interactions that would have occurred at the birth of the Universe, and find that primordial hydrogen would have been partly converted into helium. This discovery becomes known as the alpha beta gamma theory.

1941: Taylor, Joseph Hooton, Jr. Together with Russell Hulse, Taylor discovered the binary pulsar in 1974.

1941: The Maksutov telescope, an improved version of the Schmidt camera, is developed by Dmitri Maksutov. The system gives very good-quality images over a wide field of view, making it ideal for astronomical photography; however, such telescopes use correcting lens which are

1941: Japanese attack Pearl Harbor.

Birth dates of scientists	Key dates in science	Key dates in history
	difficult to manufacture for large aperture instruments, and this limits their professional applications.	
1942: Hawking, Stephen William. Has made major contributions to our understanding of black holes and the origin of the universe.	**1942**: First controlled nuclear chain reaction in a uranium 'pile' at the University of Chicago.	**1942**: Nuclear energy released.
1942: Rees, Sir Martin John. British astrophysicist who became the 15th Astronomer Royal at the beginning of 1995. His researches span almost the entire range of astrophysics, from the nature of quasars and active galaxies to the birth of the Universe, black holes, the mystery of dark matter and anthropic cosmology. He was knighted in 1992.		
1943: Bell Burnell, (Susan) Jocelyn. Discovered pulsars while still a student, in 1967.		**1943**: First electronic computer.
	1944: The terms 'Population I' and 'Population II', used to classify different forms of star, are first introduced by Walter Baade. Population I stars are those that are found chiefly in the plane of our Galaxy, especially in the spiral arms; Population II stars are those that are found in the halo of our Galaxy, especially in the globular clusters, near the Galactic centre – these are in a thick layer much fatter than, and sandwiching, the	

Birth dates of scientists	*Key dates in science*	*Key dates in history*
	plane defined by the spiral arms.	
	Carl von Weizsäcker revives the nebular hypothesis of the origin of the Solar System, which has remained the accepted model ever since.	**1945**: Atomic bomb.
	1946: Radio astronomy work begins at Jodrell Bank.	**1946**: First meeting of the United Nations.
	Cygnus A identified by Martin Ryle as the first known radio galaxy.	
	Robert Dicke and his colleagues, a team of radio astronomers looking for cool radiation from space, find no evidence of radiation with a temperature above 20 K, the limit set by their instrument.	
1947: Guth, Alan Harvey. American physicist who came up with the idea of inflation and gave the theory its name.	**1947**: Bart Bok discovers small, dark, circular clouds of material in space, which show up against the background of stars or luminous clouds. These Bok globules are thought to be stars like our Sun in the process of forming.	**1947**: First supersonic flight. India freed from British rule.
	1947: The Hale telescope is completed at the Palomar Observatory. For three decades it will be the largest telescope in the world.	
1948: Linde, Andrei Dmitrivitch. Played a major role in the development of the theory of the very early Universe known as inflation; also made important contributions to the understanding of fermions in the context of	**1948**: Ralph Alpher and Robert Herman extend the alpha beta gamma theory to predict that the Universe today must be filled with background radiation at a temperature of about 5 K.	**1948**: George Orwell writes *1984*.
	First atomic clock developed by the US	

gauge theories. In the mid-1990s, he was still developing the idea of the Universe as a self-reproducing system that sprouts other inflationary universes and is itself a sprout from another universe.

National Bureau of Standards. Miranda, one of the moons of Uranus, is discovered by Gerard Kuiper.

The largest recorded sunspot occurs, covering an area of the Sun's surface of more than 18 billion square kilometres.

1949: Fred Whipple puts forward the 'dirty snowball' model of comets.

1950: Hulse, Russell Alan. Together with Joseph Taylor, Russell Hulse discovered the binary pulsar in 1974.

1950: Jan Oort develops the idea that comets surround the Solar System in a great cloud halfway to the nearest star.

1950s: The Palomar Sky Survey, a photographic catalogue of all of the northern sky and part of the southern sky (as much as can be seen from Los Angeles), made at the Mount Palomar Observatory, using a 48-inch Schmidt camera.

First important radio telescopes making use of aperture synthesis.

Mills cross, a type of radio telescope using the principle of interferometry, made up of two lines of antennae at right angles to one another, is developed; named after the astronomer B. Y. Mills.

Large numbers of radio galaxies are discovered. These are objects far beyond the Milky Way that are in many respects similar to radio-quiet galaxies, but

1950s: Television broadcasting takes off.

1950: Korean War begins. First charge card introduced by Diner's Club.

1951: First commercially available electronic computer sold by Remington Rand.

Birth dates of scientists	Key dates in science	Key dates in history
	which produce far more radio noise than a galaxy like our own or the Andromeda galaxy.	
	Clouds of hydrogen in space are identified by their characteristic emission at a wavelength of 21 cm. These investigations provide a way to study not only the distribution of hydrogen around the galaxy, but also its motion – since the radiation is emitted at a very precisely known wavelength, the motion of the emitting clouds shows up clearly through the Doppler effect.	
	1952: Walter Baade's revision of the Cepheid distance scale doubles the size and age of the known Universe.	**1952**: Invention of the transistor radio. Mount Everest climbed.
	1953: Iosif Samuilovich Shklovskii becomes the first to suggest that radio waves and X-rays from the Crab nebula are produced by synchrotron radiation. Synchrotron emission occurs when electrons, freed from their atoms, move in spirals in a strong magnetic field – this often produces radio emission, but there is so much energy available in the Crab that the electrons also radiate visible light.	**1953**: Structure of DNA deciphered; hydrogen bomb. Invention of the maser.
	Milton Humason discovers a galaxy receding at 20 per cent of the speed of light.	**1954**: TV dinners go on sale in the USA. First sub-four-minute mile.

Birth dates of scientists	*Key dates in science*	*Key dates in history*
	1955: Smithsonian Institution moves from Washington to Harvard. First radio interferometer built in Cambridge, England.	
	1956: Neutrinos proven to exist, in experiments that monitor the flood of neutrinos emitted by a nuclear reactor at Savannah River in the United States.	**1956**: Hungarian uprising crushed by Soviet Union. First transatlantic telephone cable. Suez crisis.

	1957: Geoffrey and Margaret Burbidge, Willy Fowler and Fred Hoyle publish 'B^{2}FH' paper, describing how all the naturally occurring varieties of nuclei except primordial hydrogen and helium are built up inside stars by nucleosynthesis.
	    The first artificial satellite, Sputnik 1, is launched from the Soviet Union on 4 October.
	    The Nuffield Radio Astronomy Laboratories complete a 250-foot (76.2 m) diameter, fully steerable radio telescope, and begin operations. The telescope is eventually to be named the Lovell Telescope, in honour of Sir Bernard Lovell, in 1987.
	    Robert Dicke publishes a paper pointing out that the size of the Universe is 'not random but conditioned by biological factors'.
		1957: First artificial Earth satellite.

	1958: NASA created, to handle all non-military aspects of the US space programme.	**1958**: Fidel Castro leads revolution in Cuba.

Birth dates of scientists	Key dates in science	Key dates in history
	1959: Freeman Dyson suggests that an intelligent species with an expanding population, confined to a single planetary system like our own Solar System, will eventually rearrange the raw materials provided by those planets to build a hollow sphere around the parent star.	**1959**: First commercial photocopier. First successful hovercraft.

Birth dates of scientists | *Key dates in science* | *Key dates in history*

1959: Freeman Dyson suggests that an intelligent species with an expanding population, confined to a single planetary system like our own Solar System, will eventually rearrange the raw materials provided by those planets to build a hollow sphere around the parent star.

Lunik III obtains the first views of the far side of the Moon.

1960: Karl Schwarzschild Observatory founded.

The surface oscillations that make it possible to probe the Sun's interior by the process of helioseismology are discovered, accidentally, by researchers from Caltech planning to study random (or chaotic) movements of the hot gas at the surface of the Sun. This discovery is not exploited until the 1980s, when the technology needed to analyse the oscillations in detail is developed.

Aperture synthesis technique developed.

1960s and **1970s**: Development of quark theory.

Early **1960s**: Igor Novikov predicts the existence of background radiation.

The 'photodisintegration' of iron nuclei within a dying star is first suggested by Willy Fowler and Fred Hoyle. A dying

1959: First commercial photocopier. First successful hovercraft.

star uses fusion reactions involving silicon to stabilize it; but these reactions last only about a week. As all the silicon in the core is converted into iron-group elements, the core collapses, in a few tenths of a second, from about the size of the Sun into a lump only tens of kilometres across. During this initial collapse, gravitational energy is converted into heat, producing a flood of energetic photons which rip the heavy nuclei in the core apart, undoing the work of 11 million years of nuclear fusion.

A faint radio galaxy is found to have a redshift of 0.46, which makes it the most distant object known, at this time. By the 1990s, the record redshift will be pushed out to above 4, corresponding to looking back in time along 90 per cent of the history of the Universe since the Big Bang.

1961: Parkes Observatory, Australia's national radio astronomy observatory, completes a 64 m dish antenna, to become part of the Australia telescope.

Yuri Gagarin becomes the first man in space.

1961: First manned space-flight.

1962: The archetypal X-ray source in our Galaxy, Scorpius X-1, is discovered using a rocket-borne detector. X-ray astronomy proper

1962: Cuban missile crisis. Telstar is launched and relays the first TV pictures between the USA and Europe.

Birth dates of scientists	*Key dates in science*	*Key dates in history*

begins, as astronomers are able to lift X-ray detectors above the obscuring layers of the atmosphere on rockets, balloons and (eventually) satellites.

Green Bank Telescope starts operations.

Mariner 2 satellite reaches Venus.

Arecibo radio telescope used to bounce radar signals off Mercury.

1963: Roy Kerr discovers the Kerr solution, the mathematical description of a rotating, uncharged black hole, which is the most likely to exist in our Universe.

Hydroxyl becomes the first interstellar molecule to be detected.

Taurus X-1 becomes the second X-ray source, apart from the Sun, to be discovered.

Cyril Hazard pointed out that in 1963 the Moon would pass in front of one of many intense sources of radio noise which could not be identified with optical stars or galaxies; this occultation enabled astronomers to pinpoint the position of the source 3C 273 by timing the moment when the edge of the Moon cut off the radio noise. This led to identification of 3C 273 with a star-like object – but a 'star' with a large redshift, and an object roughly the same size as our Solar System – a quasar.

1963: John F. Kennedy assassinated.

Birth dates of scientists	Key dates in science	Key dates in history

The first in a series of Soviet spaceprobes known as Zond spaceprobes is launched.

1964: The Davis detector, designed to detect solar neutrinos, is installed 1,500 m below the ground, in the Homestake Gold Mine at Lead, South Dakota. Here, Ray Davis and colleagues will monitor the flux of solar neutrinos from the Sun. The results of this experiment will conclude that there are only about one-third of the expected number of solar neutrinos.

Ranger 7 provides 4,316 high-quality close-up pictures of the Moon.

1964: Tonkin incident leads to the start of the US–Vietnam war.

1965: Roger Penrose shows that, according to the general theory of relativity, any object which contracts within its event horizon must collapse all the way to a singularity, a point of infinite density and zero volume, where the laws of physics break down and literally anything can happen.

The first OH maser identified in the Orion nebula.

The Kerr–Newman solution, a solution to Einstein's equations of the general theory of relativity that describes a rotating, electrically charged black hole, is developed by Ezra Newman and colleagues, starting from the Kerr solution.

1965: First communications satellite.

Birth dates of scientists	Key dates in science	Key dates in history
	Arno Penzias and Robert Wilson, at the Bell Research Laboratories, accidentally find a persistent source of interference that proves to be background radiation.	
	Early Bird communications satellite placed in a geosynchronous 'Clarke Orbit'.	
	1966: John Wheeler and Franco Pacini speculate that the source of energy emitted by the Crab nebula might be a spinning neutron star.	**1966–8**: Cultural Revolution in China.
	Fred Hoyle, Bob Wagoner and Willy Fowler develop the understanding of Big Bang nucleosynthesis.	
	X-rays detected from Cygnus A.	
	Sco X-1 identified with an optical star.	
	Luna IX makes the first soft landing on the Moon.	
	Two Lunar Orbiter space-craft send back spectacular pictures of the Moon.	
	1967: Franco Pacini publishes a paper in which he points out that, if an ordinary star collapsed to form a neutron star, the collapse would make the star spin faster, and would strengthen the star's magnetic field as it was squeezed, along with the matter, into a smaller volume. Such a rotating magnetic dipole would pour out electromagnetic radiation.	

The Isaac Newton Telescope (INT) begins operations at Herstmonceux, in Sussex.

The first pulsars are discovered by Jocelyn Bell, a radio astronomer working in Cambridge under the supervision of Antony Hewish.

Mariner 9 sends back pictures of Mars.

The Davis detector, developed through the efforts of theorist John Bahcall and experimenter Ray Davis, and making use of chlorine (in the form of perchlorethylene) to monitor the arrival of neutrinos from the Sun, commences operations. This detector finds only about one-third of the expected number of solar neutrinos, leading to what is now known as the 'solar neutrino problem'.

1968: Identification of the first BL Lac object. BL Lac objects, probably closely related to quasars, are unusual, bright, compact, extremely variable sources of intense energy lying at the hearts of some galaxies.

The Maffei galaxies, two galaxies which lie almost in the plane of the Milky way and are almost hidden by interstellar dust, are discovered by Paolo Maffei.

Crab pulsar discovered.

The Vela pulsar, a pulsar associated with a supernova remnant 10,000 years old in the

1968: Soviet troops crush the 'Prague spring' in Czechoslovakia.

Birth dates of scientists	*Key dates in science*	*Key dates in history*

constellation Vela, is discovered.

The speculation that the pulses of radio noise from pulsars might be produced by a lighthouse effect as a rapidly rotating neutron star flicks its beams past is first published by Tommy Gold. This is now known as the Pacini–Gold model.

1969: The identification and investigation of the Crab pulsar establishes the Pacini–Gold model beyond any reasonable doubt.

Studies of the faint variable star associated with Sco X-1 enable Paul Feldman and John Gribbin to show that it is a neutron star surrounded by an accretion disc.

First manned Moon landing.

1969: First humans on the Moon.

1970: Charge coupled devices (CCDs) invented at the Bell Laboratories in the United States.

1970: First Boeing 747s introduced on transatlantic routes. Floppy disks used for storing computer data.

1970: Westerbork Radio Observatory opens.

X-ray astronomy is transformed by the launch of Uhuru, the first X-ray astronomy satellite, an orbiting observatory dedicated to the study of the Universe at X-ray wavelengths.

First large telescope (a 224 cm reflector) installed on Mauna Kea.

1970s: Fred Hoyle and Chandra Wickramasinghe

revive the panspermia hypothesis, suggesting that life evolved in molecular clouds in space, and was carried to Earth on board a comet, rather than on a meteorite.

Giant molecular clouds, large clouds of gas mainly composed of molecules of hydrogen, in our Galaxy, are first revealed by radio astronomy techniques. These clouds contain millions of times as much mass as our Sun, and provide the raw material from which new stars are made.

Several astronomers independently come up with the insight that each of the short-lived local vibrations which were observed to occur on the surface of the Sun could be better explained as an effect caused by millions of far smaller vibrations – sound waves trapped inside the Sun, making the surface ring like a bell.

Astronomers are surprised by the variety and complexity of complex molecules found in space.

1970s and **1980s**: Fred Hoyle and Chandra Wickramasinghe claim that some of the features seen in the radiation from molecular clouds can be explained in terms of very large organic molecules, called polymers. This idea is taken further with their proposition that life evolved in interstellar clouds over a very long time before the Solar System had even formed.

1971: An X-ray source known as Cygnus X-1 is identified with a star known as HDE 226868. As the source of the X-rays has a mass greater than the Oppenheimer–Volkoff limit, it can only be a black hole.

Mariner 9 goes into orbit around Mars. It is joined 14 days later by Mars 2. Mars 3 lands on Mars, but fails to send back data.

1972: One of the most important pioneering gamma-ray observatories, SAS II, is launched by NASA.

The Sunyaev–Zel'dovich effect, a distortion in the background radiation caused by the way the radiation is affected by the material in rich clusters of galaxies along the line of sight, is predicted by Rashid Sunyaev and Yakov Zel'dovich. This effect changes the temperature of background radiation by about 0.01 per cent.

TD-1, a European satellite, is launched. It will make the first survey of the sky in the ultraviolet.

Venera 8 lands on Venus.

1973: Announcement of the detection of 'bursters', very powerful gamma-ray sources which appear and

1971: Direct dialling introduced for telephone calls between Europe and the USA. First pocket electronic calculator put on sale by Texas Instruments.

1973: Genetic engineering techniques developed; first oil crisis increases cost of energy.

fade away again in a matter of a few seconds. Bursters were originally discovered in the late 1960s, but were held as classified information until 1973.

Smithsonian Astrophysical Observatory combines with the Harvard College Observatory, to become the Harvard–Smithsonian Center for Astrophysics.

UK Schmidt Telescope, a 1.2 m Schmidt camera located at the Anglo-Australian Observatory, begins observing.

First Skylab missions.

1974: Frank Tipler suggests (in a paper published in *Physical Review D*) that it might be possible to build a time machine. The trick involves making a naked singularity that rotates extremely rapidly. The effect of the rotation is to twist space-time, tipping it over so that one of the space dimensions is replaced by the time dimension, enabling a spaceship to make a journey which seems to be through ordinary space, but is in fact a journey through time.

Mariner 10 sends back pictures from Mercury.

Brandon Carter draws a distinction between the 'weak anthropic principle' and the 'strong anthropic principle'.

The binary pulsar is

1974: Richard Nixon resigns in the wake of the Watergate scandal.

Birth dates of scientists	Key dates in science	Key dates in history
	discovered by Russell Hulse and Joseph Taylor.	
	1975: One of the most important pioneering gamma-ray observatories, Cos B, a European satellite, is launched by the European Space Agency.	**1975**: Last US troops leave Vietnam. First personal computer available in the USA; it has a memory of 256 bytes (a quarter of a kilobyte).
	The Kuiper Airborne Observatory, a flying observatory for infrared astronomy, begins operations.	
	Two unmanned Mars Missions, known as Viking spaceprobes, are launched by NASA. Each consists of an orbiter and a lander.	**1976**: Concorde enters service. Unmanned spacecraft land on Mars.
	Vera Rubin and her colleagues discover that the Milky Way is moving through space at 500 km/sec, relative to distant galaxies.	
	Apollo–Soyuz, the first co-operative space project between the USA and the (then) USSR.	
	1977: Two spaceprobes known as Voyager are launched by NASA, to send back pictures and other data from the outer planets of the Solar System.	**1977**: Apple II personal computer launched.
	A catalogue of stars with strong emission lines is compiled by Bruce Stephenson and Nicholas Sanduleak.	
	Chiron, a supercomet that orbits between the orbits of Saturn and Uranus, is discovered by Charles Kowal.	
	Brent Tully and Richard	

Fisher discover the relationship between the width of the hydrogen line in the radio spectrum of a galaxy and the absolute magnitude of the galaxy. Through comparing these and measuring the galaxy's apparent magnitude, it is possible to work out the distance to that galaxy. This method becomes known as the Tully-Fisher method.

The Vela pulsar becomes the second pulsar to be identified as an optical pulsar.

Motion of the Milky Way measured relative to the background radiation.

1978: Charon, a satellite of the planet Pluto, is discovered by James Christy.

International Ultraviolet Explorer (IUE), a satellite, is launched by NASA to make observations of the Universe at ultraviolet wavelengths.

Joseph Taylor announces that behaviour of the binary pulsar exactly matches the predictions of Einstein's theory.

The Einstein satellite is launched by NASA. This oribiting X-ray observatory has far better resolution and sensitivity than earlier X-ray observatories.

1979: Hakucho, the first Japanese X-ray astronomy satellite, is launched.

Starlink, the computer network that links British

Birth dates of scientists	*Key dates in science*	*Key dates in history*

astronomers, is set up.

UK Infrared Telescope (UKIRT), a 3.8 m telescope for infrared astronomy, located on Mauna Kea in Hawaii and run by the Royal Observatory, Edinburgh, begins observing.

Voyager 1 visits Jupiter and its moons.

Voyager 2 visits the Jupiter system.

End of the **1970s**: The first inflationary model is developed by Alexei Starobinsky, at the L. D. Landau Institute of Theoretical Physics in Moscow, though it is not, at first, called inflation. Unfortunately, because of the difficulties Soviet scientists still had in travelling abroad or communicating with colleagues outside the Soviet sphere of influence at that time, this discovery does not spread outside the USSR. Alan Guth independently hits on a similar idea in 1979, and gives it the name 'inflation'.

1980s: Relativists attempting to prove that there is a law of nature preventing the existence of wormholes find they cannot. There is nothing in the general theory of relativity that forbids the existence of wormholes.

Helioseismology – the technique of investigating

Birth dates of scientists	*Key dates in science*	*Key dates in history*

the interior structure of the Sun by studying the way in which its surface moves – is made possible by the development of technology sufficiently advanced to analyse the surface oscillations in detail.

The Einstein ring, a gravitational lens effect predicted by Albert Einstein in the 1930s, is observed for the first time.

Researchers at the AT&T Bell Laboratories identify huge numbers of dwarf galaxies at redshifts corresponding to a look back in time of 2–3 billion years.

The Kaluza–Klein theory is revived, as investigations into the possibility of a grand unified theory increasingly suggest the existence of many extra dimensions, and string theory is increasingly considered seriously. This theory describes entities that we used to think of as point-like particles as tiny pieces, far smaller than protons, of one-dimensional 'string'. String theory also only 'works' in many dimensions, but it produces an enormous bonus – gravity.

Inflation becomes established as the standard model of the very early Universe.

1980: Very Large Array becomes operational.

First suggestions that neutrinos may have mass.

1980: Development of optical fibre communications links.

Birth dates of scientists	*Key dates in science*	*Key dates in history*
	Identification of fuzzy blobs around quasars suggests that they lie at the hearts of galaxies.	**1980s:** Development of smaller, faster, cheaper computers.
	1981: Steven Hawking suggests that the singularity found in the normal model of the Big Bang – which acts like a boundary in time to the Universe – may be meaningless; a product of equations, rather than an actual boundary. An analogy may here be drawn with the Earth. When one reaches the North Pole, one does not stop – one simply continues going, travelling south; in the same way, if one could go back in time to 'time zero', one would find oneself simply travelling through it and would then travel forwards in time.	**1981:** Introduction of the IBM PC.
	The first void – a region of the Universe that does not contain any visible bright galaxies – is discovered in the direction of the constellation Bootes. Voids take up about 98 per cent of the volume of the Universe, with bright stuff concentrated in the skin of the 'bubbles' making up the other 2 per cent. Models of the Universe based on the presence of dark matter suggest that most of the mass of the Universe may be in the form of dark stuff in the voids, while the visible bright galaxies form only in	

regions where the density of baryons is at its highest.

Voyager 1 visits Saturn, and then moves out of the solar system.

Voyager 2 visits Saturn and its rings.

1982: Andrei Linde develops the chaotic model of inflation. This presents a realization that there need not be anything special about the Planck-sized region of space-time that expanded to become our Universe. If this region was part of some larger space-time in which all kinds of scalar fields were at work, then only the regions in which those fields produced inflation could lead to the emergence of a large Universe like our own.

The first known millisecond pulsar, PSR 1937+211, is identified by astronomers from Princeton University using the Arecibo radio telescope. This pulsar has a period of just 1.6 milliseconds, which means that a star roughly as massive as our Sun is squeezed into a ball of material with the density of an atomic nucleus, spinning more than 600 times every second.

1983: EXOSAT, an X-ray astronomy satellite, is launched by the European Space Agency.

IRAS, the Infrared

1982: Space Shuttle begins operations. CD players available.

Birth dates of scientists	Key dates in science	Key dates in history
	Astronomical Satellite, is launched by NASA.	
	1984: The Isaac Newton telescope (INT), having been moved from Herstmonceux in Sussex to La Palma in the Canary Islands and given a new mirror, resumes observations. The two Vega spaceprobes are launched from the Soviet Union, to fly past Venus (dropping probes into the planet's atmosphere) and on to encounter Halley's Comet.	**1984:** Apple Macintosh introduced.
	1985: Giotto, another Halley-bound spaceprobe, is launched by the European Space Agency. Kip Thorne realizes that it would be possible to hold a wormhole open using so-called exotic matter. The critical factor is that this exotic matter has an enormous tension, sufficient to hold the wormhole open in spite of the inward tug of gravity. This exotic matter almost exactly matches the description of cosmic string, so that if anything like cosmic string existed, and could be captured, it would be capable of holding wormholes open.	**1985:** Mexico City devastated by earthquake.
	Mid-**1980s:** By now it is clear that overall our Galaxy, the Milky Way, contains up to ten times as much dark matter as the matter we can see in stars.	

Birth dates of scientists	*Key dates in science*	*Key dates in history*
	Late 1980s: Research shows that genuine time travel is not forbidden by the laws of physics, as at present understood.	
	1986: Giotto carries out a close encounter with Halley's Comet, providing confirmation of the accuracy of the 'dirty snowball' theory.	**1986**: Chernobyl nuclear disaster.
	Andrei Linde suggests that the Universe might be part of a self-reproducing system of baby universes.	
	The James Clerk Maxwell Telescope begins operations.	
	Voyager 2 visits Uranus.	
	Discovery of the 'great attractor'.	
	1987: Ginga, a Japanese X-ray astronomy satellite, launched.	
	Supernova 1987A becomes the first supernova visible from Earth to the naked eye since the one Kepler observed in 1604. This supernova is the explosion of a star about 160,000 light-years away from us, in the Large Magellanic Cloud; the precursor star was later identified on old photographic plates as a supergiant known as Sanduleak $-69°$ 202.	
	The William Herschel Telescope, a 4.2 m reflecting telescope operated by the Royal Greenwich Observatory but based at the Roque de los Muchachos	

Birth dates of scientists	*Key dates in science*	*Key dates in history*

Observatory, begins
operating.

First direct evidence of
planet-sized objects
orbiting other stars.

1988: Morris, Thorne and
Yurtsever publish their
conclusions about the
possibility of wormholes
linking different times, in
the journal *Physical Review
Letters*.

1989: Interest in the black
hole V404 Cygni is stirred
when the Japanese X-ray
astronomy satellite Ginga
finds a burst of X-rays from
an object labelled at the
time GS2023 + 338.
Astronomers soon realize
that the burst of X-rays is
too strong to be explained
in terms of the presence of
a white dwarf in a recurrent
nova, so has to be coming
from a neutron star or a
black hole.

Galileo, a NASA
spaceprobe and the first
spaceprobe intended to go
into orbit around Jupiter, is
launched.

GRANAT, a Russian
satellite, is launched to
make X-ray and gamma-
ray observations of the sky.

Hipparcos satellite
launched by the European
Space Agency, to carry out
an extremely accurate
survey of the positions of
stars. It will measure the
positions of more than
100,000 stars to an
accuracy of 0.002 arc
seconds.

Voyager 2 visits Neptune.

End of the **1980s**: Japanese researchers carry out a version of the 'double slit' experiment in which an electron gun fires electrons one at a time, on to an electron detector. This experiment confirms the wave–particle duality of electrons.

1990s: Computer simulations of the way non-spherical objects (such as spindles) collapse suggest that they may indeed form singularities that are not concealed behind event horizons – naked singularities.

GONG, the Global Oscillation Network Group, monitors the behaviour of the Sun from six observation stations spaced roughly evenly around the world. The particular aim of the project is to monitor the way in which the Sun oscillates, using helioseismology to probe the Sun's interior in a similar fashion to the way in which seismologists study earthquakes to find out the structure of the planet.

Canadian researchers develop a reflecting telescope system which uses a curved rotating dish covered with a thin layer of mercury as the mirror.

Astronomers find

Birth dates of scientists	*Key dates in science*	*Key dates in history*

evidence of a complex molecule in the form of a ring in interstellar clouds. This provides the first independent support for controversial claims made by Fred Hoyle and Chandra Wickramasinghe in the 1970s and 1980s, suggesting that some of the features observed in the radiation from molecule clouds can be explained in terms of very large organic molecules called polymers.

1990: Hubble Space Telescope, an orbiting observatory built jointly by NASA and the European Space Agency, is launched.
    ROSAT, a German/British/ American satellite for X-ray astronomy, is launched, and carries out the first complete survey of the sky in the part of the spectrum bridging the far ultraviolet and X-rays.
    Ulysses, a spaceprobe, is launched by NASA for the European Space Agency. Ulysses will fly out of the plane of the ecliptic to observe the Sun's polar regions, which cannot be seen from Earth.
    The Moon is 'discovered' as an X-ray source, by instruments on board the satellite ROSAT.

1991: Compton Gamma Ray Observatory, a satellite, is launched by NASA to study the Universe

1991: Collapse of the Soviet Union.

in the gamma-ray part of
the spectrum.

1992: Keck telescope, a 10
m diameter reflecting
telescope regarded as the
most powerful ground-
based instrument for
optical astronomy, is
completed.

The satellite COBE
(COsmic Background
Explorer) finds ripples in
the background radiation
that are exactly the right
size to conform with the
standard Big Bang model.
This is considered the
ultimate triumph of the Big
Bang theory.

1993: Flaws with the
Hubble Space Telescope
are largely corrected by
astronauts.

Gravitational lensing, in
which a dark object in our
own Galaxy passes directly
in front of a more distant
star, and makes its image
brighten briefly, is first
observed, confirming the
existence of compact dark
objects (MACHOs) in our
Galaxy. Gravitational
lenses are best described as
cosmic magnifying glasses,
produced when the gravity
of an object bends light
from a more distant object
and makes the image of the
distant object brighter to
astronomers.

1994: An amino acid
(glycine) is discovered in
an interstellar cloud, giving
credence to the theory of

Sir Fred Hoyle and Chandra Wickramasinghe that interstellar clouds are a birthplace of the molecules of life.

Astronomers at the Kitt Peak National Observatory report the first detection of a faint glow of starlight surrounding a distant galaxy and tracing out the dark halo. This confirms the reality of dark haloes.

Physicists attempting to recreate the conditions that existed in the quark–gluon plasma, by colliding massive nuclei together in particle accelerators on Earth, produce 'little bangs' which unambiguously confirm that the photons produced in these events are coming from a quark–gluon plasma, confirming the theory of the Big Bang fireball.

Ulysses makes the surprising discovery that there is no well-defined magnetic pole in the southern hemisphere of the Sun.

The last of the six members of the quark family, known as the top quark, is identified by researchers at Fermilab, in Chicago.

1995: Researchers at Los Alamos claim to have measured the mass of the neutrino, about 5 electron Volts.